MODERN HEURISTIC OPTIMIZATION TECHNIQUES

Books in the IEEE Press Series on Power Engineering

MODERN HEURISTIC OPTIMIZATION TECHNIQUES

THEORY AND APPLICATIONS TO POWER SYSTEMS

Edited by

Kwang Y. Lee and Mohamed A. El-Sharkawi

POWER ENGINEERING

Mohamed E. El-Hawary, *Series Editor*

A JOHN WILEY & SONS, INC., PUBLICATION

Disclaimer: The editors of this volume are not endorsing evolution as a scientific fact, in that species evolve from one kind to another. The term "evolutionary" in "Evolutionary Computation," or EC, simply means that the characteristics of an individual change within the population of the same species, as observed in the nature.

Library of Congress Cataloging-in-Publication Data is available.

ISBN 978-0471-45711-4

Printed in the United States of America

10 9 8 7 6 5 4 3 2 1

6   Fundamentals of Tabu Search   **101**

Alcir J. Monticelli, Rubén Romero, and Eduardo Nobuhiro Asada

9   Differential Evolution, an Alternative Approach to Evolutionary Algorithm

Kit Po Wong and ZhaoYang Dong

PART 2   SELECTED APPLICATIONS OF MODERN HEURISTIC OPTIMIZATION IN POWER SYSTEMS   235

12   Overview of Applications in Power Systems   237

Alexandre P. Alves da Silva, Djalma M. Falcão, and Kwang Y. Lee

17   Genetic Algorithms for Solving Optimal Power Flow Problems   471

Loi Lei Lai and Nidul Sinha

19 Hybrid Systems   **525**

Vladimiro Miranda

■■■■■ PREFACE

Several heuristic tools have evolved in the past decades that facilitate solving optimization problems that were previously difficult or impossible to solve. These tools include evolutionary computation, simulated annealing, tabu search, particle swarm, and so forth. Reports of applications of each of these tools have been widely published. Recently, these new heuristic tools have been combined among themselves and with knowledge elements, as well as with more traditional approaches such as statistical analysis, to solve extremely challenging problems. Developing solutions with these tools offers two major advantages: (1) development time is much shorter than when using more traditional approaches, and (2) the systems are very robust, being relatively insensitive to noisy and/or missing data.

The purpose of this book is to provide basic knowledge of evolutionary computation and other heuristic optimization techniques and how they are combined with knowledge elements in computational intelligence systems. Applications to power problems are stressed, and example applications are presented.

The book is composed of two parts: The first part gives an overview of modern heuristic optimization techniques, including fundamentals of evolutionary computation, genetic algorithms, evolutionary programming and strategies, particle swarm optimization, ant colony search algorithm, simulated annealing, tabu search, hybrid systems of evolutionary computation, and hybrid optimization of local heuristics and dynamical trajectory.

The second part of the book gives an overview of power system applications and deals with specific applications of the heuristic approaches to power system problems, such as security assessment, generation and maintenance scheduling, economic dispatch, transmission network expansion planning, generation expansion and reactive power planning, distribution system optimization, power plant and power system control, FACTS, and hybrid systems of heuristic methods.

Evolutionary Computation

Natural evolution is a hypothetical population-based optimization process. Simulating this process on a computer results in stochastic optimization techniques that can often outperform classic methods of optimization when applied to difficult real-world problems. This tutorial provides a background in the inspiration, history, and application of evolutionary computation and other heuristic optimization methods to system identification, automatic control, gaming, and other combinatorial problems.

The objectives are to provide an overview of how evolutionary computation and other heuristic optimization techniques may be applied to problems within your domain of expertise, to provide a good understanding of the design issues involved in tailoring heuristic algorithms to real-world problems, to compare and judge the efficacy of modern heuristic optimization techniques with other more classic methods of optimization, and to program fundamental evolutionary algorithms and other heuristic optimization routines.

Genetic Algorithm

A genetic algorithm (GA) is a search algorithm based on the conjecture of natural selection and genetics. The features of a genetic algorithm are different from other search techniques in several aspects. First, the algorithm is a multipath that searches many peaks in parallel, hence reducing the possibility of local minimum trapping. Second, the GA works with a coding of parameters instead of the parameters themselves. The coding of parameter will help the genetic operator to evolve the current state into the next state with minimum computations. Third, the GA evaluates the fitness of each string to guide its search instead of the optimization function. The genetic algorithm only needs to evaluate objective function (fitness) to guide its search. There is no requirement for derivatives or other auxiliary knowledge. Hence, there is no need for computation of derivatives or other auxiliary functions. Finally, the strategies employ GA explores the search space where the probability of finding improved performance is high.

Evolution Strategies and Evolutionary Programming

Evolution strategies (ES) employ real-coded variables, and, in its original form, it relied on mutation as the search operator, and a population size of one. Since then, it has evolved to share many features with the GA. The major similarity between these two types of algorithms is that they both maintain populations of potential solutions and use a selection mechanism for choosing the best individuals from the population. The main differences are ES operates directly on floating-point vectors, whereas classic GAs operate on binary strings; GAs rely mainly on recombination to explore the search space, whereas ES uses mutation as the dominant operator; and ES is an abstraction of evolution at individual behavior level, stressing the behavioral link between an individual and its offspring, whereas GAs maintain the genetic link.

Evolutionary programming (EP) is a stochastic optimization strategy similar to a GA that places emphasis on the behavioral linkage between parents and their offspring, rather than seeking to emulate specific genetic operators as observed in nature. EP is similar to ES, although the two approaches developed independently. Like both ES and GAs, EP is a useful method of optimization when other techniques such as gradient descent or direct analytical discovery are not possible. Combinatorial and real-valued function optimization in which the optimization surface or fitness landscape is "rugged," possessing many locally optimal solutions, are well suited for EP.

Differential Evolution

As a relatively new population-based optimization technique, Differential Evolution (DE) has been attracting increasing attention for a wide variety of engineering applications. Unlike the conventional evolutionary algorithms that depend on predefined probability distribution function for mutation process, DE uses the differences of randomly sampled pairs of objective vectors for its mutation process. Consequently, the object vectors' differences will pass the objective functions' topographical information toward the optimization process and therefore provide more efficient global optimization capability. DE is a stochastic direct search optimization method. It is generally considered as an accurate, reasonably fast, and robust optimization method. The main advantages of DE are its simplicity and therefore easy use in solving optimization problems requiring a minimization process with real-valued and multimodal (multiple local optima) objective functions. DE uses a nonuniform crossover that makes use of child vector parameters to guide through the minimization process. The mutation operation with DE is performed by arithmetical combinations of individuals rather than perturbing the genes in individuals with small probability compared with one of the most popular EAs, genetic algorithms (GAs). Another main characteristic of DE is its ability to search with floating point representation instead of binary representation as used in many basic EAs such as GAs. The characteristics together with other factors of DE make it a fast and robust algorithm as an alternative to EA.

Particle Swarm Optimization

Particle swarm optimization (PSO) is an exciting new methodology in evolutionary computation that is somewhat similar to a genetic algorithm in that the system is initialized with a population of random solutions. Unlike other algorithms, however, each potential solution (called a particle) is also assigned a randomized velocity and then flown through the problem hyperspace. Particle swarm optimization has been found to be extremely effective in solving a wide range of engineering problems. It is very simple to implement (the algorithm comprises two lines of computer code) and solves problems very quickly.

Ant Colony Search Algorithm

The ant colony search algorithms mimic the behavior of real ants. As is well-known, real ants are capable of finding the shortest path from food sources to the nest without using visual cues. They are also capable of adapting to changes in the environment, for example, finding a new shortest path once the old one is no longer feasible because of a new obstacle. Studies by ethnologists reveal these such capabilities are essentially due to what is called "pheromone trails," which ants use to communicate information among individuals regarding path and to decide where to go. Ants deposit a certain amount of pheromone while walking, and each ant probabilistically prefers to follow a direction rich in pheromone rather than a poorer one.

Tabu Search

Tabu search (TS) is basically a gradient-descent search with memory. The memory preserves a number of previously visited states along with a number of states that might be considered unwanted. This information is stored in a tabu list. The definition of a state, the area around it, and the length of the tabu list are critical design parameters. In addition to these tabu parameters, two extra parameters are often used: aspiration and diversification. Aspiration is used when all the neighboring states of the current state are also included in the tabu list. In that case, the tabu obstacle is overridden by selecting a new state. Diversification adds randomness to this otherwise deterministic search. If the tabu search is not converging, the search is reset randomly.

Simulated Annealing

In statistical mechanics, a physical process called annealing is often performed in order to relax the system to a state with minimum free energy. In the annealing process, a solid in a heat bath is heated up by increasing the temperature of the bath until the solid is melted into liquid, then the temperature is lowered slowly. In the liquid phase, all particles of the solid arrange themselves randomly. In the ground state, the particles are arranged in a highly structured lattice, and the energy of the system is minimum. The ground state of the solid is obtained only if the maximum temperature is sufficiently high and the cooling is done sufficiently slowly. Based on the annealing process in statistical mechanics, simulated annealing (SA) was introduced for solving complicated combinatorial optimization.

The name *simulated annealing* originates from the analogy with the physical process of solids, and the analogy between physical system and simulated annealing is that the cost function and the solution (configuration) in the optimization process correspond with the energy function and the state of statistical physics, respectively. In a large combinatorial optimization problem, an appropriate perturbation mechanism, cost function, solution space, and cooling schedule are required in order to find an optimal solution with simulated annealing. SA is effective in network reconfiguration problems for large-scale distribution systems, and its search capability becomes more significant as the system size increases. Moreover, the cost function with a smoothing strategy enables the simulated annealing to escape more easily from local minima and to reach rapidly to the vicinity of an optimal solution.

Pareto Mutiobjective Optimization

Compared with single-objective (SO) optimization problems, which have a unique solution, the solution to multiobjective (MO) problems consists of sets of tradeoffs between objectives. The goal of multiobjective optimization (MOO) algorithms is to generate these trade-offs. Exploring all these trade-offs is particularly important because it provides the system designer/operator with the ability to understand and weigh the different choices available to them. Solving MO problems has traditionally consisted of converting all objectives into a SO function. The ultimate goal is to find the solution that minimizes or maximizes this single objective while maintaining the physical constraints of the system or process. The optimization solution results in a single value that reflects a compromise between all objectives. This simple optimization process is no longer acceptable for systems with multiple conflicting objectives: System engineers may desire to know all possible optimized solutions of all objectives simultaneously. In the business world, it is known as a trade-off analysis. It is commonplace for real-world optimization problems in business, management, and engineering to feature multiple, generally conflicting objectives. This book focuses on heuristic multiobjective optimization, particularly with population-based stochastic algorithms such as evolutionary algorithms and presents the basic principles behind MOO, notably introducing the Pareto optimality concepts and formulation.

KWANG Y. LEE
MOHAMED A. EL-SHARKAWI

University Park, Pennsylvania
Seattle, Washington
November 2007

Koay Chin Aik, Ph.D., Department of Electrical and Computer Engineering, National University of Singapore, Singapore

Alexandre P. Alves da Silva, Professor, COPPE, Federal University of Rio de Janeiro, Electrical Engineering Graduate Program, Rio de Janeiro, Brazil

Eduardo Nobuhiro Asada, Professor, Department of Electrical Engineering, School of Engineering of Sao Carlos, University of Sao Paulo, Sao Carlos, Sao Paulo, Brazil

Hsiao-Dong Chiang, Professor, School of Electrical and Computer Engineering, Cornell University, Ithaca, New York

ZhaoYang Dong, Professor, School of Information Technology and Electrical Engineering, The University of Queensland, St. Lucia, Australia

Mohamed A. El-Sharkawi, Professor, Department of Electrical Engineering, University of Washington, Seattle, Washington

Djalma M. Falcão, COPPE, Federal University of Rio de Janeiro, Electrical Engineering Graduate Program, Rio de Janeiro, Brazil

David B. Fogel, Ph.D., Natural Selection, Inc., San Diego, California

Warren L. J. Fox, Ph.D., Applied Physics Laboratory, University of Washington, Seattle, Washington

Yoshikazu Fukuyama, Ph.D., General Manager, Fine Technology Components Department, Tokyo Factory, Tokyo, Japan

Hamid Ghezelayagh, Ph.D., Atmel Corporation, San Jose, California

Youngjae Jeon, Ph.D., Korea Energy Management Corporation (KEMCO), Gyeonggi-do, Korea

Ioannis N. Kassabalidis, Ph.D., Department of Informatics, School of Applied Sciences, Aristotle University of Thessaloniki, Thessaloniki, Greece

Mingoo Kim, Ph.D., Samsung Networks, Seoul, Korea

Loi Lei Lai, Professor, School of Engineering and Mathematical Sciences, City University London, St. John, London, United Kingdom

Germano Lambert-Torres, Professor, Federal University of Itajuba, Itajuba, Brazil

Jaewook Lee, Ph.D., Department of Industrial and Management Engineering, POSTECH, Pohang, Kyungbuk, Korea

Kwang Y. Lee, Professor and Chair, Department of Electrical and Computer Engineering, Baylor University, Waco, Texas

Chen-Ching Liu, Professor, Department of Electrical and Computer Engineering, Iowa State University, Ames, Iowa

Haiyan Lu, Ph.D., University of Technology Sydney, Sydney, Australia

Robert J. Marks, Ph.D., Distinguished Professor of Electrical and Computer Engineering, Baylor University, Waco, Texas

Koichi Nara, President, Fukushima National College of Technology, Iwaki, Fukushima, Japan

Vladimiro Miranda, Professor and Director of INESC Faculty of Engineering, The University of Porto, Porto, Portugal

Alcir J. Monticelli, Ph.D., Estadual de Campinas University, Campinas, Sao Paulo, Brazil

Patrick N. Ngatchou, Ph.D., Intermolecular, Inc., San Jose, California

Jong-Bae Park, Professor, Electrical Engineering Department, Konkuk University, Kwanggingu, Seoul, Korea

Rubén Romero, Professor, Department of Electrical Engineering, Estadual Paulista Julio de Mesquita Filho University, Sao Paulo, Brazil

Nidul Sinha, Professor, Department of Electrical Engineering, National Institute of Technology, Silchar, Assam, India

Yong-Hua Song, Professor and Pro-Vice-Chancellor, University of Liverpool, Liverpool, United Kingdom

Dipti Srinivasan, Professor, Department of Electrical and Computer Engineering, National University of Singapore, Singapore

John G. Vlachogiannis, Professor, Industrial and Energy Informatics Laboratory (IEI-Lab), R.S. Lianokladiou, Lamia, Greece

Kit Po Wong, Chair Professor and Head, Department of Electrical Engineering, The Hong Kong Polytechnic University, Hung Hom, Kowloon, Hong Kong

Ying Xiao, Ph.D., AREVA T&D Inc., Redmond, Washington

I.K. Yu, Professor, Changwon National University, Changwon, Kyongnam, Korea

Anahita Zarei, University of Washington, Department of Electrical Engineering, Seattle, Washington

PART 1

THEORY OF MODERN HEURISTIC OPTIMIZATION

CHAPTER 1

Introduction to Evolutionary Computation

DAVID B. FOGEL

1.1 INTRODUCTION

Darwinian evolution is intrinsically a robust search and optimization mechanism. Living organisms demonstrate optimized complex behavior at every level: the cell, the organ, the individual, and the population. The problems that biological species have solved are typified by chaos, chance, temporality, and nonlinear inter-activities. These are also characteristics of problems that have proved to be especially intractable to classic methods of optimization and appear routinely in the area of power systems. The evolutionary process can be applied to these problems, where heuristic solutions are not available or generally lead to unsatisfactory results. As a result, evolutionary algorithms have recently received increased interest, particularly with regard to the manner in which they may be applied for practical problem solving.

Evolutionary computation, the term now used to describe the field of investigation that concerns all evolutionary algorithms, offers practical advantages to the researcher facing difficult optimization problems. These advantages are multifold, including the simplicity of the approach, its robust response to changing circumstance, its flexibility, and many other facets. This chapter summarizes some of these advantages, offers a brief review of some parts of evolutionary computation theory, and introduces a new optimization technique that models swarming behavior in insects or schooling in fish. The reader who wants to further review the basic concepts of evolutionary algorithms is referred to Fogel [1–3], Bäck [4], and Michalewicz [5].

Modern Heuristic Optimization Techniques. Edited by K. Y. Lee and M. A. El-Sharkawi

3

1.2 ADVANTAGES OF EVOLUTIONARY COMPUTATION

1.2.1 Conceptual Simplicity

A primary advantage of evolutionary computation is that it is conceptually simple. The main flowchart that describes every evolutionary algorithm applied for function optimization is depicted in Fig. 1.1. The algorithm consists of initialization, which may be a purely random sampling of possible solutions, followed by iterative variation and selection in light of a performance index. This figure of merit must assign a numeric value to any possible solution such that two competing solutions can be rank ordered. Finer granularity is not required. Thus, the criterion need not be specified with the precision that is required of some other methods. In particular, no gradient information needs to be presented to the algorithm. Over iterations of

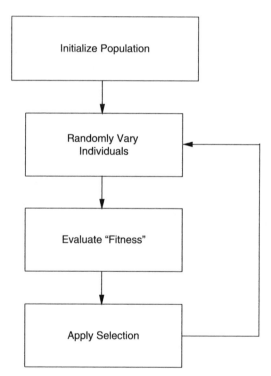

FIGURE 1.1 The main flowchart of the vast majority of evolutionary algorithms. A population of candidate solutions to a problem at hand is initialized. This often is accomplished by randomly sampling from the space of possible solutions. New solutions are created by randomly varying existing solutions. This random variation may include mutation and/or recombination. Competing solutions are evaluated in light of a performance index describing their "fitness" (or equivalently, their error). Selection is then applied to determine which solutions will be maintained into the next generation, and with what frequency. These new "parents" are then subjected to random variation, and the process iterates.

random variation and selection, the population can be made to converge asymptotically to optimal solutions (Fogel [6], Rudolph [7], and others).

The evolutionary search is similar to the view offered by Wright [8] involving "adaptive landscapes." A response surface describes the fitness assigned to alternative genotypes as they interact in an environment (Fig. 1.2). Each peak corresponds with an optimized collection of behaviors (phenotypes), and thus one or more sets of optimized genotypes. Evolution probabilistically proceeds up the slopes of the topography toward peaks as selection culls inappropriate phenotypic variants. Others (Atmar [9], Raven and Johnson [10], pp. 400–401) have suggested that it is more appropriate to view the adaptive landscape from an inverted position. The peaks become troughs, or "minimized prediction error entropy wells" (Atmar [9]). Such a viewpoint is intuitively appealing. Searching for peaks depicts evolution as a slowly advancing, tedious, and uncertain process. Moreover, there appears to be a certain fragility to an evolving phyletic line; an optimized population might be expected to quickly fall off the peak under slight perturbations. The inverted topography leaves an altogether different impression. Populations advance rapidly, falling down the walls of the error troughs until their cohesive set of interrelated behaviors is optimized. The topography is generally in flux, as a function of environmental erosion

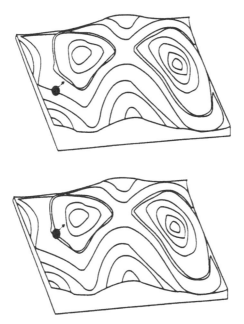

FIGURE 1.2 Evolution on an inverted adaptive topography. A landscape is abstracted to represent the fitness of alternative phenotypes and, as a consequence, alternative genotypes. Rather than viewing the individuals or populations as maximizing fitness and thereby climbing peaks on the landscape, a more intuitive perspective may be obtained by inverting the topography. Populations proceed down the slopes of the topography toward valleys of minimal predictive error.

and variation, as well as other evolving organisms, and stagnation may never set in. Regardless of which perspective is taken, maximizing or minimizing, the basic evolutionary algorithm is the same: a search for the extrema of a functional describing the objective worth of alternative candidate solutions to the problem at hand.

The procedure may be written as the difference equation

$$\mathbf{x}[t+1] = s(v(\mathbf{x}[t])), \tag{1.1}$$

where $\mathbf{x}[t]$ is the population at time t under a representation $\mathbf{x}$, v is a random variation operator, and s is the selection operator (Fogel and Ghozeil [11]). There are a variety of possible representations, variation operators, and selection methods (Bäck et al. [12]). Not more than about 10–12 years ago, there was a general recommendation that the best representation was a binary coding, as this provided the greatest "implicit parallelism" (more detail is offered later in this chapter, and see also Goldberg [13]). But this representation was often cumbersome to implement (consider encoding the solution to a traveling salesman problem as a string of symbols from $\{0, 1\}$), and empirical results did not support any necessity, or even benefit, to binary representations (e.g., Davis [14], Michalewicz [15], Koza [16]). Moreover, the suggestions that advantages would accrue from recombining alternative solutions through crossover operators and amplifying solutions based on their relative fitness also did not obtain empirical support (e.g., Fogel and Atmar [17], Bäck and Schwefel [18], Fogel and Stayton [19], and many others). Recent mathematical results have proved that there can be no best choice for these facets of an evolutionary algorithm that would hold across all problems (Wolpert and Macready [20]), and even that there is no best choice of representation for any individual problem (Fogel and Ghozeil [21]). The effectiveness of an evolutionary algorithm depends on the interplay between the operators s and v as applied to a chosen representation $\mathbf{x}$ and initialization $\mathbf{x}[0]$. This dependence provides freedom to the human operator to tailor the evolutionary approach for their particular problem of interest.

1.2.2   Broad Applicability

Evolutionary algorithms can be applied to virtually any problem that can be formulated as a function optimization task. It requires a data structure to represent solutions, a performance index to evaluate solutions, and variation operators to generate new solutions from old solutions (selection is also required but is less dependent on human preferences). The state space of possible solutions can be disjoint and can encompass infeasible regions, and the performance index can be time varying or even a function of competing solutions extant in the population. The human designer can choose a representation that follows his or her intuition. In this sense, the procedure is representation independent, in contrast with other numerical techniques that might be applicable for only continuous values or other constrained sets. Representation should allow for variation operators that maintain a behavioral link between parent and offspring. Small changes in the structure of a parent should lead to small changes in the resulting offspring, in order to facilitate an understanding of the problem space, and likewise large changes should engender gross alterations. A continuum of possible changes should be allowed such that the effective "step size" of the algorithm can be tuned, perhaps online in a self-adaptive manner (discussed later). This flexibility allows for applying essentially the same procedure to discrete combinatorial problems, continuous-valued parameter optimization problems, mixed-integer problems, and so forth.

1.2.3 Outperform Classic Methods on Real Problems

Real-world function optimization problems often (1) impose nonlinear constraints, (2) require payoff functions that are not concerned with least-squared error, (3) involve nonstationary conditions, (4) incorporate noisy observations or random processing, or include other vagaries that do not conform well to the prerequisites of classic optimization techniques. The response surfaces posed in real-world problems are often multimodal, and gradient-based methods converge rapidly to local optima (or perhaps saddle points), which may yield insufficient performance. For simpler problems, where the response surface is, say, strongly convex, evolutionary algorithms do not perform as well as traditional optimization methods (Bäck [4]). But this is to be expected as these traditional techniques were designed to take advantage of the convex property of such surfaces. Schwefel [22] has shown in a series of empirical comparisons that in the obverse condition of applying classic methods to multimodal functions, evolutionary algorithms offer a significant advantage. In addition, in the often-encountered case of applying linear programming to problems with nonlinear constraints, this offers an almost certainly incorrect result because the assumptions required for the technique are violated. In contrast, evolutionary computation can directly incorporate arbitrary linear and nonlinear constraints (Michalewicz [5]).

Moreover, the problem of defining the payoff function for optimization lies at the heart of success or failure: Inappropriate descriptions of the performance index lead to generating the right answer for the wrong problem. Within classic statistical methods, concern is often devoted to minimizing the squared error between forecast and actual data. But in practice, equally correct predictions are not of equal worth, and errors of identical magnitude are not equally costly. Consider the case of correctly predicting that a particular customer will demand 10 units of energy in a particular time period. This is typically worth less than correctly predicting that the customer will demand 100 units of energy, yet both predictions engender zero error and are weighted equally in classic statistics. Further, the error of predicting the customer will demand 10 units and having them actually demand 100 units is not of equal cost to the energy supplier as predicting the customer will demand 100 units and having them demand 10. One error leaves a missed opportunity cost and the other leaves a 90-unit oversupply. Yet again, under a squared error criterion, these two situations are treated identically. In contrast, within evolutionary algorithms, any definable payoff function can be used to judge the appropriateness of alternative behaviors. There is no restriction that the criteria be differentiable, smooth, or continuous.

1.2.4   Potential to Use Knowledge and Hybridize with Other Methods

It is always reasonable to incorporate domain-specific knowledge into an algorithm when addressing particular real-world problems. Specialized algorithms can outperform unspecialized algorithms on a restricted domain of interest (Wolpert and Macready [20]). Evolutionary algorithms offer a framework such that it is comparably easy to incorporate such knowledge. For example, specific variation operators may be known to be useful when applied to particular representations (e.g., 2-OPT on the traveling salesman problem). These can be applied directly as mutation or recombination operations. Knowledge can also be implemented into the performance index, in the form of known physical or chemical properties (e.g., van der Waals interactions: Gehlhaar et al. [23]). Incorporating such information focuses the evolutionary search, yielding a more efficient exploration of the state space of possible solutions.

Evolutionary algorithms can also be combined with more traditional optimization techniques. This may be as simple as the use of a conjugate-gradient minimization used after primary search with an evolutionary algorithm (e.g., Gehlhaar et al. [23]), or it may involve simultaneous application of algorithms (e.g., the use of evolutionary search for the structure of a model coupled with gradient search for parameter values; Harp et al. [24]). There may also be a benefit to seeding an initial population with solutions derived from other procedures (e.g., a greedy algorithm; Fogel and Fogel [25]). Furthermore, evolutionary computation can be used to optimize the performance of neural networks (Angeline et al. [26]), fuzzy systems (Haffner and Sebald [27]), production systems (Wilson [28]), and other program structures (Koza [16], Angeline and Fogel [29]). In many cases, the limitations of conventional approaches (e.g., the requirement for differentiable hidden nodes when using back propagation to train a neural network) can be avoided.

1.2.5   Parallelism

Evolution is a highly parallel process. As distributed processing computers become more readily available, there will be a corresponding increased potential for applying evolutionary algorithms to more complex problems. It is often the case that individual solutions can be evaluated independently of the evaluations assigned to competing solutions. The evaluation of each solution can be handled in parallel, and only selection (which requires at least pairwise competition) requires some serial processing. In effect, the running time required for an application may be inversely proportional to the number of processors. Regardless of these future advantages, current desktop computing machines provide sufficient computational speed to generate solutions to difficult problems in reasonable time (e.g., the evolution of a neural network for classifying features of breast carcinoma involving more than 5 million separate function evaluations requires only about 3 hours on a 200-MHz 604e PowerPC, Fogel et al. [30], or equivalently one third of an hour on a 2-GHz PC).

1.2.6 Robust to Dynamic Changes

Traditional methods of optimization are not robust to dynamic changes in the environment and often require a complete restart in order to provide a solution (e.g., dynamic programming). In contrast, evolutionary algorithms can be used to adapt solutions to changing circumstance. The available population of evolved solutions provides a basis for further improvement and in most cases it is not necessary, nor desirable, to reinitialize the population at random. Indeed, this procedure of adapting in the face of a dynamic environment can be used to advantage. For example, Wieland [31] used a genetic algorithm to evolve recurrent neural networks to control a cart-pole system comprising two poles (Fig. 1.3). The degree of difficulty depended on the relative pole lengths (i.e., the closer the poles were to each other in length, the more difficult the control problem). Wieland [31] developed controllers for a case of one pole of length 1.0 m and the other of 0.9 m by successively controlling systems where the shorter pole was started at 0.1 m and incremented sequentially to 0.9 m. At each increment, the evolved population of networks served as the basis

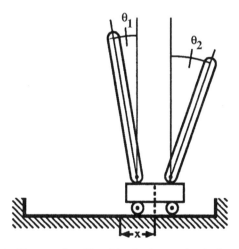

FIGURE 1.3   A cart with two poles. The objective is to maintain the cart between the limits of the track while not allowing either pole to exceed a specified maximum angle of deflection. The only control available is a force with which to push or pull on the cart. The difficulty of the problem is dependent on the similarity in pole lengths. Wieland [31] and Saravanan and Fogel [32] used evolutionary algorithms to optimize neural networks to control this plant for pole lengths of 1.0 m and 0.9 m. The evolutionary procedure required starting with poles of 1.0 m and 0.1 m and iteratively incrementing the length of the shorter pole in a series of dynamic environments. In each case, the most recent evolved population served as the basis for new trials, even when the pole length was altered.

for a new set of controllers. A similar procedure was offered in Saravanan and Fogel [32] and Fogel [33].

The ability to adapt on the fly to changing circumstance is of critical importance to practical problem solving. For example, suppose that a particular simulation provides perfect fidelity to an industrial production setting. All workstations and processes are modeled exactly, and an algorithm is used to find a "perfect" schedule to maximize production. This perfect schedule will, however, never be implemented in practice because by the time it is brought forward for consideration, the plant will have changed: machines may have broken down, personnel may not have reported to work or failed to keep adequate records of prior work in progress, other obligations may require redirecting the utilization of equipment, and so forth. The "perfect" plan is obsolete before it is ever implemented. Rather than spend considerable computational effort to find such perfect plans, a better prescription is to spend less computational effort to discover suitable plans that are robust to expected anomalies and can be evolved on the fly when unexpected events occur.

1.2.7 Capability for Self-Optimization

Most classic optimization techniques require appropriate settings of exogenous variables. This is true of evolutionary algorithms as well. However, there is a long history of using the evolutionary process itself to optimize these parameters as part of the search for optimal solutions (Reed et al. [34], Rosenberg [35], and others). For example, suppose a search problem requires finding the real-valued vector that minimizes a particular functional $f(\mathbf{x})$, where $\mathbf{x}$ is a vector in n dimensions. A typical evolutionary algorithm (Fogel [1]) would use Gaussian random variation on current parent solutions to generate offspring:

$$x_i' = x_i + \sigma_i N(0, 1), \tag{1.2}$$

where the subscript indicates the ith dimension, and σ_i is the standard deviation of a Gaussian random variable, denoted by $N(0, 1)$. Setting the "step size" of the search in each dimension is critical to the success of the procedure (Fig. 1.4). This can be accomplished in the following two-step fashion:

$$\sigma_i' = \sigma_i \exp(\tau N(0, 1) + \tau' N_i(0, 1)) \tag{1.3}$$

$$x_i' = x_i + \sigma_i' N_i(0, 1), \tag{1.4}$$

where $\tau \propto (2n)^{0.5}$ and $\tau' \propto (2n^{0.5})^{0.5}$ (Bäck and Schwefel [18]). In this manner, the standard deviations are subject to variation and selection at a second level (i.e., the level of how well they guide the search for optima of the functional $f(\mathbf{x})$). This general procedure has also been found effective in addressing discrete optimization problems (Angeline et al. [36], Chellapilla and Fogel [37], and others). Essentially,

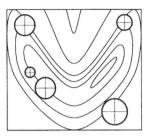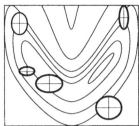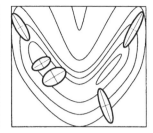

⊖ Line of equal probability density to place an offspring

FIGURE 1.4   When using Gaussian mutations in all dimensions (as in evolution strategies or evolutionary programming), the contours of equal probability density for placing offspring are depicted above (Bäck [4]). In the left panel, all standard deviations in each dimension are equal, resulting in circular contours. In the middle panel, the standard deviations in each dimension may vary, but the perturbation in each dimension is independent of the others (zero covariance), resulting in elliptical contours. In the right panel, arbitrary covariance is applied, resulting in contours that are rotatable ellipses. The method of self-adaptation described in text can be extended to adapt arbitrary covariances, thereby allowing the evolutionary algorithm to adapt to the changes in the response surface during the search for the optimum position on the surface. Similar procedures have been offered for self-adaptation when solving discrete combinatorial problems.

the effect is much like a temperature schedule in simulated annealing; however, the schedule is set by the algorithm as it explores the state space rather than *a priori* by a human operator.

1.2.8   Able to Solve Problems That Have No Known Solutions

Perhaps the greatest advantage of evolutionary algorithms comes from the ability to address problems for which there are no human experts. Although human expertise should be used when it is available, it often proves less than adequate for automating problem-solving routines. Troubles with such expert systems are well-known: the experts may not agree, may not be self-consistent, may not be qualified, or may simply be in error. Research in artificial intelligence has fragmented into a collection of methods and tricks for solving particular problems in restricted domains of interest. Certainly, these methods have been successfully applied to specific problems (e.g., the chess program Deep Blue). But most of these applications require human expertise. They may be applied impressively to difficult problems requiring great computational speed, but they generally do not advance our understanding of intelligence. "They solve problems, but they do not solve the problem of how to solve problems" (Fogel [1]). In contrast, evolution provides a method for solving the problem of how to solve problems even in the absence of human expertise. It is a recapitulation of the scientific method (Fogel et al. [38]) that can be used to learn fundamental aspects of any measurable environment.

1.3   CURRENT DEVELOPMENTS

As indicated above, evolutionary computation has a long history with several independent beginnings. Each of these beginnings, whether they occurred in the simulation of genetic systems (Fraser [39], Bremermann [40], Holland [41]), engineering optimization (Rechenberg [42]), artificial intelligence (Fogel et al. [38]), or other areas, had specific traits that were originally unique to them individually. For example, the classic genetic algorithm of Holland [41] operated on binary strings (regardless of the problem). In contrast, the classic work of Rechenberg and Schwefel (termed *evolution strategies*) relied on real-valued representations, whereas the classic evolutionary programming work of L. Fogel used finite state machines. These issues of different representations, as well as variation operators, selection methods, and other particular aspects of evolutionary algorithms, have grown to be virtually nonexistent in current evolutionary computation practice. It now makes little sense (scientifically) to speak of these originally different methods as being currently disparate: They are all plainly similar in the extreme, all relying on diverse possible representations, with many choices for variation and selection.

This similarity has grown mainly from increased communications across the groups over the past decade, as well as a recognition that early theory in evolutionary algorithms had many flaws. Some historical and current aspects of evolutionary algorithm theory are offered here.

1.3.1   Review of Some Historical Theory in Evolutionary Computation

Some early efforts (1975–1990) in evolutionary algorithm theory focused on (1) the belief that it is possible to generate superior general problem solvers, (2) the notion that maximizing implicit parallelism is useful, (3) a schema theorem to describe the propagation of components of solutions in a population, and (4) an analysis of a two-armed bandit problem that was intended to indicate an optimal sampling plan for evolutionary algorithms. Unfortunately, the conventional wisdom regarding these four focal points from even just a decade ago has been shown to be either incomplete or incorrect. As a result, the foundations of evolutionary computation have been reconsidered, and an integrated approach to the study of evolutionary computation as a whole has been undertaken.

1.3.2   No Free Lunch Theorem

It is natural to ask if there is a best evolutionary algorithm that would always give superior results across the possible range of problems. Is there some choice of variation operators and selection mechanisms that will always outperform all other choices regardless of the problem? Sadly, the answer is no: There is no best evolutionary algorithm. In mathematical terms, let an algorithm a be represented as a mapping from previously unvisited sets of points to a single new (i.e., previously unvisited) point in the state space of all possible points (solutions). Let $P(d_m^y|f, m, a)$ be the conditional probability of obtaining a particular sample d_m when algorithm a is iterated m times on cost function f. Given these preliminaries, Wolpert and Macready [20] proved the so-called no free lunch theorem:

Theorem 1.1 (No Free Lunch). For any pair of algorithms a_1 and a_2,

$$\sum_f P(d_m^y|f, m, a_1) = \sum_f P(d_m^y|f, m, a_2).$$

(See Appendix A of Wolpert and Macready [20] for the proof; English [43] showed that a similar no free lunch result holds whenever the values assigned to points are independent and identically distributed random variables.) That is, the sum of the conditional probabilities of visiting each point d_m is the same over all cost functions f regardless of the algorithm chosen. The immediate corollary of this theorem is that for any performance measure $\Phi(d_m^y)$, the average over all f of $P(\Phi(d_m^y)|f, m, a)$ is independent of a. In other words, there is no best algorithm, whether or not that algorithm is "evolutionary," and moreover whatever an algorithm gains in performance on one class of problems is necessarily offset by that algorithm's performance on the remaining problems.

This simple theorem has engendered a great deal of controversy in the field of evolutionary computation and some associated misunderstanding. There has been considerable effort expended in finding the "best" set of parameters and operators for evolutionary algorithms since at least the mid-1970s. These efforts have involved the type of recombination, the probabilities for crossover and mutation, the representation, the population size, and so forth. Most of this research involved empirical trials on benchmark functions. But the no free lunch theorem essentially dictates that the conclusions made on the basis of such sampling are in the strict mathematical sense limited to only those functions studied. Efforts to find the best crossover rate, the best mutation operator, and so forth, in the absence of restricting attention to a particular class of problems and methods are pointless.

For an algorithm to perform better than even random search (which is simply another algorithm), it must reflect something about the structure of the problem it faces. By consequence, it mismatches the structure of some other problem. Note too that it is not enough to simply identify that a problem has some structure associated with it: that structure must be appropriate to the algorithm at hand. Moreover, the structure must be specific. It is not enough to say, as is often heard, "I am concerned only with real-world problems, not all possible problems, and therefore the no free lunch theorem does not apply." What is the structure of "real-world" problems? Indeed, what is a real-world problem? The obvious vague quality of this description is immediately problematic. What constitutes a real-world problem now might not have been a problem at all, say, 100 years ago (e.g., what to watch on television on a Thursday night). Regardless, simply narrowing the domain of possible problems without identifying the correspondence between the set of problems considered and the algorithm at hand does not suffice to claim any advantage for a particular method of problem solving.

One apt example of how the match between an algorithm and the problem can be exploited was offered in De Jong et al. [44]. For a very simple problem of finding the two-bit vector **x** that maximizes the function

$$f(\mathbf{x}) = \text{integer}(\mathbf{x}) + 1,$$

where integer(**x**) returns 0 for [00], 1 for [01], 2 for [10], and 3 for [11], De Jong et al. [44] employed an evolutionary algorithm with (1) one-point crossover at either a probability of 1.0 or 0.0, (2) a constant mutation rate of 0.1, and (3) a population of size 5. In this trivial example, it was possible to calculate the exact probability that the global best vector would be contained in the population as a function of the number of generations. In this case, the use of crossover definitely increased the likelihood of discovering the best solution, and mutation alone was better than random search.

In this example, the function f assigned values $\{1, 2, 3, 4\}$ to the vectors $\{[00], [01], [10], [11]\}$, respectively, but this is not the only way to assign these fitness values to the possible strings. In fact, there are $4! = 24$ different permutations that could be used. De Jong et al. [44] showed that, in this case, the performance obtained with the 24 different permutations falls into three equivalence classes, each containing eight permutations that produce identical probability of success curves. In the second and third equivalence classes, the use of crossover was seen to be detrimental to the likelihood of success, and in fact random search outperformed evolutionary search for the first 10–20 generations in the third equivalence class. In the first equivalence class, crossover could combine the second- and third-best vectors to generate the best vector. In the second and third equivalence classes, it could not usefully combine these vectors: the structure of the problem did not match the structure of the crossover operator. Any specific search operator can be rendered superior or inferior simply by changing the structure of the problem.

Intrinsic to every evolutionary algorithm is a representation for manipulating candidate solutions to the problem at hand. The no free lunch theorem establishes that there is no best evolutionary algorithm across all problems. The fundamental result is twofold: (1) claims that evolutionary algorithms must rely on specific operators to be successful (e.g., the less-often-heard but still occasional claim that crossover is a necessary component of a successful evolutionary algorithm, as found in [13]) are not correct, and (2) efforts to make generally superior evolutionary algorithms are misguided.

1.3.3 Computational Equivalence of Representations

Holland ([41], p. 71) suggested that alternative representations in an evolutionary algorithm could be compared by calculating the number of schemata (subset templates, see below) processed by the algorithm. In order to maximize this *intrinsic parallelism*, it was recommended that representations should be chosen with the fewest "detectors with a range of many attributes." In other words, alphabets with low cardinality were to be favored because they generate more schemata. For example, six elements, each with a range of 10 values, can generate 1 million distinct representations, which is about the same as 20 elements with a range of 2 values ($2^{20} = 1,048,576$). But the number of schemata processed is 11^6 ($=1,771,561$) versus 3^{20} ($=3.49 \times 10^9$). This increased number of schemata was suggested to give a "larger information flow" to reproductive plans such as genetic algorithms. Emphasis was therefore placed on binary representations, this offering the lowest possible cardinality and the greatest number of schemata.

To review, a schema is a template with fixed and variable symbols. Consider a string of symbols from an alphabet **A**. Suppose that some of the components of the string are held fixed while others are free to vary. Following Holland [41], define a wild card symbol, $\# \in \mathbf{A}$, that matches any symbol from **A**. A string with fixed and/or variable symbols defines a schema, which is a set denoted by a string over the union of $\{\#\}$ and the alphabet $\mathbf{A} = \{0, 1\}$. Consider the schema [01##], which includes [0100], [0101], [0110], and [0111]. Holland ([41], pp. 64–74) offered that every evaluated string actually offers partial information about the expected fitness of all possible schemata in which that string resides. That is, if string [0000] is evaluated to have some fitness, then partial information is also received about the worth of sampling from variations in [####], [0###], [#0##], [#00#], [#0#0], and so forth. This characteristic was termed *intrinsic parallelism* (or *implicit parallelism*), in that through a single sample, information is gained with respect to many schemata.

Antonisse [45] offered a different interpretation of the wild card symbol # that led to an alternative recommendation regarding the cardinality of a chosen representation. Rather than view the # symbol in a string as a "don't care" character, it can be viewed as indicating all possible subsets of symbols at the particular position. If $\mathbf{A} = \{0, 1, 2\}$, then the schema [000#] would indicate the sets $\{[0000]\ [0001]\}$, $\{[0000]\ [0002]\}$, $\{[0001]\ [0002]\}$, and $\{[0000]\ [0001]\ [0002]\}$ as the # symbol indicates the possibilities of (a) 0 or 1, (b) 0 or 2, (c) 1 or 2, and (d) 0, 1, or 2. When schemata are viewed in this manner, the greater implicit parallelism comes from the use of more, not fewer, symbols.

Fogel and Ghozeil [21] showed that, in contrast with the arguments offered in both Holland [41] and Antonisse [45], there can be no intrinsic advantage to any choice of cardinality: Given some weak assumptions about the structure of the problem space and representation, equivalent evolutionary algorithms can be generated for any choice of cardinality. Moreover, the theorems in Ref. 21 indicate that there can be no intrinsic advantage to using any particular two-parent search operator (e.g., a crossover) as there always exists an alternative two-parent operator that performs the equivalent function, regardless of the chosen representation. The proofs carry considerable notation, and the reader is recommended to Ref. 21 to review them if interested.

These theorems from Ref. 21 provide an extension of the result offered in Battle and Vose [46] where it was shown that isomorphisms exist between alternative instances of genetic algorithms for binary representations (i.e., the operations of crossover and mutation on a population under a particular binary representation in light of a fitness function can be mapped equivalently to alternative similar operators for any other binary representation). They also extend the results of Vose and Liepins [47] and Radcliffe [48], where it was shown that there can be no general advantage for any particular binary representation. Although particular representations and operators may be more computationally tractable or efficient than others in certain cases, or may appeal to the designer's intuition, under the conditions studied in Ref. 21, no choice of representation, or one-point or two-point variation operator, can offer a capability not found in another choice of representation or analogous operator.

1.3.4 Schema Theorem in the Presence of Random Variation

The traditional method of selection in genetic algorithms (from, say, Holland [41]) requires solutions to be reproduced in proportion to their fitness relative to the other solutions in the population (sometimes termed *roulette wheel selection* or *reproduction with emphasis*). That is, if the fitness of the ith string in the population at time t, x_i^t, is denoted by $f(x_i^t)$, then the probability of selecting the ith string for reproduction for each available slot in the population at time $t + 1$ is given by

$$P(x_i^{t+1}) = \frac{f(x_i^t)}{\sum\limits_{i=1}^{n} f(x_i^t)}, \tag{1.5}$$

where there are n members of the population at time t. This procedure requires strictly positive fitness values.

The implementation of proportional selection leads to the well-known variant of the schema theorem (Holland [41]):

$$EP(H, t + 1) = P(H, t)\frac{f(H, t)}{\bar{f}_t}, \tag{1.6}$$

where H is a particular schema (the notation of H is used to denote the schema as a hyperplane), $P(H, t)$ is the proportion of strings in the population at time t that are an instance of schema H, $f(H, t)$ is the mean fitness of strings in the population at time t that are an instance of H, $\bar{f}_t$ is the mean fitness of all strings in the population at time t, and $EP(H, t + 1)$ denotes the expected proportion of strings in the population at time $t + 1$ that will be instances of the schema H. Equation (1.2) does not include any effects of recombination or mutation. It only describes the effects of proportional selection on a population in terms of the manner in which schemata are expected to increase or decrease over time.

The fitness associated with a schema depends on which instances of that schema are evaluated. Moreover, in real-world practice, the evaluation of a string will often include some random effects (e.g., observation error, time dependency, uncontrollable exogenous variables). That is, the observed fitness of a schema H (or any element of that schema) may not be described by a constant value, but rather by a random variable with an associated probability density function. Selection operates not on the mean of all possible samples of the schema H but only on the fitness associated with each observed instance of the schema H in the population. It is therefore of interest to assess the expected allocation of trials to schemata when their observed fitness takes the form of a random variable. Fogel and Ghozeil [49] showed that this can result in the introduction of a bias such that the expected sampling from alternative schemata will not be in proportion to their mean fitness.

The schema theorem of Holland [41] refers only to the specific realized values of competing schemata in a population; it is not intended to handle the case when the fitness of these schemata are described by random variables. The analysis in Fogel and Ghozeil [49] (cf., Rana et al. [50], see also Gillespie [51]) indicates that the expected proportion of a particular schema H in the population at the next time step is not generally governed by the ratio of the mean of that schema H to the sum of the means of schema H and schema H' (everything that is not in H). In general, there is no *a priori* reason to expect the schema theorem to adequately describe the mean sampling of alternative schemata when the fitness evaluation of those schemata is governed by random variation. This may occur in the form of observation noise on any particular string, random selection of individual strings from a particular schema, or a number of other mechanisms. Under such conditions, reliably estimating mean hyperplane performance is not sufficient to predict the expected proportion of samples that will be allocated to a particular hyperplane under proportional selection. The fundamental result is that the schema theorem of Ref. 41 cannot, in general, be used to reliably predict the average representation of schemata in future populations when schema fitness depends on random factors.

1.3.5 Two-Armed Bandits and the Optimal Allocation of Trials

Recalling the idea of sampling from alternative schemata, in order to optimally allocate trials to schemata, a loss function describing the worth of each trial must be formulated. Holland ([41], pp. 75–83) considered and analyzed the problem of minimizing expected losses while sampling from alternative schemata. The essence of this problem can be modeled as a two-armed bandit (slot machine): Each arm of the bandit is associated with a random payoff described by a mean and variance (much like the result of sampling from a particular schema where the payoff depends on which instance of that schema is sampled). The goal is to best allocate trials to the alternative arms conditioned on the payoffs that have already been observed in previous trials. Holland ([41], pp. 85–87) extended the case of two competing schemata to any number of competing schemata. The results of this analysis were used to guide the formulation of one early version of evolutionary algorithms, so it is important to review this analysis and its consequences. This is particularly true because the formulation has recently been shown mathematically to be flawed (Rudolph [52]; Macready and Wolpert [53]).

Holland ([41], pp. 75–83) examined the two-armed bandit problem where there are two random variables, RV_1 and RV_2 (representing two slot machines) from which samples may be taken (much like schemata represent samples from a solutions space). The two random variables are assumed to be independent and possess some unknown means (μ_1, μ_2) and unknown variances (σ_1^2, σ_2^2). A trial is conducted by sampling from a chosen random variable; the result of the trial is the payoff. Suppose some number of trials has been devoted to each random variable (n_1 and n_2, respectively) and that the average payoff (the sum of the individual payoffs divided by the number of trials) from one random variable is greater than for the other. Let RV_{high} be the random variable for which the greater average payoff has been observed (not necessarily the random variable with the greater mean), and let RV_{low} be the other random variable. The objective in this problem is to allocate trials to each random variable so as to minimize the expected loss function

$$L(n_1, n_2) = [q(n_1, n_2)n_1 + (1 - q(n_1, n_2))n_2] \cdot |\mu_1 - \mu_2|, \tag{1.7}$$

where $L(n_1, n_2)$ describes the total expected loss from allocating n_1 samples to RV_1 and n_2 samples to RV_2, and

$$q(n_1, n_2) = \begin{cases} \Pr(\dot{x}_1 > \dot{x}_2) & \text{if } \mu_1 < \mu_2 \\ \Pr(\dot{x}_1 > \dot{x}_2) & \text{if } \mu_1 > \mu_2 \end{cases},$$

where x_1 and x_2 designate the mean payoffs (sample means) for having allocated n_1 and n_2 samples to RV_1 and RV_2, respectively. Note that specific measurements and the number of samples yield explicit values for the sample means.

As summarized in Goldberg ([13], p. 37), Holland ([41], pp. 77–78) proved the following (notation consistent with Holland [41]):

Theorem 1.2   Given N trials to be allocated to two random variables with means $\mu_1 > \mu_2$ and variances σ_1^2 and σ_2^2, respectively, and the expected loss function described in Eq. (1.7), the minimum expected loss results when the number of trials allocated to the random variable with the lower observed average payoff is

$$n^* \cong b^2 \ln \left[\frac{N^2}{8 \pi b^4 \ln N^2} \right], \tag{1.8}$$

where $b = \sigma_1 / (\mu_1 - \mu_2)$. The number of trials to be allocated to the random variable with the higher observed average payoff is $N - n^*$.

If the assumptions associated with the proof (Holland [41], pp. 75–83) held, and if the mathematical framework were correct, the above analysis would apply equally well to a k-armed bandit as to a two-armed bandit.

Unfortunately, as offered in Macready and Wolpert [53], the error in Holland ([41], pp. 77–85) stems from considering the unconditioned expected loss for allocating $N - 2n$ trials after allocating n to each bandit, rather than the expected loss conditioned on the information available after the $2n$ pulls. Macready and Wolpert [53] showed that a simple greedy strategy that pulls the arm that maximizes the payoff for the next pull based on a Bayesian update from prior pulls outperforms the strategy offered in Holland ([41], p. 77). The trade-off between exploitation (pulling the best Bayesian bandit) and exploration (pulling the other bandit in the hopes that the current Bayesian information is misleading) offered in Holland [41] is not optimal for the problem studied. Macready and Wolpert [53] remarked that the algorithms proposed in Holland [41] "are based on the premise that one should engage in exploration, yet for the very problem invoked to justify genetic exploration, the strategy of not exploring at all performs better than n^* exploring algorithms...."

The ramifications of this condition are important: One of the keystones that has differentiated genetic algorithms from other forms of evolutionary computation has been the notion of optimal schema processing. Until recently, this was believed to have a firm theoretical footing. It now appears, however, that the basis for claiming that the n^* sampling offered in Ref. [41] is optimal for minimizing expected losses to competing schemata is lacking. In turn, neither is the use of proportional selection optimal for achieving a proper allocation of trials (recall that proportional selection was believed to achieve an exponentially increasing allocation of trials to the observed best schemata, this in line with the mathematical analysis of the two-armed bandit). And as yet another consequence then, the schema theorem under proportional selection, which describes the expected allocation of trials to schemata in a single generation, lacks fundamental importance (cf. [13]). It is simply a formulation for describing one method of performing selection but not a generally *optimal* method.

1.4 CONCLUSIONS

Although the history of evolutionary computation dates back to the 1950s and 1960s (Fogel [3]), only within the past decade have evolutionary algorithms become practicable for solving real-world problems on desktop computers (Bäck et al. [12]). As computers continue to deliver accelerated performance, these applications will only become more routine. The flexibility of evolutionary algorithms to address general optimization problems using virtually any reasonable representation and performance index, with variation operators that can be tailored for the problem at hand and selection mechanisms tuned for the appropriate level of stringency, gives these techniques an advantage over classic numerical optimization procedures. Moreover, the two-step procedure to self-adapt parameters that control the evolutionary search frees the human operator from having to handcraft solutions, which would often be time consuming or simply infeasible. Evolutionary algorithms offer a set of procedures that may be usefully applied to problems that have resisted solution by common techniques and can be hybridized with such techniques when such combinations appear beneficial. Moreover, the notion of modeling natural systems can be brought to fruition in other forms, including particle swarm methods as indicated here, as well as ant colony optimization and other techniques not mentioned. If this brief introduction serves to spur the imagination of the reader into finding new ways to model natural systems, particularly as they may be applied to problems in power systems, it will have succeeded immensely.

ACKNOWLEDGMENTS

The author would like to thank the editors for inviting this presentation, IEEE/Wiley for permissions to reprint sections of the author's prior publications, and T. Bäck for permission to reprint his figure from Bäck [4].

REFERENCES

1. Fogel DB. Evolutionary computation: Toward a new philosophy of machine intelligence. 2nd ed. Piscataway, NJ: IEEE Press; 2000.

2. Fogel DB. What is evolutionary computation? IEEE Spectrum 2000; February:26–32.

3. Fogel DB, ed. Evolutionary computation: The fossil record. Piscataway, NJ: IEEE Press; 1998.

4. Bäck T. Evolutionary algorithms in theory and practice. New York: Oxford University Press; 1996.

5. Michalewicz Z. Genetic algorithms + data structures = evolution programs. 3rd ed. Berlin: Springer; 1996.

6. Fogel DB. Asymptotic convergence properties of genetic algorithms and evolutionary programming: analysis and experiments. Cybern Syst 1994; 25:389–407.

7. Rudolph G. Convergence analysis of canonical genetic algorithms. IEEE Trans Neural Networks 1994; 5:96–101.

8. Wright S. The roles of mutation, inbreeding, crossbreeding, and selection in evolution. Proc. 6th Int. Cong. Genetics. Vol. 1. Genetic Society of America, Ithaca. 1932. p. 356–366.

9. Atmar W. The inevitability of evolutionary invention. 1979 (unpublished manuscript).

10. Raven PH, Johnson GB. Biology. St. Louis: Times Mirror; 1986.

11. Fogel DB, Ghozeil A. Using fitness distributions to design more efficient evolutionary computations. Proc. of 1996 IEEE Conf. on Evol. Comp. Keynote Lecture. New York: IEEE Press; 1996. p. 11–19.

12. Bäck T, Fogel DB, Michalewicz Z. eds. Handbook of evolutionary computation. New York: Oxford University Press; 1997.

13. Goldberg DE. Genetic algorithms in search, optimization and machine learning. Reading, MA: Addison-Wesley; 1989.

14. Davis L, ed. Handbook of genetic algorithms. New York: Van Nostrand Reinhold; 1991.

15. Michalewicz Z. Genetic algorithms + data structures = evolution programs. Berlin: Springer; 1992.

16. Koza JR. Genetic programming. Cambridge, MA: MIT Press; 1992.

17. Fogel DB, Atmar JW. Comparing genetic operators with Gaussian mutations in simulated evolutionary processes using linear systems. Biol Cybern 1990; 63:111–114.

18. Bäck T, Schwefel H-P. An overview of evolutionary algorithms for parameter optimization. Evol Comp 1993; 1:1–24.

19. Fogel DB, Stayton LC. On the effectiveness of crossover in simulated evolutionary optimization. BioSystems 1994; 32:171–182.

20. Wolpert DH, Macready WG. No free lunch theorems for optimization. IEEE Trans Evol Comp 1997; 1:67–82.

21. Fogel DB, Ghozeil A. A note on representations and variation operators. IEEE Trans Evol Comp 1997; 1:159–161.

22. Schwefel H-P. Evolution and optimum seeking. New York: John Wiley & Sons; 1995.

23. Gehlhaar DK, Verkhivker GM, Rejto PA, Sherman CJ, Fogel DB, Fogel LJ, Freer ST. Molecular recognition of the inhibitor AG-1343 by HIV-1 protease: Conformationally flexible docking by evolutionary programming. Chem Biol 1995; 2:317–324.

24. Harp SA, Samad T, Guha A. Towards the genetic synthesis of neural networks. In: Schaffer JD, ed. Proc. of the 3rd Intern. Conf. on Genetic Algorithms. San Mateo, CA: Morgan Kaufmann; 1989. p. 360–369.

25. Fogel DB, Fogel LJ. Using evolutionary programming to schedule tasks on a suite of heterogeneous computers. Comp Oper Res 1996; 23:527–534.

26. Angeline PJ, Saunders GM, Pollack JB. An evolutionary algorithm that constructs recurrent neural networks. IEEE Trans Neural Networks 1994; 5:54–65.

27. Haffner SB, Sebald AV. Computer-aided design of fuzzy HVAC controllers using evolutionary programming. In: Fogel DB, Atmar W, eds. Proc. of the 2nd Ann. Conf. on Evolutionary Programming. La Jolla, CA: Evolutionary Programming Society; 1993. p. 98–107.

28. Wilson SW. Classifier fitness based on accuracy. Evol Comp 1995; 3:149–175.

29. Angeline PJ, Fogel DB. An evolutionary program for the identification of dynamical systems. In: Rogers SK, Rock D, eds. Aerosence 97, Symp. on Neural Networks. Vol. 3077. Orlando, FL: SPIE 1997. p. 409–417.

30. Fogel DB, Wasson EC, Boughton EM, Porto VW. A step toward computer-assisted mammography using evolutionary programming and neural networks. Cancer Lett 1997; 119:93–97.

31. Wieland AP. Evolving controls for unstable systems. In: Touretzky DS, Elman JL, Sejnowski TJ, Hinton GE, eds. Connectionist models: Proceedings of the 1990 Summer School. San Mateo, CA: Morgan Kaufmann; 1990. p. 91–102.

32. Saravanan N, Fogel DB. Evolving neurocontrollers using evolutionary programming. In: IEEE Conf. on Evol. Comp. Vol. 1. Piscataway, NJ: IEEE Press; 1994. p. 217–222.

33. Fogel DB. A 'correction' to some cart-pole experiments. In: Fogel LJ, Angeline PJ, Bäck T, eds. Evolutionary programming VI. Cambridge, MA: MIT Press; 1996. p. 67–71.

34. Reed J, Toombs R, Barricelli NA. Simulation of biological evolution and machine learning. J Theor Biol 1967; 17:319–342.

35. Rosenberg R. Simulation of genetic populations with biochemical properties. Ph.D. dissertation, University of Michigan, Ann Arbor. 1967.

36. Angeline PJ, Fogel DB, Fogel LJ. A comparison of self-adaptation methods for finite state machines in a dynamic environment. In: Fogel LJ, Angeline PJ, Bäck T, eds. Evolutionary programming V. Cambridge, MA: MIT Press; 1996. p. 441–449.

37. Chellapilla K, Fogel DB. Exploring self-adaptive methods to improve the efficiency of generating approximate solutions to traveling salesman problems using evolutionary programming. In: Angeline PJ, Reynolds RG, McDonnell JR, Eberhart R, eds. Evolutionary Programming VI. Berlin: Springer; 1997. p. 361–371.

38. Fogel LJ, Owens AJ, Walsh MJ. Artificial intelligence through simulated evolution. New York: John Wiley & Sons; 1996.

39. Fraser AS. Simulation of genetic systems by automatic digital computers. I. Introduction. Australian J Biol Sci 1957; 10:484–491.

40. Bremermann HJ. Optimization through evolution and recombination. In: Yovits MC, Jacobi GT, Goldstein GD, eds. Self-organizing systems—1962. Washington, DC: Spartan Books; 1962. p. 93–106.

41. Holland JH. Adaptation in natural and artificial systems. Ann Arbor, MI: Univeristy of Michigan Press. 1975.

42. Rechenberg I. Cybernetic solution path of an experimental problem. Royal Aircraft Establishment; Library Translation 1122, 1965.

43. English TM. Evaluation of evolutionary and genetic optimizers: no free lunch. In: Fogel LJ, Angeline PJ, Bäck T, eds. Evolutionary programming V: Proc. of the 5th Annual Conference on Evolutionary Programming. Cambridge, MA: MIT Press; 1996. p. 163–169.

44. De Jong KA, Spears WM, Gordon DF. Using Markov chains to analyze GAFOs. In: Whitley LD, Vose MD, eds. Foundations of genetic algorithms 3. San Mateo, CA: Morgan Kaufmann; 1995. p. 115–137.

45. Antonisse J. A new interpretation of schema notation that overturns the binary encoding constraint. In: Schaffer JD, ed. Proceedings of the Third Int. Conf. on Genetic Algorithms. San Mateo, CA: Morgan Kaufmann; 1989. p. 86–91.

46. Battle DL, Vose MD. Isomorphisms of genetic algorithms. Artificial Intelligence 1993; 60:155–165.

47. Vose MD, Liepins GE. Schema disruption. In: Belew RK, Booker LB, eds. Proceedings of the Fourth International Conference on Genetic Algorithms. San Mateo, CA: Morgan Kaufmann; 1991. p. 237–240.

48. Radcliffe NJ. Non-linear genetic representations. In: Männer R, Manderick B, eds. Parallel problem solving from nature. 2. Amsterdam: North-Holland; 1992. p. 259–268.

49. Fogel DB, Ghozeil A. Schema processing under proportional selection in the presence of random effects. IEEE Trans Evol Computat 1997; 1(4): p. 290–293.

50. Rana S, Whitley LD, Cogswell R. Searching in the presence of noise. In: Voigt H-M, Ebeling W, Rechenberg I, Schwefel HP, eds. Parallel problem solving from nature— PPSN IV. Berlin: Springer; 1996. p. 198–207.

51. Gillespie H. Natural selection for variances in offspring numbers: a new evolutionary principle. Am Naturalist 1977; 111:1010–1014.

52. Rudolph G. Reflections on bandit problems and selection methods in uncertain environments. In Bäck T, ed. Proc. of 7th Intern. Conf. on Genetic Algorithms, San Francisco: Morgan Kaufmann; 1997. p. 166–173.

53. Macready WG, Wolpert DH. Bandit problems and the exploration/exploitation tradeoff. IEEE Trans Evol Computat 1998; 2(1):2–22.

CHAPTER 2

Fundamentals of Genetic Algorithms

ALEXANDRE P. ALVES DA SILVA and DJALMA M. FALCÃO

2.1 INTRODUCTION

Research on genetic algorithms (GAs) has shown that the initial proposals are incapable of solving hard problems in a robust and efficient way. Usually, for large-scale optimization problems, the execution time of first-generation GAs increases dramatically whereas solution quality decreases. The aim of this chapter is to point out the main design issues in tailoring GAs to large-scale optimization problems. Important topics such as encoding schemes, selection procedures, and self-adaptive and knowledge-based operators are discussed.

2.2 MODERN HEURISTIC SEARCH TECHNIQUES

Optimization is the basic concept behind the application of GAs, or any other evolutionary algorithm [1–3], to any field of interest. Over and above the problems in which optimization itself is the final goal, it is also a way for achieving (or the main idea behind) modeling, forecasting, control, simulation, and so forth.

Traditional optimization techniques begin with a single candidate and search iteratively for the optimal solution by applying static heuristics. On the other hand, the GA approach uses a population of candidates to search several areas of a solution space, simultaneously and adaptively.

Evolutionary computation allows precise modeling of the optimization problem, although not usually providing mathematically optimal solutions. Another advantage of using evolutionary computation techniques is that there is no need for having an explicit objective function. Moreover, when the objective function is available, it does not have to be differentiable. Genetic algorithms have been most commonly applied to solve combinatorial optimization problems. Combinatorial optimization usually involves a huge number of possible solutions, which makes the use of

Modern Heuristic Optimization Techniques. Edited by K. Y. Lee and M. A. El-Sharkawi

enumeration techniques (e.g., cutting plane, branch and bound, or dynamic programming) hopeless.

Thermal unit commitment, hydrothermal coordination, expansion planning (generation, transmission, and distribution), reactive compensation placement, maintenance scheduling, and so forth, have the typical features of a large-scale combinatorial optimization problem. In problems of this kind, the number of possible solutions grows exponentially with the problem size. Therefore, the application of optimization methods to find the optimal solution is computationally impracticable. Heuristic search techniques are frequently employed in this case for achieving high-quality solutions within reasonable run time.

Among the heuristic search methods, there are the ones that apply local search (e.g., hill climbing) and the ones that use a nonconvex optimization approach, in which cost-deteriorating neighbors are accepted also. The most popular methods that go beyond simple local search are GAs [4–7] (and other evolutionary techniques, such as evolutionary programming, evolutionary strategies, etc.), simulated annealing (SA) [8], and tabu search (TS) [9]. Particle swarm [10] is another optimization technique that has shown great potential lately. However, more experience is still necessary to indicate its efficiency and robustness.

Simulated annealing uses a probability function that allows a move to a worse solution with a decreasing probability as the search progresses. With GAs, a pool of solutions is used, and the neighborhood function is extended to act on pairs of solutions. Tabu search uses a deterministic rather than stochastic search. Tabu search is based on a neighborhood search with local optima avoidance. In order to avoid cycling, a short-term adaptive memory is used in TS. Genetic algorithms have a basic distinction when compared with other methods based on stochastic search. They often use a coding (genotypic space) for representing the problem. The other methods often solve the optimization problem in the original representation space (phenotypic).

The most rigorous global search methods have asymptotic convergence proof (also known as convergence in probability); that is, the optimal solution is guaranteed to be found if infinite time is available. Among SA, GA, and TS algorithms, simulated annealing and genetic algorithms are the only ones with proof of convergence. However, there is no such proof for the canonical GA [11], that is, the one with proportional selection (Section 2.6.1.6) and crossover/mutation with constant probabilities (Section 2.6.4), which in fact is divergent.

Although all the mentioned algorithms have been applied successfully to real-world problems, several of their crucial parameters have been selected empirically. Theoretical knowledge of the impact of these parameters on convergence is still an open problem. In fact, there is only a beginning of theoretical results for tabu and particle swarm searches.

The choice of representation for a GA is fundamental to achieving good results. Encoding allows a kind of *tunneling* in the original search space. That means a particle has a nonzero probability of passing a potential barrier even when it does not have enough energy to jump over the barrier. The tunneling idea is that rather than escaping from local minima by random uphill moves, escape can be achieved with the quantum tunnel effect. It is not the height of the barrier that determines the rate of escape from a local optimum, but rather its width relative to current population variance.

The main shortcoming of the standard SA procedure is the slow asymptotic convergence with respect to the *temperature* parameter T. In the standard SA algorithm, the cooling schedule for asymptotic global convergence is inversely proportional to the logarithm of the number of iterations (k); that is, $T(k) = c/(1 + \log k)$. The constant c is the largest depth of any local minimum that is not the global minimum. Convergence in probability cannot be guaranteed for faster cooling rates (e.g., lower values for c).

Tabu search owes its efficiency to an experience-based fine-tuning of a large collection of parameters. Tabu search is a general search scheme that must be tailored to the details of the problem at hand. Unfortunately, as mentioned before, there is little theoretical knowledge for guiding this tailoring process.

Heuristic search methods utilize different mechanisms in order to explore the state space. These mechanisms are based on three basic features:

- The use of memoryless search (e.g., standard SA and GA) or adaptive memory (e.g., TS);
- The kind of neighborhood exploration used, that is, random (e.g., SA and GAs) or systematic (e.g., TS); and
- The number of current solutions taken from one iteration to the next (GAs, as opposed to SA and TS, take multiple solutions to the next iteration).

The combination of these mechanisms for exploring the state space determines the search diversification (global exploration) and intensification (local exploitation) capabilities. The standard SA algorithm is notoriously deficient with respect to the diversification aspect. On the other hand, the standard GA is poor in intensification.

When the objective function has very many equally good local minima, wherever the starting point is, a small random disturbance can avoid the small local minima and reach one of the good ones, making this an appropriate problem for SA. However, SA is less suitable for a problem in which there is one global minimum that is much better than all the other local ones. In this case, it is very important to find that valley. Therefore, it is better to spend less time improving any set of parameters and more time working with an ensemble to examine different regions of the space. This is what GAs do best. Hybrid methods have been proposed in order to improve the robustness of the search.

2.3 INTRODUCTION TO GAs

Genetic algorithms operate on a population of individuals. Each individual is a potential solution to a given problem and is typically encoded as a fixed-length binary string (other representations have also been used, including character-based and real-valued encodings, etc.), which is an analogy with an actual chromosome. After an initial population is randomly or heuristically generated, the algorithm evolves the population through sequential and iterative application of three operators: selection, crossover, and mutation. A new generation is formed at the end of each iteration.

For large-scale optimization problems, the initial population can incorporate prior knowledge about solutions. This procedure should not drastically restrict the population diversity, otherwise premature convergence could occur. Typical population sizes vary between 30 and 200. The population size is usually set as a function of the chromosome length.

The execution of a GA iteration is basically a two-stage process. It starts with the current population. Selection is applied to create an intermediate population (mating pool). Then, crossover and mutation are applied to the intermediate population to create the next generation of potential solutions. Although a lot of emphasis has been placed on the three above-mentioned operators, the coding scheme and the fitness function are the most important aspects of any GA, because they are problem dependent.

The original explanation about how GAs could result in robust search relied on the argument of hyperplane sampling. In order to understand this idea, assume a problem encoded with 3 bits. The search space is represented by a cube with one of its vertices at the origin 000. For example, the upper surface of the cube contains all the points of the form $*1*$, where $*$ could be either 0 or 1.

A string that contains the symbol $*$ is referred to as a schema. It can be viewed as a (hyper)plane representing a set of solutions with common properties. The order of a schema is the number of fixed positions present in the string. The defining length is the distance between the first and last fixed positions of a particular schema. Building blocks are highly fit strings of low defining length and low order.

The true fitness of a hyperplane partition corresponds with the average fitness of all strings that lie in that hyperplane. Genetic algorithms use the population as a sample for estimating the fitness of that hyperplane partition. After the initial generation, the pool of new strings is biased toward regions that have previously contained strings that were above average with respect to previous populations. In order to further explore the search space, crossover and mutation generate new sample points while partially preserving the distribution of strings that is observed after selection. Recently, the widespread belief that GAs are robust by virtue of their schema processing has been proved to be false [12].

In the following sections, several important design stages of a GA are presented. Section 2.4 shows different possibilities for encoding. It emphasizes the importance of the encoding scheme on GA convergence. Section 2.5 treats the formulation of the fitness function. Section 2.6 presents different propositions for the selection, crossover, and mutation operators. Parameter control in GAs is addressed in Section 2.6, too. This chapter concludes with a short presentation of niching methods and parallel GAs.

2.4 ENCODING

In order to apply a GA to a given problem, the first decision one has to make is the kind of genotype the problem needs. That means a decision must be taken on how the parameters of the problem will be mapped into a finite string of symbols, known as genes (with constant or dynamic length), encoding a possible solution in a given problem space. The issue of selecting an appropriate representation is crucial for the search. The symbol alphabet used is often binary, though other representations have also been used, including character-based and real-valued encodings.

Many GA applications use a binary alphabet, and their length is constant during the evolutionary process. Also, all the parameters decode to the same range of values and are allocated the same number of bits for the genes in the string. A problem occurs when a gene may only have a finite number of discrete valid values if a binary representation is used. If the number of values is not a power of 2, then some of the binary codes are redundant (i.e., they will not correspond to any valid gene value). The most popular compromise is to map the invalid code to a valid one.

Another shortcoming of binary encoding is the so-called Hamming cliffs (e.g., although the integers 3 and 4 are neighbors in decimal representation, the Hamming distance between the corresponding binary representation, i.e., [0 1 1] and [1 0 0], respectively, is three [different bits]). It is worthwhile to mention that Gray coding, although frequently recommended as a solution to Hamming cliffs, because adjacent numbers differ by a single bit, has an analogous drawback for numbers at the opposite extremes of the decimal scale (e.g., the minimum and maximum gene values differ by only 1 bit, too). Binary encoding can also introduce an additional nonlinearity, thus making the combined objective function (the one in the genotype space) more multimodal than the original one (in the phenotype space).

At the beginning of GA research, the binary representation was recommended because it was supposed to give the largest number of schemata (plural of schema), therefore providing the highest degree of implicit parallelism. However, new interpretations have shown that high-cardinality alphabets (e.g., real numbers) can be more effective due to the higher expression power and low effective cardinality [13–15]. Complex applications suggest nonbinary alphabets. Integer or continuous-valued genes are typically used in large-scale function optimization problems. Another advantage of nonbinary representations, particularly the real-valued one, is the easy definition of problem-specific operators.

When using binary coding, the positions of the genes in the chromosome are extremely important for a successful GA design, unless uniform crossover is applied (Section 2.6.2). A bad choice can make the problem harder than necessary. Therefore, correlated binary genes should be coded together in order to form building blocks, thus diminishing the disruptive effects of crossover. However, this information is usually unavailable beforehand.

Epistasis is a possible measure of problem difficulty for GAs. It represents the interaction among different genes in a chromosome. This depends on the extent to which the change in chromosome fitness resulting from a small change in one gene varies according to the values of other genes. The higher the epistasis level, the harder the problem is. This is obviously also true when applying uniform crossover or real-valued encoding. As mentioned earlier, a possibility for making the gene ordering irrelevant is to apply uniform crossover, because the result of this operation is not affected by the positions of the genes. The same goal can be achieved with real-valued encoding and recombination operators that also turn the genes positions irrelevant (Section 2.6.2). However, making the gene ordering irrelevant does not necessarily mean an easier way to a good solution.

One possible answer for the binary gene position problem is to use an operator called inversion. This is implemented by extending every gene by adding the position it occupies in the string. Inversion is interesting because it can freely mix the genes of the same string in order to put together the building blocks, automatically, during evolution (e.g., [(2 1) (3 0) (1 0) (4 1)], where the first number is a bit tag that indexes the bit and the second one represents the bit value; i.e., (3 0) means that the third bit is equal to zero). At first sight, the inversion operator looks very useful when the correlated parameters are not known *a priori*. With the association of a position to every gene, the string can be correctly reordered before evaluation. However, for large-scale problems, inversion has not demonstrated any utility. Reordering greatly expands the search space, making the problem much more difficult to solve.

Therefore, the very hard encoding problem still remains in the hands of the designer. In order to achieve good performance for large tasks, GAs must be matched to the search problem at hand. The only way to succeed is by using domain-specific knowledge to select an appropriate representation.

2.5 FITNESS FUNCTION

Each string is evaluated and assigned a fitness value after the creation of an initial population. It is useful to distinguish between the objective function and the fitness function used by a GA. The objective function provides a measure of performance with respect to a particular set of gene values, independently of any other string. The fitness function transforms that measure of performance into an allocation of reproductive opportunities (i.e., the fitness of a string is defined with respect to other members of the current population). After decoding the chromosomes (i.e., applying the genotype to phenotype transformation), each string is assigned a fitness value. The phenotype is used as input to the fitness function. Then, the fitness values are employed to relatively ponder the strings in the population.

The specification of an appropriate fitness function is crucial for the correct operation of a GA [16]. As an optimization tool, GAs face the task of dealing with problem constraints [17]. Crossover and mutation, that is, the perturbation (variation) mechanism of GAs, are general operators that do not take into account the feasibility region. Therefore, infeasible offspring appear quite frequently. There are four basic techniques for handling constraints when using GAs.

The simplest alternative is the rejecting technique in which infeasible chromosomes are discarded all over the generations. A different strategy is the repairing procedure, which uses a converter to transform an infeasible chromosome into a feasible one. Another possible technique is the creation of problem-specific genetic operators to preserve feasibility of chromosomes.

The previous procedures do not generate infeasible solutions. This is not usually an advantage. In fact, for large-scale, highly constrained optimization problems, this is certainly a great drawback. Particularly for power system problems, where the optimal solutions usually are on the boundaries of feasible regions, the above-mentioned techniques for handling constraints often lead to poor solutions. One possible way for overcoming this drawback is to apply the repairing procedure only to a fraction (10%, for instance) of the infeasible population.

It has been suggested that constraint handling for such types of optimization problem should be performed allowing search through infeasible regions. Penalty functions allow the exploration of infeasible subspaces [18]. An infeasible point close to the optimum solution generally contains much more information about it than a feasible point far from the optimum. On the other hand, the design of penalty functions is difficult and problem dependent. Usually, there is no *a priori* information about the distance to optimal points. Therefore, penalty methods consider only the distance from the feasible region. Penalties based on the number of violated constraints do not work well.

There are two main possible forms to build a fitness function with penalty term: the addition and multiplication forms. The former is represented as $g(x) = f(x) + p(x)$; where for maximization problems $p(x) = 0$ for feasible points, and $p(x) < 0$ otherwise. The maximum absolute $p(x)$ value cannot be greater than the minimum absolute $f(x)$ value for any generation, in order to avoid negative fitness values. The multiplication form is represented as $g(x) = f(x)p(x)$; where for maximization problems $p(x) = 1$ for feasible points, and $0 \leq p(x) < 1$ otherwise.

The penalty term should vary not only with respect to the degree of constraint violations, but also with respect to the GA iteration count. Therefore, besides the amount of violation, the penalty term usually contains variable penalty factors, too (one per violated constraint). The key for a successful penalty technique is the proper setting of these penalty factors. Small penalty factors can lead to infeasible solutions, and very large ones totally neglect infeasible subspaces. In average, the absolute values of the objective and penalty functions should be similar. At least in theory, the parameters of the penalty functions can, also, be encoded as GA parameters. This procedure creates an adaptive method, which is optimized as the GA evolves toward the solution.

In summary, the main problems associated with the fitness function specification are the following:

- Dependence on whether the problem is related to maximization or minimization;
- When the fitness function is noisy for a nondeterministic environment [19];
- The fitness function may change dynamically as the GA is executed;
- The fitness function evaluation can be so time consuming that only approximations to fitness values can be computed;
- The fitness function should allocate very different values to strings in order to make the selection operator work easier (Section 2.6.1.6);

- It must consider the constraints of the problem; and
- It could incorporate different subobjectives.

The fitness function is a black box for the GA. Internally, this may be achieved by a mathematical function, a simulator program, or a human expert that decides the quality of a string. At the beginning of the iterative search, the fitness function values for the population members are usually randomly distributed and widespread over the problem domain. As the search evolves, particular values for each gene begin to dominate. The fitness variance decreases as the population converges. This variation in fitness range during the evolutionary process often leads to the problems of premature convergence and slow finishing.

2.5.1   Premature Convergence

A frequent problem with GAs, known as deception, is that the genes from a few comparatively highly fit (but not optimal) individuals may rapidly come to dominate the population, causing it to converge on a local maximum or stagnate somewhere in the search space. Once the population has converged, the ability of the GA to continue searching for better solutions is nearly eliminated. Crossover (Section 2.6.2) of almost identical chromosomes generally produces similar offspring. Only mutation (Section 2.6.3), with its random perturbation mechanism, remains to explore new regions of the search space.

The schema theorem, which was proved erroneous a few years ago [20], says that reproductive opportunities should be given to individuals in proportion to their relative fitnesses. However, by doing that, premature convergence occurs because the population is not infinite (basic hypothesis of the theorem). This is due to genetic drift (Section 2.7). In order to make GAs work effectively on finite populations, the (proportional) way individuals are selected for reproduction must be modified. Different ways of performing selection are described in Section 2.6.1 The basic idea is to control the number of reproductive opportunities each individual gets. The strategy is to compress the range of fitnesses, without loosing selection pressure (Section 2.5.2), and avoid any super-fit individual from suddenly dominating the population.

2.5.2   Slow Finishing

This is the opposite problem of premature convergence. After many generations, the population has almost converged, but it is still possible that the global maximum (or a high-quality local one) has not been found. The average fitness is high, and the difference between the best and the average individuals is small. Therefore, there is insufficient variance in the fitness function values to localize the maxima.

The same techniques used to tackle premature convergence are used also for fighting slow finishing. An expansion of the range of population fitnesses is produced, instead of a compression. Both procedures are prone to bad remapping (underexpansion or overcompression) due to super-poor or super-fit individuals.

2.6 BASIC OPERATORS

In this section, several important design issues for the selection, crossover, and mutation operators are presented. Selection implements the survival of the fittest according to some predefined fitness function. Therefore, high-fitness individuals have a better chance of reproducing, whereas low-fitness ones are more likely to disappear. Selection alone cannot introduce any new individuals into the population (i.e., it cannot find new points in the search space). Crossover and mutation are used to explore the solution space.

Crossover, which represents mating (recombination) of two individuals, is performed by exchanging parts of their strings to form two new individuals (offspring). In its simplest form, substrings are exchanged after a crossover point is randomly determined. The crossover operator is applied with a certain probability, usually in the range [0.5, 1.0]. This operator allows the evolutionary process to move toward promising regions of the search space. It is likely to create even better individuals by recombining portions of good individuals. The new offspring created from mating, after being subject to mutation, are put into the next generation.

The purpose of the mutation operator is to maintain diversity within the population and inhibit premature convergence to local optima by randomly sampling new points in the search space. The GA stopping criterion may be specified as a maximal number of generations or as the achievement of an appropriate level for the generation average fitness (stagnation).

2.6.1 Selection

Selection, more than crossover and mutation, is the operator responsible for determining the convergence characteristics of GAs [21, 22]. Selection pressure is the degree to which the best individuals are favored [23]. The higher the selection pressure, the more the best individuals are favored. The selection intensity of GAs is the expected change of average fitness in a population after selection is performed. Analyses of selection schemes show that the change in mean fitness at each generation is a function of the population fitness variance.

The convergence rate of a GA is determined largely by the magnitude of the selection pressure. Higher selection pressures imply higher convergence rates. If the selection pressure is too low, the convergence rate will be slow, and the GA will unnecessarily take longer to find a high-quality solution. If the selection pressure is too high, it is very probable that the GA will converge prematurely to a bad solution. In fact, selection schemes should also preserve population diversity, in addition to providing selection pressure. One possibility to achieve this goal is to maximize the product of selection intensity and population fitness standard deviation. Therefore, if two selection methods have the same selection intensity, the method giving the higher standard deviation of the selected parents is the best choice.

Many selection schemes are currently in use. They can be classified in two groups: proportionate selection and ordinal-based selection. Proportionate-based procedures select individuals based on their fitness values relative to the fitness of the other individuals in the population. Ordinal-based procedures select individuals not based on their fitness but based on their rank within the population.

An ordinal selection scheme has a fundamental advantage over a proportional selection one. The former is translation and scale invariant (i.e., the selection pressure does not change when every individual's fitness is multiplied and added by a constant. The selection intensity of proportionate selection is the only one that is sensitive to the current population distribution [24]. However, conclusive statements about the performance of rank-based selection schemes are difficult to make because, by suitable (but tricky!) adjustment, proportionate selection can give similar performance.

2.6.1.1 Tournament Selection   This selection scheme is implemented by choosing some number of individuals randomly from the population, copying the best individual from this group into the intermediate population, and repeating it until the mating pool is complete. Tournaments are frequently held only between two individuals. Bigger tournaments are also used with arbitrary group sizes (not too big in comparison with the population size). Tournament selection can be implemented very efficiently because no sorting of the population is required.

The tournament procedure selects the mating pool without remapping the fitnesses. By adjusting the tournament size, the selection pressure can be made arbitrarily large or small. Bigger tournaments have the effect of increasing the selection pressure, because below-average individuals do not have good chances of winning a competition.

2.6.1.2 Truncation Selection   In truncation selection, only a subset of the best individuals is chosen to be possibly selected, with the same probability. This procedure is repeated until the mating pool is complete. As a sorting of the population is required, truncation selection has a greater time complexity than tournament selection. As in tournament selection, there is no fitness remapping in truncation selection.

2.6.1.3 Linear Ranking Selection   The individuals are sorted according to their fitness values, and the last position is assigned to the best individual, whereas the first position is allocated to the worst one. The selection probability is linearly assigned to the individuals according to their ranks. All individuals get a different selection probability, even when equal fitness values occur.

2.6.1.4 Exponential Ranking Selection   Exponential ranking selection differs from linear ranking selection only in that the probabilities of the ranked individuals are exponentially weighted.

2.6.1.5 Elitist Selection   Preservation of the elite solutions from the preceding generation ensures that the best solutions known so far will remain in the population and have more opportunities to produce offspring. Elitist selection is used in combination with other selection strategies.

2.6.1.6 Proportional Selection   This is the first selection method proposed for GAs. The probability that an individual will be selected is simply proportionate to its fitness value. The time complexity of the method is the same as in tournament selection. This mechanism works only if all fitness values are greater than zero. The selection probabilities strongly depend on the scaling of the fitness function. In fact, most of the scaling procedures described in the next sections have been proposed to keep proportional selection working. One big drawback of proportional selection is that the selection intensity is usually low, because a single individual, either the fittest or the worst, dictates the degree of compression of the range of fitnesses. This is quite common even during the early stage of the search, when the population variance is high. Negative selection intensity is also possible.

Notice that in ordinal-based selection schemes, the effect of extreme individuals is negligible, irrespective of how much greater or smaller their fitnesses are than the rest of the population. Therefore, despite its popularity inside the power system research community, proportional selection (i.e., roulette wheel) is usually an inferior scheme. There are different scaling operators that help in separating the fitness values in order to improve the work of the proportional selection mechanism. The most common ones are linear scaling, sigma truncation, and power scaling.

2.6.1.6.1 Linear Scaling   Linear scaling (i.e., $f' = af + b$) works well except when most populations members are highly fit, but a few very poor individuals are present. The coefficients a and b are usually chosen to enforce equality of the objective and fitness functions average values and also cause maximum scaled fitness to be a specified multiple (usually two) of the average fitness. These two conditions ensure that average population members receive one offspring copy on average, and the best receives the specified multiple number of copies. Notice that proportional selection with linear scaling is not the same as linear ranking selection.

2.6.1.6.2 Sigma Truncation   In order to overcome the presence of super-poor individuals, the use of population variance information has been suggested to preprocess objective function values before scaling. This procedure subtracts a constant from the objective function values; $f' = \max[0, f - (\bar{f} - d\sigma)]$, where $\bar{f}$ is the mean objective function value in the population. The constant d is chosen as a multiple (between 1 and 3) of the population standard deviation, and negative results are arbitrarily set to zero.

2.6.1.6.3 Power Scaling   Another possibility is power scaling (i.e., $f' = f^p$). In general, the p value is problem dependent and may require adaptation during a run to expand or compress the range of fitness function values. The problem with all fitness scaling schemes is that the degree of compression can be determined by a single extreme individual, degrading the GA performance.

[1 1| 0   1   0   1   0   1] [1 1   0   1   1   0   0   0]

⇒

[1 0| 0   1   1   0   0   0] [1 0   0   1   0   1   0   1]

FIGURE 2.1   Example of one-point crossover.

2.6.2 Crossover

Crossover is a very controversial operator due to its disruptive nature (i.e., it can split important information). In fact, the usefulness of crossover is problem dependent. The traditional GA uses one-point crossover (Fig. 2.1), where the two parents are each cut once at specific points and the segments located after the cuts exchanged. The positions of the bits in the schema determine the likelihood that these bits will remain together after crossover. Obviously, an order-1 schema is not affected by recombination, because the critical bit is always inherited by one of the offspring.

The crossover operator presented above can be generalized in order to apply multiple-point crossover. However, more than two crossover points, although giving a better exploration capacity, can be too disruptive. The crossover mechanism can be better visualized treating strings as rings. In Fig. 2.2, two-point crossover is applied to the example shown in Fig. 2.1. Each offspring takes one ring segment, in between adjacent cut points, from each parent. The contiguous ring segment(s) is taken from the other parent. For more than two crossover points, this procedure is repeated until the last segment is filled. An extra cut is assumed at the beginning of the string (i.e., between genes g8 and g1) for an odd number of cut points.

From the linear string point of view, the elements in between the two crossover points are swapped between two parents to form two offspring (Fig. 2.2). One-point crossover can be represented by the ring geometry as a two-point crossover

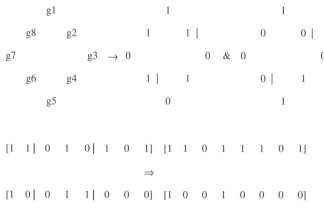

[1 1| 0   1   0| 1   0   1] [1 1   0   1   1   1   0   1]

⇒

[1 0| 0   1   1| 0   0   0] [1 0   0   1   0   0   0   0]

FIGURE 2.2   Ring representation and two-point crossover.

[1 1 0 1 0 1 0 1]

↑ ↓ ↓ ↑ ↓ ↑ ↑ ↓ ⇒ [1 0 0 1 1 1 0 0]

[1 0 0 1 1 0 0 0]

FIGURE 2.3 Example of uniform crossover, where each arrow points to the randomly picked gene value.

with the first cut point always between genes g8 and g1. For multiple-point crossover, the cut points can be anywhere, as long as they are not the same.

Uniform crossover is another important recombination mechanism [25]. Offspring is created by randomly picking each bit from either of the two parent strings (Fig. 2.3). This means that each bit is inherited independently from any other bit. Uniform crossover has the advantage that the ordering of the genes is irrelevant in terms of splitting building blocks.

Uniform crossover is more disruptive than two-point crossover. On the other hand, two-point crossover performs poorly when the population has largely converged because of the inability to promote diversity. For small populations, which is not usually the case for large-scale problems, more disruptive crossover operators such as uniform or m-point ($m > 2$) may perform better because they help overcome the limited amount of information.

Reduced surrogates can be used to improve two-point crossover exploration ability. It is highly recommended for large-scale problems. The idea is to ignore all bits that are equivalent in the two parent strings (Fig. 2.4). Afterwards, crossover is applied on the reduced surrogates (i.e., only one possible cut is considered between any pair of nonequivalent bits).

Notice that the reduced surrogate form implements the original crossover operation in an unbiased way. For example, the cut points between genes 2|3, 3|4, and 4|5 produce the same effect on offspring. Therefore, two-point reduced surrogate crossover considers these cut points as one single possible cross point.

The crossover operator can be redefined for real-valued encoding. Different combinations have been utilized (e.g., a convex combination such as $\lambda_1 \underline{x}_1 + \lambda_2 \underline{x}_2$, where $\lambda_1, \lambda_2 \in i^+$ and $\lambda_1 + \lambda_2 = 1$). One possibility is to take the average of the two corresponding parent genes. The square-root of the product of the two values can also be used. Another possibility is to take the difference between the two values and add it to the higher or subtract it from the lower.

[1 1 0 1 0 1 0 1]   [– 1 – – 0 1 – 1]

reduced surrogates ⇒

[1 0 0 1 1 0 0 0]   [– 0 – – 1 0 – 0]

FIGURE 2.4 Implementation of reduced surrogates to improve crossover exploration capability.

2.6.3 Mutation

The GA literature has reflected a growing recognition of the importance of mutation in contrast with viewing it as responsible for reintroducing inadvertently lost gene values. The mutation operator is more important at the final generations when the majority of the individuals present similar quality. As is shown in Section 2.6.4, a variable mutation rate is very important for the search efficiency. Its setting is much more critical than that of crossover rate.

In the case of binary encoding, mutation is carried out by flipping bits at random, with some small probability (usually in the range [0.001; 0.05]). For real-valued encoding, the mutation operator can be implemented by random replacement (i.e., replace the value with a random one). Another possibility is to add/subtract (or multiply by) a random (e.g., uniformly or Gaussian distributed) amount. Mutation can also be used as a hill-climbing mechanism.

2.6.4 Control Parameters Estimation

Typical values for the population size, crossover, and mutation rates have been selected in the intervals [30, 200], [0.5, 1.0], and [0.001, 0.05], respectively. Fixed crossover and mutation operators do not provide enough search power for tackling large-scale optimization problems. Tuning parameters manually is common practice in GA design. Often, one parameter is tuned at a time in order to avoid the challenging task of simultaneous estimation. However, as they strongly interact in complex forms, this tuning procedure is prone to suboptimality.

In fact, any static set of parameters is inappropriate, regardless of how they are tuned. The GA search technique is an adaptive process, which requires continuous tracking of the search dynamics. Therefore, the use of constant parameters leads to inferior performance. For example, it is obvious that large mutations can be helpful during early generations to improve the GA exploration capability. This is not the case for the end of the search, when small mutation steps are needed to fine-tune suboptimal solutions.

The proper way for dealing with this problem is by using parameters that are functions of the number of generations. Deterministic rules are frequently applied for implementing this idea. However, besides being very difficult to define, they fail to take into account the actual progress of the population performance. Adaptive rules based on population variance, or even the search for optimal parameters as part of the GA processing (i.e., including parameters as part of the chromosomes), seem to be more promising [26, 27].

2.7 NICHING METHODS

Two agents cause the reduction of population fitness variance at each generation. The first, selection pressure, multiplies copies of the fitter individuals. The other agent is independent of fitness. It is called genetic drift [28] and is due to the stochastic nature of the selection operator (i.e., bias on the random sampling of the population). When there is lack of selection pressure, genetic drift is responsible for premature convergence. The GA still ends up on a single peak, even when there are several ones of equal fitness.

Therefore, even when multiobjective optimization is not the main goal, the identification of multiple optima is beneficial for the GA performance. Niching methods extend standard GAs by creating stable subpopulations around global and local optimal solutions. Niching methods maintain population diversity and allow GAs to explore many peaks simultaneously. They are based on either fitness sharing or crowding schemes [29].

Fitness sharing decreases each element's fitness proportionately to the number of similar individuals in the population (i.e., in the same niche). The similarity measure is based on either the genotype (e.g., Hamming distance) or the phenotype (e.g., Euclidian distance). On the contrary, crowding schemes do not require the setting of a similarity threshold (niche radius). Crowding implicitly defines neighborhood by the application of tournament rules. It can be implemented as follows. When an offspring is created, one individual is chosen, from a random subset of the population, to disappear. The chosen one is the element that most closely resembles the new offspring.

Another idea used by niching methods is restricted mating. This mechanism avoids the recombination of individuals that do not belong to the same niche. Highly fit, but not similar, parents can produce highly unfit offspring. Restricted mating is based on the assumption that if similar parents (i.e., from the same niche) are mated, then offspring will be similar to them.

It is important to notice that similarity of genotypes does not necessarily imply similarity of the corresponding phenotypes. The hypothesis that highly fit parents generate highly fit offspring is valid only under the occurrence of building blocks and low epistasis. When the genes strongly interact, there is no guarantee that these offspring will not be lethals.

2.8 PARALLEL GENETIC ALGORITHMS

In several cases, the application of GAs to actual problems in business, engineering, and science requires long processing time (hours, days). Most of this time is usually spent in thousands of fitness function evaluations that may involve long calculations. Fortunately, these evaluations can be performed quite independently of each other, which makes GAs very adequate for parallel processing.

The many ways parallel GAs can be implemented depends on how the population is dealt with and on the computer hardware available. A single population or multiple populations may be considered. In the last case, the populations may be isolated or communicate by exchanging individuals or some other information. The computer hardware, on the other hand, may range from a simple cluster of PCs loosely coupled by a local area network to massive parallel single instruction, multiple data (SIMD) computers, which execute the same single instruction on all the processors.

The usual classification of parallel GAs found in the literature is [30, 31]:

- *Single-population master–slave GAs*: Conceptually, they are the same as the conventional GAs, with the difference that one master node executes the basic GA operations (selection, crossover, and mutation), and the fitness evaluations are distributed among several slave processors. The only gain in this type of parallel GA is in processing speed as the GA itself performs exactly in the same way as the sequential one.

- *Multiple-population GAs*: This type of GA consists of several subpopulations that exchange individuals occasionally. This operation is called *migration* and may be implemented according to several different strategies. Besides possible speedup if implemented in a parallel computer platform, this type of parallel GA may exhibit a better performance than the conventional GA as far as the quality of the solutions found. This type of parallel GA is also known as distributed GA because it is usually implemented in distributed-memory parallel computers.

- *Fine-grained GAs*: These consist of a single spatially structured population, usually in a two-dimensional rectangular grid, with one individual per grid point. The fitness evaluation is performed simultaneously for all the individuals, and selection and mating are usually restricted to a small neighborhood around each individual. This type of parallel GA is well suited for massively parallel multiple instruction, multiple data (MIMD) computers.

Other types of parallel GAs may be found in the literature but are less used than the above-mentioned or are a combination of them. Also, the parallel GAs may be implemented in a synchronous or asynchronous way.

Single-population master–slave GAs are easy to implement and usually present high speedup in inexpensive hardware like PC clusters. Multiple-population and fine-grained GAs, on the other hand, introduce new parameters such as the number of populations and their sizes, the topology of communications (e.g., each population is connected to all the others), and the migration rate. Although many implementations of parallel GAs have been described in the literature, the effect of these new parameters on the quality of the search is still under analysis [32].

2.9   FINAL COMMENTS

This tutorial on GAs has pointed out the main topics on their design. The focus on the essential topics helps one to not miss the forest for the trees. The first generation of GAs, based on the canonical algorithm, considering proportional selection and crossover/mutation with constant probabilities, was not originally proposed for solving static optimization problems [33]. Almost three decades of research has adapted the original proposal [34] to deal with this type of problem.

The first applications to power systems appeared after 1991 [35]. Since then, GAs have been applied not only to pure optimization problems in power systems but also to model identification, control, and neural network training. After a necessary period of maturing, GAs are being used now, frequently in combination with conventional optimization techniques, for solving large-scale problems.

ACKNOWLEDGMENTS

This work was supported by the Brazilian Research Council (CNPq) and by the State of Rio de Janeiro Research Foundation (FAPERJ).

REFERENCES

1. Fogel DB. An introduction to simulated evolutionary optimization. IEEE Trans Neural Networks 1994; 5(1):3–14.

2. Fogel DB. Evolutionary computation—Toward a new philosophy of machine intelligence. Piscataway, NJ: IEEE Press; 1995.

3. Bäck T, Hammel U, Schwefel H-P. Evolutionary computation: Comments on the history and current state. IEEE Trans Evol Computat 1997; 1:3–17.

4. Goldberg DE. Genetic algorithms in search, optimization and machine learning. Reading, MA: Addison-Wesley; 1989.

5. Beasley D, Bull DR, Martin RR. An overview of genetic algorithms: Part 1, fundamentals. University Computing 1993; 15(2):58–69.

6. Beasley D, Bull DR, Martin RR. An overview of genetic algorithms: Part 2, research topics. University Computing 1993; 15(4):170–181.

7. Whitley D. A genetic algorithm tutorial. Statistics Computing 1994; 4:65–85.

8. Aarts E, Korst J. Simulated annealing and Boltzmann. New York: John Wiley & Sons; 1989.

9. Glover F, Laguna M. Tabu search. Boston: Kluwer Academic; 1997.

10. Kennedy J, Eberhart RC. Swarm intelligence. San Mateo, CA: Morgan Kaufmann; 2001.

11. Rudolph G. Convergence analysis of canonical genetic algorithms. IEEE Trans Neural Networks 1994; 5(1):96–101.

12. Vose MD, Wright AH. Form invariance and implicit parallelism. Evol Computat 2001; 9(3):355–370.

13. Antonisse HJ. A new interpretation of schema notation that overturns the binary encoding constraint. In: Proc. 3rd Int. Conf. on Genetic Algorithms. San Mateo, CA: Morgan Kaufmann; 1989. 86–91.

14. Goldberg DE. The theory of virtual alphabets. In: Parallel problem solving from nature 1. Lecture Notes in Computer Science, Vol. 496. New York: Springer; 1991. 13–22.

15. Eshelman LJ, Schaffer JD. Real-coded genetic algorithms and interval-schemata. In: Foundations of genetic algorithms 2. San Mateo, CA: Morgan Kaufmann; 1993. 187–202.

16. Radcliffe NJ, Surry PD. Fitness variance of formae and performance prediction. In: Foundations of genetic algorithms 3. San Mateo, CA: Morgan Kaufmann; 1995. 51–72.

17. Michalewicz Z, Schoenauer M. Evolutionary algorithms for constrained parameter optimization problems. Evol Computat 1996; 4(1):1–32.

18. Gen M, Cheng R. A survey of penalty techniques in genetic algorithms. In: Proc. 3rd IEEE Conf. on Evolutionary Computation. Piscataway, NJ: IEEE Press; 1996. 804–809.

19. Miller BL, Goldberg DE. Genetic algorithms, selection schemes, and the varying effects of noise. University of Illinois at Urbana-Champaign; IlliGAL Report, No. 95009. 1995.

20. Macready WG, Wolpert DH. Bandit problems and the exploration/exploitation tradeoff. IEEE Trans Evol Computat 1998; 2(1):2–22.

21. Goldberg DE, Deb K. A comparative analysis of selection schemes used in genetic algorithms. In: Foundations of genetic algorithms. San Mateo, CA: Morgan Kaufmann; 1991. 69–93.

22. Blickle T, Thiele L. A comparison of selection schemes used in genetic algorithms. Swiss Federal Institute of Technology; TIK-Report, Nr. 11, 2nd Version. 1995.

23. Bäck T. Selective pressure in evolutionary algorithms: A characterization of selection mechanisms. In: Proc. 1st IEEE Conf. on Evolutionary Computation. Piscataway, NJ: IEEE Press; 1994. 57–62.

24. Whitley LD. The GENITOR algorithm and selection pressure: Why rank-based allocation of reproductive trials is best. In: Proc. 3rd Int. Conf. on Genetic Algorithms. San Mateo, CA: Morgan Kaufmann; 1989. 116–121.

25. Syswerda G. Uniform crossover in genetic algorithms. In: Proc. 3rd Int. Conf. on Genetic Algorithms. San Mateo, CA: Morgan Kaufmann; 1989. 2–9.

26. Saravanan N, Fogel DB, Nelson KM. A comparison of methods for self-adaptation in evolutionary algorithms. BioSystems 1995; 36:157–166.

27. Eiben AE. Hinterding R, Michalewicz Z. Parameter control in evolutionary algorithms. IEEE Trans Evol Computat 1999; 3(2):124–141.

28. Rogers A, Prügel-Bennet A. Genetic drift in genetic algorithm selection schemes. IEEE Trans Evol Computat 1999; 3(4):298–303.

29. Sareni B, Krähenbühl L. Fitness sharing and niching methods revisited. IEEE Trans Evol Computat 1998; 2(3):97–106.

30. Cantú-Paz E. Efficient and accurate genetic algorithms. Boston: Kluwer; 2000.

31. Alba E, Troya JM. A survey of parallel distributed genetic algorithms. Complexity 1999; 4(4):31–52.

32. Cantu-Paz E. Markov chain models of parallel genetic algorithms. IEEE Trans Evol Computat 2000; 4(3):216–226.

33. De Jong KA. Genetic algorithms are NOT function optimizers. In: Foundations of genetic algorithms 2. San Mateo, CA: Morgan Kaufmann; 1993. 5–17.

34. Holland JH. Adaptation in natural and artificial systems. University of Michigan Press; Ann Arbor, MI; 1975.

35. Miranda V, Srinivasan D, Proença LM. Evolutionary computation in power systems. Electrical Power & Energy Systems 1998; 20(2):89–98.

CHAPTER 3

Fundamentals of Evolution Strategies and Evolutionary Programming

VLADIMIRO MIRANDA

3.1 INTRODUCTION

This chapter is devoted to the description of a branch of techniques in evolutionary computing (EC) that, for historical reasons, was kept divided for many years, although the techniques may be seen in fact as the same generic approach: evolutionary programming (EP) and evolution strategies (ES). This does not happen by chance or coincidence: today, we have difficulties in explaining to newcomers the differences between the two approaches.

Instead of two distinct paradigms, what we find is a collection of variations over the same theme, and perhaps it is about time to accept calling all of them by a family name such as "phenotypic methods" inside evolutionary methods . . . . Nevertheless, history records that evolutionary programming and evolution strategies were created independently, and for this reason, these converging approaches received distinct names to the present day.

The evolutionary programming concept is attributed to Lawrence J. Fogel in the early 1960s, in parallel with the proposals of John Holland on adaptive systems, and came to light in a number of technical papers, although the landmark publication is widely considered to be the book *Artificial Intelligence Through Simulated Evolution* (1966). Whereas the works of Holland led to development of the genetic algorithm (GA) approach, Fogel successfully persisted in a line that led to what is now known as EP.

Evolution strategies is a method developed by I. Rechenberg and H.-P. Schwefel, who report the first developments at the Technical University of Berlin in 1963. They stimulated an active group that produced a remarkable set of practical and theoretical results, always with a sense of generalization of a family of methods (thus the plural "strategies").

Modern Heuristic Optimization Techniques. Edited by K. Y. Lee and M. A. El-Sharkawi
Copyright © 2008 the Institute of Electrical and Electronics Engineers, Inc.

These two communities evolved separately, gathered interested researchers around them and organized independent series of workshops and conferences. It is fair to remember that in the 1960s, neither computing power nor telecommunications and the Internet were a reality as we know them today, and information did not flow as easily and readily as today, when we often fall into the error of taking it for granted. Furthermore, the groups were publishing in different languages, in different media targeting different audiences with distinct application interests. Regular contacts were only established since the end of the 1980s. This division, resulting from historical reasons, seems however not useful in the present day and instead may be confusing to newcomers in the field. With a sense of independence, we would rather see this dual name disappear and be replaced by a unifying designation.

As in any evolutionary process, ES and EP rely on the definition of a fitness function, which sets up the *environment* and establishes the way to measure the quality of each solution (called *individual*). The fitness function is just like the objective function of an operations research problem; as such, it may include penalties for the violation of constraints.

To be fair, the concept of fitness function may include a loose definition of what a function is. In fact, the only real requirement is a process that allows a ranking of alternatives in the solution space, such that this ranking is in agreement with the preferences defined by the decision maker. This process can, therefore, include rules, besides mathematical expressions. But it is true that the traditional models have analytical expressions as their fitness function.

Both EP and ES share with GA this "selection-by-fitness-function" principle. It is worth noticing this, because this is not the only process that may ignite and drive an evolving process. For instance, "selection-by-arms-race" is another possible mechanism; it would require at least two distinct populations, one of them possibly predating the other. But, in fact, GA, ES, and EP are all based on the same algorithmic approach, where individuals are not confronted with one another but are measured against an external common selection function.

But ES and EP move away from GA in the way they represent alternatives, solutions, or individuals in a population. While genetic algorithms rely on the power of a discrete genetic representation to generate new offspring with higher survival chances, EP and ES rely on the power of a direct representation of individuals. This is often a source of confusion and misunderstanding, and we will try to make our point of view more clear—with a warning to the reader that this is not at all consensual and that many passionate discussions gravitate around this topic.

A phenotype is an individual (or the description of an individual); a genotype is the coding (for a program) to make an individual. These concepts are often mistaken, but their understanding forms perhaps one of the clearest bases to distinguish what is "genetic" from what cannot be rigorously called "genetic."

The phenotypic representation of an individual may often be complex: and, as any representation, it also demands some rule of interpretation for an observer. But the basic assumption for an individual representation is that it is biunivocal; that is, that given the representation coding, one may derive the individual and vice versa. This means that there is no limitation for the form of representation, and this includes binary representations as well—if there is a one-to-one matching between a form of representation and an individual, we will sustain that such representation may be justly called phenotypic. Let us make this point clear: changing a real value representation of a variable into a binary representation of the same value does not, in our view, change the representation from phenotypic to genotypic nor the technique from EP or ES to GA. A transformation in the phenotype representation can always be accompanied by a suitable transformation of variation operators, and, therefore, the choice of suitable phenotype representation is a matter of simplicity and practical interest and does not reflect substantial theoretical differences.

Other authors, however, will argue that a phenotypic representation relates only to the use of the same variables upon which the environment/fitness function exerts its action directly. This leads us to distinguish among three forms of representation of an individual, in evolutionary computation:

- Pure phenotypic: the individual is represented by the same variables that are used by the fitness function;
- Transformed phenotypic: the coding representing an individual is composed of variables that, under a suitable transform, may be mapped into the variables used by the fitness function;
- Genetic: the coding representing an individual must be interpreted as a program that allows building a phenotypic representation of the individual.

One may say that a pure phenotypic representation is a string of data whose interpretation is straightforward. The variables of the problem may even be represented as binary numbers, but this does not make this representation "genetic," in our definition. A transformed phenotypic representation is a string of data requiring an external program to allow the extraction of the values of the natural variables of the problem. But the program that allows the interpretation of the string of data is still completely external. On the contrary, the name *genetic* should have been reserved to representations where the instructions to build an individual are also coded (but, in the literature, the application of the term *genetic* has had a wider and uncontrolled adoption).

Also, one should not confuse the representation of an individual with the behavior of an individual. For instance, a possible representation of a neural network (with a given architecture) is by the value of its weights; these parameters may receive a real-valued representation, or a binary representation—it is indifferent to our case, because both are phenotypic representations: given the weights, you have the network, and given the network, you have the weights. This has nothing to do with the *behavior* of an individual, which is dictated as much by its inner constitution (phenotype) and by the environment it is submitted to—the fitness function. In our virtual world, the individual is its representation—and a transform applied to the variables does not change this.

If one has a problem where one evaluates the behavior of competing programs for some task, then a program is an individual and its representation is at phenotypic level, however odd this may seem. A genotypic representation of such an individual would be under the form of a "program to build the program/individual." Having programs as individuals is certainly something not alien to EP, especially to the early works of L. Fogel, but the popular view of EP and ES is based on problems described by variables, and in this chapter we will stick to that (perhaps limited) understanding.

In GA, one must establish a univocal mapping between the space of the genes and the space of the phenotypic variables. This may be a one-way mapping, from a domain of programs (to build individuals) to the domain of representation of individuals. The variation introduced by crossover and mutation is generated at the gene level, and the phenotypic consequences are evaluated against the fitness function at the problem variable level.

In ES and EP, there is no real gene level and no need for a mapping process between genes (as programs) and problem variables. In the most widespread and known models of ES and EP (the ones we will deal with in this chapter), each solution is represented by its own variables, with real or integer values, within their feasible domains. Therefore, one can say that variation is introduced directly at the level of the phenotype.

Variation and/or diversity are essential to make selection useful. They allow the coverage of the search space. *Variation* refers to how different are offspring from ancestor generations, whereas *diversity* refers to how different are individuals from one another within a generation. When evolution depends heavily on diversity, the loss of diversity leads usually to an early termination of evolutionary algorithms, either at suboptimal solutions or at local optima. When the evolution depends heavily on variation, this must also be controlled in order to allow progress from generation to generation: too little variation may trap the progress in local optima, and too much variation may disrupt the evolving process completely.

Both EP and ES have used a variety of hybrids of variation and diversity to accelerate convergence. In classic pure EP (as a model of species evolution), variation is introduced solely by mutation, but early works related to finite state machines already used recombination processes to originate new individuals. In ES, which started by using mutation, processes similar to crossover were later introduced to originate variation and new phenotypic expressions. But, as we shall see, there is no real need for the distinction between the two approaches and they have actually converged in conceptual terms.

Contrary to many approaches to the theme, we will start with the description of ES. When we constrain ourselves to real-valued problems, in what refers to EP and ES, it seems to us that it is a more didactic approach. Nevertheless, the readers are invited to search in the literature for other evolutionary models where EP has been applied with success.

3.2 EVOLUTION STRATEGIES

Under the general designation of evolution strategies, the ideas started by Rechenberg and Schwefel were explored and resulted in a remarkable legacy of theoretical and experimental work.

The geographical independence of the development of EP (in the United States) and ES (in Europe) and subsequent creation of schools of followers, reinforced by the factors of lack of communication referred to in the previous section, are probably the best explanation why we still today find a distinction in the literature. As we shall see, the most fundamental reason that one sees claimed, to distinguish EP from ES, is the fact that "pure" EP does not make use of recombination of individuals as a means to generate offspring and diversity and relies only on mutation. However, this claim is not accurate, and, as previously mentioned, both EP and ES have used different mixes of mutation and recombination methods.

Evolution strategies is a name that covers a wide family of related algorithms. These algorithms follow the general biological paradigm of exploring a search space by means of processes mimicking mutation, recombination, and selection. They are distinct from GA models by the fact that there is no distinction between genotype and phenotype (i.e., there is not a separation of worlds where variation is generated in one of them while selection acts on the effects felt in another one). In ES, the problem and the alternatives are represented in their natural variables—the heaviest work of ES has been, in fact, over problems with real-valued variables.

The problem addressed in the early 1960s by Rechenberg and Schwefel was related to finding optimal shapes presenting minimum drag in a wind tunnel—a typical engineering problem. Between 1963 and 1974, these researchers developed the foundations of a theory for evolution strategies. The results, however interesting, remained within the knowledge of a closed community, perhaps especially composed of civil or structural engineers. It seems that a reason exists for this: there are many structural or technical optimization problems for which no mathematical analytical closed form for an objective function exists. Therefore, engineers had to rely on their intuition and professional judgment, instead of using analytical solutions available at the time.

Nowadays, there is a variety of models or versions of ES. In the following sections, we will discuss some aspects of what could be called a canonical model—the $(\mu, \kappa, \lambda, \rho)$ES model—using the notation of Ref. 1.

3.2.1 The General $(\mu, \kappa, \lambda, \rho)$ Evolution Strategies Scheme

The designation "$(\mu, \kappa, \lambda, \rho)$ES model" has been proposed by Schwefel [1] and has the following parameters:

- μ: number of parents in a generation
- κ: number of generations of survival or maximum number of reproduction cycles of an individual
- λ: number of offspring or children created in one generation
- ρ: number of direct ancestors for each individual

In this light, a simple canonical evolutionary programming trusting only on two ancestors per individual and mutation is like a $(\mu, \infty, \mu + \mu, 1)$ evolution strategy.

This means that a whole family of processes can be started, depending on the choice of the above parameters. Some of the varieties have been researched in depth, and some are still an open field for research. Of course, the simplest ones have been the most investigated, and this effort has brought insight into the mechanisms that power the evolution strategies and make them so successful (or that provoke divergence and lack of success in some cases).

The aim of this chapter is to explain to power system researchers and engineers the basics on how ES and EP are built and work. Therefore, our didactic strategy will be to start with simple models and progressively increase them in complexity. After all, it will be like retracing the story of the theoretical development of ES.

The first ES models had less degrees of freedom than the $(\mu, \kappa, \lambda, \rho)$ model admits. The first approach became known as the $(1 + 1)$ES model: it had, in each generation, only one parent, only one descendent was generated, and the selection acted upon the set constituted by parent and child. Later, one spoke of an opposition of $(\mu + \lambda)$ES against (μ, λ)ES. In the $(\mu + \lambda)$ES, the μ survivors in each generation were selected from a population formed by the union of the sets of μ parents and λ children. This meant that an individual had the possibility, in theory, of living forever. According to this notation, the first experiments obeyed a $(1 + 1)$ES strategy.

On the other hand, in a (μ, λ)ES with $\lambda \geq \mu \geq 1$, the new μ future parents are selected from the λ offspring only, no matter how good their parents might be. It has been demonstrated that this strategy risks divergence in some cases, if the solution "best-so-far" is not stored externally or at least preserved within the generation cycle (this deterministic preservation of the best is called *elitism*). The first models of this kind have been called the $(1, \lambda)$ES.

The (μ, λ)ES implies that an individual can have children only once, and that its life duration is of one generation, as opposed to the $(\mu + \lambda)$ES where there is no limit for the life span of an individual.

The $(\mu, \kappa, \lambda, \rho)$ES introduces new degrees of freedom in defining an evolution strategy. With the variable κ defining a life span for each individual, one can now test a variety of strategies and look at the (μ, λ)ES and the $(\mu + \lambda)$ES as the extreme cases of such variety. Furthermore, by recognizing the role of the parameter ρ, which defines how many parents an individual has, one explicitly introduces the operation of recombination as one major factor conditioning the development of an evolution strategy. But in contemporary ES, the $(\mu, \kappa, \lambda, \rho)$ are not the only parameters to take into account. We can list some more:

- P: the (start) population
- **mut**: the mutation operator
- p_m: mutation probability
- **rec**: the recombination operator
- p_r: recombination probability
- **sel**: selection operator

- ζ: number of stochastic tournament participators
- Other parameters may be recognized, some of them associated with the operators **rec**, **mut**, and **sel** adopted

It is usual to distinguish in the representation of an individual two types of parameters: *object* parameters (OP) and *strategy* parameters (SP). Say individual c is represented by $c(OP, SP)$, such that

$$OP = (o_1, o_2, \ldots, o_{no}) \quad \text{and} \quad SP = (s_1, s_2, \ldots, s_{ns}),$$

given "no" object parameters and "ns" strategy parameters.

The object parameters may be $OP = (\theta, x_1, \ldots, x_n)$. The x are the classic n variables of the problem (or the phenotypic variables) and are the only ones that enter the fitness function. The parameter θ counts the remaining life span measured in number of iterations (reproduction cycles). Of course, at the birth of an individual, $\theta = \kappa$.

The strategy parameters usually refer to the standard deviations σ for mutations, which can be global or in each of the n dimensions or variables of an individual, and to parameters α establishing correlation between mutations in distinct variables (sometimes called "angle" parameters). It is fair to say that strategy parameters have a *genetic* flavor, because they could be seen as coding for the way to build a new individual, instead of representing phenotypic traits of the individual carrying them. We will not discuss this point further here.

The general algorithm for evolution strategies could be something like this:

```
Procedure ES
(BEGIN)
      define parameters and operators
set μ, κ, λ, ρ and other parameters;
set operators (rec, mut, sel);
      start the generation counter
g := 0;
      initialize a random population P of μ elements
Initpopulation P[g];
      evaluate the fitness of all individuals of the
population
Evaluate P[g];
while not done do
            reproduction - generate λ offspring...
            ... by recombination
      P'[g] := recombine(P[g])
            ... by mutation - introduce stochastic
perturbations in the new population
      P̃[g]:= Mutate (P'[g]);
```

```
            evaluation - calculate the fitness of the
new individuals
      Evaluate P̃[g];
            selection - of μ survivors for the next
generation, based on fitness value
      P[g+1]:= select (P[g] ∪ P̃[g]);
            test for termination criterion (based on
fitness, number of generations, etc.)
      If test is positive then done:= TRUE;
            increase the generation counter
      g:= g + 1;
End while
(End ES)
```

3.2.2 Some More Basic Concepts

Although there is still much work to be done in order to establish on solid grounds a general theory about generalized evolution strategies, there are some achievements made over some simplified models that allowed insight into the way ES work. Although this text is not meant to organize all theory behind ES and is instead oriented to give an introduction to the topic to power engineers and researchers, we will nevertheless introduce some basic concepts that have been used by researchers in the field. In trying to develop a formal description of the behavior of ES, researchers have worked mostly on the so-called spherical model, in order to gain insight into how local convergence properties of algorithms behave comparatively, and have introduced the concept of *progress rate*.

The progress rate φ is defined as the *expectation* of the change in the (Euclindean) distance, from one generation to the following, between the optimum (wherever it is) and the average location of the population. The spherical model consists of an isotropic fitness landscape defined by

$$F_1(\mathbf{y}) = c_0 + \sum_{i=1}^{n} c_i \left(y_i - y_i^*\right)^2,$$

with $c_i = 1, \forall i = 1, \ldots, n$, and the symbol $*$ denoting *optimum*. The y is the object parameter or search variable.

This is an interestingly workable model because it has radial symmetry. Therefore, the fitness function may be also written as $F(\mathbf{y}) = F(\mathbf{y}^* - \mathbf{R}) = Q(\mathbf{R})$, where $\mathbf{R}$ is the *distance-to-optimum* vector; but due to the radial symmetry of the spherical model, $Q(\mathbf{R})$ is dependent only on $R = \|\mathbf{R}\|$, the length of $\mathbf{R}$ (i.e., $Q = Q(R)$ and has $\mathbf{y}^*$ as symmetry center).

A mutated individual is therefore defined as

$$\tilde{\mathbf{y}} = \mathbf{y} + \mathbf{Z},$$

where $\mathbf{Z}$ is a random vector. Observing Fig. 3.1, we can understand that the rate of progress φ may be defined as the expectation

$$\varphi = E[R - r],$$

where r is the distance of a mutated individual to the optimum.

This has been the basic model adopted by researchers like Beyer [2] to analyze, in the spherical model, the progress rate of an ES and to derive laws about the probability of success (i.e., the probability of the mutated $\tilde{\mathbf{y}}$ being inside the circle defined by R around the optimum), and about how to achieve an optimal progress rate (i.e., the fastest progress possible toward the optimum).

3.2.3 The Early (1 + 1)ES and the 1/5 Rule

The first ES models experimented with in the 1960s used only one parent and one child per generation. In what is now called a $(1 + 1)$ES, or the *plus* strategy, one parent generates one offspring per generation by applying normally distributed mutations (this means that smaller changes are more likely than major changes); if one child has a better performance than a parent, measured in terms of the optimization objective, it replaces the parent (becomes selected), otherwise the child is discarded and a new offspring is generated from the parent individual by mutation.

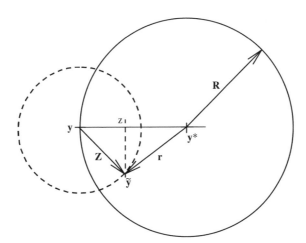

FIGURE 3.1 Representation of the projection of a fitness landscape in $n = 2$ dimensions, with identification of the optimum $\mathbf{y}^*$ and a point $\mathbf{y}$ discovered at a generation g; the circle with radius R defines the domain of success of mutations $\mathbf{Z}$ added to $\mathbf{y}$, for which $r < R$.

No recombination is used, the mutation scheme is kept unchanged along the generations, and there is an exogenous control of the step size.

We see that this is a simple scheme with elitist selection. So simple, in fact, that it allowed the derivation of some theoretical results for this strategy, namely about convergence velocity and step size. This model will have an individual in a generation represented as a set of variables (without loss of generality; let's admit for the moment that we are dealing only with real-valued variables), such as in Fig. 3.2.

The mutation scheme may be described as follows: a mutation step is carried out at generation g by adding a random perturbation $\mathbf{Z}$ to the parent individual $\mathbf{X}^{(g)}$, creating a new individual $\tilde{\mathbf{X}}$:

$$\tilde{\mathbf{X}} := \mathbf{X}^{(g)} + \mathbf{Z}, \tag{3.1}$$

where

$$\mathbf{Z} = \sigma(N_1(0, 1), \ldots, N_n(0, 1))^t. \tag{3.2}$$

The individual $\mathbf{X}$ is a vector of "n" object variables. $N_n(0, 1)$ corresponds with the Gaussian distribution with zero mean and unit variance of the jth variable, and σ is the mutation strength (sometimes incorrectly called *step size* when, in fact, it is the square root of the variance of the probability distribution defining the production of mutations). $\mathbf{Z}$ is a random vector and, given the definition above, the mutation distribution is isotropic.

The global mutation strength σ cannot be maintained constant if convergence is to be achieved. A useful rule-of-thumb has been proposed by Rechenberg [3], the so-called $1/5$ success rule: in order to have an optimum convergence velocity, the "success rate" $S(h)$, or the ratio between successful mutations, and all mutations, should approach the value of $1/5$. Therefore, the mutation variance should be increased if the observed success rate is greater than $1/5$ and decreased if it is less than $1/5$. $S(h)$ is the success rate in the last h generations.

According to Rechenberg, there is an optimal value for h given by multiplying by 10 the number n of variables or dimensions of the search space. So, if σ defines the initial mutation variance for a problem with 10 variables, then one should evaluate the success rate in a set of $h = 100$ generations and σ may change to $a\sigma$ if $S(h) > 1/5$ and to σ/a if $S(h) < 1/5$. The value of a should be chosen such that $a \in [1, 2]$.

This result has been obtained for plane approximations of strongly convex functions of the spherical model type. This spherical model became important in many subsequent theoretical advancements, because in many situations it is locally a good approximation of a given fitness landscape. In the case of spherical models

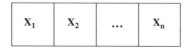

FIGURE 3.2   Representation of an individual i with n real-valued variables.

and similar, a linear convergence order for convergence velocity has been proved. But for models distinct from the spherical model, the value of $1/5$ may be replaced by other numerical values to approximate an optimal behavior of the algorithms. The "good" value to use will depend, therefore, on the topology of the search space.

However, this "$1/5$ rule" may also reduce the effectiveness of the algorithm in finding an optimum; it may accelerate the discovery of the optimum, but the probability of actually reaching it becomes reduced, because it tends to get trapped in local optima; from then on, if may become difficult to find improvements in the neighborhood and then, after a number of generations, the application of the $1/5$ rule will further reduce the amplitude of the probability density function regulating mutations, making it even more difficult to escape.

3.2.4 Focusing on the Optimum

Mutations leading to important variations in the individuals are usually beneficial to the procedure in the beginning, because they allow new individuals to jump away from parents and thus to probe vast regions of the feasible domain of the problem. However, at a later stage, large perturbations drive individuals away from the region of the optimum (Fig. 3.3).

Common sense tells us that when we have solutions neighboring a possible optimum, the spread of the probability distribution that regulates mutation should become narrower. This allows a fine adjustment of the solutions and is part of the rational behind Rechenberg's rule. This rule introduces or externally defines rules for reducing the spread of the probability distributions with the increase in generations. Proposed as such, this naïve scheme is mechanical, deterministic, and rigid and against the very spirit of evolutionary processes.

One technique instead gained popularity, because it uses the very same principles of evolution in a sort of meta-evolutionary scheme. This scheme associates with each individual an extra variable, which represents precisely the variance of its mutation distribution. Although we will further elaborate on this topic later, we can immediately state that this strategy is quite successful in many cases and that it has theoretical background to support it.

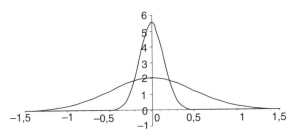

FIGURE 3.3 Two mutation Gaussian distributions with distinct variances. Their mean value is the present value of a variable. A smaller variance allows greater probability for smaller perturbations in the value of a variable.

3.2.5   The $(1, \lambda)$ES and σSA Self-Adaptation

In this strategy, one parent $\mathbf{X}$ generates λ offspring individuals, and selection acts among these to obtain a single survivor to become a parent in the following generation. Therefore, parent and children do not compete against one another.

This model allows us to introduce a dynamic control of the mutation strength, as opposed to the Rechenberg 1/5 rule, which was seen by many as a concept alien to the ES spirit: the mutation strength was controlled externally by some deterministic rule. Therefore, the research effort was directed toward achieving a dynamic control of the mutation strength under principles of evolution and self-adaptation—this means that a mutation strength parameter would also be subject to mutation and selection in order to adapt the progress of the algorithm to an optimal progress rate.

A successful family of models in this line of reasoning is the σSA, or σ-self adaptation, strategy, originally developed by Schwefel [4, 5]. The central idea is that each individual is governed by object parameters and by evolvable strategy parameters, and if an individual is selected with respect to its fitness, the corresponding set of strategic parameters survives as well. These strategic parameters, if optimal, should drive the individuals into a regime of optimum gain (i.e., of maximal expected fitness improvement per generation).

In the $(1, \lambda)$ strategy, we have only one evolvable strategy parameter—a mutation strength σ. An individual at a generation g can therefore be represented as in Fig. 3.4. The hope in a self-adaptation approach is that the problem will "learn" the value of an optimal strategic parameter σ^* whose adoption would lead to the maximum possible progress rate. Unfortunately, this is a value that is not available beforehand, as its calculation would depend on the knowledge of the location of the optimum that one is precisely seeking—it is an unknown theoretical value to be approached. However, there will always be random fluctuations around the exact value that will cause some loss in efficiency.

As in $(1 + 1)$ES, the object parameters are subject to mutation such that, departing from a generation g, λ offspring are generated by

$$\tilde{\mathbf{X}}_k = \mathbf{X}^{(g)} + \mathbf{Z}_k, \quad k = 1 \text{ to } \lambda, \tag{3.3}$$

where

$$\mathbf{Z}_k = \sigma_k^{(g+1)}[N(0, 1), \ldots, N(0, 1)]^t, \tag{3.4}$$

FIGURE 3.4   Representation of an individual with n real-valued variables; it has an extra variable related to the variance of the Gaussian distribution commanding the mutations in its offspring.

but now the mutation strength σ has been itself subject to mutation such that

$$\sigma_k^{(g+1)} = \Pi\left[\sigma^{(g)}\right].$$

The $\Pi[\cdot]$ mutation operator performs multiplicative mutations. This can be done by the multiplication of the parental $\sigma^{(g)}$ by a random number ξ such that

$$\sigma_k^{(g+1)} = \xi\sigma^{(g)}, \quad k = 1 \text{ to } \lambda. \tag{3.5}$$

The expectation of ξ must not deviate too much from 1, that is,

$$E[\xi] \approx 1.$$

There are some distributions of the random variable ξ that have found practical use. One of the most important is the lognormal distribution, originally proposed by Schwefel, which has the property that some value will have the same probability of being doubled or of becoming divided in half:

$$p_\sigma(\xi) = \frac{1}{\tau\sqrt{2\pi}}\frac{1}{\xi}e^{-(1/2)(\ln \xi/\tau)^2}.$$

For practical purposes, the random variable ξ can easily be generated from the Gaussian $N(0, 1)$ by an exponential transformation such as

$$\xi = e^{\tau N(0,1)}. \tag{3.6}$$

These expressions introduce the external value τ—the learning parameter, which will soon be discussed. The learning parameter τ conditions the speed and accuracy of the σSA evolution strategy. Therefore, the question on how to choose good values for τ remains for the moment.

Another mutation rule used in practice depends on the symmetrical two-point distribution. For practical purposes, the mutated value of $\sigma_k^{(g+1)}$ is given by

$$\sigma^{(g+1)} = \begin{cases} \sigma^{(g)}(1 + \beta), & \text{if } U(0, 1) \leq 0.5 \\ \sigma^{(g)}/(1 + \beta), & \text{if } U(0, 1) > 0.5, \end{cases}$$

with $U(0, 1)$ being a sampling from the uniform distribution in $[0, 1]$. Also, the application of this distribution depends on a learning parameter β, sometimes appearing under the form of $\alpha = 1 + \beta$.

It has been proved that, given τ and β sufficiently small, the effects of these two mutation schemes become comparable. It has been demonstrated that there is an equivalence between the two approaches, given the correspondence $\tau = \beta\sqrt{1 - \beta}$, if τ is sufficiently small.

3.2.6   How to Choose a Value for the Learning Parameter?

It has been demonstrated, within the hyperspherical model, which is usually a good local model, that large values for β or τ should be avoided. The objective is to find a compromise value that leads the algorithm to a near-optimal performance, measured in terms of *rate of progress* toward the optimum. The Schwefel's rule establishes that τ should be chosen proportional to $1/\sqrt{n}$, where n is the dimension of the search space. Practical rules suggested by Beyer [6] are the following:

$$\text{For } \lambda \geq 10, \ \tau \approx \frac{1}{\sqrt{n}} c_{1,\lambda}$$

$$\text{For } 4 < \lambda < 10, \ \tau \approx \frac{1}{\sqrt{n}} \frac{c_{1,\lambda}}{\sqrt{2c_{1,\lambda}^2 + 1 - 2d_{1,\lambda}^{(2)}}}$$

where $c_{1,\lambda}$ is called the *progress coefficient* and $d_{1,\lambda}^{(2)}$ is called the *second-order progress coefficient* [7]. Table 3.1 is a table of coefficients extracted from Ref. [6], calculated by numerical integration from a theoretical model.

Below $\lambda = 4$, one cannot adopt the second formula, because it yields an imaginary result; a value larger than $c_{1,\lambda}$ may be used instead. The σSA algorithm cannot adapt itself in such a way to obtain a theoretical optimum rate of progress, but it still self-adapts the mutation strength.

As a general indication, it must be said that there is a theoretical value for τ that maximizes the progress rate. However, this maximum is not symmetrical with respect to τ, and one has a much stronger risk of degrading the performance of

TABLE 3.1   Coefficients $c_{\mu,\lambda}$ to Adopt in a $(1, \lambda)$ES

λ	$c_{1,\lambda}$	$d_{1,\lambda}^{(2)}$	λ	$c_{1,\lambda}$	$d_{1,\lambda}^{(2)}$
2	0.5642	1.000	30	2.0428	4.4187
3	0.8463	1.2757	40	2.1608	4.8969
4	1.0294	1.5513	50	2.2491	5.2740
5	1.1630	1.8000	60	2.3193	5.5856
6	1.2672	2.0217	70	2.3774	5.8512
7	1.3522	2.2203	80	2.4268	6.0827
8	1.4236	2.3995	90	2.4697	6.2880
9	1.4850	2.5626	100	2.5076	6.4724
10	1.5388	2.7121	200	2.7460	7.7015
20	1.8675	3.7632	300	2.8778	8.4610

the algorithm by choosing a τ value too small than from using $\tau > c_{1,\lambda}/\sqrt{n}$. Also, it has been observed (as naturally expected) that a transient period precedes the establishment of a steady state in the progress of the algorithm. The transient behavior time is proportional to n. This is not a serious problem if the dimension of the problem is not too large, say, $n < 200$, which is realistic for a number of practical problems.

The magnitude of the learning parameter influences the duration of the transient period, whose duration in generation number is inversely proportional to τ^2. If τ is chosen according to the rule that makes it proportional to $1/\sqrt{n}$, then the transient phase duration becomes proportional to the space dimension n. If n is very large, say, $n > 1000$, this may be a serious problem, and then it is advisable to keep a fixed $\tau = 0.3$ during an initial period before starting to apply the rules above.

The choice of the learning parameter according to Schwefel's rule leads to a nearly optimum progress rate of the algorithm, once the transient phase is finished. A σSA algorithm with a learning parameter chosen according to the rules above exhibits a *linear convergence order*. There are always fluctuations, and it is not possible to attain the theoretical optimum rate of progress of the algorithm just by manipulating τ. Therefore, other mechanisms have been tried, like keeping a memory of the past values of the mutation rate in order to act upon some kind of moving average instead of upon the most recent value.

3.2.7 The (μ, λ)ES as an Extension of $(1, \lambda)$ES

The emergence of a (μ, λ)ES model with (μ parents $\leq \lambda$ offspring) is a natural development of evolution strategies. In (μ, λ), there is a population of μ individuals evolving in the parameter space; they generate λ offspring by randomly selecting one parent and mutating it, and doing this λ times. The μ individuals of the following generation are selected as the best among the λ individuals generated by mutation—an elitist strategy.

Beyer, in [8], developed theoretical work in order to explain the progress of a (μ, λ)ES as a generalized model of $(1, \lambda)$ES. He managed to derive a formula for the progress rate dependent on a single $c_{\mu,\lambda}$ progress coefficient parameter, generalizing the result obtained for the $(1, \lambda)$ES. In Table 3.2, we reproduce the table included in Ref. 8, which gives the value of $c_{\mu,\lambda}$ to consider when developing an evolution strategy model of this kind.

One of the consequences of having μ parents is that one can keep a number μ of distinct σ strategic parameters, each associated with a parent and mutated according to the rules explained above. It has been demonstrated, however, that this (μ, λ)ES risks divergence, and therefore Schwefel has recommended the adoption of elitist strategies, such as keeping or preserving the best individuals to the following generation. But if one is going to follow this elitist approach, then why not just adopt a $(\mu + \lambda)$ES and naturally keep the best individual in the successive generations?

TABLE 3.2   Coefficients $c_{\mu,\lambda}$ to Adopt for a Diversity of (μ, λ) Evolution Strategies

$\mu\backslash\lambda$	5	10	20	30	40	50	100	200	300
1	1.16	1.54	1.87	2.04	2016	2025	2.51	2.75	2.88
2	0.92	1.36	1.72	1.91	2.04	2.13	2.40	2.65	2.79
3	0.68	1.20	1.60	1.80	1.93	2.03	2.32	2.57	2.71
4	0.41	1.05	1.49	1.70	1.84	1.95	2.24	2.51	2.65
5	0.00	0.91	1.39	1.62	1.77	1.87	2.18	2.45	2.60
10		0.00	0.99	1.28	1.46	1.59	1.94	2.24	2.40
20			0.00	0.76	1.03	1.20	1.63	1.97	1.15
30				0.00	0.65	0.89	1.41	1.79	1.99
40					0.00	0.57	1.22	1.65	1.86
50						0.00	1.06	1.53	1.75
100							0.00	1.07	1.36

3.2.8   Self-Adaptation in (μ, λ)ES

Departing from the basic ideas of self-adaptation, tested, examined, and theoretically explained for $(1, \lambda)$ES, some variants have been considered and developed for the (μ, λ)ES, in what we could call the σSA (μ, λ)ES, which follows the lines of σSA $(1, \lambda)$ES.

In a σSA (μ, λ)ES, we must consider again object parameters (the variables of the problem) and strategic parameters—in this case, a mutation rate σ_k associated with each individual to be mutated. The object parameters (variable values) are mutated as usual, by having

$$\tilde{\mathbf{X}}_k = \mathbf{X}_k^{(g)} + \mathbf{Z}_k, \quad k = 1 \text{ to } \lambda. \tag{3.7}$$

where

$$\mathbf{Z}_k = \sigma_k^{(g+1)}[N(0, 1), \ldots, N(0, 1)]^t.$$

In order to approximate an optimal progress rate, the σ_k are mutated according to

$$\tilde{\sigma}_k = \sigma_k e^{z_0} e^{z_k} = \sigma_k e^{z_0 + z_k}.$$

As usual, the symbol $\sim$ denotes a mutated variable.

According to Schwefel [1], the mutating factors should be given by Gaussian distributions dependent on learning parameters τ such that

$$z_0 \in N(0, \tau_0^2), \quad \tau_0^2 = K^2 \frac{1}{2n}$$

$$z_k \in N(0, \tau_k^2), \quad \tau_k^2 = K^2 \frac{1}{2\sqrt{n}}.$$

For $1 < \mu < \lambda$, large n and λ and not too small μ, one may have a direct relation of K with the progress rate, and then

$$K \approx \frac{1}{2} \mu \ln \frac{\lambda}{\mu}.$$

Having uncorrelated distinct mutation strengths associated with the variables of the problem allows the evolution to adapt to an anisotropic shape of the fitness landscape; however, the search proceeds much along the coordinate axes of the search space, as illustrated in Fig. 3.5. This could be recognized in searching for the optimum in a function tested by Schwefel [9] as simple as

$$F(\mathbf{X}) = \sum_{i=1}^{n} i x_i^2,$$

where each variable is differently scaled—self-adaptation demands the learning of the scaling of n distinct σ_i.

It was verified that the self-adaptation scheme was very successful, after examining the results of a series of experiments of a σSA (μ, 100)ES for μ varying between 1 and 30. Furthermore, it was discovered that the most successful scheme was with $\mu = 12$, and that both smaller and larger values would cause loss of convergence speed.

The interpretation given is that for self-adaptation to work properly and efficiently, it requires a certain degree of diversity, represented in a number of parents. Furthermore, it has been discovered that having $\lambda > \mu$ is important, as well as having a limited life span of individuals (not allowing them to survive for more

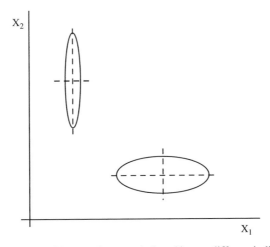

FIGURE 3.5 Illustration of the search pattern induced in two different individuals by distinct mutation rates affecting the distinct variables.

than a given number κ of generations), and also the application of recombination in the strategy parameters are all prerequisites for a successful self-adaptation scheme, which then can be made to approach theoretical optimum convergence speed.

3.3 EVOLUTIONARY PROGRAMMING

3.3.1 The $(\mu + \lambda)$ Bridge to ES

Instead of a (μ, λ)ES, one may have a $(\mu + \lambda)$ES. In this case, the μ parents of a $(g + 1)$ generation are chosen among the μ parents from generation (g) plus the λ offspring created by mutation from those μ parents. The practical indications on the values of the parameters to adopt in a basic $(\mu + \lambda)$ES follow the general trend of the (μ, λ)ES.

It is interesting to notice that a $(\mu + \lambda)$ES, with $\mu = \lambda$, is similar and can be assimilated to some formulations of evolutionary programming—and this is the bridge we were looking for in order to present EP, if we keep in mind that the body of EP work is broader than this. There is only one traditional difference that is minor but that has been inflated in many publications targeting a "general public" to sustain the idea that evolutionary programming and evolution strategies are two well-separate methods—it is the form of selection. Whereas in ES it was traditional to have an elitist selection (the best at each generation would be selected to the next one), in EP the tradition had preferred selection by stochastic tournament. We say tradition, on purpose: tournament and elitist selection have been used by both schools, of ES and EP—and by GA and GP (genetic programming) for that matter.

We recall that the most simple stochastic tournament is $T(1, 2)$, the one that, in successive operations, randomly samples two individuals from the parent population and, with a given probability fixed externally, selects the one with better fitness to be included in the next generation. This is done as many times as necessary until the required number λ of offspring is generated. Other kinds of stochastic tournaments can be conducted, such as $T(m, n)$, where the best m out of a sample of n parents is selected. Still, the first tournament method that was introduced into EP, in 1988, is different. In the words of D. Fogel, "it makes each solution compete with a random collection of some size of other solutions in pair-wise comparisons. A win is awarded to the solution that proves better or is awarded probabilistically according to a chosen formula. Then the solutions with most wins become the new parents."

In some models, however, the stochastic nature of the selection is abandoned, and pure elitist processes are adopted. For instance, one simply examines the fitness of all individuals in the parent population and selects the best λ individuals to form the next generation.

In parallel with the ES community, the evolutionary programming followers also developed a self-adaptive strategy. Applied to EP, this process was introduced in

1992 as *meta-EP* [10]. The mutation process governing the evolution of the mutation strength parameter in each individual from generation g to generation $g + 1$ is given by

$$\sigma^{(g+1)} = \xi \sigma^{(g)}, \tag{3.8}$$

where ξ is a random number given by

$$\xi = 1 + \tau N(0, 1), \tag{3.9}$$

and τ is the *learning parameter*, fixed externally. One may observe that the mutations in the mutation strength parameter are still of multiplicative type, like in ES, while the mutations in the phenotypic variables are additive.

Under this light, we can observe that the mutation operator used in EP, in the variant called meta-EP and discussed above, can be derived from the lognormal operator presented for ES in Section 3.2.5 by taking the Taylor expansion to its linear term, which gives precisely

$$\xi = 1 + \tau N(0, 1). \tag{3.10}$$

The optimal value of the learning parameter τ has been the object of empirical and theoretical studies. In many practical models, we found that this value has been fixed by trial and error, but the conclusions derived for ES are perfectly valid for the meta-EP. Therefore, the "meta-EP" proposed by Fogel [11] with its Gaussian approach, if τ is sufficiently small, becomes included in the class of models of $(\mu + \lambda)$ES and exhibits the same behavior. It can rightly be related, therefore, to this set of evolution strategies.

3.3.2 A Scheme for Evolutionary Programming

A typical EP model, as any model in the EC family, requires the definition of a fitness function and of a population of individuals. Each individual is represented by its variables in their natural domains. If a solution requires the representation of structural or topological aspects, these can also be represented as naturally as possible, namely by discrete variables.

Mutations act directly on variable values of an individual. Real-valued variables are subject to a zero mean multivariate Gaussian perturbation in each generation. This means that minor variations in an offspring become highly probable, whereas substantial variations become increasingly unlikely. The Gaussian scheme, however, does not prevent them. This procedure allows real-valued variables to converge to a possible optimum continuously, avoiding the discrete nature of a genetic binary coding. Also, this allows, sometimes, a new area in the search space to be explored, by an individual that suffered an important successful mutation. For discrete variables, often associated with topology features, mutations may follow Poisson distributed deletions or additions.

So here is the pseudo-code for a general simple EP algorithm:

```
Procedure EP
(BEGIN)
        start the generation counter
g:= 0;
        initialize a random population P
Initpopulation P[g];
            evaluate the fitness of all individuals of the
population
Evaluate P[g];
while not done do
                reproduction - duplicate the population
        P'[g]:= P[g]
                mutation - introduce stochastic
                perturbations in the new population,
                including in the strategic parameter σ
        P̃[g]:= Mutate (P'[g]);
                evaluation - calculate the fitness of the
                new individuals
        Evaluate P̃[g];
                selection - of the survivors for the next
                generation, based on the fitness
        value and on stochastic tournament
        P[g+1]:= select(P[g]∪P̃[g]);
                test for termination criterion (based on
                fitness, no. generations, etc.)
        If test is positive then done := TRUE;
            increase the generation counter
        g:= g + 1;
End while
(End EP)
```

Observing this piece of pseudo-code, we can see that the selection procedure acts upon a generation that is composed of "parents" and "sons"—$P[g]$ and $P'[g]$. This helps preserving the best individuals so that they may allow the exploration of promising regions by giving place to "good" mutations. In the spirit of evolutionary programming, the selection procedure should be stochastic; that is, the best individuals should be selected to the following generation with a given (usually high) probability. However, it is usual to find also, in practical applications and in the present day, procedures for deterministic selection, where the best are always selected.

3.3.3 Other Evolutionary Programming Variants

One cannot be fair to EP without mentioning that, besides modeling problems with real-valued variables, EP concepts have been applied to other types of problems. Historically, in fact, the first EP models were applied to evolving algorithms (machines) that would predict symbols emerging from a sequence of symbols. The work of L. Fogel was based on five-state machines running prediction experiments, with a population of three in a process similar to a $(3 + 3)$ES, where parent and child population would be simultaneously competing.

The fact that an important line in EP research has been devoted to evolving machines led to a sort of specialization of EP in mutation variants, techniques, and operators, and this is one of the sources of the common perception that EP is mainly focused on mutation as an engine for evolution. In fact, in many EP models more than one mutation process is simultaneously applied in complex combinations, adapted to the type of applications under study. We will not develop this topic further in this chapter, particularly devoted to problems modeled with real-valued variables, but the attention of the reader is called upon this line of development, which has produced extremely fruitful work, namely in areas connected with machine learning

3.4 COMMON FEATURES

3.4.1 Enhancing the Mutation Process

After having experiments with a global evolving mutation strength, and defining a global learning factor, as we have discussed so far, researchers tried to decouple mutations in one individual, so that the distinct variables could undergo distinct evolution processes. This meant that, for an individual, one would set not a single mutation strength but instead n mutation strengths for the n objective parameters or n variables defining an individual.

This scheme has been tried with some success. Going back to the *spherical model*, one has postulated that it was a good local approximation in many cases. However, it assumed an isotropic topology of the search space. Therefore, allowing distinct mutation strengths according to distinct coordinate directions (the variables of the problem) would in principle allow a more accurate approximation of regions with a sort of ellipsoid topology, instead of spherical. A scheme with n mutation strengths allows, therefore, a decoupling of the mutation rate evolution according to the axial directions of the search space. In many cases, this will be enough to enhance the performance of a self-adaptive evolution strategy or meta-EP.

But, in some cases, this is not enough. Correlations must be established between evolution along some direction and along some other direction. Otherwise, slow progress or even divergence may occur.

The original scheme with one single mutation strength assumed an evolution in an isotropic space—say, the length R of a vector $\mathbf{R}$, or $\|\mathbf{R}\|$, is given by $\|\mathbf{R}\| = \mathbf{R}^t\mathbf{R}$. Decoupling variable mutation strengths is equivalent to assuming a diagonal metrics matrix in the search space; therefore, the length R of a vector $\mathbf{R}$ would be given by $\|\mathbf{R}\| = \mathbf{R}'\mathbf{DR}$, with $\mathbf{D}$ being a diagonal matrix. This has been illustrated in Fig. 3.5.

Recognizing this, ES has incorporated correlation between mutations as strategic variables. This is equivalent to considering a Mahalanobis metric in space—the length of a vector $\mathbf{R}$ will be given by $\|\mathbf{R}\| = \mathbf{R}'\mathbf{TR}$, where $\mathbf{T}$ is a full matrix. Figure 3.6 illustrates the effect of having nonzero covariances between variables, allowing the exploration of the search space along directions not aligned with the coordinate axes.

ES followers have adopted a formal mathematical representation of the possible covariances of the mutation distribution in the several directions in space. The basic concept is the one of "inclination angle α" as the departure point to defining linearly dependent mutation correlations for the object variables. Given an angle α_j, a basic covariance matrix between directions p and q may be defined by the transformation matrix $\mathbf{T}$, given by

$$
\mathbf{T}_{pq}(\alpha) =
\begin{bmatrix}
1 & 0 & & & & & & & 0 & & 0 \\
0 & 1 & & & & & & & 0 & & 0 \\
& & \ddots & & & & & & & & \\
& & & \cos\alpha_j & & & -\sin\alpha_j & & & & \\
& & & & 1 & 0 & & & & & \\
& & & & & \ddots & & & & & \\
& & & & 0 & 1 & & & & & \\
& & & \sin\alpha_j & & & \cos\alpha_j & & & & \\
0 & 0 & & & & & & & 1 & & 0 \\
& & & & & & & & & \ddots & \\
0 & 0 & & & & & & & 0 & & 1
\end{bmatrix},
$$

where only lines and columns p and q have elements distinct from 0 or 1.

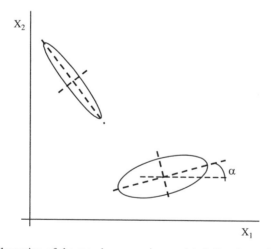

FIGURE 3.6   Illustration of the search process in correlated directions of two distinct individuals with different linear correlations between variables. The angle α becomes one strategic variable subject to mutation (and recombination).

The product of all $\mathbf{T}_{pq}$ matrices according to all combinations of p and q gives the matrix $\mathbf{C}$ of covariances. This allows the calculation of an actual mutation $\tilde{\mathbf{X}}$ to a given individual $\mathbf{X}$ by

$$\tilde{\mathbf{X}} = \mathbf{X} + \mathbf{CZ}, \tag{3.11}$$

where $\mathbf{Z} = (z_1, \dots, z_n)$ and $z_i \in N(0, \sigma_i^2)$. The vector $\mathbf{CZ}$ is, therefore, a random vector with normally distributed and eventually correlated components, as a function of the α_i and the σ_i.

The strategic variables that establish the correlations or the covariances (filling up of nonzeros the elements out of the main diagonal of $\mathbf{C}$) are the inclination angles α; one can readily see that if these angles are set to 0, then all matrices $\mathbf{T}_{pq}$ become identity matrices, and therefore mutations will develop independently in all dimensions of the search space. These angles α are, therefore, taken as strategic variables and may also be mutated and subject to a self-adaptive procedure.

In summary, to establish correlated mutations, one may proceed step by step as follows:

1. Mutate the σ_i
2. Mutate the α_k
3. Calculate and apply the matrix $\mathbf{C}$ to obtain a new mutated individual

The angles α_k should be mutated with

$$\tilde{\alpha}_k = \alpha_k + z_k \quad \text{with} \quad z_k \in N(0, \beta^2).$$

Values of β come from experimental work, and Schwefel [1] has recommended something around $5°$ or 0.0873 rad as giving good results in practice.

One more thing may be said about mutation processes and the search for an effective way of making them effective. In EP, it is not rare that several mutation operators, with different effects, be used to generate offspring. This operation is often ruled by probabilities and may remain constant or be changed from generation to generation [12].

3.4.2 Recombination as a Major Factor

So far, we have only discussed mutation as a factor of progress in an evolutionary process. However, this is an incorrect or at least an incomplete view. In fact, recombination may play an important, even dominant role in accelerating the progress to the optimum and enhancing the chances of success of the search procedure. To be fair, the idea of recombination was already present in the early works of L. Fogel [13], where the generation of offspring in evolving finite state machines was produced by such a kind of mechanism. In the ES community, this idea was adopted at a later stage, especially for problems with real-valued variables. In the EP community, namely dealing with this type of problem, recombination was not favored until recently as a major tool or not at all.

This distinction between ES and EP (on the use of recombination) is recorded, for instance, in the Glossary—Evolutionary Algorithms—Terms and Definitions available online on the Internet (http://ls11-www.cs.uni-dortmund.de/people/beyer/EA-glossary/) and organized by H.G. Beyer and other German researchers.

Using the notation that Schwefel and Rechenberg proposed, as we have been doing so far, we have now to consider $(\mu/\rho, \lambda)$ evolution strategies. In this variant, the parameter ρ determines the number of parents that recombine to form one new individual. The biological construction is based on $\rho = 2$, but when building an evolution strategy or designing an evolutionary programming approach, we do not need to be limited to that option and may experiment with strategies with $\rho > 2$.

Recombination is a word that designates a number of distinct procedures that share the property of building an individual departing from a set of parents. Here are some possible recombination schemes that have been used in evolution strategies:

Uniform crossover: In this variant, the value for each variable in the newly formed individual is obtained by randomly selecting one of the ρ parents to "donate" its value. In the case of $\rho = 2$, it is traditional to randomly generate a bit string with length equal to the number of variables of the individuals and then to use such string to command the recombination procedure: if a bit associated with a variable is 1, the value from the first parent is selected; if it is 0, then the value from the second parent is selected.

Intermediary recombination: In this variant, the value of any variable in the offspring receives a contribution from all parents. This could result either from averaging the values of all parents (global intermediary recombination) or from averaging the values from a subset of the parents only, randomly chosen (local intermediary recombination). In these processes, one may still choose to average values with equal weights or to randomly define weights for a weighted average. In the case of $\rho = 2$, one could have the value of a variable given by

$$x_k^{\text{new}} = u_k x_{k,\,j1} + (1 - u_k)x_{k,\,j2},$$

where the indices $j1$ and $j2$ denote the two parent individuals and u_k sampled from a uniform distribution in $[0, 1]$.

Point crossover: In this variant, parallel to the one adopted in genetic algorithms, first one randomly defines $\gamma \, (< n)$ crossover points, common to all individuals in the set of parents, and then the offspring successively receives a part from each parent, in turn. Experience has demonstrated the power of recombination to greatly accelerate the convergence of evolution strategies. This being so, some theoretical explanations were sought.

In the genetic algorithm community, the *building block theory* became popular. It stated that recombination allowed good blocks from each parent to join together in a better descendant. But in the ES community, other mathematical descriptions allowed distinct views to emerge. Beyer, for instance, argued differently [14], based on his developments of the concept of progress rate: he suggested that recombination acted as a sort of "genetic repair" mechanism, compensating the disruptive effects of mutation. Therefore, larger mutation strengths or larger learning parameters were allowed, contributing to a higher progress rate than in an ES without recombination.

Furthermore, under some assumptions, Beyer also demonstrated that the highest impulse from recombination was achieved when all μ parents contributed to form one new individual. He justified this assertion with a mathematical demonstration and called his model the $(\mu/\mu, \lambda)$ES. Recombination, thus, plays a major role in modern evolution strategies and is not a secondary technique.

3.4.3 Handling Constraints

Unlike GA, EP allows a very natural way of handling constraints in a problem. Because each individual is coded in its original or phenotypic variables, it is usually easy to enforce constraints.

One way to do that is during the mutation phase—each time a new individual is mutated, it can be checked for feasibility, discarded if the constraints are not met, and a replacement generated until one is found that respects constraints. Furthermore, during the mutation phase, mutations can be in many cases conditioned so that there is no possibility for an unfeasible descendant to be created. This was the original ES scheme for handling constraints—but sometimes it may be a very time-consuming procedure.

The other possibility is to handle constraints during the selection phase, by attributing a low fitness value to individuals that violate constraints.

3.4.4 Starting Point

To start an evolution strategy process, one has to generate a first initial population of μ individuals. This can be typically done in two ways:

By randomly generating the coordinates for the μ individuals, or

By generating mutations from a seed or starting individual.

It was traditional in the community of evolution strategies to make sure that the initial population was composed of feasible individuals. However, this may not be mandatory if an adequate method of penalties and handling constraints is adopted.

3.4.5 Fitness Function

The fitness function is usually represented by the objective function to be evaluated, representative of the problem to be solved. To this fitness function, one may add the effect of penalties to represent the undesired violation of constraints. This is an approach adopted in all evolutionary algorithms variants.

One simple and yet claimed as effective way of adding the effect of the violation of constraints, in a problem of maximization, is to count the number of violated constraints or add up the amount of violations, and attribute the fitness value according to the following rule:

If no violations occur, $\text{Fitness}(X) = F(X)$.

If violations occur, $\text{Fitness}(X) = -\Sigma$ violation values i, $i \in$ constraint set.

3.4.6 Computing

One may find many software sources allowing the implementation of evolution strategies. A few examples are mentioned below to help beginners in the search in the internet.

One well known possibility is **evoC**, available from the Bionics and Evolution Department of the Technical University of Berlin, Germany—it is an application written in C that may be used in a variety of platforms, from MS-DOS to LINUX. It is free but not in the public domain and could until recently be obtained from ftp://ftp-bionic.fb10.tu-berlin.de under the directory /pub/software/evoC.

A set of MATLAB tools with a user-friendly interface developed at the University of Magdburg, Germany, can be requested and obtained from bihn@infaut.et.uni-magdburg.de (Dr. Bihn).

There is also a set of demonstration programs with a relevant didactic interest that were developed and are available from the ftp server of the Technical University of Berlin at ftp://ftp-bionic.fb10.tu-berlin.de. These demonstrations are available also on the Internet, and have the support of a technical report [15] available from the same server.

GAGS ia a public domain object-oriented library for evolutionary computation, and although it may be used for GA models, it may also serve for other EC paradigms such as ES/EP. It may be obtained from http://kal-el.ugr.es/gags.html.

A rather large EA library written in Java 1.3 is available from the site of EvaLife in Denmark, http://www.evalife.dk/. Last but not least, we must refer to EvoWeb, http://evonet.dcs.napier.ac.uk/, the Web site of EvoNet (the European Network of Excellence in Evolutionary Computing).

3.5 CONCLUSIONS

Evolution strategies is the designation of a broad family of evolutionary algorithms that have in common the representation of solutions in the space of problem variables, instead of using any sort of coding like the binary coding adopted in genetic algorithms.

The ES school of thought had its birth in Germany and remained confined to a closed community, perhaps due to the fact that many of the early publications were made in the German language. However, it is now obvious that ES is a rich and fruitful field. Detached from the historical processes that give birth to new ideas, one may also with no difficulty recognize that evolutionary programming, which had an independent development, converged with evolution strategies for real-valued variable problems.

There is a substantial theoretical work justifying the success of evolution strategy and evolutionary programming algorithms and giving indications on how to tune an algorithm in order to obtain the maximum possible efficiency, measured by the rate of progress toward the optimum. This theoretical basis clearly indicates that recombination is a major operator inducing fast evolutionary progress. Under this light, pure evolutionary programming models, relying only on mutation, may seem to be more limited.

Theory and experiments also suggest that good strategies should use a number of parents μ generating a larger offspring $\lambda > \mu$, and that in many cases generalized recombination processes, using all μ parents to generate each offspring, offer the faster rates of progress or algorithm efficiency.

Finally, it is evident today that self-adaptation schemes are usually very effective and offer the best chances of reaching the absolute optimum while exogenously controlled mutation strengths, even if allowing in some cases a fast progress toward the optimum, risk becoming trapped in local optima.

REFERENCES

1. Schwefel H-P, Rudolph G. Contemporary evolution strategies. In: Morán F, Moreno A, Merelo JJ, Chacón P, eds. Advances in artificial life. 3rd International Conference on Artificial Life. Vol. 929 of Lecture Notes in Artificial Intelligence. Berlin: Springer; 1995. pp. 893–907.

2. Beyer H-G. Towards a theory of evolution strategies: Some asymptotical results from the $(1, +\lambda)$-theory. Evol Computat 1993; 1(2):165–188.

3. Rechenberg I. Evolutionsstrategie-Optimierung technischer Systeme nach Prinzipen der biologischen Evolution. Stuttgart: Frommann-Holzboog; 1973.

4. Schwefel H-P. Adaptive Mechanismen in der biologischen Evolution und ihr Einfluß auf die Evolutionsgeschwindigkeit. Technical report, Technical University Berlin; 1974.

5. Schwefel H-P. Evolution and optimum seeking. New York: John Wiley & Sons; 1995.

6. Beyer H-G. Toward a theory of evolution strategies: Self-adaptation. Evol Computat 1996; 3(3):311–347.

7. Beyer H-G. Towards a theory of evolution strategies: Progress rates and quality gain for $(1, +\lambda)$ strategies on (nearly) arbitrary fitness functions. In: Davidor Y, Männer R, Schwefel H-P, eds. Parallel problem solving from nature, 3. Heidelberg: Springer-Verlag; 1994. 58–67.

8. Beyer H-G. Toward a theory of evolution strategies: The (μ, λ) theory. Evol Computat 1995; 2(4): 381–407.

9. Schwefel H-P. Natural evolution and collective optimum-seeking. In: Sydow A, ed. Computational system analysis: Topics and trends. Amsterdam: Elsevier; 1992. 5–14.

10. Fogel DB. Evolving artificial intelligence. Ph.D. thesis, University of California San Diego; 1992.

11. Fogel DB, Fogel LJ, Atmar JW. Meta-evolutionary programming. In: Chen RR, ed. Proceedings of the 25th Asilomar Conference on Signals, Systems and Computers. San Jose, CA: Maple Press; 1991. 540–545.

12. Fogel DB. Evolutionary computation: Toward a new philosophy of machine intelligence. 2nd ed. Piscataway, NJ: IEEE Press; 2000.

13. Fogel LJ, Owens AJ, Walsh MJ, Artificial intelligence through simulated evolution. New York: John Wiley & Sons; 1966.

14. Beyer H-G. Toward a theory of evolution strategies: on the benefits of sex—the $(\mu/\mu, \lambda)$ theory. Evol Computat 1995; 3(1):81–111.

15. Herdy M, Patone G. Evolution strategy in action. In: Proceedings of the International Conference on Evolutionary Computing—the Third Conference on Parallel Problem Solving from Nature (PPSN III), Lecture Notes in Computer Science, Vol. 866, Springer-Verlag, 1994.

CHAPTER 4

Fundamentals of Particle Swarm Optimization Techniques

YOSHIKAZU FUKUYAMA

4.1 INTRODUCTION

Natural creatures sometimes behave as a swarm. One of the main streams of artificial life research is to examine how natural creatures behave as a swarm and reconfigure the swarm models inside a computer. Reynolds developed *boid* as a swarm model with simple rules and generated complicated swarm behavior by computer graphic animation [1]. Boyd and Richerson examined the decision process of human beings and developed the concept of *individual learning and cultural transmission* [2]. According to their examination, human beings make decisions using their own experiences and other persons' experiences.

A new optimization technique using an analogy of swarm behavior of natural creatures was started in the beginning of the 1990s. Dorigo developed ant colony optimization (ACO) based mainly on the social insect, especially ant, metaphor [3]. Each individual exchanges information through pheromones implicitly in ACO. Eberhart and Kennedy developed particle swarm optimization (PSO) based on the analogy of swarms of birds and fish schooling [4]. Each individual exchanges previous experiences in PSO. These research efforts are called *swarm intelligence* [5, 6]. This chapter focuses on PSO as one of the *swarm intelligence* techniques.

Other evolutionary computation (EC) techniques such as genetic algorithms (GAs), utilize multiple searching points in the solution space like PSO. Whereas GAs can treat combinatorial optimization problems, PSO was aimed to treat nonlinear optimization problems with continuous variables originally. Moreover, PSO has been expanded to handle combinatorial optimization problems and both discrete and continuous variables as well. Efficient treatment of mixed-integer nonlinear optimization problems (MINLPs) is one of the most difficult problems in practical optimization. Moreover, unlike other EC techniques, PSO can be realized with only a small

Modern Heuristic Optimization Techniques. Edited by K. Y. Lee and M. A. El-Sharkawi
Copyright © 2008 the Institute of Electrical and Electronics Engineers, Inc.

program; namely, PSO can handle MINLPs with only a small program. This feature of PSO is one of its advantages compared with other optimization techniques.

This chapter is organized as follows: Section 4.2 explains the basic PSO method and Section 4.3 explains variations of PSO. Section 4.4 shows some applications of PSO, and Section 4.5 concludes this chapter with some remarks.

4.2 BASIC PARTICLE SWARM OPTIMIZATION

4.2.1 Background of Particle Swarm Optimization

Swarm behavior can be modeled with a few simple rules. Schools of fishes and swarms of birds can be modeled with such simple models. Namely, even if the behavior rules of each individual (agent) are simple, the behavior of the swarm can be complicated. Reynolds utilized the following three *vectors* as simple rules in the researches on *boid*.

1. Step away from the nearest agent
2. Go toward the destination
3. Go to the center of the swarm

The behavior of each agent inside the swarm can be modeled with simple *vectors*. The research results are one of the basic backgrounds of PSO.

Boyd and Richerson examined the decision process of humans and developed the concept of *individual learning and cultural transmission* [2]. According to their examination, people utilize two important kinds of information in decision process. The first one is their *own experience*; that is, they have tried the choices and know which state has been better so far, and they know how good it was. The second one is *other people's experiences*; that is, they have knowledge of how the other agents around them have performed. Namely, they know which choices their neighbors have found most positive so far and how positive the best pattern of choices was. Each agent decides its decision using its own experiences and the experiences of others. The research results are also one of the basic background elements of PSO.

4.2.2 Original PSO

According to the above background of PSO, Kennedy and Eberhart developed PSO through simulation of bird flocking in a two-dimensional space. The position of each agent is represented by its x, y axis position and also its velocity is expressed by vx (the velocity of x axis) and vy (the velocity of y axis). Modification of the agent position is realized by the position and velocity information.

Bird flocking optimizes a certain objective function. Each agent knows its best value so far (pbest) and its x, y position. This information is an analogy of the *personal experiences* of each agent. Moreover, each agent knows the best value so far in the group (gbest) among pbests. This information is an analogy of *the knowledge of how the other agents around them have performed.* Each agent tries to modify its position using the following information:

The current positions (x, y),

The current velocities (vx, vy),

The distance between the current position and pbest

The distance between the current position and gbest

This modification can be represented by the concept of velocity (modified value for the current positions). Velocity of each agent can be modified by the following equation:

$$v_i^{k+1} = wv_i^k + c_1\text{rand}_1 \times (\text{pbest}_i - s_i^k) + c_2\text{rand}_2 \times (\text{gbest} - s_i^k) \quad (4.1)$$

where v_i^k is velocity of agent i at iteration k, w is weighting function, c_j is weighting coefficients, rand is random number between 0 and 1, s_i^k is current position of agent i at iteration k, pbest_i is pbest of agent i, and gbest is gbest of the group.

Namely, velocity of an agent can be changed using three *vectors* such like *boid*. The velocity is usually limited to a certain maximum value. PSO using (4.1) is called the Gbest model.

The following weighting function is usually utilized in (4.1):

$$w = w_{\max} - \frac{w_{\max} - w_{\min}}{\text{iter}_{\max}} \times \text{iter}, \quad (4.2)$$

where $w_{\max}$ is initial weight, $w_{\min}$ is final weight, $\text{iter}_{\max}$ is maximum iteration number, and iter is current iteration number.

The meanings of the right-hand side (RHS) of (4.1) can be explained as follows [7]. The RHS of (4.1) consists of three terms (*vectors*). The first term is the previous velocity of the agent. The second and third terms are utilized to change the velocity of the agent. Without the second and third terms, the agent will keep on "flying" in the same direction until it hits the boundary. Namely, it tries to explore new areas and, therefore, the first term corresponds with *diversification* in the search procedure. On the other hand, without the first term, the velocity of the "flying" agent is only determined by using its current position and its best positions in history. Namely, the agents will try to converge to the their pbests and/or gbest and, therefore, the terms correspond with *intensification* in the search procedure. As shown below, for example, $w_{\max}$ and $w_{\min}$ are set to 0.9 and 0.4. Therefore, at the beginning of the search procedure, *diversification* is heavily weighted, while *intensification* is heavily weighted at the end of the search procedure such like simulated annealing (SA). Namely, a certain velocity, which gradually gets close to pbests and gbest, can be calculated. PSO using (4.1), (4.2) is called *inertia weights approach* (IWA).

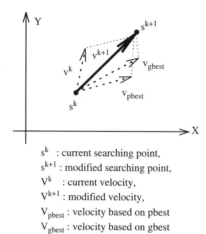

s^k : current searching point,
s^{k+1} : modified searching point,
V^k : current velocity,
V^{k+1} : modified velocity,
V_{pbest} : velocity based on pbest
V_{gbest} : velocity based on gbest

FIGURE 4.1 Concept of modification of a searching point by PSO.

The current position (searching point in the solution space) can be modified by the following equation:

$$s_i^{k+1} = s_i^k + v_i^{k+1}. \tag{4.3}$$

Figure 4.1 shows a concept of modification of a searching point by PSO, and Fig. 4.2 shows a searching concept with agents in a solution space. Each agent changes its current position using the integration of vectors as shown in Fig. 4.1.

The general flowchart of PSO with IWA can be described as follows:

Step 1. Generation of initial condition of each agent. Initial searching points (s_i^0) and velocities (v_i^0) of each agent are usually generated randomly within the allowable range. The current searching point is set to pbest for each agent. The best-evaluated value of pbest is set to gbest, and the agent number with the best value is stored.

Step 2. Evaluation of searching point of each agent. The objective function value is calculated for each agent. If the value is better than the current pbest of the agent,

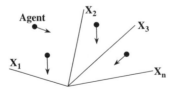

FIGURE 4.2 Searching concept with agents in a solution space by PSO.

the pbest value is replaced by the current value. If the best value of pbest is better than the current gbest, gbest is replaced by the best value and the agent number with the best value is stored.

Step 3. Modification of each searching point. The current searching point of each agent is changed using (4.1), (4.2), and (4.3).

Step 4. Checking the exit condition. The current iteration number reaches the predetermined maximum iteration number, then exits. Otherwise, the process proceeds to step 2.

Figure 4.3 shows the general flowchart of PSO. The features of the searching procedure of PSO can be summarized as follows:

(a) As shown in (4.1), (4.2), and (4.3), PSO can essentially handle continuous optimization problems.

(b) PSO utilizes several searching points, and the searching points gradually get close to the optimal point using their pbests and the gbest.

(c) The first term of the RHS of (4.1) corresponds with diversification in the search procedure. The second and third terms correspond with intensification in the search procedure. Namely, the method has a well-balanced mechanism to utilize diversification and intensification in the search procedure efficiently.

(d) The above concept is explained using only the x, y axis (two-dimensional space). However, the method can be easily applied to n-dimensional problems. Namely, PSO can handle continuous optimization problems with continuous state variables in a n-dimensional solution space.

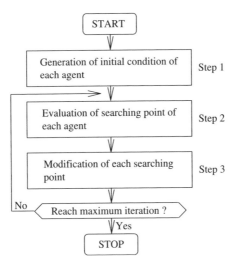

FIGURE 4.3   A general flowchart of PSO.

Shi and Eberhart tried to examine the parameter selection of the above parameters [7, 8]. According to their examination, the following parameters are appropriate and the values do *not* depend on problems:

$$c_i = 2.0, \qquad w_{max} = 0.9, \qquad w_{min} = 0.4.$$

The values are also proved to be appropriate for power system problems [9, 10]. The basic PSO has been applied to a learning problem of neural networks and Schaffer f6, a famous benchmark function for GA, and the efficiency of the method has been observed [4].

4.3   VARIATIONS OF PARTICLE SWARM OPTIMIZATION

4.3.1   Discrete PSO

The original PSO described in Section 4.2 treats nonlinear optimization problems with continuous variables. However, practical engineering problems are often formulated as combinatorial optimization problems. Kennedy and Eberhart developed a discrete binary version of PSO for these problems [11]. They proposed a model wherein the probability of an agent's deciding yes or no, true or false, or making some other decision is a function of personal and social factors as follows:

$$P(s_i^{k+1} = 1) = f(s_i^k, v_i^k, \text{pbest}_i, \text{gbest}). \tag{4.4}$$

The parameter v, an agent's tendency to make one or the other choice, will determine a probability threshold. If v is higher, the agent is more likely to choose 1, and lower values favor 0 choice. Such a threshold requires staying in the range [0, 1]. One of the functions accomplishing this feature is the sigmoid function, which is usually utilized in artificial neural networks.

$$\text{sig}(v_i^k) = \frac{1}{1 + \exp(-v_i^k)}. \tag{4.5}$$

The agent's tendency should be adjusted for success of the agent and the group. In order to accomplish this, a formula for each v_i^k that will be some function of the difference between the agent's current position and the best positions found so far by itself and by the group should be developed. Namely, like the basic continuous version, the formula for the binary version of PSO can be described as follows:

$$v_i^{k+1} = v_i^k + \text{rand}_1 \times (\text{pbest}_i - s_i^k) + \text{rand}_2 \times (\text{gbest} - s_i^k) \tag{4.6}$$

$$\rho_i^{k+1} < \text{sig}(v_i^{k+1}) \quad \text{then } s_i^{k+1} = 1;$$
$$\text{else } s_i^{k+1} = 0, \tag{4.7}$$

where rand is a positive random number drawn from a uniform distribution with a predefined upper limit, and ρ_i^{k+1} is a vector of random numbers of [0.0, 1.0].

These formulas are iterated repeatedly over each dimension of each agent. The second and third term of the RHS of (4.6) can be weighted like the basic continuous version of PSO. v_i^k can be limited so that $\text{sig}(v_i^k)$ does not approach too closely to 0.0 or 1.0. This ensures that there is always some chance of a bit flipping. A constant parameter $V_{\max}$ (limited value of v_i^k) can be set at the start of a trial. In practice, $V_{\max}$ is often set in $[-4.0, +4.0]$. The entire algorithm of the binary version of PSO is almost the same as that of the basic continuous version except for the above state equations.

4.3.2 PSO for MINLPs

Engineering problems often pose both discrete and continuous variables using nonlinear objective functions. Kennedy and Eberhart discussed the integration of binary and continuous versions of PSO [6]. Fukuyama et al. presented a PSO for MINLPs by modifying the continuous version of PSO [9].

The method of Ref. 9 can be described briefly as follows. Discrete variables can be handled in (4.1) and (4.3) with little modification. Discrete numbers instead of continuous numbers can be used to express the current position and velocity. Namely, a discrete random number is used for rand in (4.1), and the whole calculation of RHS of (4.1) is discretized to the existing discrete number. Using this modification for discrete numbers, both continuous and discrete numbers can be handled in the algorithm with no inconsistency. In Ref. 9, the PSO for MINLPs was applied successfully to a reactive power and voltage control problem in electric power systems with promising results. The details can be found in Chapter 16.

4.3.3 Constriction Factor Approach (CFA)

The basic system equation of PSO [(4.1), (4.2), and (4.3)] can be considered as a kind of difference equation. Therefore, the system dynamics, that is, the search procedure, can be analyzed using eigenvalues of the difference equation. Actually, using a simplified state equation of PSO, Clerc and Kennedy developed CFA of PSO by eigenvalues [6, 12]. The velocity of the constriction factor approach (simplest constriction) can be expressed as follows instead of (4.1) and (4.2):

$$v_i^{k+1} = K[v_i^k + c_1 \times \text{rand}_1 \times (\text{pbest}_i - s_i^k)$$
$$+ c_2 \times \text{rand}_2 \times (\text{gbest} - s_i^k)] \qquad (4.8)$$

$$K = \frac{2}{|2 - \varphi - \sqrt{\varphi^2 - 4\varphi}|}, \quad \text{where } \varphi = c_1 + c_2, \ \varphi > 4, \qquad (4.9)$$

where φ and K are coefficients.

For example, if $\varphi = 4.1$, then $K = 0.73$. As φ increases above 4.0, K gets smaller. For example, if $\varphi = 5.0$, then $K = 0.38$, and the damping effect is even more pronounced. The convergence characteristic of the system can be controlled by φ. Namely, Clerc et al. found that the system behavior can be controlled so that the system behavior has the following features:

(a) The system does not diverge in a real-valued region and finally can converge.
(b) The system can search different regions efficiently by avoiding premature convergence.

The whole PSO algorithms by IWA and CFA are the same except that CFA utilizes a different equation for calculation of velocity [(4.8) and (4.9)]. Unlike other EC methods, PSO with CFA ensures the convergence of the search procedures based on mathematical theory. PSO with CFA can generate higher-quality solutions for some problems than PSO with IWA [10, 13]. However, CFA only considers dynamic behavior of only one agent and studies on the effect of the interaction among agents. Understanding the effects of pbests and gbest in the system dynamics remains for future work.

4.3.4 Hybrid PSO (HPSO)

Angeline developed HPSO by combining the selection mechanism of EC techniques with PSO [14]. The number of highly evaluated agents is increased whereas the number of lowly evaluated agents is decreased at each iteration by HPSO. Because the PSO search procedure depends greatly on pbest and gbest, the searching area is limited by pbest and gbest. However, the effect of pbest and gbest is gradually vanished by the selection mechanism, and a broader area search can be realized by HPSO. Agents with low evaluation values are replaced by those with high evaluation values using the selection. The replaced rate is called selection rate (Sr). It should be noted that the pbest information of each agent is maintained even if the agent position is replaced to another agent's position. Therefore, intensive search in a current attractive area and dependence on the past high-evaluation position are realized. Figure 4.4 shows a general flowchart of HPSO. As shown in the figure, only step 3 is added to the original PSO. Figure 4.5 shows the concepts of steps 2, 3, and 4 of the general flowchart.

The original PSO changes the current search point using the state equations (4.1), (4.2), and (4.3) step by step. Therefore, it sometimes takes time to get into an *attractive area* in the solution space. On the contrary, HPSO moves the lowly evaluated agents to the attractive area directly using the selection method, and concentrated search especially in the current attractive area is realized. However, a jump to the attractive area does not always lead to effective search, and a more attractive area may exist between the current searching point and the jumped area. As Angeline pointed out in Ref. 14, HPSO does not always work better than the original PSO. HPO with IWA calculates velocity and new searching points using (4.1), (4.2), and (4.3), whereas HPSO with CFA calculates them using (4.8), (4.9), and (4.3).

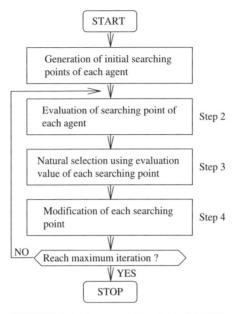

FIGURE 4.4   A general flowchart of HPSO.

4.3.5 Lbest Model

Eberhart and Kennedy also developed an *lbest model* [6]. In the model, agents have information only of their own and their nearest array neighbors' bests (lbests), rather than that of the entire group. Namely, in (4.1), gbest is replaced by lbests in the model. For example, lbest of agent no. 3 can be determined as the best pbest among pbests of agents no. 2, 3, and 4.

$$v_i^{k+1} = wv_i^k + c_1 \text{rand}_1 \times (\text{pbest}_i - s_i^k) + c_2 \text{rand}_2 \times (\text{lbest}_i - s_i^k), \tag{4.10}$$

where lbest$_i$ is local best of agent i.

The whole PSO algorithms by lbest model and gbest model are the same except that lbest model utilizes a different equation for calculation of velocity [(4.10)].

4.3.6 Adaptive PSO (APSO)

The following points are improved to the original PSO with IWA.

1. The search trajectory of PSO can be controlled by introducing the new parameters (P_1, P_2) based on the probability to move close to the position of (pbest, gbest) at the following iteration.
2. The wv_i^k term of (4.1) is modified as (4.12). Using the equation, the center of the range of particle movements can be equal to gbest.

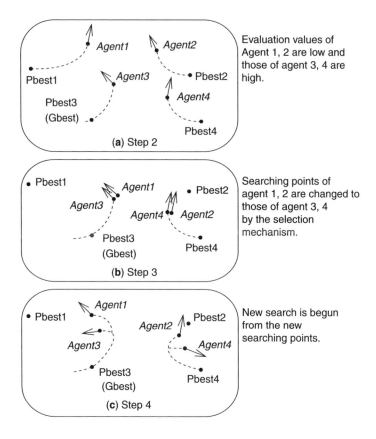

FIGURE 4.5   Concept of searching process by HPSO.

3. When the agent becomes gbest, it is perturbed. The new parameters (P_1, P_2) of the agent are adjusted so that the agent may move away from the position of (pbest, gbest).

4. When the agent is moved beyond the boundary of feasible regions, pbests and gbest cannot be modified.

5. When the agent is moved beyond the boundary of feasible regions, the new parameters (P_1, P_2) of the agent are adjusted so that the agent may move close to the position of (pbest, gbest).

The new parameters are set to each agent. The weighting coefficients are calculated as follows:

$$c_2 = \frac{2}{P_1}, \qquad c_1 = \frac{2}{P_2} - c_2, \qquad (4.11)$$

The search trajectory of PSO can be controlled by the parameters (P_1, P_2). Concretely, when the value is enlarged more than 0.5, the agent may move close to the position of pbest/gbest.

$$w = \text{gbest} - (\{c_1(\text{pbest} - x) + c_2(\text{gbest} - x)\}/2 + x). \qquad (4.12)$$

Namely, the velocity of the improved PSO can be expressed as follows:

$$v_i^{k+1} = w_i + c_1\text{rand}_1 \times (\text{pbest}_i - s_i^k) + c_2\text{rand}_2 \times (\text{gbest} - s_i^k). \qquad (4.13)$$

The improved PSO can be expressed as follows (steps 1 and 5 are the same as PSO):

1. *Generation of initial searching points*: Basic procedures are the same as PSO. In addition, the parameters (P_1, P_2) of each agent are set to 0.5 or higher. Then, each agent may move close to the position of (pbest, gbest) at the following iteration.
2. *Evaluation of searching points*: The procedure is the same as PSO. In addition, when the agent becomes gbest, it is perturbed. The parameters (P_1, P_2) of the agent are adjusted to 0.5 or lower so that the agent may move away from the position of (pbest, gbest).
3. *Modification of searching points*: The current searching points are modified using the state equations (4.13), (4.3) of adaptive PSO.

4.3.7 Evolutionary PSO (EPSO)

The idea behind EPSO [16, 17] is to grant a PSO scheme with an explicit selection procedure and with self-adapting properties for its parameters. At a given iteration, consider a set of solutions or alternatives that we will keep calling particles. The general scheme of EPSO is the following:

1. REPLICATION: each particle is replicated R times
2. MUTATION: each particle has its weights mutated
3. REPRODUCTION: each mutated particle generates an offspring according to the particle movement rule
4. EVALUATION: each offspring has its fitness evaluated
5. SELECTION: by stochastic tournament the best particles survive to form a new generation.

The movement rule for EPSO is the following: given a particle s_i^k, a new particle s_i^{k+1} results from

$$s_i^{k+1} = s_i^k + v_i^{k+1} \qquad (4.14)$$

$$v_i^{k+1} = w_{i0}^* v_i^k + w_{i1}^* (\text{pbest}_i - s_i^k) + w_{i2}^* (\text{gbest}^* - s_i^k). \qquad (4.15)$$

So far, this seems like PSO—the movement rule keeps its terms of inertia, memory, and cooperation. However, the weights undergo mutation

$$w_{ik}^* = w_{ik} + \tau N(0, 1), \tag{4.16}$$

where $N(0, 1)$ is a random variable with Gaussian distribution, 0 mean, and variance 1; and the global best gbest is randomly disturbed to give

$$\text{gbest}^* = \text{gbest} + \tau' N(0, 1). \tag{4.17}$$

The τ, τ' are learning parameters (either fixed or treated also as strategic parameters and therefore also subject to mutation).

This scheme benefits from two "pushes" in the right direction: the Darwinistic process of selection and the particle movement rule, and therefore it is natural to expect that it may display advantageous convergence properties when compared with ES or PSO alone. Furthermore, EPSO can also be classified as a self-adaptive algorithm, because it relies on the mutation and selection of strategic parameters, just as any σ-SA evolution strategy.

4.4 RESEARCH AREAS AND APPLICATIONS

References 18 to 65 show other PSO-related papers. Most of these papers are related to PSO itself and its modification and comparison with other EC methods. PSO is a new EC technique, and there are a few applications. Table 4.1 shows early applications of PSO in general fields. The last eight applications are in power and energy system fields. Detailed descriptions of applications to reactive power and voltage control [9, 60] can be found in Chapter 16. Optimal operation of cogeneration [31] has actually been operated in some factories of an automobile company in Japan.

TABLE 4.1 PSO Applications

Application Field	Ref. No.
Neural network learning algorithm	[29, 51]
Human tremor analysis	[23]
Rule extraction in fuzzy neural network	[32]
Battery pack state-of-charge estimation	[6]
Computer numerically controlled milling optimization	[59]
Reactive power and voltage control	[9, 60, 64]
Distribution state estimation	[10]
Power system stabilizer design	[18]
Fault state power supply reliability enhancement	[46]
Dynamic security assessment	[61]
Economic dispatch	[62, 63]
Short-term load forecasting	[65]
Optimal operation of cogeneration	[31]

However, applications of PSO to various fields are at the early stage. More applications can be expected.

4.5 CONCLUSIONS

This chapter presented fundamentals of PSO techniques. Whereas many evolutionary computation techniques have been developed for combinatorial optimization problems, PSO was developed originally for nonlinear optimization problems with continuous variables. PSO has several variations, including integration with selection mechanism and hybridization for handling both discrete and continuous variables. Moreover, a recently developed constriction factor approach is based on mathematical analysis and useful for obtaining high-quality solutions. Advanced PSOs such as adaptive PSO (APSO) and evolutionary PSO (EPSO) have been developed. A few applications have already appeared using PSO including practical applications. PSO can be an efficient optimization tool for nonlinear continuous optimization problems, combinatorial optimization problems, and mixed-integer nonlinear optimization problems (MINLPs).

REFERENCES

1. Reynolds C. Flocks, herds, and schools: A distributed behavioral model. Computer Graphics 1987; 21(4):25–34.
2. Boyd R, Richerson PJ. Culture and the evolutionary process. Chicago: University of Chicago Press; 1985.
3. Colorni A, Dorigo M, Maniezzo V. Distributed optimization by ant colonies. Proceedings of First European Conference on Artificial Life. Cambridge, MA: MIT Press; 1991. p. 134–142.
4. Kennedy J, Eberhart R. Particle swarm optimization. Proceedings of IEEE International Conference on Neural Networks (ICNN'95) Perth, Australia: IEEE Press; 1995. Vol. IV. p. 1942–1948.
5. Bonabeau E, Dorigo M, Theraulaz G. Swarm intelligence: From natural to artificial systems. Oxford: Oxford University Press; 1999.
6. Kennedy J, Eberhart R. Swarm intelligence. San Mateo, CA: Morgan Kaufmann; 2001.
7. Shi Y, Eberhart R. A modified particle swarm optimizer. Proceedings of IEEE International Conference on Evolutionary Computation (ICEC'98). Anchorage: IEEE Press; 1998. p. 69–73.
8. Shi Y, Eberhart R. Parameter selection in particle swarm optimization. Proceedings of the 1998 Annual Conference on Evolutionary Programming. San Diego: MIT Press; 1998.
9. Fukuyama Y. et al. A particle swarm optimization for reactive power and voltage control considering voltage security assessment. IEEE Trans Power Systems 2000; 15(4):1232–1239.
10. Naka S, Genji T, Yura T, Fukuyama Y. A hybrid particle swarm optimization for distribution state estimation. IEEE Trans Power Systems 2003; 18(1):60–68.

11. Kennedy J, Eberhart R. A discrete binary version of the particle swarm optimization algorithm. Proceedings of the 1997 Conference on Systems, Man, and Cybernetics (SMC'97). IEEE Press; 1997. p. 4104–4109.

12. Clerc M, Kennedy J. The particle swarm—explosion, stability, and convergence in a multidimensional complex space. IEEE Trans Evol Computat 2002; 6(1):58–73.

13. Eberhart R, Shi Y. Comparing inertia weights and constriction factors in particle swarm optimization. Proceedings of the Congress on Evolutionary Computation (CEC2000). IEEE Press; 2000. p. 84–88.

14. Angeline P. Using selection to improve particle swarm optimization. Proceedings of IEEE International Conference on Evolutionary Computation (ICEC'98). Anchorage: IEEE Press; 1998.

15. Ide A, Yasuda K. The improvement of search trajectory of particle swarm optimization. National Convention Record I.E.E. Japan: IEEE of Japan; 2003. p. 41–42. (in Japanese.)

16. Miranda V, Fonseca N. New evolutionary particle swarm algorithm (EPSO) applied to voltage/var control. Proceedings of PSCC'02—Power System Computation Conference. Seville, Spain, Butterworth-Heinemann; June 24–28; 2002.

17. Miranda V, Fonseca N. EPSO—Best of two world of meat-heuristic applied to power system problems. Proceedings of the 2002 Congress of Evolutionary Computation: IEEE Press; 2002.

18. Abido MA. Particle swarm optimization for multimachine power system stabilizer design. Proceedings of IEEE Power Engineering Society Summer Meeting: IEEE Press; 2001.

19. Angeline P. Evolutionary optimization versus particle swarm optimization: philosophy and performance differences. Proceeding of the Seventh Annual Conference on Evolutionary Programming: IEEE Press; 1998.

20. Carlisle A, Dozier G. Adapting particle swarm optimization to dynamic environments. Proceedings of International Conference on Artificial Intelligence. Monte Carlo Resort, Las Vegas, Nevada Computer Science Research, Education, and Applications Press; 2000.

21. Carlisle A, Dozier G. An off-the-shelf particle swarm optimization. Proceedings of the Workshop on Particle Swarm Optimization. Indianapolis, IN: Purdue School of Engineering and Technology, 2001.

22. Eberhart R, Kennedy J. A new optimizer using particle swarm theory. Proceedings of the Sixth International Symposium on Micro Machine and Human Science. Nagoya, Japan. Piscataway, NJ: IEEE Service Center; 1995. p. 39–43.

23. Eberhart R, Hu X. Human tremor analysis using particle swarm optimization. Proceedings of Congress on Evolutionary Computation (CEC'99). Washington, DC. Piscataway, NJ: IEEE Service Center; 1999. p. 1927–1930.

24. Eberhart R, Shi Y. Comparison between genetic algorithms and particle swarm optimization. Proceedings of the Seventh Annual Conference on Evolutionary Programming; IEEE Press; 1998.

25. Eberhart R, Shi Y. Evolving artificial neural networks. Proceedings of International Conference on Neural Networks and Brain. Beijing, P.R.C., PL5-PL13; Publishing House of Electronics Industry; 1998.

26. Eberhart R, Shi Y. Comparison between genetic algorithms and particle swarm optimization. In: Porto VW, Saravanan N, Waagen D, Eiben AE, eds. Evolutionary programming VII: Proceedings 7th Annual Conference on Evolutionary Programming. San Diego, CA. Berlin: Springer-Verlag; 1998.

27. Eberhart R, Shi Y. Tracking and optimizing dynamic systems with particle swarms. Proceedings of Congress on Evolutionary Computation (CEC2001), Seoul, Korea. Piscataway, NJ: IEEE Service Center; 2001.

28. Eberhart R, Shi Y. Particle swarm optimization: developments, applications and resources. Proceedings of Congress on Evolutionary Computation (CEC2001). Seoul, Korea, Piscataway, NJ: IEEE Service Center; 2001.

29. Eberhart R, Simpson P, Dobbins R. Computational intelligence PC tools. Boston: Academic Press Professional; 1996.

30. Fan H-Y, Shi Y. Study of Vmax of the particle swarm optimization algorithm. Proceedings of the Workshop on Particle Swarm Optimization. Indianapolis, IN: Purdue School of Engineering and Technology; 2001.

31. Fukuyama Y, et al. Particle swarm optimization for optimal operational planning of a cogeneration system. Proceedings of the 4th IASTED International Conference on Modelling, Simulation, and Optimization (MSO2004); ACTA Press; 2004.

32. He Z, Wei C, Yang L, Gao X, Yao S, Eberhart R, Shi Y. Extracting rules from fuzzy neural network by particle swarm optimization. Proceedings of IEEE International Conference on Evolutionary Computation (ICEC'98). Anchorange, Alaska, IEEE Press; May 4–9; 1998.

33. Ismail A, Engelbrecht AP. Training product units in feedforward neural networks using particle swarm optimization. Proceedings of the International Conference on Artificial Intelligence. Durban, South Africa; CSREA Press; 1999. p. 36–40.

34. Kennedy J. The particle swarm: Social adaptation of knowledge. Proceedings of International Conference on Evolutionary Computation (CEC'97). Indianapolis, IN: Piscataway, NJ: IEEE Service Center; 1997. p. 303–308.

35. Kennedy J. Minds and cultures: Particle swarm implications. Socially intelligent agents: Papers from the 1997 AAAI Fall Symposium. Technical report FS-97-02, 67–72. Menlo Park, CA: AAAI Press; 1997.

36. Kennedy J. Methods of agreement: Inference among the elementals. Proceedings of International Symposium on Intelligent Control. Piscataway, NJ: IEEE Service Center; 1998.

37. Kennedy J. The behavior of particles. In: Porto VW, Saravanan N, Waagen D, Eiben AE, eds. Evolutionary programming VII: Proceedings 7th Annual Conference on Evolutionary Programming Conference. San Diego, CA. Berlin: Springer-Verlag; 1998. p. 581–589.

38. Kennedy J. Thinking is social: Experiments with the adaptive culture model. J Conflict Resolution 1998; 42:56–76.

39. Kennedy J. Small worlds and mega-minds: Effects of neighborhood topology on particle swarm performance. Proceedings of Congress on Evolutionary Computation (CEC'99), 1931–1938. Piscataway, NJ: IEEE Service Center; 1999.

40. Kennedy J. Stereotyping: Improving particle swarm performance with cluster analysis. Proceedings of the 2000 Congress on Evolutionary Computation (CEC2000). San Diego, CA. Piscataway, NJ: IEEE Press; 2000.

41. Kennedy J. Out of the computer, into the world: externalizing the particle swarm. Proceedings of the Workshop on Particle Swarm Optimization. Indianapolis, IN: Purdue School of Engineering and Technology; 2001.

42. Kennedy J, Eberhart R. The particle swarm: Social adaptation in information processing systems. In: Corne D, Dorigo M, Glover F, eds. New ideas in optimization. London: McGraw-Hill; 1999.

43. Kennedy J, Spears W. Matching algorithms to problems: An experimental test of the particle swarm and some genetic algorithms on the multimodal problem generator. Proceedings of IEEE International Conference on Evolutionary Computation (ICEC'98). Anchorage, Alaska, IEEE Press; May 4–9; 1998.

44. Løvbjerg M, Rasmussen T, Krink T. Hybrid particle swarm optimization with breeding and subpopulations. Proceedings of the Third Genetic and Evolutionary Computation Conference (GECCO-2001); Morgan Kaufmann Press; 2001.

45. Mohan C, Al-kazemi B. Discrete particle swarm optimization. Proceedings of the Workshop on Particle Swarm Optimization. Indianapolis, IN: Purdue School of Engineering and Technology; 2001.

46. Nara K, Mishima Y. Particle swarm optimisation for fault state power supply reliability enhancement. Proceedings of IEEE International Conference on Intelligent Systems Applications to Power Systems (ISAP2001). Budapest; IEEE Press; 2001.

47. Ozcan E, Mohan C. Analysis of a simple particle swarm optimization system. Intelligent Engineering Systems Through Artificial Neural Networks 1998; 8:253–258.

48. Ozcan E, Mohan C. Particle swarm optimization: surfing the waves. Proceedings of 1999 Congress on Evolutionary Computation (CEC'99). Washington, DC, Institute of Electrical & Electronics Enginee, July 6–9, 1999.

49. Parsopoulos K, Plagianakos V, Magoulas G, Vrahatis M. Stretching technique for obtaining global minimizers through particle swarm optimization. Proceedings of the Workshop on Particle Swarm Optimization. Indianapolis, IN: Purdue School of Engineering and Technology; 2001.

50. Ray T, Liew KM. A swarm with an effective information sharing mechanism for unconstrained and constrained single objective optimization problems. Proceedings of the 2001 Congress on Evolutionary Computation (CEC2001). Seoul, Korea; IEEE Press; 2001.

51. Salerno J. Using the particle swarm optimization technique to train a recurrent neural model. Proceedings of 9th International Conference on Tools with Artificial Intelligence (ICTAI'97). IEEE Press, 1997.

52. Secrest B, Lamont G. Communication in particle swarm optimization illustrated by the traveling salesman problem. Proceedings of the Workshop on Particle Swarm Optimization. Indianapolis, IN: Purdue School of Engineering and Technology; 2001.

53. Shi Y, Eberhart R. Parameter selection in particle swarm optimization. In: Evolutionary programming VII: Proceedings EP98. New York: Springer-Verlag; 1998. p. 591–600.

54. Shi Y, Eberhart R. Empirical study of particle swarm optimization. Proceedings of the 1999 Congress on Evolutionary Computation (CEC'99). Piscataway, NJ: IEEE Service Center; 1999. p. 1945–1950.

55. Shi Y, Eberhart R. Experimental study of particle swarm optimization. Proceedings of SCI2000 Conference. Orlando, FL; IIIS Press; 2000.

56. Shi Y, Eberhart R. Fuzzy adaptive particle swarm optimization. Proceedings of Congress on Evolutionary Computation (CEC2001). Seoul, South Korea. Piscataway, NJ: IEEE Service Center; 2001.

57. Shi Y, Eberhart R. Particle swarm optimization with fuzzy adaptive inertia weight. Proceedings of the Workshop on Particle Swarm Optimization. Indianapolis, IN: Purdue School of Engineering and Technology; 2001.

58. Suganthan P. Particle swarm optimiser with neighbourhood operator. Proceedings of the 1999 Congress on Evolutionary Computation (CEC'99). Piscataway, NJ: IEEE Service Center; 1999. p. 1958–1962.

59. Tandon V. Closing the gap between CAD/CAM and optimized CNC end milling. Master's thesis, Purdue School of Engineering and Technology, Indiana University Purdue University Indianapolis; 2000.

60. Yoshida H, Kawata K, Fukuyama Y, Nakanishi Y. A particle swarm optimization for reactive power and voltage control considering voltage stability. In: Torres GL, Alves da Silva AP, eds. Proceedings of International Conference on Intelligent System Application to Power Systems (ISAP'99), Rio de Janeiro, Brazil; IEEE Press; 1999; p. 117–121.

61. Kassabalidis I, El-sharkawi M, et al. Dynamic security border identification using enhanced particle swarm optimization. IEEE Trans Power Systems 2002; 17(3):723–729.

62. Gaing Z-L. Particle swarm optimizatin to solving the economic dispatch considering the generator constraints. IEEE Trans Power Systems 2003; 18(3):1187–1195.

63. Park J-B, Lee KY, et al. A particle swarm optimization for economic dispatch with non-smooth cost functions. IEEE Trans Power Systems 2005; 20(1):34–42.

64. Zhao B, et al. A multiagent-based particle swarm optimization approach for optimal reactive power dispatch. IEEE Trans Power Systems 2005; 20(20):1070–1078.

65. Huang C, et al. A particle swarm optimization to identifying the ARMAX model from short-term load forecasting. IEEE Trans Power Systems 2005; 20(2):1126–1133.

CHAPTER 5

Fundamentals of Ant Colony Search Algorithms

YONG-HUA SONG, HAIYAN LU, KWANG Y. LEE, and I.K. YU

5.1 INTRODUCTION

The ant colony optimization (ACO) is a meta-heuristic approach for combinatorial optimization problems, which can be regarded as a paradigm for all ant colony search algorithms that are inspired by the foraging behavior of the social insects, especially the ants [1]. From a broader perspective, the ACO algorithms belong to the class of model-based search (MBS) algorithm according to Dorigo [2]. MBS algorithms have become increasingly popular methods for solving combinatorial optimization problems. An MBS algorithm is characterized by the use of a (parameterized) probabilistic model that is used to generate solutions to the problem under consideration. The MBS algorithms can be classified into two categories based on how the probabilistic model is used: (i) the algorithms use a given probabilistic model without changing the model structure during run-time; and (ii) the algorithms use and change the probabilistic model in alternating phases. The ACO algorithms fall into the first category. At run-time, ACO algorithms will update the parameters' values of the probabilistic model in such a way that there will be more chances to generate high-quality solutions over time.

Ant colony search algorithms (ACSAs) have recently been introduced as powerful tools to solve a diverse set of optimization problems, such as the traveling salesman problem (TSP) [3, 4], the quadratic assignment problem [4, 5], and optimization problems in power systems [6–15, 23, 24], such as the generation scheduling problem, unit commitment problem (economic dispatch problem) [16–18], and web usage mining problem [19]. This chapter focuses on the fundamentals of the ACSAs. The applications of ACSAs in power systems will be covered in depth in Chapter 20.

Modern Heuristic Optimization Techniques. Edited by K. Y. Lee and M. A. El-Sharkawi
Copyright © 2008 the Institute of Electrical and Electronics Engineers, Inc.

5.2 ANT COLONY SEARCH ALGORITHM

The first ACS algorithm was proposed by Dorigo in the early 1990s. The ACS belongs to biologically inspired heuristic algorithms. It was developed mainly based on the observation of the foraging behavior of a real ant. It will be useful to understand how ants, which are almost blind animals with very simple individual capacities acting together in a colony, can find the shortest route between the ant nest and a source of food. Section 5.2.1 describes the behavior of real ants, followed by the presentation of the two most (experimentally) successful ant colony search algorithms, with their progenitor in Section 5.2.2. The major characteristics of the ACSAs are summarized in the Section 5.2.3.

5.2.1 Behavior of Real Ants

As is well-known, real ants are capable of finding the shortest path from food sources to the nest without using visual cues. They are also capable of adapting to changes in the environment, for example, finding a new shortest path once the old one is no longer feasible due to a new obstacle. Figure 5.1 clearly illustrates these phenomena. In Fig. 5.1a, ants are moving on a straight line that connects a food source (A) to the nest (E). Once an obstacle appears as shown in Fig. 5.1b, the path is cut off. Shortly, the ants can establish a new shortest path as shown in Fig. 5.1c.

The studies by ethnologists reveal that such capabilities that ants have are essentially due to what is called *pheromone trails*, which ants use to communicate information among individuals regarding the walking path or the decision about where to go when they are foraging. According to Blum and Dorigo [2], real ants initially explore the area surrounding their nest in a random manner when searching for

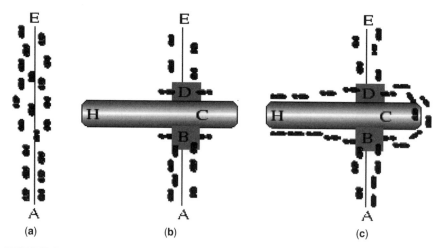

FIGURE 5.1 The behavior of real ants. (**a**) Ants travel the shortest path; (**b**) an obstacle breaks the path; (**c**) ants choose the shorter path.

food. As soon as an ant finds a food source, it carries some of the found food to the nest. While it is walking, the ant deposits a chemical pheromone trail on the ground. The pheromone trails deposited on the ground will guide other ants to the food source. Each ant probabilistically prefers to follow a direction rich in pheromone rather than a poorer one. The indirect communication between the ants via the pheromone trails allows them to find the shortest paths between their nest and food sources.

The process illustrated in Fig. 5.1 can then be explained as follows: In Fig. 5.1a, the path between the food source and the nest is clear. The ants carrying the food walk back to the nest in a random fashion initially. Supposing that all ants walk at approximately the same speed, the shortest path will have most ants visited on average. Therefore, the pheromone on the shortest path will accumulate faster than on any other paths. After a short transitory period, the difference in the amount of pheromone on the paths is sufficiently noticeable to the new ants coming into the system. From that point of time, new ants will preferably walk along the shortest path, which is a straight line that connects a food source to the nest. When an obstacle blocks an established path, as shown in Fig. 5.1b, the pheromone trail is interrupted. Those ants that are just in front of the obstacle cannot continue to follow the pheromone trail and therefore they would choose to turn right or turn left randomly as they do not have any clue about which is the best choice. In Fig. 5.1c, those ants that choose by chance the shorter path around the obstacle will more rapidly establish the interrupted pheromone trail compared with those that choose the longer path as there are more ants walking along the shorter path at each time unit. Hence, the shorter path will have a higher amount of pheromone deposited on average, and this will in turn cause a higher number of ants to choose the shorter path. Due to this positive feedback (autocatalytic) process, very soon all ants will choose the shorter path.

5.2.2 Ant Colony Algorithms

The foraging behavior of real ants has inspired the ant colony algorithms, which are algorithms in which a set of artificial ants cooperate to find the solution to a combinatorial optimization problem by changing information via pheromone deposited on the artificial paths.

For the ease of description of the ant colony algorithms, we look at a discrete optimization problem $\wp = (S, f, \Omega)$ characterized [1] as

- A finite set $C = \{c_1, c_2, \ldots, c_{Nc}\}$ of solution components.
- A finite set $L = \{l_{i,j} | (c_i, c_j) \in \tilde{C}\}$ of possible connections among the elements of $\tilde{C}$, which is a subset of $C \times C$ (the Cartesian product of C and C), where $l_{i,j}$ indicates the connection between components c_i and c_j. The number of possible connections is denoted by $|L| \leq N_C^2$.
- A finite set $\Omega \equiv \Omega(C, L)$ of constraints assigned over the elements of C and L.
- A finite set χ of states of the problem, defined in terms of all possible sequences $s = \langle c_i, c_j, \ldots, c_k, \ldots \rangle$ over the elements of C. The number of components in the

sequence is denoted by $|s|$. The maximum length of a sequence is bounded by a positive constant number $N < +\infty$.

- A set $\tilde{\chi}$ of all states that are feasible with respect to the constraints $\Omega(C, L)$, $\tilde{\chi} \subseteq \chi$.
- A set $S = \{s | s \in \tilde{\chi}\}$ of (candidate) solutions, $S \subseteq \chi$.
- $J_{i,j}$ is a connection cost associated with each $l_{i,j} \in L$.
- $J_s(L)$ is a cost associated with the solution s, which can be calculated from all the costs $J_{i,j}$ of all the components belonging to the solution s.
- A neighborhood structure of components: Given two components, c_1 and c_2, the component c_2 is the component c_1 if both components are in C, and the component c_2 can be reached from the component c_1 in one logical step. The set of all neighbor components of the component c is denoted by N_c.
- Each connection $l_{i,j}$ is weighted by $\tau_{i,j}$.

In order to solve the above discrete problem, it is often useful to map the problem on a graph $G = (C, L)$ and refer to this graph as the construction graph. The solutions to the optimization problem can be expressed in terms of feasible paths on the graph G, and the optimal solution will be the path that yields minimum connection cost.

There are a number of experimental successful ant algorithms published in the literature that can be used to solve the above problem [20]. Here we describe two of the most successful ant algorithms, namely, the ant colony system (ACS) and the max-min ant system (MMAS), with their progenitor, the ant system (AS) for the sake of application of ant colony search algorithms in solving combinatorial optimization problems facing power systems.

These three algorithms share the same algorithmic skeleton as shown in Fig. 5.2. The skeleton consists of three major phases, which are the initialization phase, the solution construction phase, and the pheromone updating phase and other optional phases, such as a phase that locally updates the pheromone trails, as shown in the dashed-line block in the figure. The initialization phase does almost the same tasks as listed below across all three algorithms:

- Set up a construction graph, which is a completely connected and weighted graph $\varsigma = (C, L, T)$, where the vertices are the components C, the set L fully connects the components C, and T is a vector whose components representing the so-called pheromone trail strength, τ, which are considered to be associated with possible connections (or edges) only. This task is often the most difficult task when applying an ant colony search algorithm to an optimization problem.
- Initialize the pheromone trail strength for all the edges. Except for the MMAS in which the initial pheromone trail strength for all the edges is set to be a large value, the AS and ACS initialize these values to be a small value, in general.
- Set the number of artificial ants (ants for abbreviation) in a colony as m, and put each ant on a randomly chosen vertex.

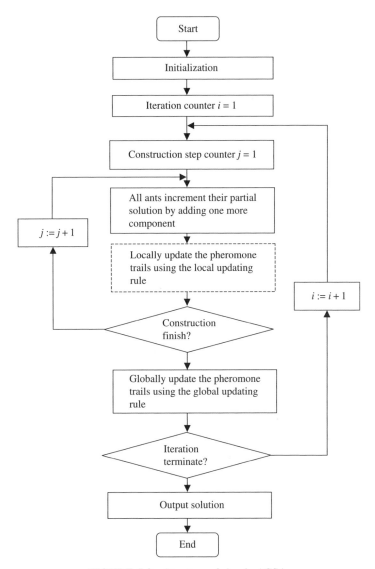

FIGURE 5.2 Structure of simple ACSA.

- Set up the termination criteria for the iteration looping, which may be that the iteration number exceeds a predefined maximum number of solution construction steps or that the computation time has exceeded a given CPU-time limit.

When the solution construction phase starts, m ants have been put on randomly chosen vertices on the construction graph, and their paths consist of their initial vertices. Within each construction step, all the ants construct their feasible paths (partial solutions) by moving to the next vertex based on a probabilistic decision according to the so-called state transition rule. The state transition rules are different among various ant search algorithms. However, they all serve the purpose of guiding the ants to make a more cost-effective move from their current positions. After all the ants have made one move, their current feasible paths may be improved by applying local pheromone updating rule depending on the ant search algorithm under consideration. The solution construction phase is repeated until all ants have completed their feasible paths. Then the pheromone trails will be globally updated using the global pheromone updating rule. The global updating rule enforces two things: one is the pheromone evaporation, which stops pheromone trails from unlimited accumulation, and the other is the pheromone reinforcement, which makes the favorite edges have stronger pheromone trails. The global updating rule can be different from one algorithm to another, but they all aim to fade out the pheromone trails on the edges that are not used by ants exponentially and to enforce the favorite edges so that they can attract more ants in following construction steps.

In the rest of this subsection, we will describe how these three algorithms work with emphasis on the state transition rules and the pheromone updating rules they are applying.

5.2.2.1 *The Ant System* The ant system (AS) was the first ant colony search algorithm proposed as a novel heuristic method for solving combinatorial problems in the early 1990s [1, 21, 22]. Informally, an AS can be described as follows: After the initialization phase, m ants are put on their initial vertex on the construction graph, and all pheromone trails are initialized to be a small value (i.e., within the range from 0 to 1) for all the edges. In each solution construction step, each ant incrementally constructs a solution by adding one solution component to the partial solutions it has constructed so far respectively. Suppose that the kth ant is at the vertex c_i in the construction step t. The kth ant performs a randomized walk from the vertex c_i to the next vertex to construct a feasible solution incrementally in such a way that the next vertex is chosen stochastically according to the transition probabilities of candidate vertices with respect to the current vertex of the kth ant. The probability with which the kth ant at the vertex i chooses to move to the vertex j can be determined by the random-proportional state transition rule [3] as shown below:

$$p_k(i,j) = \begin{cases} \dfrac{[\tau(i,j)]^\alpha \cdot [\eta(i,j)]^\beta}{\sum_1^t [\tau(i,u)]^\alpha \cdot [\eta(i,u)]^\beta} & j, u \in N_{k,i} \\ 0 & \text{otherwise,} \end{cases} \tag{5.1}$$

where $\tau(i,j)$ represents the pheromone trail associated with $l_{i,j}$, which is the connection between vertices i and j, $\eta(i,j)$ a heuristic value, called the desirability of adding connection $l_{i,j}$ to the solution under construction and can be determined according to the optimization problem under consideration. It is usually set to be the reverse of the connection cost, $J_{i,j}$, associated with the edge $l_{i,j}$, that is, $1/J_{i,j}$, $N_{k,i}$ is the feasible neighbor components of the kth ant at the vertex c_i with respect to the problem constraints Ω, $N_{k,i} \subseteq N_i$, α and β are two parameters that determine the relative importance of pheromone versus the heuristic value ($\alpha, \beta > 0$).

Once all ants have completed their solutions, the pheromone trails are updated on all edges according to the pheromone global updating rule as shown below:

$$\tau(i, j) \leftarrow (1 - \alpha) \cdot \tau(i, j) + \sum_{k=1}^{m} \Delta\tau_k(i, j), \tag{5.2}$$

where α is a pheromone decay parameter with $0 < \alpha \leq 1$, and

$$\Delta\tau_k(i,j) \begin{cases} 1/J_{i,j} & (i, j) \in \varphi \\ 0 & \text{otherwise,} \end{cases} \tag{5.3}$$

with φ being the set of moves done by the kth ant.

The introduction of the heuristic value $\eta(i, j)$ in (5.1) is mainly to favor the edges that are cost-effective and also have a greater amount of pheromone. The first term in (5.2) models the pheromone evaporation, in which the pheromone trails are lowered by the pheromone decay factor. The pheromone trails for the unfavorite edges will be evaporated exponentially. The second term in (5.2) models the reinforcement of pheromone trails. It makes the edges that have been visited by ants become more desirable for future ants. Ants deposit an amount of pheromone related to the quality of the solutions that they produced so that the more cost-effective solution generated by an ant, the greater the amount of pheromone it deposits on the edges that it used to produce the solution. According to Dorigo and Gambardella [3], the pheromone updating rule was meant to simulate the change in the amount of pheromone due to both the addition of new pheromone deposited by ants on the visited edges and the pheromone evaporation.

Although the AS performance was not competitive with the state-of-the-art algorithms for some benchmark problems, such as the traveling salesman problem (TSP), it gave rise to a whole set of successful ant colony search algorithms that have recently been unified in a novel meta-heuristic called ant colony optimization (ACO).

5.2.2.2 The Ant Colony System

The ant colony system (ACS) was developed after the AS. The ACS can outperform the AS due to the following three modifications [3]: (i) the state transition rule provides a direct way to balance exploration of new edges and exploitation of a found solution and accumulated knowledge about the problem; (ii) the global pheromone updating rule is applied only to the edges that belong to the best journey; and (iii) while ants construct a solution, a local pheromone updating rule is applied for each construction step.

Informally, the ACS algorithm can be described as follows: m ants are initially put on randomly chosen vertices on the construction graph. Each connection or edge on the graph is associated with a pheromone trail that is initialized to be a small positive value. Each ant walks randomly on the graph to construct its solution by repeatedly adding one solution component to its partial solution following the state transition rule, which is different from the one used in AS. Suppose that the kth ant is at the vertex c_i and the construction has been up to the tth step. The kth ant will move to the next vertex to increment its solution by one component. Its next vertex, say c_j, can be determined by

$$
c_j = \begin{cases} \arg \max_{u \in N_{k,i}} \{ [\tau(i, u)]^a \cdot [\eta(i, u)]^\beta \} & q \leq q_0 \\ c^* & \text{otherwise,} \end{cases} \tag{5.4}
$$

where q is a uniformly distributed random number within $[0, 1]$, q_0 is a parameter $(0 < \alpha < 1)$, which determines the relative importance of local exploitation versus global exploration, and c^* is a random variable selected according to the probability distribution given in (5.1). The state transition rule resulting from (5.1) and (5.4) is called *pseudo-random-proportional rule* [3]. This transition rule favors moves to vertices connected by more cost-effective edges and with denser pheromone trails. By applying the *pseudo-random-proportional rule*, for each construction step, an ant at the vertex c_i has to choose a vertex c_j to move to; at that moment, a uniformly distributed random number $q_0 \in [0, 1]$ is generated. If $q \leq q_0$, then the next vertex will be the vertex determined by (5.4), which yields the best move, otherwise, it will be the component c^* chosen according to (5.1). While ants construct their solutions, for each step, the pheromone trails for the edges the ants have visited will be changed by applying a locally update rule shown below:

$$
\tau(i, j) \leftarrow (1 - \rho) \cdot \tau(i, j) + \rho \cdot \Delta\tau(i, j), \tag{5.5}
$$

where ρ with $0 < \rho < 1$, is a parameter, and $\Delta\tau(i, j)$ is the amount of pheromone that kth ant puts on the edge $l_{i,j}$. The pheromone update $\Delta\tau(i, j)$ can be set to be the value of the initial pheromone trails for the simple ACS algorithm [3].

Once all the ants have completed their solutions, the pheromone trails will be updated by the pheromone global updating rule shown below:

$$
\tau(i, j) \leftarrow (1 - \alpha) \cdot \tau(i, j) + \alpha \cdot \Delta\tau_k(i, j), \tag{5.6}
$$

where α is a pheromone decay parameter with $0 < \alpha \leq 1$, and

$$
\Delta\tau_k(i, j) = \begin{cases} (J_{gb})^{-1} & l_{i,j} \in s^* \\ 0 & \text{otherwise,} \end{cases} \tag{5.7}
$$

where J_{gb} is the cost of the globally best ant solution, s^*, from the beginning of the trial.

5.2.2.3 The Max-Min Ant System

According to Stützle and Hoos [4], the max-min ant system (MMAS) is an ACO algorithm derived from AS and is one of the best ACO algorithms for the TSP and the quadratic assignment problem (QAP), which is a typical benchmark problem for challenging any combinatorial optimization algorithms.

The MMAS differs from the AS in following three key aspects [4]:

- To exploit the best solution found during an iteration or during the run of the algorithm, after each iteration only one single ant adds pheromone. This ant may be the one that found the best solution in the current iteration (iteration-best ant) or the one that found the best solution among all the iterations from the beginning of the trial (global-best ant), which is the same one used in the state transition rule of ACS.
- To avoid stagnation of the search, the range of possible pheromone trails (edges in the construction graph) on each solution component is limited to an interval $[\tau_{min}, \tau_{max}]$.
- Initialize the pheromone intensity for all edges to be τ_{max} to achieve a higher exploration of solution at the beginning of trails.

Informally, the MMAS algorithm can be described as follows: m ants are initially positioned at randomly chosen vertices on the construction graph. The pheromone trails for all the connections (or edges) are initialized to be an arbitrarily large value such that all pheromone trails are the upper limit of all possible pheromone trail values. Each ant walks randomly on the graph to construct its solution by repeatedly adding one solution component to its current partial solution following the state transition rule, which is the same as the one used in the AS. Suppose that the kth ant is at the vertex c_i and the construction has been up to the tth step. The kth ant will move to the next vertex to increment its partial solution by one component. The next vertex is chosen stochastically according to the transition probabilities of candidate vertices with respect to the current vertex of the kth ant. The probability with which the kth ant at the vertex i chooses to move to the vertex j can be determined by the random-proportional state transition rule [3] as shown in (5.1).

Once all ants complete their solutions, in other words, the program completes one iteration, the pheromone trails consisting of the best solution will be updated by applying a global update rule shown below:

$$\tau(i, j) \leftarrow (1 - \alpha) \cdot \tau(i, j) + \Delta\tau^{best}(i, j), \tag{5.8}$$

where α with $0 < \alpha \leq 1$ is a pheromone decay parameter, and $\Delta\tau^{best}(i, j) = 1/f(s^{best})$ and $f(s^{best})$ denotes the solution cost of either the iteration-best solution (s^{ib}) or the global-best solution (s^{gb}). The MMAS mainly uses the iteration-best solution for $f(s^{best})$ whereas the ACS uses only the global-best solution in global pheromone update.

After updating the pheromone trails, the pheromone trail constraint $\tau_{min} \leq \tau(i, j) \leq \tau_{max}$, where $\tau(i, j)$ represents the pheromone trails for the connection $l_{i,j}$, will be enforced such that if $\tau(i, j) < \tau_{min}$, $\tau(i, j)$ will be re-set to be τ_{min} and if $\tau(i, j) > \tau_{max}$, $\tau(i, j)$ will be re-set to be τ_{max}.

The iterations continue until the termination criterion is met and one run completes.

5.2.3   Major Characteristics of Ant Colony Search Algorithms

There are some attractive properties of ant search algorithms when compared with other search methods:

5.2.3.1   Distributed Computation: Avoid Premature Convergence
Conventionally, scientists choose to work on a system simplified to a minimum number of components in order to observe essential information. An ant search algorithm often simplifies as much as possible the components of the system, for the purpose of taking into account their large number. The power of the massive parallelism in ant algorithms is able to deal with incorrect, ambiguous, or distorted information, which is often found in nature. The computational model contains the dynamics that is determined by the nature of local interactions between many elements (artificial ants).

5.2.3.2   Positive Feedback: Rapid Discovery of Good Solution   The ant search algorithms make use of a unique indirect communication means to share global information while solving a problem. Occasionally, the information exchanged may contain errors and should alter the behavior of the ants receiving it. As the search proceeds, the new population of ants often containing the states of higher fitness will affect the search behavior of others and will eventually gain control over other agents while at the same time actively exploiting inter-ant communication by mean of the pheromone trail laid on the path. The artificial ant foraging behavior dynamically reduces the prior uncertainty about the problem at hand. As ants doing a task can be either "successful" or "unsuccessful," they can switch between these two according to how well the task is performed. Unsuccessful ants also have a certain chance to switch to be inactive, and successful ants have a certain chance to recruit inactive ants to their task. Therefore, the emerging collective effect is in a form of autocatalytic behavior, in that the more ants following a particular path, the more attractive this path becomes for the next ants that meet. It can give rise to global behavior in the colony.

5.2.3.3   Use of Greedy Search and Constructive Heuristic Information: Find Acceptable Solutions in the Early Stage of the Process   Based on the available information collected from the path (pheromone trail level and heuristic information), the decision is made at each step as a constructive way by the artificial ants, even if each ant's decision always remains probabilistic. It tends to evolve a group of initial poorly generated solutions to a set of acceptable solutions through successive generations. It uses objective function to guide the search only and does not need any other auxiliary knowledge. This greatly reduces the complexity of the problem. The user only has to define the objective function and the infeasible regions (or obstacle on the path).

5.3 CONCLUSIONS

This chapter focuses the fundamentals of ant colony search algorithms by presenting two major ACSAs, which are the ACS and the MMAS. The individual ants are rather simple; however, the entire colony foraging toward the bait site can exhibit complicated dynamics resulting in a very attractive search capability. The ACSAs consist of three major phases: (i) the initialization phase, in which the construction graph is built based on the combinatorial optimization problem at hand and related parameters are initialized; (ii) the solution construction phase, in which each ant constructs its own solution to the problem incrementally following the state transition rule until all ants complete their solutions; (iii) the pheromone update phase, in which the pheromone trails corresponding with each connection in the construction graph will be updated based on the global pheromone trail updating rule. The state transition rule and the pheromone trail updating rule can be different from one algorithm to the other. The ACSAs have attractive features compared with other heuristic search methods. The important features are (i) distributed computation in nature, (ii) autocatalytic process, and (iii) use of greedy search and constructive heuristic information. The experimental studies on the typical benchmark problems, such as the TSP and QAS, have shown that the ACSAs are promising in solving combinatorial optimization problems.

REFERENCES

1. Dorigo M, Di Caro G. Ant colony optimization: a new meta-heuristic. Proceedings of the 1999 Congress on Evolutionary Computation, 1999 (CEC'99), the IEEE; Vol. 2. 6–9 July 1999. p. 1470–1477.

2. Blum C, Dorigo M. The hyper-cube framework for ant colony optimization. IEEE Trans Systems Man Cybernetics B 2004; 34(2):1161–1172.

3. Dorigo M, Gambardella LM. Ant colony system: A cooperative learning approach to the traveling salesman problem. IEEE Trans Evol Computat 1997; 1(1):53–66.

4. Stützle T, Hoos HH. MAX–MIN ant system. Future Generation Computer Systems 2000; 16(8):889–914.

5. Maniezzo V, Colorni A. The ant system applied to the quadratic assignment problem. IEEE Trans Knowledge Data Eng 1999; 11(5):769–778.

6. Song YH, ed. Modern optimisation techniques in power systems. Boston: Kluwer Academic Publishers; 1999.

7. Irving MR, Song YH. Optimisation techniques for electrical power systems. IEE Power Eng J 2000; 14(5):245–254.

8. Song YH, Irving MR. Optimisation techniques for electrical power systems. Part 2: Heuristic optimisation techniques. IEE Power Eng J 2001; 15(3):151–160.

9. Song YH, Chou CS. Application of ant colony search algorithms in power system optimisation. IEEE Power Eng Rev 1998; 18(12):63–64.

10. Song YH, Chou CS, Min Y. Large-scale economic dispatch by artificial ant colony search algorithms. Electric Machines and Power Systems 1999; 27(5):679–690.

11. Yu IK, Song YH. A novel short-term generation scheduling technique of thermal units using ant colony search algorithms. Electric Power and Energy Systems 2001; 23:471–479.

12. Song YH, Chou CS, Stonham TJ. Combined heat and power economic dispatch by improved ant colony search algorithm. Electric Power Systems Research 1999; 52:115–121.

13. Yu I-K, Chou CS, Song YH. Application of the ant colony search algorithm to short-term generation scheduling problem of thermal units. Proceedings of 1998 International Conference on Power System Technology (POWERCON '98), the CSEE; Vol. 1. 18–21 Aug. 1998; p. 552–556.

14. Huang S-J. Enhancement of hydroelectric generation scheduling using ant colony system based optimization approaches. IEEE Trans Energy Conversion 2001; 16(3):296–301.

15. Shi L, Hao J, Zhou J, Xu G. Short-term generation scheduling with reliability constraint using ant colony optimization algorithm. Fifth World Congress on Intelligent Control and Automation, 2004 (WCICA 2004), the IEEE; Vol. 6. 15–19 June 2004; p. 5102–5106.

16. Sisworahardjo NS, El-Keib AA. Unit commitment using the ant colony search algorithm. Large Engineering Systems Conference on Power Engineering 2002 (LESCOPE '02), the IEEE; 26–28 June 2002; p. 2–6.

17. Sum-im T, Ongsakul W. Ant colony search algorithm for unit commitment. 2003 IEEE International Conference on Industrial Technology, the IEEE; Vol. 1. 10–12 Dec. 2003; p. 72–77.

18. Hou Y-H, Wu Y-W, Lu L-J, Xiong X-Y. Generalized ant colony optimization for economic dispatch of power systems. Proceedings of International Conference on Power System Technology, 2002 (PowerCon 2002), the CSEE; Vol. 1. 13–17 Oct. 2002; p. 225–229.

19. Abraham A, Ramos V. Web usage mining using artificial ant colony clustering and linear genetic programming. The 2003 Congress on Evolutionary Computation, 2003 (CEC '03), the IEEE; Vol. 2. 8–12 Dec 2003; p. 1384–1391.

20. Dorigo M, Bonabeau E, Theraulaz G. Ant algorithm and stigmergy. Future Generation Computer Systems 2000; 16(8):851–871.

21. Dorigo M. Optimization learning and natural algorithm. PhD dissertation; Politecnico di Milano, 1992.

22. Dorigo M, Maniezzo V, Colorni A. The ant system: Optimisation by a colony of co-operating agents. IEEE Trans Systems Man Cybernetics B 1996; 26(1):29–41.

23. Teng J-H, Liu Y-H. A novel ACS-based optimum switch relocation method. IEEE Trans Power Syst 2003; 18(1):113–120.

24. Jeon Y-J, Kim J-C, Yun S-Y, Lee KY. Application of ant colony algorithm for network reconfiguration in distribution systems. Proceedings, IFAC Symposium for Power Plants and Power Systems Control, Seoul, South Korea, The IFAC; June 9–12, 2003; p. 266–271.

CHAPTER 6

Fundamentals of Tabu Search

ALCIR J. MONTICELLI, RUBÉN ROMERO, and EDUARDO NOBUHIRO ASADA

6.1 INTRODUCTION

Tabu search (TS) consists of a meta-heuristic procedure used to manage heuristic algorithms that perform local search. Meta-heuristics consists of advanced strategies that allow the exploitation of the search space by providing means of avoiding being entrapped into local optimal solutions. As it happens with other combinatorial approaches, TS carries out a number of transitions in the search space aiming to find the optimal solutions or a range of near-optimal solutions. The name *tabu* is related to the fact that in order to avoid revisiting certain areas of the search space that have already been explored, the algorithm turns these areas tabu (or forbidden). It means that for a certain period of time (the tabu tenure), the search will not consider the examination of alternatives containing features that characterize the solution points belonging to the area declared *tabu*.

Tabu search was developed from concepts originally used in artificial intelligence. Unlike other combinatorial approaches such as genetic algorithms and simulated annealing, its origin is not related to biological or physical optimization processes [1–4]. TS was originally proposed by Fred Glover in the early 1980s and has ever since been applied with success to a number of complex problems in science and engineering. Applications to electric power network problems are already significant and growing. These include, for example, the long-term transmission network expansion problem and distribution planning problems such as the optimal capacitor placement in primary feeders.

6.1.1 Overview of the Tabu Search Approach

Compared with simulated annealing and genetic algorithms, tabu search explores the solution space in a more aggressive way (i.e., it is greedier than those algorithms). Tabu search algorithms are initialized with a configuration (or a set of configurations

Modern Heuristic Optimization Techniques. Edited by K. Y. Lee and M. A. El-Sharkawi

when the search is performed in parallel), which becomes the current configuration. At every iteration of the algorithm, a neighborhood structure is defined for the current configuration; a move is then made to the best configuration in this neighborhood (i.e., in a minimization problem, the algorithm switches to the configuration presenting the smallest cost). Normally, only the most promising neighbors are evaluated, otherwise the problem could become intractable. Unlike gradient type algorithms used for local search, the neighborhood in tabu search is updated dynamically. Another difference is that transitions to configurations with higher cost are allowed (this gives the method the ability to move out of local minimum points). An essential feature of tabu search algorithms is the direct exclusion of search alternatives temporarily classed as forbidden (tabu). Consequently, the use of memory becomes crucial in these algorithms: one has to keep track of the tabu's restrictions.

Other mechanisms of tabu search are the *intensification* and *diversification*: by the intensification mechanism, the algorithm does a more comprehensive exploration of attractive regions that may lead to a local optimal point; by the diversification mechanism, on the other hand, the search is moved to previously unvisited regions, something that is important in order to avoid local minimum points. Tabu search consists of a set of principles (or functions) applied in an integrated way to solve a complex problem in an intelligent manner. According to Glover [5]:

> Tabu search is based on the premise that problem solving, in order to qualify as intelligent, must incorporate adaptive memory and responsive exploration. The use of adaptive memory contrasts with "memoriless" designs, such as those inspired by metaphors of physics and biology, and with "rigid memory" designs, such as those exemplified by branch and bound and its AI-related cousins. The emphasis on responsive exploration (and hence purpose) in tabu search, whether in a deterministic or probabilistic implementation, derives from the supposition that a bad strategic choice can yield more information than a good random choice.

The principal features (or functions) of tabu search are summarized in [3, 5] as follows:

1. Adaptive memory
2. Selectivity (including strategic forgetting)
3. Abstraction and decomposition (through explicit and attributive memory)
4. Timing (which means both recency and frequency of events and differentiation between short-term and long-term)
5. Quality and impact (meaning the relative attractiveness of alternative choices and magnitude of changes in structure or constraining relationships)
6. Context (including both regional, structural, and sequential interdependences)
7. Sensible exploration
8. Strategically imposed constraints and inducements (or, tabu conditions and aspiration levels)

9. Concentrated focus on good regions and good solution features (intensification process)
10. Characterizing and exploring promising new regions (diversification process)
11. Non-monotonic search patterns (strategic oscillations)
12. Integrating and extending solutions (path relinking)

We can formulate different tabu search algorithms by combining the functions above to solve specific problems. Of course, the way the actual implementation is made depends on the problem characteristics and on the degree of sophistication needed in a particular application. Although the set of functions listed above can be expanded and/or modified, it is worth noting that the approach was originally proposed, and tested successfully in a number of problems, with only a reduced set of such functions (tabu search with short-term memory with tabu lists and aspiration criteria).

6.1.2 Problem Formulation

Generally speaking, TS algorithms solve problems formulated as follows:

$$\text{Min } f(x)$$
$$\text{Subject to } x \in X, \tag{6.1}$$

where x is a configuration (or a decision variables), $f(\cdot)$ is the objective function, and X is the search space. Notice also that no assumptions are made regarding the convexity of $f(x)$ and X or about the differentiability of $f(x)$. A variety of combinatorial optimization problems can be represented as the minimization of an objective function subject to a set of algebraic constraints, as above. There are cases, however, that the constraints cannot be represented easily as algebraic constraints. This is the case, for example, of the radiality constraints in certain distribution system operation and planning problems. This type of constraint, which may be a nuisance for certain mathematical approaches, is easily handled by TS as normally TS does not work directly with the algebraic constraints; configurations are represented by a coding instead. Tabu search solves a problem (Eq. 6.1) by first applying a local heuristic search in which, given a configuration x (a solution), the neighborhood of x is defined as the set of all configurations $x' \in N(x)$ that can be obtained by the same transition mechanism applied to x (Fig. 6.1). The conditions required for x' being a neighbor of x defines the structure of the neighborhood of x. The local search algorithm finds the transition, which leads to the configuration x' presenting the largest decrement in the objective function (in the same way as a steepest gradient algorithm). The repetition of this procedure eventually leads to a local optimal solution.

Tabu search differs from the simple local search algorithm above in at least two essential aspects (Fig. 6.1):

1. Transition leading to configurations for which the objective function is actually greater than it is for the current solution is allowed (we are considering a minimization problem such as problem 6.1).

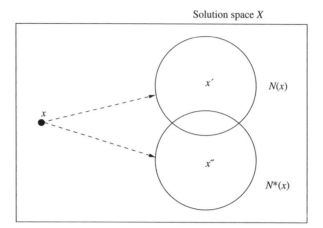

FIGURE 6.1 Illustration of a transition in tabu search.

2. The neighborhood of x, that is, $N(x)$, is not static; it can change both in size as well as in structure. A modified neighborhood $N^*(x)$ is determined in different ways, for example:

   a. Using a tabu list, which contains attributes of configurations that are forbidden. In this case $N^*(x) \subset N(x)$, as noted above, this is useful to avoid cycling and escape from local optima.

   b. Using strategies to reduce the size of the neighborhood in order to speed up the local search. As in the previous case, the reduced neighborhood is such that $N^*(x) \subset N(x)$.

   c. Using the so-called elite configurations to perform path relinking. In this case, it is not necessarily true that $N^*(x) \subset N(x)$.

   d. Redefining $N(x)$ during the optimization process; this is normally done in order to profit from the specific properties of the problem.

6.1.3 Coding and Representation

As it happens with other combinatorial approaches (e.g., simulated annealing and genetic algorithms), representation and coding are key issues in the tabu search. The definitions of feasible and unfeasible solutions, as well as the characterization of the objective function, are directly connected to the type of representation and coding used to model the problem being solved. Another critical issue affected by the method used in representation and coding is the characterization of the neighborhood of a given configuration.

Regarding the basic functions of tabu search, three aspects are of interest: (a) the way the transitions are made in the solution space; (b) how each solution in the search space is characterized; and (c) how the neighborhood of a given configuration is defined. These concepts are best understood with the help of the graphical illustration

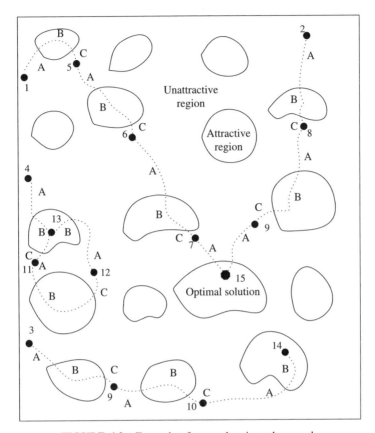

FIGURE 6.2 Example of a search using tabu search.

discussed in the following. For instance, the search space can be visualized as shown in Fig. 6.2, where *attractive* and *unattractive* regions are indicated including an optimal solution. The tabu search starts at a point such as point 1 in the figure and moves toward the optimal solution.

6.1.4 Neighborhood Structure

Figure 6.3 illustrates a generic point x with its neighborhood. Given a transition mechanism, a neighbor of x is any point that can be reached from x by means of a single transition. There are both feasible and unfeasible neighbors. Among the feasible ones, usually only the attractive configurations are of interest. Figure 6.4 includes in the neighborhood of x configurations that are interesting although unfeasible. The reason is that the temporary transition through an attractive unfeasible configuration in certain circumstances may be the shortest way leading to the desired attractive feasible solution. Figure 6.5 shows the complementary transition, which takes the configuration from an attractive unfeasible point to an attractive feasible one. This type of strategy has been successfully used in the transmission expansion planning

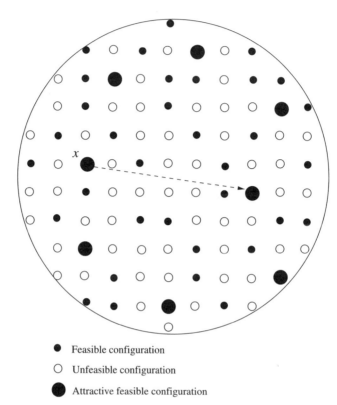

● Feasible configuration

○ Unfeasible configuration

● Attractive feasible configuration

FIGURE 6.3 Illustration of a neighborhood of a given configuration.

problem, which is a good example of complex system with plenty of local optimal solutions [6, 7]. The transition through unfeasible points is facilitated by the inclusion of penalty terms in the objective function to represent the cost of unfeasibility; hence, if the decrement in the actual cost more than compensates for the cost component due to the unfeasibility, then the transition is allowed.

Example 6.1

Consider now the well-known n-queens problem that consists of placing n queens on an $n \times n$ chessboard so that no two queens can capture each other [8]. This problem can be seen as an optimization problem for which an optimal solution is such that no two queens can be placed in the same row, same column, or same diagonal. Figure 6.6a shows one topology for the n-queens problem (with $n = 7$) in which there are four collisions, or four possibilities for attack. Figure 6.6b shows an optimal solution (i.e., a solution in which no two queens can capture each other). The examples 1 through 6 which will be presented in the following draws upon, with minor modifications, the material originally developed by Laguna in [8] and are an excellent guide for understanding the basic workings of tabu search, mainly regarding the use of tabu lists and aspiration criteria. An efficient coding for this

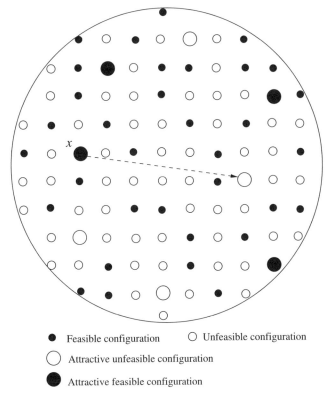

● Feasible configuration   ○ Unfeasible configuration

◯ Attractive unfeasible configuration

⬤ Attractive feasible configuration

FIGURE 6.4 Example of feasible ⇒ unfeasible transition.

problem consists in describing a configuration by an $n \times 1$ vector P whose ith element, $P(i)$, represents the column of the chessboard where the ith queen is placed; the ith queen is assumed to be placed in row i. With this coding, the topology given in Fig. 6.6a is described as follows as shown Fig. 6.7 where queen 1 is in row 1 and column 4, queen 2 is in row 2 and column 5, and so forth. With this coding, no two queens are placed in the same row or in the same column, and so part of the problem is already solved. The remaining problem can then be formulated as the minimization of diagonal collisions.

Next, we have to establish the objective function for the n-queens problem. In order to do that, the concepts of positive and negative diagonals are introduced, as illustrated in Fig. 6.8a and Fig. 6.8b. In order to find the number of collisions of a configuration, it is then necessary to go through the positive and negative diagonals and check for collisions. This task is made easier noticing that the positive diagonals are characterized by the fact that the difference $i - j$ is constant, whereas for the negative diagonal $i + j$ is constant, as illustrated in Fig. 6.9a and Fig. 6.9b. For example, for the configuration of Fig. 6.6a, collisions occur in positive diagonals 5, -2 and -3 and in negative diagonal 7. Thus, it can be seen that the evaluation of the objective function for this type of coding can be easily performed. Notice that although the

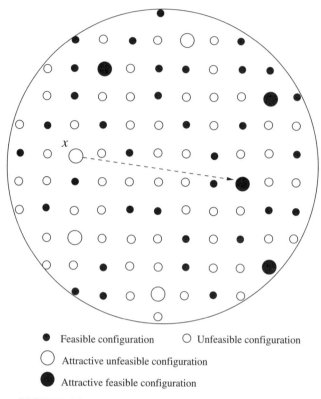

● Feasible configuration   ○ Unfeasible configuration

◯ Attractive unfeasible configuration

● Attractive feasible configuration

FIGURE 6.5   Example of unfeasible ⇒ feasible transition.

n-queens problem could have been mathematically formulated as a $0 - 1$ problem, in tabu search such mathematical modeling is in fact unnecessary.

6.1.5 Characterization of the Neighborhood

A critical issue in tabu search is the need for evaluating the configurations in a neighborhood. The number of configurations in a neighborhood may be very large and the quality of these configurations may vary a lot. At each iteration of a basic TS algorithm, it is normally necessary to identify the best configuration in the neighborhood of the current solution taking into account that the new solution has no tabu attributes or, if it happens to have, the aspiration criterion is satisfied (i.e., the new configuration is good enough to justify relaxing the tabu constraint). One of the main strategies of tabu search consists in moving to the best configuration in the neighborhood of the current configuration. Usually, the sizes of neighborhoods are much larger than can be evaluated by the algorithm, and thus only the most attractive part of a neighborhood is actually explored. Having the means of finding efficiently such attractive parts is critical to TS methodology. Recent work on tabu search recommends general-purpose strategies for reducing neighborhood sizes. These strategies however do not

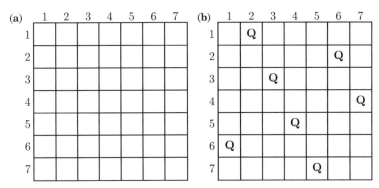

FIGURE 6.6 Seven-queens problem: (**a**) an initial topology; (**b**) an optimal topology.

$$P_1 = \boxed{4\,|\,5\,|\,3\,|\,6\,|\,7\,|\,1\,|\,2} \;\leftarrow\; \text{row (queen)}$$
$$\qquad\qquad \leftarrow \text{column}$$

FIGURE 6.7 Coding of topology shown in Figure 6.6(a).

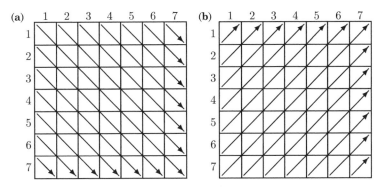

FIGURE 6.8 (**a**) Positive diagonals; (**b**) negative diagonals.

(a)

	1	2	3	4	5	6	7
1	0	-1	-2	-3	-4	-5	-6
2	1	0	-1	-2	-3	-4	-5
3	2	1	0	-1	-2	-3	-4
4	3	2	1	0	-1	-2	-3
5	4	3	2	1	0	-1	-2
6	5	4	3	2	1	0	-1
7	6	5	4	3	2	1	0

(b)

	1	2	3	4	5	6	7
1	2	3	4	5	6	7	8
2	3	4	5	6	7	8	9
3	4	5	6	7	8	9	10
4	5	6	7	8	9	10	11
5	6	7	8	9	10	11	12
6	7	8	9	10	11	12	13
7	8	9	10	11	12	13	14

FIGURE 6.9 Characterization of diagonals: (**a**) positive diagonals; (**b**) negative diagonals.

TABLE 6.1 Neighbor Configurations for the Topology of Fig. 6.6

No.	Swap	v	Δv	No.	Swap	v	Δv	No.	Swap	v	Δv
1	1–7	2	−2	8	3–5	3	−1	15	1–4	5	1
2	2–4	2	−2	9	3–6	3	−1	16	2–3	5	1
3	2–6	2	−2	10	4–7	3	−1	17	3–7	5	1
4	5–6	2	−2	11	6–7	3	−1	18	4–6	5	1
5	1–5	3	−1	12	2–5	4	0	19	5–7	5	1
6	1–6	3	−1	13	1–2	5	1	20	4–5	6	2
7	2–7	3	−1	14	1–3	5	1	21	3–4	7	3

necessarily work in all problems, and problem specific strategies have to be developed.

Example 6.2

A movement in tabu search consists in moving from the current configuration to the best neighboring configuration. In order to perform a move, it is necessary then to know the neighborhood of the current configuration. A popular choice for the n-queens problem is as follows: a neighbor configuration is a configuration that can be found by swapping the columns occupied by any two queens. For example, for the configuration of Fig. 6.6a, swapping queens 2 and 6 obtain a neighbor of the current configuration. The corresponding configurations are as shown in Fig. 6.10.

For this neighborhood structure, the current configuration has $n(n-1)/2$ neighbors; for $n = 7$, the number of neighbors is 21. For large values of n, however, the number of neighbors can be prohibitive, and an appropriated simplification of the neighborhood will have to be considered. Table 6.1 shows the objective function in the neighborhood of the current configuration as defined above. There are four neighbors with cost reduction of -2, which means that the corresponding moves would reduce the number of collisions from 4 to 2.

6.2 FUNCTIONS AND STRATEGIES IN TABU SEARCH

A tabu search strategy is an algorithm, which normally forms part of a more general tabu search procedure. The basic functions of tabu search are intensification, diversification, strategic oscillation, elite configurations, and path relinking. This section describes how these functions can be combined to build a range of tabu search strategies and operators, and how these can be used in effective tabu search programs.

6.2.1 Recency-Based Tabu Search

Recency-based memory is an important feature of tabu search: it is a type of *short-term memory* that keeps track of solution attributes that have changed during the most recent moves made by the algorithm. The information contained in this memory allows labeling as tabu-active selected attributes of recently visited

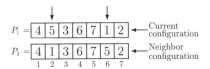

FIGURE 6.10 Generation of a neighbor configuration by swapping elements.

solutions; this feature avoids revisiting a solution already visited in the recent past. A number of practical applications reported in the literature are direct implementations of this basic tabu search algorithm.

This is the most basic type of tabu search algorithm and is based on a list of forbidden attributes and an aspiration criterion. The main objectives of the tabu list are (a) to avoid cycling (i.e., revising already visited solutions) and (b) reducing the size of neighborhoods by excluding from consideration configurations labeled as tabu. The main disadvantage of the use of a tabu list is that a forbidden attribute may be part of an attractive solution of a neighborhood that has not been visited so far. To cope with this problem, an aspiration criterion is used such that if the cost associated with a tabu configuration is smaller than the costs of the last kp transitions, or is inferior to the cost of the incumbent solution, then the tabu constraint is relaxed and the transition is allowed.

Figure 6.2 illustrates the working of a short-term memory tabu search algorithm of the type describe above. Four different processes, or paths, are shown in the figure: paths $1-5-6-7-15$ and $2-8-9-15$ lead to the optimal solution; path $3-9-10-14$ is entrapped into a local optimal solution; and path $4-11-12-13$ produces cycling. Even when tabu restraints are enforced, cycling may occur if the number of moves k during which a tabu is active is relatively small. Excessively large values of k, on the other hand, can turn the search inefficient as the visit to certain attractive configurations may be delayed. An alternative path not shown in Fig. 6.2 may have an optimal solution as an intermediate point. In this case, the solution process passes through the optimal solution and continues until it stops at a nonoptimal solution. This does not represent a serious problem because the optimal solution is kept in memory as an incumbent solution.

Example 6.3

This example illustrates the use of attributes to implement tabu constraints. Consider again the n-queens problem with $n = 7$. Assume that the current solution is the configuration shown in Fig. 6.6a. This configuration is codified as shown in Fig. 6.11.

The 21 neighbors of this configuration are summarized in Table 6.1. The best neighbor corresponds with the move that swaps queens 1 and 7. This move becomes tabu and

$$P_1 = \boxed{4\;5\;3\;6\;7\;1\;2}$$

row (queen) — over columns 1 2 3 4 5 6 7; ← column

FIGURE 6.11 Coding of 7 queen problem (initial topology).

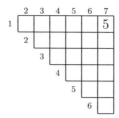

FIGURE 6.12 Attribute storage for the n-queens problem.

can be stored as illustrated in Fig. 6.12, which indicates that the swap 1–7 will stay forbidden for the next five moves. The same arrangement can be used for other possible moves. The arrangement is updated after every move: for example, when the next move is performed, the number 5 in position (1, 7) is decremented to 4 to take into account that the corresponding tabu tenure has decreased. Alternatively, rather than storing the tabu tenure for each tabu constraint, one can store the iteration where the tabu is activated: for example, if the tabu is activated in iteration 237, this number is entered in position (1, 7) of the storage arrangement of Fig. 6.12.

6.2.2 Basic Tabu Search Algorithm

A tabu search algorithm with short-term memory has the following characteristics: (1) It is a process of k_T moves in the search space, which comprises both feasible and unfeasible solutions. k_T is either predefined or determined adaptively. (2) The neighborhood of the current configuration is searched for a configuration with minimum cost. The move is validated if the move found is not tabu, of even being tabu, if an aspiration criterion is met. Moves to configurations with cost higher than that of the current configuration are allowed. (3) At each iteration, the list of tabu attributes is updated.

Example 6.4

A possible algorithm for the n-queens problem is as follows:

1. Define the tabu tenure and the aspiration criterion.
2. Define neighborhood structure.
3. Find the initial topology.
4. Compute and order the objective function for the entire neighborhood.
5. Move to the best neighbor if it is not tabu.
6. Move to the best neighbor if it is tabu but satisfies the aspiration criterion.
7. Update the tabu list.
8. Repeat steps 4 to 7 until a topology with zero cost is found.

The n-queens problem can be solved considering [3]: (1) tabu tenure of three iterations; (2) aspiration criterion that accepts a new (tabu) solution if its cost is lower

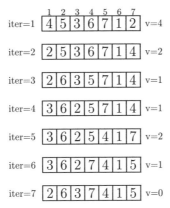

FIGURE 6.13 Transitions of tabu search algorithm.

than that of the incumbent solution; (3) the neighborhood is formed by the topologies obtained by exchanging the positions of any two queens; (4) the initial topology is that of Fig. 6.6(a).

The application of this algorithm yields the following sequence of topologies leading to an optimal solution as shown in Fig. 6.13.

The best moves and the corresponding objective functions and tabu lists at each iteration are summarized in the following:

- 1; move $= 1-7$; $v = 2$; tabu $= 1-7(3)$
- 2; move $= 2-4$; $v = 1$; tabu $= 1-7(2)$, $2-4(3)$
- 3; move $= 1-3$; $v = 1$; tabu $= 1-7(1)$, $2-4(2)$, $1-3(3)$
- 4; move $= 5-7$, $v = 2$; tabu $= 2-4(1)$, $1-3(2)$, $5-7(3)$
- 5; move $= 4-7$, $v = 1$; tabu $= 1-3(1)$, $5-7(2)$, $4-7(3)$
- 6; move $= 1-3$, $v = 0$; tabu $= 5-7(1)$, $4-7(2)$, $1-3(3)$

Remarks

- The optimal solution was found after six iterations (moves). In the two first iterations, the objective function was reduced. In the third iteration, the objective function remained constant.
- In the fourth move, there was no neighbor configuration with quality better than that of the current solution. Although there were two topologies with attributes $1-3$ and $1-7$ that lead to topologies with the same objectives these are forbidden (the attribute $1-3$ would lead to a topology already visited, whereas the attribute $1-7$ would lead to a new topology, since the aspiration criterion is not satisfied, the move is not performed).
- In the fourth move, the objective function is in fact increased.

- In the fifth iteration, there was a decrease in the objective function.
- Finally, in the sixth iteration, the attribute $1-3$ is used because the aspiration criterion is satisfied (the objective function is smaller than that of the incumbent).

6.2.2.1 *Candidate List Strategies*

Once the neighborhood is defined, it is evaluated. Because normally either the neighborhood is excessively large or the evaluation of each alternative is time consuming, a screening to limit the search to the most attractive neighbors has to be performed. In power system applications, for example, evaluation may imply the need for solving either a linear program or a nonlinear program or a power flow problem. Four strategies have been used in the literature to perform the screening of a neighborhood: (1) aspiration plus; (2) elite candidate list; (3) successive filter strategy; and (4) sequential fan candidate list.

The *aspiration plus* technique defines a minimum quality level for a configuration to be considered a new solution candidate. This method requires that the potential neighbors be put in an ordered list and analyzed one by one until a specified threshold is passed. The process stops after a few more configurations are evaluated after the threshold is hit (*value plus*). In order to avoid a reduced or an excessive number of neighbors being analyzed, upper and lower bounds have to be satisfied. There are at least two alternatives for defining the threshold: (1) the value of objective function of the best configuration visited in the last k moves, and (2) the value of objective function of incumbent solution.

The elite candidate list technique starts with a master list formed by the n_p best elements of the neighborhood of the initial configuration. Then a series of moves are performed considering that the set of top n_p neighbors remain the same. When either a configuration satisfying a given objective is found or a maximum number of moves is performed, a new master list is built and the process is repeated.

The successive filter strategy technique is normally used in cases in which the neighborhood structure is defined by swapping attributes. This is what happens, for example, in network transmission planning, where a neighbor is obtained by the addition of a circuit and the removal of another circuit (a swap of candidate circuits). A reduced neighborhood is then obtained by defining two short lists, one with elements that can be added and another with elements that can be removed from the current configuration.

The *sequential fan candidate list* technique is similar to the concept of population used in genetic algorithms. Given an initial configuration, the entire neighborhood is evaluated, and a reduced list of the n_p best neighbors is formed. These configurations are then called current configurations (a population). Next, a reduced number of neighbors of each current configuration is evaluated and a successor is found for each current configuration.

6.2.2.2 *Tabu Tenure*

Tabu tenure is the number of tabu search iterations (i.e., number of moves or transitions) an attribute remains forbidden. In a typical application, several tabu lists can be maintained simultaneously each with a different tenure. Thus, not only single attributes but also combinations of attributes can be forbidden. Very seldom, a tabu is specified by giving the complete information about the undesired configuration; normally, only certain attributes of the configuration are put in the tabu list. The tabu tenure can be either static (predefined) or dynamic (determined on the fly). Dynamic tenure can be implemented in two different ways: (1) random dynamic tabu tenure or (2) systematic dynamic tabu tenure. The random dynamic tabu tenure is implemented using bounds t_{min} and t_{max}; a randomly chosen tenure $t_{min} \leq t \leq t_{max}$ when a new tabu attribute is established. There are two alternatives for implementing this technique: in the first one, the value t is kept for a certain time αt_{max}, where α is a parameter. After that time, a new t is determined and so forth; the second alternative consists in determining a new t at each move, so that each tabu attribute will remain forbidden for different periods of time.

As with the random dynamic tabu tenure, the systematic dynamic tenure can be implemented using two different approaches. The first alternative is a simple variation of the random dynamic tabu tenure in which the random choice is replaced by a systematic choice: for example, assuming bounds $t_{min} = 4$ and $t_{max} = 9$, the systematic choice is made choosing t cycling through the sequence 4; 5; 6; 7; 8; 9 [3]. The second alternative is the so-called *moving gap* technique, where the tabu list is partitioned into two halves, one static and one dynamic. For example, in a scheme with eight iterations, all attributes remain in the tabu list for the first four iterations, whereas the last four iterations are variable, as described in the following. In the case of the *right gap*, the attributes initially remain in the tabu list for four iterations, then they are dropped from the list for two iterations, and come back to the list for two additional iterations. In the *middle gap*, the attributes initially remain in the tabu list for five iterations (four plus one), then they are dropped from the list for two iterations, and come back to the list for an additional iteration. And in the *left gap*, the attributes initially remain in the tabu list for six iterations (four plus two), then they are dropped from the list for the two remaining iterations.

6.2.2.3 *Aspiration Criteria*

Typically, a tabu test begins with the determination of a trial solution x' in the neighborhood of the current solution x. Then, attributes of x changed in the move from x to x' are identified. If among the attributes there are tabu-active attributes, the move still can be validated if x' satisfies an aspiration level (i.e., if x' is good enough for justifying the relaxation of the tabu restraint).

6.2.3 The Use of Long-Term Memory in Tabu Search

Although a significant part of the literature on tabu search deals with algorithms based exclusively on short-term memory techniques, more complex, sophisticated applications require the use of long-term memory. There are several different ways to use long-term memory. We can (1) reinitialize the search from a high-quality configuration; (2) redefine the neighborhood structure based on a high-quality solution; (3) redefine the objective function to penalize certain attributes of a high-quality solution; and (4) change the search strategy based on knowledge acquired during previous searches.

In practice, normally a combination of short-term and long-term memory is adopted: the short-term memory is usually implemented as a subroutine of a more general long-term algorithm. The long-term algorithm is normally based on three techniques: (1) frequency-based memory; (2) intensification; and (3) diversification.

6.2.3.1 Frequency-Based Memory

Frequency information is stored in order to be used to change future search strategies. Two principal types of frequency memory are used in practice: the *residence frequency* and the *transition frequency*.

By the *residence frequency* technique, the number of times an attribute occurs in a predefined set of configurations is kept in memory. This set of configurations can be the set of all configurations visited so far, or a set of elite configuration or a set of low-quality solutions, and so forth. For example, regarding the set of elite configurations, the frequency of a given attribute may indicate it is highly desirable; and the opposite happens if we are dealing with a set of low-quality solutions. In the first case (elite configuration), these attributes can be used both in intensification and in path relinking. On the other hand, if the set of configurations is diverse and the residence frequency of a certain attribute is high, it may indicate that this attribute is limiting the search space thus should be penalized (become tabu).

By the *transition frequency* technique, the number of times an attribute occurs in transitions is stored. The frequent occurrence of these attributes does not necessarily indicate that they will form part of the optimal solution. The information regarding these attributes can be used to change the search strategy by means of diversification; these attributes are penalized or become tabu. In transmission expansion planning, for example, this occurs with low-cost circuits, which are frequently used as crack fillers (i.e., they are temporarily used in moving between two alternative configurations and then are removed from the current solution).

Example 6.5

Figure 6.14 illustrates the use of memory in a tabu search algorithm in the 7-queens problem discussed in the previous examples. In this case, the lower triangle of the 7×7 matrix is used to store the transition frequency of each possible move, whereas the upper triangle is used for short-term memory, as in the previous example. In the case summarized in Fig. 6.14, 25 moves have been performed; for example, five of these moves were performed by swapping the positions of queens 1 and 3 or two moves involving queens 1 and 5, and so forth.

The information of the frequency-based memory can be used in different ways. From the current topology, for which the number of collisions is equal to one, a diversification process can be started. For example, the attributes that appear with high frequency are penalized during the diversification process.

6.2.3.2 Intensification

Both intensification and diversification are considered *advanced* functions of tabu search, in the sense that they can be added to a basic

Current solution

1	2	3	4	5	6	7
1	3	2	6	7	5	4

No. of collisions = 4

Tabu structure

	1	2	3	4	5	6	7	
						3		1
								2
	5					2		3
		1	3				1	4
	2	1	3	4				5
	1				1			6
		3	1					7

FIGURE 6.14 Transition frequency-based memory (lower triangle).

tabu search algorithm that uses short-term memory with tabu lists and aspiration criteria. Thus, the search normally begins with the basic algorithm that is then followed by intensification or diversification. Tabu search algorithms explore intensification in a systematic way. Configurations found during the search are stored and their neighborhoods are then explored more thoroughly; the local optimal solutions closest to these solutions are found in the intensification phase. Intensification can be performed starting from any elite solutions. In this case, intensification can be implemented using basically the same algorithm used in the short-term memory algorithm but with some modification in the neighborhood structure, as normally, the intensification is restricted to the neighborhood of an elite configuration and aims to find a local optimal solution in that neighborhood. In the transmission expansion problem, for example, this objective can be achieved performing simple operations such as single circuit additions or removals as well as circuit swaps.

Intensification can also be based on building blocks that are present in elite solutions stored in memory. Through a mechanism called path relinking, we can produce new attractive configurations utilizing such building blocks. In fact, *path relinking* is a means of implementing both intensification and diversification, depending on the amplitude of the changes introduced by this mechanism in the current configuration.

Decomposition is an alternative way to implement intensification. In this case, severe constraints are imposed on the problem structure in order to intensify the search in a restricted region. This process keeps certain problem variables constant at predefined values, and a restricted search is then performed. For example, in the traveling salesman problem, a certain number of partial paths are kept constant and the search is performed in the resulting reduced space to find the best way to connect and complement these paths.

6.2.3.3 *Diversification* The objective of diversification is to move the search to unvisited regions of the search space as well as to avoid cycling. For example, in the

TABLE 6.2 Actual Objective Function and Modified One for the Five Best Neighbors

Queens	v	v_p
3–4	1	4
1–6	2	3
2–5	2	3
2–6	2	2
2–4	3	4

capacitor placement problem, this is carried out by temporarily changing the rules for finding new configurations: for instance, certain capacitor banks are removed from a configuration and become tabu-active (i.e. they will not be allowed to be reintroduced for a certain period of time). The removal of key capacitor banks usually changes the architecture of a solution in such a way as to force the algorithm to visit unexplored regions.

Diversification can be implemented either by *restarting* the search from a new configuration or by *modifying the choice rules*. A restarting configuration can be obtained, for example, via path relinking or by using attributes of elite configurations or attributes found by the residence frequency mechanism. Modified choice rules, on the other hand, involve the change of objective function to take into account information obtained via residence frequency (neighbor configurations with high residence frequency are penalized so that the algorithm will tend to search in a different neighborhood).

Example 6.6

Diversification can be applied to the configuration shown in Fig. 6.14. The penalized objective function

$$v_p = v + \alpha n_f, \tag{6.2}$$

where v_p is the modified objective function, v is the original objective function (number of collisions), α is the penalty factor, and n_f is the number of times a given attribute was used to perform a transition. For $\alpha = 1$, the modified objectives for the five best neighbors of the current topology are given in Table 6.2.

Hence, without penalties, the attribute 3–4 would be chosen, in which case the objective function would be $v = 1$, whereas when the penalties are taken into account, the attribute 2–6 is selected instead, with modified cost of $v_p = 2$, which then causes the diversification of the search, as the attribute 2–6 has not been used so far.

6.2.4 Other TS Strategies

In the following, three important strategies that can be used both in connection with short-term memory as well as long-term memory are discussed: path relinking, strategic oscillations, and elite configurations.

6.2.4.1 *Path Relinking*   Two or more elite configurations are used to generate a new configuration. The configurations used in path relinking are called reference solutions. Reference solutions can be classified as *initiating solutions* or *guiding solutions*. For example, a good topology can be generated taking configuration A as the initiating solution and considering configurations B and C as guiding solutions; this is an intensification process (new solutions are sought in the neighborhood of the elite configurations). An alternative way to implement path relinking, called *constructive neighborhood*, consists in finding a single configuration and initiating intensification or diversification from there. A few attributes of the initiating solution are selected, and at each step, a constructive algorithm adds an attribute of a guiding configuration to the new configuration; such added attributes are selected either to improve quality or to restore feasibility. If a large number of attributes of the initiating solution is present, then we have an intensification process. Otherwise, new configuration will be used as the basis for diversification.

6.2.4.2 *Strategic Oscillation*   Strategic oscillations is based on three different techniques that are used alternately: (1) a strategy for unfeasible regions in which the objective is to reach the boundary of a feasible region and then enter in the feasible region; (2) a strategy for searching in the feasible region for a local optimal solution; and (3) a strategy for leaving the feasible region and entering the unfeasible region. Strategic oscillation is more efficiently used in problems where the size of the unfeasible region is relatively large, as is the case of the transmission expansion planning problem.

"Good solutions at one level are likely to be found close to good solutions at an adjacent level" [5]. This phrase describes the basis for the so-called proximate optimality principle (POP). In strategic oscillation and other constructive-destructive processes, it is worthwhile to explore a level before moving to the next one. Thus, the search process is forced to stay at the current level for *npop* iterations (transitions) before moving to the next level; the transition is then taken from the best configuration found in the *npop* transitions.

6.2.4.3 *Elite Configurations*   Elite configurations is a list of the best configurations found during the search process of tabu search algorithm. These configurations are stored separately from other solutions and can be used for restarting the search process in intensification and diversification stages. However, they are most used in the path relinking as *initiating solutions* or as *guiding solutions*. The elite configurations are efficiently employed when besides the good quality they also present diversity among the solutions. Therefore, elite configurations list is formed taking into account the quality and diversity of these topologies (it also implies that a measure of the diversity must be specified).

6.3 APPLICATIONS OF TABU SEARCH

The initial applications of tabu search occurred in the late 1980s and were limited to the operation research area and covered classic problems such as the traveling salesman problem and graph coloring. Applications to complex problems in other fields appeared in the early 1990s. A survey made by the authors of this chapter based on the database of the Institute for Scientific Information (ISI) has listed 1348 publications related to tabu search and its applications, most of them in the period 1998–2005. As for applications to power system problems, according to the ISI about 48 papers have been published in the specialized literature up to April 2005, most of them in journals such as the *IEEE Transactions on Power Systems* (15), *IEE Proceedings Generation, Transmission and Distribution* (9), *Electric Power Systems Research* (14), and *International Journal of Electrical Power & Energy Systems* (6). The list of applications presented in the following is not complete and is biased by the previous experience of the authors in this field.

The traveling salesman problem is one of the combinatorial problems that has received a great deal of attention in the literature. The main difficulty associated with this problem is to find optimal, or even near-optimal solutions, for systems with high dimensions. Typical TS applications in this area are described in Refs. 9 and 10. The vehicle routing problem (VRP) is another problem in operations research that has been explored with TS techniques [11]. Assignment problems such as the quadratic assignment problem (QAP), generalized assignment problem (GAP), and multilevel generalized assignment problem (MGAP) were also studied with TS algorithms [12–14]. Other problems wherein TS has been applied include multidimensional knapsack $0-1$ problem [15], fixed charge transportation problem [16], graph partitioning [17], and telecommunication network problems [18]. An interesting application in the power system area is the use of a TS and hybrids to the unit commitment problem [19, 20]. Another application to a very complex problem is the case of the transmission network expansion planning problem described in Refs. [7, 21, 22]. Other applications are described in Refs. [7, 21, 22].

6.4   CONCLUSIONS

This chapter has presented the basic features of tabu search. Like other iterative techniques used to solve complex combinatorial optimization problems, TS has the ability to escape from local optimal solutions by accepting uphill moves; that is, moves that deteriorate the current value of the objective function. TS differs from the other techniques in the use of memory, which is crucial to the successful implementation of tabu search. As the TS algorithm traverses the solution space, it stores relevant findings in short-term and long-term memories, which are then subsequently used to redirect search and modify the local search algorithms that form part of the TS meta-heuristic. Both simple tabu search algorithms based on short-term memory with tabu lists and aspiration criteria as well as more advanced techniques such as intensification, diversification, path relinking, elite configurations, and strategic oscillations have been described. Examples were used throughout the chapter to illustrate the basic concepts.

REFERENCES

1. Glover F, Kelly JP, Laguna M. Genetic algorithms and tabu search: Hybrids for optimization. Computers Operations Res 1994; 22(1):111–134.

2. Glover F, Taillard E, de Werra D. A users guide to tabu search. Ann Operations Res 1993; 41:3–28.

3. Glover F, Laguna M. Tabu Search. Boston: Kluwer Academic Publishers; 1997.

4. Reeves CR. Modern heuristic techniques for combinatorial problems. New York: McGraw-Hill; 1995.

5. Glover F. Tabu search fundamental and uses. Graduate School of Business, University of Colorado; 1995.

6. Garver LL. Transmission network estimation using linear programming. IEEE Trans Power App Syst 1970; PAS-89:1688–1697.

7. Gallego RA, Romero R, Monticelli A. Tabu search algorithm for network synthesis. IEEE Trans Power Systems 2000; 15(2):490–495.

8. Laguna M. A guide to implementing tabu search. Investigacion Operativa 1994; 4(1):5–25.

9. Knox J. Tabu search performance on the symmetrical traveling salesman problem. Computers Operations Res 1994; 21(8):867–876.

10. Fiechter CN. A parallel tabu search algorithm for large traveling salesman problems. Discrete Appl Math 1994; 51(3):243–267.

11. Barbarosoglu G, Ozgur D. A tabu search algorithm for the vehicle routing problem. Computers Operations Res 1999; 26(3):255–270.

12. Skorinkapov J. Extensions of a tabu search adaptation to the quadratic assignment problem. Computers Operations Res 1994; 21(8):855–865.

13. Kelly JP, Laguna M, Glover F. A study of diversification strategies for the quadratic assignment problem. Computers Operations Res 1994; 21(8):885–893.

14. Osman IH. Heuristic for the generalized assignment problem—simulated annealing and tabu search approaches. OR Spektrum 1995; 17(4):211–225.

15. Hanafi S, Freville A. An efficient tabu search approach for the 0-1 multidimensional knapsack problem. Eur J Operational Res 1998; 106(2–3):659–675.

16. Sun M, Aronson JE, McKeown PG, Drinka D. A tabu search heuristic procedure for'the'fixed charge transportation problem. Eur J Operational Res 1998; 106(2–3):441–456.

17. Rolland E, Pirkul H, Glover F. Tabu search for graph partitioning. Ann Operations Res 1996; 63:209–232.

18. Costamagna E, Fanni A, Giacinto G. A tabu search algorithm for the optimization of "telecommunication networks. Eur J Operational Res 1998; 106(2–3):357–372.

19. Mantawy AH, Abdel-Magid YL, Selim SZ. Unit commitment by tabu search. IEE Proc Generation Transmission Distribution 1998; 145(1):56–64.

20. Mantawy AH, Abdel-Magid YL, Selim SZ. Integrating genetic algorithms, tabu search and simulated annealing for the unit commitment problem. IEEE Trans Power Systems 1999; 14(3):829–836.

21. Wen FS, Chang CS. Transmission network optimal planning using the tabu search method. Electric Power Systems Res 1997; 42(2):153–163.

22. Gallego RA, Monticelli A, Romero R. Comparative studies of non-convex optimization methods for transmission network expansion planning. IEEE Trans Power Systems 1998; 13(3):822–828.

23. Wang YC, Yang HT, Huang CL. Solving the capacitor placement problem in a radial'distribution system using tabu search. IEEE Trans Power Systems 1996; 11(4):1868–1873.

24. Gan DQ, Qu ZH, Cai HZ. Large-scale VAR optimization and planning by tabu search. Electric Power Systems Res 1996; 39(3):195–204.

25. Wen FS, Chang CS. A tabu search approach to fault section estimation in power systems. Electric Power Systems Res 1997; 40(1):63–73.

26. Wen FS, Chang CS. Tabu search approach to alarm processing in power systems. IEE Proc Generation Transmission Distribution 1997; 144(1):31–38.

27. Gallego RA, Monticelli A, Romero R. Optimal capacitor placement in radial distribution networks. IEEE Trans Power Systems 2001; 16(4):630–637.

28. da Silva EL, Ortiz J.M.D, De Oliveira GC, Binato S. Transmission network expansion planning under a tabu search approach. IEEE Trans Power Systems 2001; 16(1):62–68.

CHAPTER 7

Fundamentals of Simulated Annealing

ALCIR J. MONTICELLI, RUBÉN ROMERO, and EDUARDO NOBUHIRO ASADA

7.1 INTRODUCTION

Simulated annealing (SA) is one of the most flexible techniques available for solving hard combinatorial problems. The main advantage of SA is that it can be applied to large problems regardless of the conditions of differentiability, continuity, and convexity that are normally required in conventional optimization methods. In this chapter, first the principles of simulated annealing are presented on an intuitive basis. Next, the basic theory of simulated annealing along with a sequential version of the method is presented. A parallel version of the algorithm is then introduced. Detailed discussions about the application of the approach for two problems are given, namely the traveling salesman problem and the transmission network expansion problem. As happens with other combinatorial techniques, the coding of solutions, the neighborhood definition of a given configuration, the evaluation function, and the transition mechanisms are critical to the success of practical implementations of simulated annealing.

Annealing is the process of submitting a solid to high temperature, with subsequent cooling, so as to obtain high-quality crystals (i.e., crystals whose structure form perfect lattices) [1]. Simulated annealing emulates the physical process of annealing and was originally proposed in the domain of statistical mechanics as a means of modeling the natural process of solidification and formation of crystals. During the cooling process, it is assumed that thermal equilibrium (or quasi-equilibrium) conditions are maintained. The cooling process ends when the material reaches a state of minimum energy, which, in principle, corresponds with a perfect crystal. It is known that defect-free crystals (i.e., solids with minimum energy) are more likely to be formed under a slow cooling process. The two main features of the simulated annealing process are (1) the transition mechanism between states and (2) the cooling schedule. When applied to combinatorial optimization, simulated

Modern Heuristic Optimization Techniques. Edited by K. Y. Lee and M. A. El-Sharkawi

annealing aims to find an optimal configuration (or state with minimum "energy") of a complex problem.

Simulated annealing was originally proposed by Metropolis in the early 1950s as a model of the crystallization process. It was only in the 1980s, however, that independent research done by Kirkpatrick, Gelatt, and Vecchi [2] and Cerny [3] noted the similarities between the physical process of annealing and some combinatorial optimization problems. They noted that there is a correspondence between the alternative physical states of the matter and the solution space of an optimization problem. It was also observed that the objective function of an optimization problem corresponds with the free energy of the material. An optimal solution is

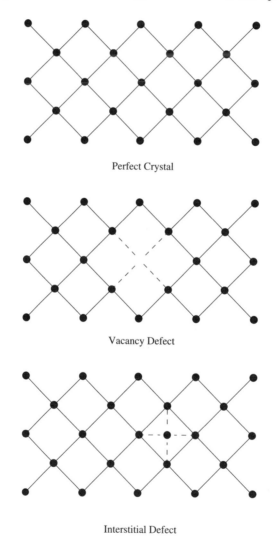

Perfect Crystal

Vacancy Defect

Interstitial Defect

FIGURE 7.1 Solidification of crystals and common defects.

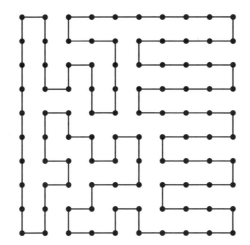

FIGURE 7.2   Symmetrical TSP with 100 cities.

associated with a perfect crystal, whereas a crystal with defects corresponds with a local optimal solution (Fig. 7.1) [4]. The analogy is not complete, however, because in the annealing process there is a physical variable that is the temperature, which under proper control leads to the formation of a perfect crystal. When simulated annealing is used as an optimization technique, the "temperature" becomes simply a control parameter that has to be properly determined in order to achieve the desired results.

In this chapter, two complex problems are used to illustrate the workings of the simulated annealing algorithm: (1) the traveling salesman problem (TSP) and (2) the transmission network expansion planning problem. Cerny [2] has used a very simple version of SA to solve the traveling salesman problem for cases in which the cities are supposed to be arranged in a symmetrical array. Figure 7.2 shows a symmetrical configuration with 100 cities. Such symmetrical structures normally involve a number of alternative optimal solutions where the horizontal and vertical distance between adjacent cities is one unit of length. An optimal tour in this case has a total length of $v = 100$.

7.2   BASIC PRINCIPLES

This section summarizes the theoretical fundamentals of simulated annealing. A more detailed description of the theory of simulated annealing can be found in Refs. 1 and 14.

7.2.1   Metropolis Algorithm

The original idea behind the simulated annealing algorithm is the Metropolis algorithm that models the microscopic behavior of sets of large numbers of particles, as in a solid, by means of Monte Carlo simulation. In a material, the individual particles have different levels of energy, according to a certain statistical distribution. The possible lowest level of energy, known as the fundamental level, corresponds with the state where all particles stand still and occurs at temperature 0 K. For temperatures above that level, the particles will occupy different levels of energy, such that the number of particles in each level decreases as the energy level increases (i.e., the maximum number of particles is found in the fundamental level). The distribution of the particles in the various levels varies with the temperature; for $T = 0$ K, for example, all particles are in the fundamental level; as the temperature increases, more particles are found in higher energy levels but always as a decreasing function of the energy level.

The Metropolis algorithm generates a sequence of states of a solid as follows: giving a solid in state S_i, with energy E_i, the next state S_j is generated by a transition mechanism that consists of a small perturbation with respect to the original state, obtained by moving one of the particles of a solid chosen by the Monte Carlo method. Let the energy of the resulting state, which also is found probabilistically, be E_j; if the difference $E_j - E_i$ is less than or equal to zero, the new state S_j is accepted. Otherwise, in case the difference is greater than zero, the new state is accepted with probability

$$\exp\left(\frac{E_i - E_j}{k_B T}\right), \tag{7.1}$$

where T is the temperature of the solid and k_B is the Boltzmann constant. This acceptance rule is also known as Metropolis criterion and the algorithm summarized above is the Metropolis algorithm [14]. The temperature is assumed to have a rate of variation such that thermodynamic equilibrium is reached for the current temperature level, before moving to the next level. This normally requires a large number of state transitions of the Metropolis algorithm.

7.2.2 Simulated Annealing Algorithm

For a combinatorial optimization problem to be solved by simulated annealing, it is formulated as follows: let G be a finite, although perhaps very large, set of configurations and v the cost associated with each configuration of G. The solution to the combinatorial problem consists of searching the space of configurations for the pair (G, v) presenting the lowest cost. The SA algorithm starts with an initial configuration G_0 and an initial "temperature" $T = T_0$ and generates a sequence of configurations $N = N_0$. Then the temperature is decreased; the new number of steps to be performed at the temperature level is determined, and the process is then repeated. A candidate configuration is accepted if its cost is less than that of the current configuration. If the cost of the candidate configuration is bigger than the cost of the current configuration, it still can be accepted with a certain probability. This ability to perform uphill moves allows simulated annealing to escape from local optimal configurations. The entire process is controlled by a *cooling schedule* that determines how the temperature is decreased during the optimization process.

Simulated Annealing;
Begin
   Initialize (T_0, N_0);
   $k := 0$;
   Initial configuration S_i
   Repeat procedure
   do $L := 1$ **to** N_k
      generate $(S_j \; from \; S_i)$;
      if $f(S_j) \leq f(S_i)$ **do** $S_i := S_j$;
      otherwise
      if $\exp\left(\frac{f(S_i)-f(S_j)}{T_k}\right) > \mathrm{random}[0,1]$ **do** $S_i := S_j$;
      end do;
   $k := k+1$;
   Calculation of the length (N_k);
   Determine control parameter (T_k);
   Stopping criterion
end;

FIGURE 7.3 The algorithm simulated annealing (Aarts and Korst [1]).

Figure 7.3 summarizes the simulated annealing algorithm, which consists of two basic mechanisms: the generation of alternatives and an acceptance rule. T_k is the control parameter that corresponds with the temperature in physical annealing, and N_k is the number of alternatives generated in the kth temperature level (this corresponds with the time that the system stays at a given temperature level and should be big enough for allowing the system to reach a state that corresponds with "thermal equilibrium"). Initially, when T is large, larger deterioration in the cost function is allowed; as the temperature decreases, the simulated annealing algorithm becomes greedier, and only smaller deteriorations are accepted; and finally when T tends to zero, no deteriorations are accepted.

From the current state S_i, with cost $f(S_i)$, a neighbor solution S_j, with cost $f(S_j)$, is generated by the transition mechanism. The following probability is calculated in performing the acceptance test:

$$P_T\{\text{Accept } S_j\} = \begin{cases} 1 & \text{if } f(S_j) \leq f(S_i) \\ \exp\left(\dfrac{f(S_i) - f(S_j)}{T_k}\right) & \text{if } f(S_j) > f(S_i). \end{cases} \qquad (7.2)$$

7.3 COOLING SCHEDULE

The cooling schedule is the control strategy used from the beginning until the convergence of the simulated annealing algorithm. It is characterized by the following four parameters:

1. Initial temperature T_0;
2. Final temperature T_f (stopping criterion);

3. Number of transitions, N_k, at temperature T_k.
4. Temperature rate of change given by $T_{k+1} = g(T_k)T_k$, where $g(\cdot)$ is a function that controls the temperature.

The efficiency of the algorithm, regarding the quality of the final solutions as well as the number of iterations, will depend on the choice of these parameters. The procedures used in the calculation of parameters are based on the idea of thermal equilibrium and are detailed in the following sections.

7.3.1   Determination of the Initial Temperature T_0

There are several ways of determining the initial temperature T_0 for the simulated annealing algorithm. One alternative consists of carrying out a constructive experimental process that simulates the first temperature level of the algorithm. In Ref. [1], the following procedure is suggested:

$$T_0 = \frac{\overline{\Delta V}^+}{\ln\left(\dfrac{m_2}{m_2 X_0 - m_1(1 - X_0)}\right)}. \tag{7.3}$$

$\overline{\Delta V}^+$ is the average value of differences (Δv) in the objective function considering only the raising values within m_0 tries ($m_0 = m_1 + m_2$). m_1 corresponds with the number of moves with decreasing costs, and m_2 is the number of moves with increasing costs. X_0 corresponds with the acceptance ratio of new configurations. In the literature, a commonly used value is $X_0 = 0.85$, which means that, at the initial temperature, 85% of transition tests are accepted.

An alternative way of determining T_0 was proposed in [5]:

$$T_0 = \frac{\mu}{-\ln \phi} f(x^0), \tag{7.4}$$

where it is assumed that $\phi\%$ of the uphill moves, which are $\mu\%$ worse than the initial solution $f(x^0)$, are accepted at the initial temperature level T_0.

Remarks

The result expressed by Eq. (7.4) can be verified as follows: Consider a configuration with cost $f(x^c)$ is $\mu\%$ worse than the cost $f(x^0)$ of the initial configuration; that is,

$$f(x^c) = (1 + \mu)f(x^0) = f(x^0) + \mu f(x^0) \quad \Rightarrow \mu f(x^0) = f(x^c) - f(x^0).$$

From the definition of ϕ, it follows that

$$\phi = \exp\left[-\frac{f(x^c) - f(x^0)}{T_0}\right].$$

From the two previous equations, it follows that

$$\ln\phi = -\frac{f(x^c) - f(x^0)}{T_0} = -\frac{\mu f(x^0)}{T_0} \Rightarrow T_0 = \frac{\mu}{-\ln\phi}f(x^0).$$

The determination of T_0 from Eq. (7.4) has the advantage of being simple and direct, although it depends on the estimate $f(x^0)$, which is not always easily determined.

Example 7.1

In a given optimization problem, one wishes to accept $\phi = 13\%$ of the uphill moves with costs up to 1% bigger than the cost of the initial solution, which is of $f(x^0) = 100,000$. The initial temperature T_0 is determined as follows (Eq. 7.4):

$$T_0 = \frac{0.01}{-\ln(0.13)}(100,000) = 490.$$

7.3.2 Determination of N_k

The number of moves performed at each temperature level should be such that the condition of thermal near-equilibrium is guaranteed. Hence, the value of this parameter is closely related to the rate of temperature reduction. Most algorithms use a value of N_k that depends on the problem dimension (number of decision variables). Two of the proposals that appear in the literature are summarized below.

Constant N_k:

$$N_{k+1} = N_0. \tag{7.5}$$

Variable N_k: $\rho \geq 1.0$.

$$N_{k+1} = \rho N_k. \tag{7.6}$$

N_0 is the number of transition tests at the initial temperature level, and ρ is a user-supplied parameter.

Remarks

In Ref. 6, both versions above have been implemented and compared; although the second alternative is more demanding in terms of computational effort, it normally leads to results that in general are better than the ones obtained with the first approach.

7.3.3 Determination of Cooling Rate

There are a number of ways to execute the temperature reduction in the simulated annealing. All methods, however, are based on the fact that thermal equilibrium should be reached before the temperature is reduced. Three alternatives for calculating T_{k+1} from the current temperature T_k are summarized below:

- *Constant cooling rate*: $\beta \in [0.50; 0.99]$

$$T_{k+1} = \beta T_k.\tag{7.7}$$

- *Variable cooling rate*: $\delta \in [0.01; 0.20]$

$$T_{k+1} = \frac{T_k}{\left[1 + \dfrac{\ln(1 + \delta)T_k}{3\sigma(T_k)}\right]},\tag{7.8}$$

where $\sigma(T_k)$ is the standard deviation of the costs of the configurations generated at the previous temperature level T_k.

- *Variable cooling rate*: $\lambda \leq 1.0$

$$T_{k+1} = \frac{T_k}{\exp\left(\dfrac{\lambda T_k}{\sigma(T_k)}\right)}.\tag{7.9}$$

The next value T_{k+1} also depends on the performance in the same way as in Eq. (7.8).

Remarks

The performance of the various cooling schedules is highly problem dependent. For the network transmission expansion planning problem, for example, methods Eq. (7.8) and Eq. (7.9) present nearly the same performance as the method of Eq. (7.7) as far as the quality of the solutions is concerned, although the number of iterations to convergence is normally much higher for the methods of Eq. (7.8) and Eq. (7.9). In Refs. 6 and 7, the method of Eq. (7.7), with proper calibration, has been applied with success.

7.3.4 Stopping Criterion

Stopping criteria vary a lot in degree of complexity and sophistication. Both predefined and adaptive criteria have been suggested in the literature. Some of the most common strategies are summarized below: (1) Define a constant number of temperature reductions normally between 6 and 50. (2) Use the rate of improvement of the cost function to define the stopping criterion; hence, if the incumbent solution or the cost of the best solution so far does not improves after a series of temperature reductions, it is assumed that convergence has been reached and the process is stopped. (3) Define the number of uphill moves that should be accepted. The process stops whenever the number of acceptances becomes less than the specified value.

In problems such as the network transmission expansion planning, and in a number of other applications of SA to power networks, auxiliary problems have to be solved in a number of times to verify the solution feasibility. (Notice that this is not the case with several important operation research problems such as the traveling salesman problem.) In the case of the network expansion problem, the solutions of linear programs representing the entire network are required. For these situations, more elaborate stopping criteria may be necessary in order to reduce the overall computational effort. In Refs. 6 and 7, the following criterion has been tested with success: (1) Stop the process if the number of LP solutions exceeds a specified limit; or, (2) stop the process if the incumbent solution does not improve after a specified number of iterations.

The algorithm can be further improved by adding an improvement phase with a greedy algorithm that employs a modified neighborhood structure. Next, two examples of simulated annealing are discussed: the traveling salesman problem and the transmission network expansion problem.

7.4   SA ALGORITHM FOR THE TRAVELING SALESMAN PROBLEM

The traveling salesman problem (TSP) is formulated as follows: given a set of n cities, $V = \{1, 2, 3, \ldots, n\}$, a set of edges connecting the cities, $(i, j) \in A$, and the distances (costs) d_{ij} between cities i and j (with $d_{ij} = d_{ji}$ in the symmetrical case). A traveling salesman has to make a tour starting from one of the cities, visiting each one of the other cities only once, and returning to the original city. The objective is to find the tour with minimum length. The following algorithm draws upon and improves the original algorithm presented by Cerny [3].

7.4.1   Problem Coding

The coding for the TSP can be made very easily by using a vector in which each index i corresponds with a city, and the i element of the vector contains the next city in the tour.

Consider for example the solution shown in Fig. 7.4. This solution can be coded as the vector p_1 in Fig. 7.5, which indicates that the cities are visited in the following order: 1, 9, 10, 7, 4, 6, 8, 5, 3, 2, 1. The vector p_2 in the figure was obtained by swapping the contents of positions 3 and 7 of the vector p_1. The tour corresponding with the modified vector p_2 is shown in Fig. 7.6. The optimal tour is shown in Fig. 7.7.

The coding described above has the advantage that it always guarantees a feasible solution. Also, the size of the vector is the same as the number of cities (i.e., n) and

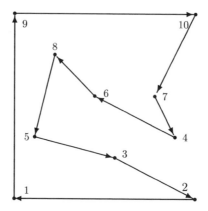

FIGURE 7.4   Traveling salesman problem with 10 cities ($v_1 = 53.72$).

then increases linearly with the size of the problem. Moreover, good neighborhood structures can be defined without difficulty.

Remarks

There are several different ways of encoding the solution of the traveling salesman problem. The coding given above is a popular choice in both simulated annealing and tabu search applications.

7.4.2 Evaluation of the Cost Function

In each step of the simulated annealing algorithm, an acceptance test has to be performed. This test requires the evaluation of the change produced in the cost function by a perturbation. For the neighborhood structure described above, the evaluation can be easily performed as follows. Consider for example that the current topology is the one given in Fig. 7.4 and that the perturbed topology is the one of Fig. 7.6. In this case, the variation in the cost function is given by:

$$\Delta v = v_2 - v_1 = d_{9,8} + d_{10,5} - d_{9,10} - d_{8,5}$$
$$\Delta v = 2\sqrt{2} + 10 - 9 - \sqrt{17} = -0.295.$$

Because there is an actual reduction in the cost, the new topology is accepted and becomes the new current solution. The new current topology is the one represented by

$$p_1 = \boxed{1\ |\ 9\ |\ 10\ |\ 7\ |\ 4\ |\ 6\ |\ 8\ |\ 5\ |\ 3\ |\ 2}$$

$$p_2 = \boxed{1\ |\ 9\ |\ 8\ |\ 6\ |\ 4\ |\ 7\ |\ 10\ |\ 5\ |\ 3\ |\ 2}$$

FIGURE 7.5   Example of neighbor configuration obtained by swapping.

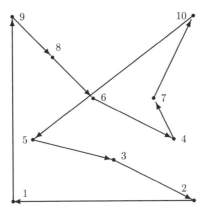

FIGURE 7.6 Another tour option obtained from the configuration of Fig. 7.4 by swapping elements 3 and 7.

encoding vector p_2, with the following associated cost:

$$v_2 = v_1 + \Delta v = 53.72 - 0.29 = 53.43.$$

7.4.3 Cooling Schedule

As mentioned earlier in the chapter, the cooling schedule is characterized by (1) determining the initial temperature, (2) the number of trials at each temperature level, (3) the rate of cooling, and (4) the stopping criterion.

The initial temperature T_0 may be calculated from Eq. (7.4). Thus, for example, if one wishes to accept $\phi = 30\%$ of the uphill moves (moves with increasing costs) with

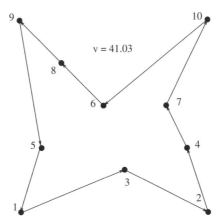

FIGURE 7.7 Optimal solution of the TSP presented in Fig. 7.4 and Fig. 7.6.

costs up to $\mu = 15\%$ worse than the initial configuration, which is of $v_1 = 53.72$ (the initial configuration represented in Fig. 7.5), the initial temperature T_0 is given by

$$T_0 = \frac{0.15}{-\ln(0.30)}(53.72) = 6.69.$$

The number of trials at each temperature level is assumed to remain constant throughout the cooling process, that is, $N_{k+1} = N_0$, where N_0 is normally found according to the number of cities. For example, for $n = 10$, one can choose $N_0 = 4n = 4(10) = 40$ trials. The rate of temperature reduction β can be assumed to be constant as well, for example, $\beta = 0.98$. Hence, for updating the temperature, one simply uses:

$$T_{k+1} = 0.98T_k.$$

The stopping criterion is based on the number of temperature levels for which the cost of the incumbent solution (the optimum solution so far) does not change (for example, five consecutive temperature levels). The SA characterized by these parameters easily solves the TSP with the initial topology given in Fig. 7.4, with $n = 10$, and finds the optimal solution shown in Fig. 7.7.

7.4.4 Comments on the Results for the TSP

The simulated annealing algorithm described above works particularly well for the traveling salesman problem. The algorithm as described above easily finds the optimal solution for the symmetrical problems of the type illustrated in Fig. 7.2 with $n = 100$. More elaborate SA algorithms are able to find a solution for cases with thousands of cities.

An important feature of the algorithm discussed above is the way the neighborhood structure was defined (k-opt, with $k = 2$ in the case shown in the figure), in which case all neighboring configurations are feasible. Notice that this is not the case with genetic algorithms and with evolutive algorithms in general, where infeasibilities may occur. Another characteristic of the simulated annealing approach as applied to the traveling salesman problem is the relatively small effort spent in computing the variations of the cost function caused by a move. This is normally less than what is required in computing the probability $p = \exp(\Delta v/T)$, and it suggests that it may be advantageous to use approximations in evaluating the exponential function.

7.5 SA FOR TRANSMISSION NETWORK EXPANSION PROBLEM

Given the network configuration for a certain year and the peak generation/demand for the next year (along with other data such as network operating limits, costs, and investment constraints), one wants to determine the expansion plan with minimum cost, (i.e., one wants to determine *where* and *which* type of new equipment should be installed). Of course, this is a subproblem of a more general case, called *dynamic* expansion planning, where, in addition to the questions *which* and *where*, one wants to know *when* to install new pieces of equipment. In this section, we focus on the initial stages of the expansion planning studies when the basic topology of the future network is determined. Network topologies synthesized by the proposed approach will then be further analyzed and improved by testing their performances using other analysis tools such as power flow, short circuit, transient and dynamic stability analysis. In the following, the static transmission expansion planning problem is formulated as a mixed-integer nonlinear programming problem in which the power network is represented by a DC power flow model.

$$
\begin{aligned}
\min v \quad &= \quad \sum_{(i,j)\in\Omega} c_{ij}n_{ij} + \sum_{k\in\Gamma} \alpha_k r_k \\
\text{subject to} \quad & \\
& \mathbf{Sf} + \mathbf{g} + \mathbf{r} = \mathbf{d} \\
& f_{ij} - \gamma_{ij}(n_{ij}^0 + n_{ij})(\theta_i - \theta_j) = 0 \\
& |f_{ij}| \le (n_{ij}^0 + n_{ij})\bar{f}_{ij} \\
& \mathbf{0} \le \mathbf{g} \le \bar{\mathbf{g}} \\
& \mathbf{0} \le \mathbf{r} \le \mathbf{d} \\
& 0 \le n_{ij} \le \bar{n}_{ij} \\
& n_{ij} \text{ integer}; \ f_{ij} \text{ and } \theta_j \text{ unbounded} \\
& (i, j) \in \Omega,
\end{aligned}
\tag{7.10}
$$

where c_{ij} is the cost of the addition of a circuit in branch $i–j$, n_{ij} is the number of circuits added in right-of-way $i–j$, γ_{ij} is the susceptance of the new circuit $i–j$, θ_j is the voltage angle of node j, $\mathbf{S}$ is the branch-node transposed incidence matrix, $\mathbf{f}$ is the vector of power flow, $\mathbf{d}$ is the vector of liquid demand, $\mathbf{g}$ is the generation vector, $\bar{\mathbf{g}}$ is the vector of maximum generation capacity, v is the total cost of the system for an investment proposal n_{ij}, $\mathbf{r}$ is the vector of artificial generators, and Γ represents the set of load buses.

Problem (7.10) is a typical hard combinatorial problem prone to combinatorial explosion as the number of decision variables increases. An extra complication regards the fact that there are cases in which planning does not simply mean the reinforcement of an existing network. Sometimes one has to start from scratch, at least for parts of the network, because of the addition of new load, transition and generation buses. In certain cases, the simple addition of one or two circuits (lines or transformers) will not be enough to guarantee network connectivity: the situation is not uncommon in which entire paths have to be built to put all the pieces together in a single network. Under these circumstances, the combinatorial burden is even heavier than it would be in simpler, reinforcement-only type of problems.

For a given set of specified variables n_{ij}, Problem Eq. (7.10) becomes a linear programming problem:

$$\min w \quad = \quad \sum_{k \in \Gamma} \alpha_k r_k$$

subject to

$$
\begin{aligned}
&\mathbf{Sf} + \mathbf{g} + \mathbf{r} = \mathbf{d} \\
&f_{ij} - \gamma_{ij}(n_{ij}^0 + n_{ij})(\theta_i - \theta_j) = 0 \\
&|f_{ij}| \le (n_{ij}^0 + n_{ij})\bar{f}_{ij} \\
&\mathbf{0} \le \mathbf{g} \le \bar{\mathbf{g}} \\
&\mathbf{0} \le \mathbf{r} \le \mathbf{d} \\
&f_{ij} \text{ and } \theta_j \text{ unbounded,}
\end{aligned}
\tag{7.11}
$$

which is solved for testing the adequacy of a candidate solution; adequacy is indicated by zero load shedding. Notice that Problem Eq. (7.10) is always feasible due to the presence of the load shedding factor $\sum_{k \in \Gamma} \alpha_k r_k$ in the objective function; thus, whenever a tentative solution set n_{ij} is inadequate, feasibility is achieved by the use of artificial generators (which represents the load shedding). It has been observed that this feature facilitates the optimization process to escape from local minima by temporarily moving through regions of inadequate solutions but still keeping the feasibility of the mathematical problem.

7.5.1 Problem Coding

A coding for this problem was suggested in Ref. 10, where only the integer variables (the number of circuits that can be added in a right-of-way) are coded; the real variables, such as the angle voltages, being determined from the solution of the linear program formulated in Eq. (7.11).

A configuration is then characterized by an n-vector, where n is the number of right-of-ways where new circuit additions are allowed. Figure 7.8 shows the initial configuration for the 6-bus network (the complete data set can be found in [8, 9]). New circuits can be added to all right-of-ways (a total of 15 right-of-ways); the number of additions per right-of-way is not limited. The coding for this configuration shown in Fig. 7.9, clearly indicates that no circuit additions have been carried out so far. All other relevant variables, including the loss shedding ($w = 545$ MW), are given in Fig. 7.8.

7.5.2 Determination of the Initial Solution

Of course, one can always determine the initial solution randomly; in this case, the elements of the coding vector are randomly selected in the range $[0, n_{ij}^{max}]$, where n_{ij}^{max} is the maximum number of additions in right-of-way i–j. This, however, normally generates a solution with an excessive number of added circuits. To cope with this problem, the allowed addition can be set, for example to 20% of the total

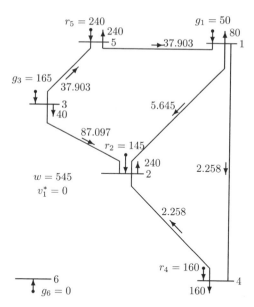

FIGURE 7.8 Basic configuration of the 6-bus network (Garver).

number of right-of-ways. Another alternative consists in initializing the configuration vector with no circuit additions; that is, assuming that initially only the already existing circuits are present (a solution with zero investment cost). A constructive heuristic algorithm can also be used to determine a high-quality initial solution. For example, the Garver's algorithm based on the transportation model can be used to generate such initial configuration [9]. Because the transportation model takes into account only Kirchhoff's current law and not the voltage law, this solution is a solution to a relaxed problem, which can be a good initialization for the DC power flow model used in the formulation of Eq. (7.10).

The evaluation of the cost of a trial configuration in the network expansion planning problem is more complicated than it is in the traveling salesman problem. This is due to the need to run a linear program in order to determine the corresponding load shedding, which forms part of the cost function. This type of difficulty is common in problems with multiple constraints on both integer and real variables such as the transmission network expansion problem.

The cost associated with the initial configuration of Fig. 7.10 for $\alpha = 5$ can be found as follows:

$$v = v_1 + v_2 = 0 + 5(545) = 2725,$$

	1-2	1-3	1-4	1-5	1-6	2-3	2-4	2-5	2-6	3-4	3-5	3-6	4-5	4-6	5-6
$n_o =$	0	0	0	0	0	0	0	0	0	0	0	0	0	0	0

FIGURE 7.9 Coding of topology shown in Figure 7.8.

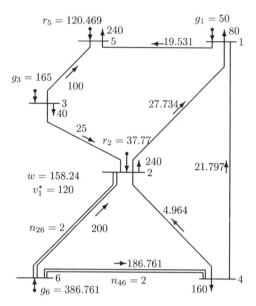

FIGURE 7.10   A topology with load shedding.

where the load shedding of $w = 545$ MW is determined from the solution of a linear program as in Eq. (7.11). As for the topology shown in Fig. 7.10, the cost is given by:

$$v = v_1 + v_2 = 4(30) + 5(158.24) = 911.2.$$

7.5.3   Neighborhood Structure

For the coding described above, there are a number of different ways of defining the structure of the neighborhood of the current solution. A simple alternative consists in adapting the method of Ref. 10, which has been used in connection with the application of SA to the knapsack multiconstraint problem. This technique has been extended to the network expansion planning problem in Ref. 6, where the neighborhood structure is formed by the configurations that result from the current configuration by (a) adding a new circuit, (b) removing a previously added circuit, and (c) by swapping two circuits (adding a new one and removing another one that has been added before).

Once the neighborhood is defined, a neighbor is randomly chosen and the acceptance test is carried out to check whether the candidate configuration will become the new current configuration or not. The technique proposed in Ref. 10 goes as follows:

1. While the current topology presents load shedding, the generation of neighbors is made in the order addition–swap–removal, where the swapping and the

	1-2	1-3	1-4	1-5	1-6	2-3	2-4	2-5	2-6	3-4	3-5	3-6	4-5	4-6	5-6
$n_1 =$	0	0	0	0	0	0	0	0	2	0	0	0	0	2	0

FIGURE 7.11 Coding of topology shown in Figure 7.10.

removal of circuits is performed only when the configuration that results from the addition of a new circuit does not pass the acceptance test.

2. When the current topology does not present load shedding, the generation of neighbors follows the order removal–swap–addition, where the swap or addition of circuits is tried only if the removal fails.

For example, consider that the current topology is the one of Fig. 7.10, which is a solution with load shedding and whose coding is as shown in Fig. 7.11.

A circuit addition is randomly chosen, for example the circuit 2–6. The coding of the new topology is as shown in Fig. 7.12.

7.5.4 Variation of the Objective Function

Most of the computational effort spent in the network expansion planning problem is concentrated on the acceptance test where the solution of the linear program formulated in Eq. (7.11) is required. This is a contrast with what happens in the traveling salesman problem where the feasibility is automatically guaranteed by the coding that has been adopted.

Consider for example that the current topology for the network expansion problem is the one of Fig. 7.10 and that the trial configuration is obtained after the addition of one circuit in 2–6. The cost associated with topology n_2 is given by

$$v = v_1 + v_2 = 5(30) + 5(99.84) = 649.2,$$

where the load shedding $w = 99.84$ was obtained from a LP (Linear Programming) solution. The cost variation is then given by

$$\Delta v = 649.2 - 911.2 = -262.0.$$

The resulting configuration also presents load shedding, although its total cost (the cost of investment plus the penalties associated with the load shedding) is less than

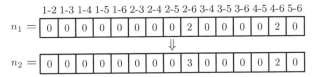

FIGURE 7.12 Generation of a neighbor topology.

the current topology, which means that the new topology is accepted and becomes the new current solution.

7.5.5 Cooling Schedule

Consider that, for example, one wishes to accept $\phi = 40\%$ of the uphill moves with costs up to $\mu = 30\%$ worse than that of the initial configuration, which is of $v_1 = 2725$ (the initial configuration is shown in Fig. 7.8). In this case, T_0 is given by (7.4):

$$T_0 = \frac{0.30}{-\ln(0.40)}(2725) = 892.2.$$

The number of trials at each temperature level is assumed to remain constant throughout the cooling process, that is, $N_{k+1} = N_0$, where N_0 is normally found according to the number of right-of-ways where new circuit additions are allowed. For example, for $n = 15$, one can choose $N_0 = 3n = 3(15) = 45$ trials.

The rate of temperature reduction β can be assumed to be constant as well, for example, $\beta = 0.90$. Hence, for updating the temperature, one simply uses:

$$T_{k+1} = 0.90 T_k.$$

The stopping criterion takes into consideration the number of LP solutions performed during the optimization process. Hence, the search stops when either (1) the maximum number of LP solutions is reached (1200 LPs in the case of the 6-bus network), or (2) the incumbent solution does not change for the last k_p trials, where k_p is a parameter set by the user ($k_p = 200$ in the case of the 6-bus network).

A simulated annealing algorithm with the parameters defined above was able to solve the network expansion planning problem of Fig. 7.8 leading to the optimal solution with $v = 200$ with additions in $n_{2-6} = 4$, $n_{3-5} = 1$, and $n_{4-6} = 2$.

7.6 PARALLEL SIMULATED ANNEALING

Notwithstanding all its good characteristics, one has to recognize that simulated annealing is extremely greedy regarding computation time requirements. To cope with this limitation, parallelization has been suggested [1, 11]. Parallelization of simulated annealing, however, is not a simple task. The main difficulty is the fact that simulated annealing works as a Markov chain, which is a typical sequential entity: a Markov chain models the simulated annealing process as a sequence of trials where the probability of the outcome of a given trial depends only on the outcome of the current trial (the current configuration, solution, or topology), and does not depend on the trials in the sequence that came before the current trial. Hence, the parallelization should be carried out in such a way that this sequential structure is not affected when the simulated annealing algorithm is mapped onto a parallel architecture.

In the simulated annealing algorithm, N_k trials are performed in each temperature level T_k. Each trial involves the following steps:

1. From the neighborhood of the current configuration, select a trial configuration.
2. Calculate the difference between the costs of the current solution and that of the trial configuration.
3. Decide whether the trial configuration should be accepted or not.
4. Replace the current configuration by the new one if this is the case.

The following remarks should be considered in performing parallelization:

1. The three first steps are independent and can be performed in parallel without affecting the sequential nature of the algorithm. Running step 4 in parallel, however, would alter the sequential nature of the algorithm.
2. The number of times the steps 1 through 3 are executed does not vary with the temperature level, although the frequency with which the step is performed varies as the temperature is reduced.
3. The computations involved in these four steps are highly problem dependent. For example, in the traveling salesman problem, the computations involved in step 2 are much smaller than those involved in the same step in the network expansion planning problem, as in the second case a LP solution is required.

7.6.1 Division Algorithm

Let np denote the number of processors in the parallel machine. A simple, effective way of parallelizing the SA algorithm consists in dividing the effort in generating the corresponding Markov chain over the np processors: each processor performing N_k/n_p trials. In order to keep the main characteristic of the SA algorithm, at each temperature level, when all processors finish processing their individual tasks, the incumbent optimal solutions are sent to the master node (node $p = 0$), which then selects the best one and broadcasts the results to all other processors ($p = 1, \ldots,$ $np - 1$).

The communication requirements of this algorithm are relatively small, which means that the potential for efficiency is high. The quality of the solution depends on the number of processors used in the parallel computation. For larger numbers of processors, the number of configurations studied by each processors can become too small, which may not allow the system to reach thermal equilibrium. To cope with this problem, the number of trials performed per temperature level can be increased, which is then compensated by an increase in the cooling rate (a smaller parameter β is used in this case).

Figure 7.13 illustrates the workings of the division algorithm, according to Ref. 1. There are eight processors in the figure. The vertical column illustrates

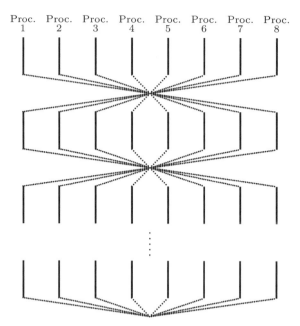

FIGURE 7.13   Division algorithm.

the working of each processor, which runs a number of trials per temperature level, and in a determined moment (when they finish processing the tasks assigned to them), they communicate to the master processor in order to receive the new global incumbent, and then they restart from there.

Remarks

When node-0 receives the partial solutions from the other processors, it also checks for thermal equilibrium conditions, as in the sequential algorithm (it verifies whether the solutions arriving from processors $p = 1, \ldots, np - 1$ coincide). This condition is normally satisfied before reaching the minimum specified temperature level, which means a faster convergence for the parallel algorithm.

7.6.2   Clustering Algorithm

In this case, in contrast with the division algorithm discussed above, the sequential nature of the SA algorithm is strictly observed [1, 7], as the np processors evaluate the N_k trials in a cooperative way, which means that all processors always work with the same current solution. Thus, whenever one processor accepts a new incumbent, it is communicated to all the other processors. This parallelization approach is less efficient at high-temperature levels where the frequency with which new current solutions are accepted is relatively high. The opposite happens at lower temperature levels where very few acceptances are performed and thus the algorithm presents best performance.

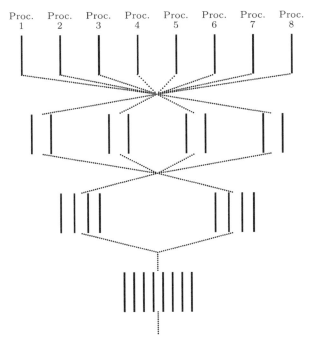

FIGURE 7.14 Hybrid division/clustering algorithm.

A good alternative consists in using a hybrid division/clustering algorithm. The process is started as a division algorithm and then switches to the clustering algorithm, which is executed for lower temperature levels. The point at which the switching to the clustering algorithm occurs is based on the observed acceptance rate, as discussed above. The way the hybrid algorithm works is graphically illustrated in Fig. 7.14.

Remarks

In Ref. 7, both versions of the parallel simulated annealing algorithm have been implemented and tested. Normally, the solutions obtained by the parallel versions have better quality when compared with the ones given by the sequential algorithm.

7.7 APPLICATIONS OF SIMULATED ANNEALING

The initial applications of simulated annealing to operation research problems occurred in the late 1980s (traveling salesman problem, quadratic assignment problem, etc.). The first important application in engineering was in very-large-scale integration (VLSI) placement and was carried out by Kirkpatrick, one of the creators of the SA technique. In the 1990s, the number of applications in engineering increased substantially, and two important books were published [1, 4]. Additional reference of theory and applications of simulated annealing can be found in Refs. [12–17]. A survey made by the authors of this paper based on the database of the Institute for Scientific Information (ISI) revealed 2500 publications about simulated annealing and its applications to optimization problems in technical journals only. The unconstrained search about simulated annealing revealed about 6000 publications on the topic. As for applications to power system problems, according to the ISI database, about 104 papers have been published in them specialized literature up to April 2005, most of them in journals such as the *IEEE Transactions on Power Systems* (39), *IEEE Transactions on Power Delivery* (12), *IEE Proceedings Generation, Transmission and Distribution* (20), *Electric Power Systems Research* (15), and *International Journal of Electrical Power & Energy Systems* (18). (Certainly, the list of applications presented in the following is not complete and is biased by the experience of the authors in this field.) SA has been extensively used to solve a variety of power systems problems. One of the most explored problems is unit commitment; a hybrid SA approach has been used in Refs. [18–20]. Another complex problem that has been tackled by SA is the transmission network expansion problem; both sequential and parallel versions have been developed [6–8, 11]. Other applications include the optimal allocation of capacitors in primary distribution feeders [21], the optimal reconfiguration of distribution networks [22], phase balancing [23], and reactive planning [24, 25]. Finally, an excellent reference to meta-heuristic methods in operation research problems can be found in Ref. 26.

7.8 CONCLUSIONS

This chapter presented the basic facts about simulated annealing, a widely used Monte Carlo type iterative technique for solving combinatorial optimization problems. SA tries to emulate the process of cooling a metal and the consequent formation of crystals. SA has the ability to escape from local optimum by performing uphill moves, that is, moves that deteriorate the current value of the objective function. As the cooling process progresses, however, the probability of accepting such uphill moves is reduced, and the SA algorithm becomes greedier regarding the value of the objective function. It can be proved that under proper cooling conditions, a global optimal solution can be reached. Notwithstanding this fact, the simulated annealing algorithm is very greedy regarding computation time requirements. This characteristic has motivated the development of parallel implementations of the simulated annealing algorithm aiming to speed up simulation as well as improve solution quality.

REFERENCES

1. Aarts E, Korst J. Simulated annealing and Boltzmann machines. New York: John Wiley & Sons; 1989.
2. Kirkpatrick S, Gelatt CD, Jr, Vecchi M. Optimization by simulated annealing. Science 1983; 220(4598):498–516.

3. Cerny V. Thermodynamical approach to the Traveling Salesman Problem: An efficient simulation algorithm. J Optim Theory Appl 1985; 45(1):41–51.

4. Van Laarhoven PJM, Aarts EH. Simulated annealing: theory and applications. Holland: Dordrecht, D. Reidel Publishing Company; 1987.

5. Reeves CR. Modern heuristic techniques for combinatorial problems. New York: John Wiley & Sons; 1993.

6. Romero R, Gallego RA, Monticelli A. Transmission system expansion planning by simulated annealing. IEEE Trans Power Systems 1996; 11(1):364–369.

7. Gallego RA, Alves AB, Monticelli A, Romero R. Parallel simulated annealing applied to long term transmission network expansion planning. IEEE Trans Power Systems 1997; 12(1):181–188.

8. Gallego RA, Monticelli A, Romero R. Comparative studies of non-convex optimization methods for transmission network expansion planning. IEEE Trans Power Systems 1998; 13(3):822–828.

9. Garver LL. Transmission network estimation using linear programming. IEEE Trans Power App Syst 1970; PAS-89:1688–1697.

10. Drexl A. A simulated annealing approach to the multiconstraint zero-one knapsack problem. Computing 1988; 40:1–8.

11. Gallego RA. Planejamento a Longo Prazo de Sistemas de Transmissão Usando Técnicas de Otimização Combinatorial. *PhD thesis, UNICAMP*; 1997.

12. Hajek B. Cooling schedules for optimal annealing. Math Operations Res 1988; 13:311–329.

13. Sait SM, Youssef H. Iterative computer algorithms with applications in engineering. IEEE Computer Society Press, Los Alamitos, CA, USA; 1999.

14. Aarts EHL, Korst JHM, Van Laarhoven PJM. A quantitative analysis of the simulated annealing algorithm: A case study for the traveling salesman problem. J Statistical Phys 1988; 50(1–2):187–206.

15. Ram DJ, Sreenivas TH, Subramaniam KG. Parallel simulated annealing algorithms. J Parallel Distributed Comput 1996; 37(2):207–212.

16. Osman IH. Heuristics for the generalized assignment problem—simulated annealing and tabu search approaches. OR Spektrum 1995; 17(4):211–225.

17. Koulmas C, Antony SR, Jaen R. A survey of simulated annealing applications to operations research problems. Omega Int J Management Sci 1994; 22(1):41–56.

18. Wong KP, Wong YW. Combined genetic algorithms/simulated annealing/fuzzy set approach to short-term generation scheduling with take-or-pay fuel contract. IEEE Trans Power Systems 1996; 11(1):128–135.

19. Mantawy AH, Abdel-Magid YL, Selim SZ. Integrating genetic algorithms, tabu search and simulated annealing for the unit commitment problem. IEEE Trans Power Systems 1999; 14(3):829–836.

20. Zhuang F, Galiana FD. Unit commitment by simulated annealing. IEEE Trans Power Systems 1990; 5(1):311–318.

21. Chiang HD, Wang JC, Cockings O, Shin HD. Optimal capacitor placement in distribution systems. Part I: A new formulation and the overall problem. IEEE Trans Power Delivery 1990; 5(2):634–642.

22. Chiang HD, Jumeau RJ. Optimal network reconfigurations in distributions systems. Part I: Formulation and a solution methodology. IEEE Trans Power Delivery 1990; 5(4):1902–1909.

23. Zhu JX, Bilbro G, Chow MY. Phase balancing using simulated annealing. IEEE Trans Power Systems 1999; 14(4):1508–1513.

24. Liu CW, Jwo WS, Liu CC, Hsiao YT. A fast global optimization approach to VAR planning for the large scale electric power systems. IEEE Trans Power Systems 1997; 12(1):437–442.

25. Hsiao YT, Chiang HD. Applying network window schema and a simulated annealing technique to optimal VAR planning in large scale power systems. Int J Electrical Power Energy Systems 2000; 22(1):1–8.

26. Glover F, Kochenberger GA. Handbook of metaheuristics. Boston: Kluwer Academic Publishers; 2003.

CHAPTER 8

Fuzzy Systems

GERMANO LAMBERT-TORRES

8.1 MOTIVATION AND DEFINITIONS

8.1.1 Introduction

Typically, engineering requires the use of exact or mathematical statements. These statements correspond with precise information, such as "$x = 3.0$," "$2 \leq \theta \leq 6$," or "$y = 3t + 24$." The value of "$x = 3$" has a grade membership comparable with 100% ($=1$); for all other values (2.8, 2.9, 3.1, 3.2), the grades of membership in the solution is zero. In the case of real-world values, however, this grade of membership is not true because of the imprecision of tools, the influence of the observer, and so forth.

Additionally, information from the real world is often imprecise (i.e., inexact statements such as "the value of x is not big" or "the temperature is about 4"). Therefore, a theory to correctly express the grade of membership is desirable.

Lofti A. Zadeh [1] proposed such a theory to make a rapprochement between the precision of classic mathematics and the imprecise information from the real world. The theory is called fuzzy sets theory and works with grades of membership of x in A, that is, $\mu_A(x)$, taken from the set M, which must be a lattice structure [2]. Usually, this set M is taken in the interval [0, 1]. Goguem [3] proposed a further generalization of the theory using values of the membership taken from the set $L = (-\infty, \infty)$ or ([-1, 1] for a normalized set).

Formally, let X be a set of objects, called the universe, whose elements are denoted x, and the membership in a subset A of X is the membership function, μ_A from X to the real interval M. The set A is called a fuzzy set and is a subset of X, and $\mu_A(x)$ is the grade of membership x in A. In this way, the set A is completely characterized by the set of pairs:

$$A = \{(x, \mu_A(x)), x \in X)\}.$$

Modern Heuristic Optimization Techniques. Edited by K. Y. Lee and M. A. El-Sharkawi
Copyright © 2008 the Institute of Electrical and Electronics Engineers, Inc.

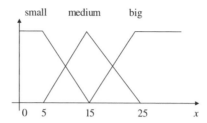

FIGURE 8.1   Examples of fuzzy terms.

When the value of $\mu_A(x)$ is closer to 1, the more x belongs to A; otherwise, when the value of $\mu_A(x)$ is closer to 0, the more x does not belong to A.

The fuzzy sets theory allows manipulation of both exact and inexact (fuzzy) statements. This is very important for load forecasting and expert system, as there are so many fuzzy factors that are difficult to characterize by a number. An instance of this could be temperature, for example: "the temperature this afternoon will be about 4." Temperature is an important factor in load forecasting, but it is not easy to characterize by an exact numerical quantity. In addition, the linguistic hedges (such as about, little, big, small, medium, large) can be modified by other linguistic hedges (such as not, very, very very) [4, 5]. For example, "value of x is not big," where *big* is a linguistic hedge and not a modifier. Figure 8.1 shows fuzzy terms.

8.1.2   Typical Actions in Fuzzy Systems

In a conventional system, input variables are continuously read and manipulated (computed); thus, a controller continuously generates values for the system output, updating the process given an answer. In a fuzzy system, the steps mentioned above are present and, in addition, two more steps are incorporated: fuzzification and defuzzification [6, 7]. The fuzzification process is the first step of the fuzzy controller and is done after reading the input variable values. Its mission is to transform "actual values" of the variables to "fuzzy values" (linguistic values), which are then manipulated by the controller. The defuzzification process is the last step of the fuzzy controller and it transforms "fuzzy values" back to "actual values." Figure 8.2 presents these steps of the control process.

A partial example of the fuzzification process is shown in the first part of Fig. 8.3, where inputs $x = 10$ and $y = 8$ generate values for the memberships "small" and

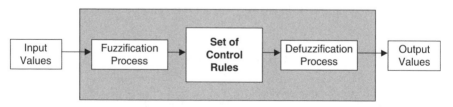

FIGURE 8.2   Fuzzy system scheme.

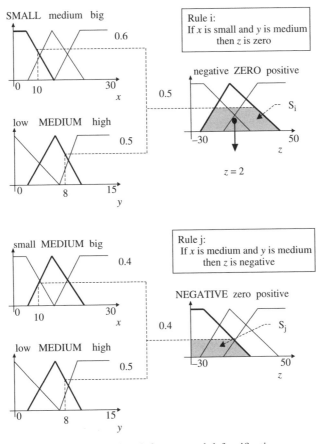

FIGURE 8.3 Fuzzification, inference, and defuzzification processes.

"medium" equal to 0.6 and 0.5, respectively. An example of defuzzification process is presented in the next item.

A fuzzy inference process executes each rule; that is, (a) the input variables are transformed into the fuzzy statements, and (b) an output value is computed. The execution of each rule is made using *modus ponens*, which means that the premise of the rule produces the degree of membership for the conclusion of the same rule. This membership degree is a function of the fuzzy values of the input variables and of the conjunctions used among them. Let's consider the example of Fig. 8.3, where the two following fuzzy control rules are shown:

Rule i: If x is small and y is medium, then z is zero.
Rule j: If x is medium and y is medium, then z is negative.

Notice that during the fuzzification process for the rule i, the values obtained are 0.6 and 0.5, as mentioned above. Because the liaison element used between the fuzzy statement is the conjunction "and" (which represents the "minimum" operator), the conclusion of the rule (value "zero") is limited by the minimum value of the premise, that is, $\min(0.6, 0.5) = 0.5$.

Each rule produces a limited area (as illustrated by the shaded area in Fig. 8.3) according to the resultant value from the premise and the output membership function of the conclusion. In the same way, for the rule j, the limitation of the value "negative" is 0.4. Of course, it is possible that the value that comes from the premise is zero. In this case, the rule has no influence on the computation of the final output value.

After the execution of all the rules, the defuzzification process begins. Each produced area is computed. Then the maximum operation is applied to define the biggest one. In Fig. 8.3, Si is bigger than Sj, that is, $\max(Si, Sj) = Si$. The final actual output value is computed using the center of gravity method of Si. The value of the abscissa found is the actual value of the output variable. (Figure 8.5 shows an example of this process, where the centroid produces a value of $z=2$.)

8.2 INTEGRATION OF FUZZY SYSTEMS WITH EVOLUTIONARY TECHNIQUES

8.2.1 Integration Types of Hybrid Systems

The artificial intelligence techniques may integrate in several forms and with several hybridization levels. For academic purposes, it is possible to divide these integrations into three main types: stand-alone systems, weak integration systems, and fused systems [8, 9]. The major difference between them comes from the number of information exchanges that happen during the problem solution.

8.2.1.1 Stand-alone Systems In this type of integration, there is no information exchange between the systems. They operate in a parallel and competitive way. This system type allows comparing the results obtained by the techniques regarding both the result quality and the processing time. An example of the use of this integration type would be the use of genetic algorithms for the optimization of load-dispatch of a set of hydrothermal power stations versus a dispatch based on numerical techniques of optimization driven by a fuzzy system.

Another type of stand-alone system is one in which one technique is preferably used. If results different to what was expected by the system are obtained (whether by the answering time or because no answer is obtained), another technique is used in sequence looking for a solution. In this way, the techniques would be acting in sequence. One example for the use of this integration type would be a search process for ways of system restoration. Initially, it is possible to use the genetic algorithms technique for the search optimization, and if it does not present a result in time, an expert system could be used to find a possible solution, even though not optimized.

8.2.1.2 Weak Integration Systems In this type of integration, the information exchange happens whether in sequence or in hierarchic form.

The system sequence form of weak interaction is quite similar to the sequence form described before; however, in this case, technique 1 provides one result that works as input data for technique 2 to continue the processing of the problem solution. This integration type is very much in use in technical literature. One example of this integration type would be the system that uses expert systems to set up operations in a substation after a blackout. Having the operations list, a system based in genetic algorithm techniques could optimize this list.

In one hierarchic system, the upper-level technique will drive the lower-level techniques aiming to have a feedback or the best possible feedback. For that purpose, it may drive one or many times the existent techniques of the lower level.

Actually, this system type is similar to the other described system, and the difference comes from the fact that the second system acts as input for the first system. One example would be a fuzzy expert system (joint fuzzy system and expert system techniques) that would act in the upper level aiming at the integration of several numeric and intelligent tools to provide the system operator with a support tool for the decision take process for contingency analysis. One modulus could be a genetic algorithm for the redispatch of the power-generating plants.

Normally, a characteristic of this integration type is the information exchange through data basis, which reduces the total processing velocity of the system solution and makes its use unfeasible for frame where the information exchange is very high.

8.2.1.3 Fused Systems In these systems, also called strongly integrated systems, the information exchange between the systems happens in a very intensive way. Therefore, for the system to have a processing time adequate to the user requirements, two integration forms may exist.

When this system type uses a hierarchic system, as the one shown before, the information is not exchanged only between the upper-level technique and the lower-level technique, but also between them. Thus, other decision levels may be set up and the information exchanged between the several techniques. Usually, this system type shows a blackboard, that is to say, a working area common to all the techniques, where the generated information required for making a determined technique is written and read.

Another integration type is the fuse. In this type, a technique is embedded into another, generating almost a new hybrid technique, it being very difficult to find the limit where one begins and the other ends. An example of this type of system could be the intelligent adaptation control of one process, where a fuse fuzzy expert system has had its pertinence functions adjusted by a genetic algorithm as a function of changes happening in the process.

8.2.2 Hybrid Systems in Evolutionary Techniques

Up to now, the major contribution of the evolutionary techniques is tight up to the capacity in optimizing a problem solution. The evolutionistic techniques look to copy the reproduction biologic system through self-organization and adaptation to some environments. A genetic system will try the conditioning of its evolution through the information coming from the environment (or a problem to be solved), adapting itself and producing new genes generations. The central idea of a genetic algorithm is to find the optimum solution for a given problem, without the need for an exhaustive search. These algorithms have been used successfully in complex problems where the search space is very large, a fact that may become nonviable with the use of conventional techniques.

8.2.3 Evolutionary Algorithms and Fuzzy Logic

The evolutionary algorithms, such as genetic algorithms or particle swarm intelligence, also have much use together with the fuzzy logic. These techniques have some common characteristics (e.g., dealing with nonlinear systems quite well) and some complementary characteristics. The fuzzy systems have the capacity of storing knowledge but have some difficulties in the process of self-learning, which may be obtained with the use of evolutionary techniques. References 10–29 present some examples of this type of integration.

(a) An example of integration of these techniques would be the support that the evolutionary algorithm could provide for a set of fuzzy rules, making adjustment of its membership functions. This could happen by representing some of the characteristics from these functions in each gene. For example, if the functions will be of triangle type, it is possible to represent them by a central value and an opening value. Thus, each function is characterized by two values.

The appraisal of the importance of each gene would be calculated by the model provided with the existing membership functions in the gene. The crossover and mutation process is changed according the representation form of each existing data in the gene. One example for the illustration of this process is found in the next section of this chapter.

(b) Another integration form would be to use fuzzy rules set to work as an evaluation function. The advantage of this procedure is the possibility for placing some heuristics into the process of solution search, driving it for a site of more interest.

(c) Another integration form would be the fusion of both techniques, building up one fuzzy genetic algorithm or fuzzy particle swarm algorithms. This is done through the representation of each gene element as being a linguistic value from a fuzzy variable. Fuzzy distances could also be used as an evaluation function in this hybrid system, and new crossing and mutation forms could be set up taking into consideration elements from the fuzzy set theory.

8.3 AN ILLUSTRATIVE EXAMPLE OF A HYBRID SYSTEM

The major aim of a software package is to park a vehicle in a garage, starting from any initial position. To go through this task, the user should begin the development of a fuzzy control set of rules and pertinence functions that will define the vehicle route. Several windows and numeric routines are available in the program for helping the users to set up such rules. The fuzzification and defuzzification processes are performed by program with no interference from the user.

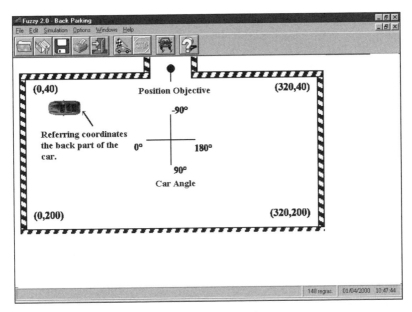

FIGURE 8.4   Program basic screen.

To represent the vehicle parking problem, the program has a basic screen that is shown in Fig. 8.4. It shows the garage position, the existing hurdles (actually the walls), and the limit coordinate values. Also, the entry variables involved in the problem are shown, namely (x, y), measured from the vehicle back central point and the car angle (ϕ).

This program basic screen also comprises a set of menus, its functions indicated by the attribute names. By starting (through keyboard or mouse) the attribute File, it will be possible to charge a set of rules and pertinence functions previously developed, to save a new set, or to start its creation. Also, it is through this attribute that there is the route for the impressions of several types and outing of the program.

8.3.1   Parking Conditions

Some conditions are set up for parking. They belong to two types: linked to the computer package and logic. The ones linked to the package are refered to as physical limitations:

- Input variables limits:
  - position (x, y): $0 < x < 32$ and $0 < y < 20$ (m) (parking limitations)
  - Vehicle angle: $-90° \leq \phi \leq 270°$
  - Vehicle direction: ahead or back
- Output variable limit:
  - Vehicle wheel angle: $-30° \leq \theta \leq 30°$ (actual model limitation)

With respect to the logic limitations, they may be of different types according to the strategy that will be used. Beside others, some examples are

- Minimizing the number of direction changes (ahead or back);
- Minimizing the vehicle travel;
- Do not use part of garage by the parking time.

The conditions for the vehicle displacement are acceleration up to 1 (m/s^2) and maximum speed of 1 (m/s). All the displacements are made taking both of these values as a reference. There are three possibilities for changing the direction of displacement, which are

- A crash against the wall: when the system detects that in the next simulation step, the vehicle will crash against the wall;
- A rule that enforces turn-back: when because of a rule the order for reversing be used, or,
- Lack of outputs: when the control does not use any rules, that is to say, if the output area were void (zero).

8.3.2 Creation of the Fuzzy Control

The user of this computer package may define a new system by creating the pertinence functions and the control rules. Figure 8.5 shows the window Create Fuzzy Sets defining the number of pertinence functions for each variable. Pertinence

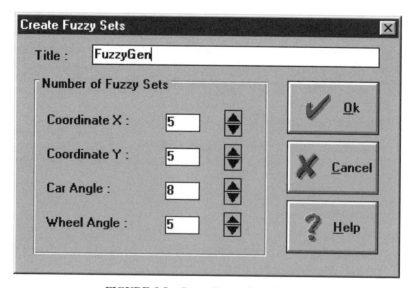

FIGURE 8.5 Create Fuzzy Sets window.

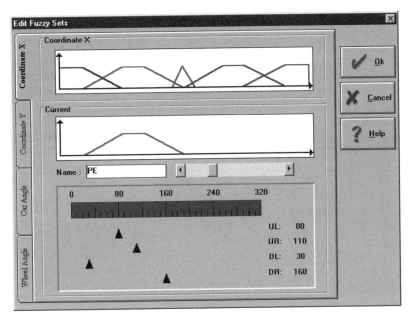

FIGURE 8.6 Edit Fuzzy Sets window.

functions are equally spaced in the variable's control surface. The user changes these pertinence functions through the window Edit Fuzzy Sets. Figure 8.6 shows this window.

The user provides the pertinence functions of the input and output variables of the program. This can be made through files that have been generated before or editing the new pertinence functions. Figure 8.6 presents an example of editing for the input variable y.

For defining the control rules, that is to say, how the functions will be grouped, there exists the window Edit Rules. Figure 8.7 presents an example of this edit window for filling up the rules output values with premise values equal to $x =$ LC, $y =$ YT, angle VE, and ahead (forward) movement. The program provides set-up for possible combinations of the pertinence function input variables, once the user fills up the rule's conclusion value. Figure 8.7 shows two regions of concern. In the first there are the possibilities for selection of the direction (ahead or back) and of the coordinate corresponding with the car angle. The second region of concern has the filling of the rule's conclusion. This can be made through the selection of one of the output values (or none of them for a nondefined rule or to reverse). So, this rule is

IF x is LC **AND** y is YT **AND** *car angle* is VE **AND**
Displacement Direction is ahead
THEN *modified angle* is NS.

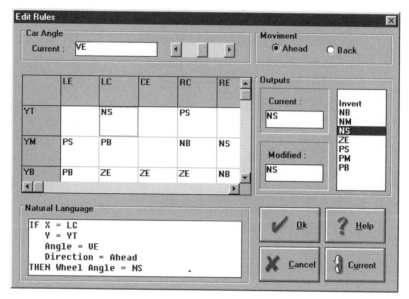

FIGURE 8.7 Edit Rules window.

8.3.3 First Simulations

The computing pack learning process used is that of trial-and-error type. The user makes the membership functions, supplies a set of rules, and performs several tests checking the control quality. It is known that this learning process (trial-and-error) may not provide the expected results because many interpretation mistakes may happen.

Figure 8.8a shows an example of control where the vehicle starts from an initial position X equal to 30, Y equal to 120, and an angle of 250°, making 449 iterations before parking. Another example is shown in Fig. 8.8b, where the vehicle starts from initial position X equal to 80, Y equal to 195, and angle of 50°, making 328 iterations before parking. If we want to reduce the iterations number (i.e. to minimize the route made by the vehicle), we have to change the rules, introduce new membership functions, or adjust the existing ones. To define the best values for membership functions by hand is difficult and takes a lot of time.

A training model using genetic algorithms was developed in this work. The control is optimized through automatic adjustment of existing membership functions, with no action required from the user. This modulus was introduced in the computing package.

8.3.4 Problem Presentation

If we want to shorten the iteration number, that is to say, to minimize the vehicle turning space, we need to change the rules or introduce new pertinence functions

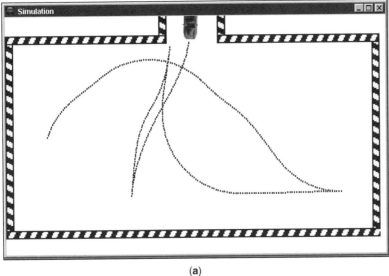

(a)

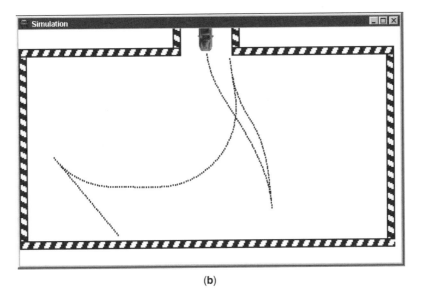

(b)

FIGURE 8.8 Simulation examples.

or adjust the existing ones. To define the best values for the pertinence functions by hand is difficult and takes a long time.

In this work, a training modulus using genetic algorithms was introduced. The control optimization is made through the automatic adjustment of the existing pertinence functions, making the user's actions no longer necessary. This modulus was incorporated into the system and has earned a new menu, the Genetic Training menu, which will be discussed in the following section.

8.3.5 Genetic Training Modulus Description

This modulus has an adjusting function for the pertinence functions, using genetic algorithms, from a control previously fostered by the user. For this purpose, we have three options from the menu Genetic Training: Starting Positions for Training, Genetic Training, and Best Results. In a general way, the integration between genetic algorithms and fuzzy control was set up as follows:

1. The chromosome was defined as being a linkage between pertinence function adjustment values.
2. The parameters are the centers and the widths of each fuzzy set. These parameters make up the chromosome's genes.
3. From a possible parameter's initial values range, the fuzzy system is rolled up to check whether it works well.
4. This information is used for set-up of the adjustment of each chromosome (adaptability) and establishing in this way a new population.
5. The cycle repetition happens until completion of the generation's number defined by the user. At every generation, the best set of values is found for the pertinence functions parameters.

8.3.6 The Option to Define the Starting Positions

For the genetic training, it is possible to define the initial positions that the vehicle will depart from to evaluate each chromosome that represents the set of values for the pertinence function's parameter, in this way looking for a control optimization not only with respect to one sole route, but also for all possible initial positions for starting the vehicle, and so the parking happens.

Through the edit window, the initial positions are edited. Thus, it is possible to define a new position, to edit an existing position, to exclude and qualify or disqualify one position so that it may or may not be used for the genetic training. This is made through the option To Use Position.

8.3.7 The Option Genetic Training

Through this option, the user makes the adjustments of the genetic training parameters. Through another window, the user may set up the adjustment value for the pertinence functions, also to define the parameters (population, generations, crossover tax, and mutation); that is, how much the function will be displaced to the right or left and how much it will shrink or expand. After start training, it is possible to follow the genetic training to each generation. For each chromosome that comprises the set of adjustment parameters of the pertinence functions, we have the number of total iterations generated for parking the vehicle, starting from all the initial positions setup.

After ending all the generations, it is possible to choose the best result found between all the generations through the button Best. This operation allows evaluating the proposed solutions for the genetic algorithms. For that purpose, the user should choose one generation and click the button Adjust.

After the adjustment, the pertinence functions are redefined according to the parameters of the chromosome corresponding with the chosen generation. The system will make the control based on these new functions.

The original pertinence functions may be restored through the option Best Options from the Genetic Training menu, clicking the Restore Original button. The mechanisms used for the generation of very good pertinence functions in the middle of fuzzy control used through the genetic algorithms will now be presented.

To describe each pertinence function of the fuzzy control introduced by the computer package, four parameters are defined. They are IE (down left), ID (down right), SE (upper left), and SD (upper right). Figure 8.6 shows the pertinence function PE parameters, variable x. For this function, the IE value is equal to 30, ID 160, SE 80, and SD 110. The adjustments of pertinence functions were defined for each of them in the following equations:

$$IE = (IE + k_i) - w_i$$
$$ID = (ID + k_i) + w_i$$
$$SE = (SE + k_i)$$
$$SD = (SD + k_i),$$

where k_i and w_i are adjustments coefficients. The k_i makes each pertinence function move to the right or left without losing its shape. The coefficient w_i makes the pertinence function shrink or expand. These coefficients take each integer negative or positive value according to the adjustment value defined by the user in the Genetics Parameters window.

Figure 8.9 shows an adjustment example in PE's function with values for IE equal to 30, ID 160, SE 80, and SD 110. With $k = -8$ and $w = 3$, the pertinence function will have the following adjustment:

$$IE' = (30 + (-8)) - 3 = 19$$
$$ID' = (160 + (-8)) + 3 = 155$$
$$SE' = (80 + (-8)) = 72$$
$$SD' = (110 + (-8)) = 102.$$

The genetic algorithms are used to find the optimum values, according the strategy and initial points used, from k_i and w_i for the pertinence functions.

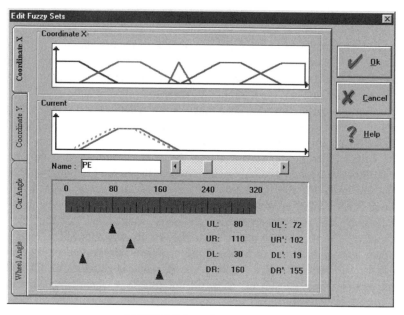

FIGURE 8.9 Adjust example.

8.3.7.1 Genetic Representation of Solutions

Usually, a possible solution for a problem is associated with a chromosome p by a vector with m positions $p = \{x_1, x_2, x_3, \ldots, x_m\}$, where each component x_i represents a gene. Among chromosomes representation, the better known ways are the binary representation and the representation by integers. The binary representation is a classic one.

However, in this paper, the representation by integers was used, because the genes of each chromosome are composed by the adjust coefficients $k_i - w_i$, which are integer values. With regard to chromosome size (i.e., how many genes each chromosome will have), this will depend on the number of membership functions defined by a user. For a fuzzy control with a group of 18 membership functions, for example, there will be chromosomes with 36 genes. This is because for each function we have two adjust coefficients: $k_i - w_i$. A 36-position vector then represents the chromosome.

8.3.7.2 Evaluation Function

This function deals with the evaluation of the aptitude level (adaptation) of each chromosome generated by algorithms. For the present problem, the aim is to minimize the vehicle-parking route. For this case, the evaluation function is given by:

$$FA = \frac{1}{1 + I},$$

where I is the total iterations up to parking, regarding the adjustment made for each chromosome in the membership function. According to this function, each chromosome aptitude will be inversely proportional to the iterations number.

8.3.7.3 Genetic Operators: Crossover and Mutation

The genetic operators came from the natural evolution principle and govern the chromosomes renewal. The genetic operators are necessary for the population diversification, and the acquired adaptation characteristics from anterior generations are preserved.

The crossover process or recombination comprises a random cut upon father chromosomes in which genes will be changed, generating two descendants. For example, let us consider two-father chromosomes $p_1 - p_2$. A crosspoint is randomly chosen. The former information to this point in one of the fathers is linked to the latter information to this point into the other father.

The mutation process consists of making the changes according to the adjustment value defined by the user in the values of one or more chromosome genes. For an adjustment value of 10, for example, the change of a selected gene will fall in some value in the interval $[-10, 10]$.

8.3.7.4 Renovation and Selection Criteria

The renovation and selection criteria introduced in this work through the algorithms were the reproduction one. Reproduction is a process in which the chromosomes copies are used for the next generation according to the evaluation function values. Chromosomes with high aptitude value will contribute one or more exactly equal descendants for the next generation.

The example of Table 8.1 shows a population where $N = 4$ with the evaluation function value of each chromosome $f_1 = f_2 = 16, f_3 = 48$, and $f_4 = 80$ into the current population. As $\Sigma f_i = 160$, the partial aptitude of each chromosome is $16/160 = 0.1$ (10%), $16/160 = 0.10$ (10%), $48/160 = 0.30$ (30%), and $80/160 = 0.50$ (50%), respectively. Starting from the partial aptitudes, the number of expected descendants from each chromosome in the next generation is calculated. In the example, for each chromosome, it is expected that $0.1 \times 4 = 0.4$, $0.1 \times 4 = 0.4$, $0.3 \times 4 = 1.2$, and $0.5 \times 0.4 = 2$, respectively.

The next generation effectively reproduced chromosomes number is given by the whole part of the expected descendants from each chromosome. Then, for the example, we have a reproduction of x_3 and two reproductions from x_4. The selection of one more chromosome for reproduction to complete the population of four chromosomes is made through proportional selection to the adjustment.

TABLE 8.1 Selection Criteria

Population	$f_i(x)$ evaluation	Partial aptitude (%)	Expected descendants	Reproduced chromosomes
x_1	16	10	0.40	0
x_2	16	10	0.40	0
x_3	48	30	1.20	1
x_4	80	50	2.00	2
Soma	160	100	4	3

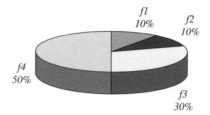

FIGURE 8.10 Roulette representation.

This process was introduced through the roulette technique, where the chromosomes that show a bigger adaptation have greater probability for being selected. Being $\Sigma f_i = 160$, the roulette for this example is represented by Fig. 8.10.

For practical purposes, the roulette is represented by vector v of M elements (ordered from $\{1, \ldots, N\}$ and one random index r, $r = 1, \ldots, M$. Then, $v(r)$ corresponds with what chromosome i was selected for. For example: With $M = 10$, for the previous example we have $7v = \{1, 2, 3, 3, 3, 4, 4, 4, 4, 4\}$. If $r = 4$, then the selected chromosome is the number 3.

Finally, the roulette turns a determinate number of times, depending on the chromosomes number necessary for the population completion, which are chosen as chromosomes that will take part in the next generation, the ones drawn by the roulette.

8.3.7.5 Stop Criteria
The stop criteria used was the one that defines the maximum number of generations to be produced. When the generation number is completed by genetic algorithms, the new populations generating process is finished, and the best solution is the one among the individuals better adapted to the evaluation function.

8.3.7.6 Algorithm Presentation
With G the generation number, P the population, P_c the crossover rate (recombination), P_m the mutation rate, and VA the allowed adjust value for membership functions, the algorithm shown below generates as a start the vector **s** with G positions. Each one of the vector elements is the best chromosome of one generation.

Step 1: Generate initial population P with genes in the interval $[-VA, +VA]$.
Step 2: Evaluate population P. Store in vector **s** the best chromosome.
Step 3: If completed in the generation number G, go to Step 13.
Step 4: Calculate the aptitude regarding population P.
Step 5: Calculate the expected descendants from population P.
Step 6: Draw descendants from population P' starting from population P.
Step 7: Compose the grouping of population P'.
Step 8: Draw the crosspoint for the father chromosomes of population P'.
Step 9: Make the crossing for population P' according to P_c.
Step 10: Make the mutation in each chromosome according to P_m.

Step 11: Evaluate the population P'. Store in vector **s** the best chromosome. Make $P = P'$.

Step 12: Go back to Step 3.

Step 13: End.

8.3.8 Tests

In this section, we present tests made with a fuzzy control, which membership functions were adjusted through genetic algorithms. These tests show those mechanisms' efficiency, providing an objective evaluation of the found results.

The originals membership functions are shown in Fig. 8.11. This control training was made starting from three initial positions as shown in Table 8.2. In this table, we also have the vehicle's number of generated iterations until parking, using the original membership functions. Figure 8.12 shows the vehicle in each starting position.

Figure 8.13 shows the courses referred to in each initial position. These positions were chosen according to the points where the vehicle does not follow a good route until parking, consequently generating an excessive number of iterations.

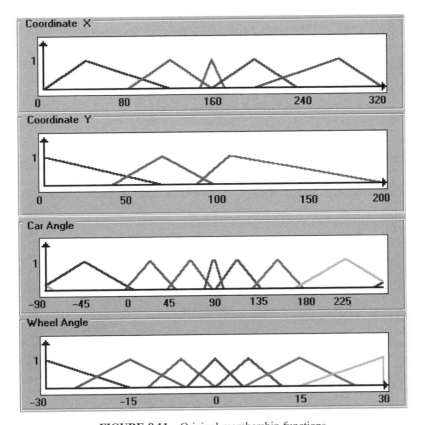

FIGURE 8.11 Original membership functions.

TABLE 8.2 Starting Positions for Training

Position	x	y	Car angle	Iterations without training
1	25	120	180	330
2	160	130	−90	888
3	275	160	−40	655

The definition of several initial positions will not only minimize the routes referred to these points but also for other points, resulting in a global minimization of traveled space. The defined genetic parameters for the training are shown in Table 8.3.

The generated results by genetic algorithms are shown in Table 8.4 and Fig. 8.14. As shown in Table 8.4, a reduction of 932 iterations (49.75%) was made for parking the vehicle starting from the initial positions set-up for training. Further, we will present simulation results made starting from initial positions not used in the training. Figure 8.15 shows the membership functions after adjustment.

Table 8.5 shows the results obtained from simulations made with 30 positions randomly chosen for the vehicle parking. The results demonstrate an average reduction of iterations number for a vehicle to reach the final position of 21% for the genetic algorithm trained control. These values represent a global reduction of the vehicle route starting from positions not used in genetic training.

It is possible to notice that in some positions, the iterations number is bigger than the ones generated by the original control (without training). In position 29 from Table 8.5, for example, the original control generates 235 iterations to park the vehicle compared with the trained control, which generates 355 iterations. This

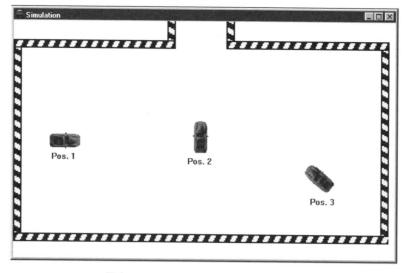

FIGURE 8.12 Initial training positions.

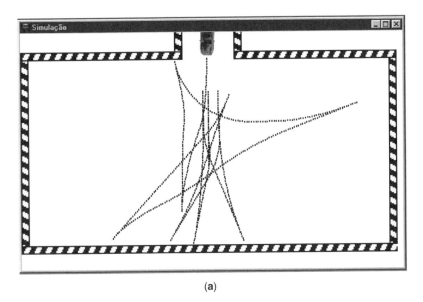

(a)

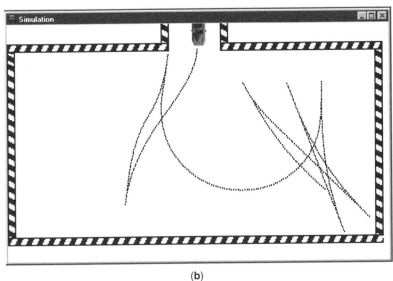

(b)

FIGURE 8.13 Routes without training: (**a**) position 2, (**b**) position 3.

TABLE 8.3 Genetic Parameters

Population size	14
Generations number	30
Crossover probability	90%
Mutation probability	1%

TABLE 8.4 Iterations After the Genetic Training

Position	Iterations without training	Iterations with training
1	330	280
2	888	384
3	655	277
Total	1873	941
Average	624.33	331.67

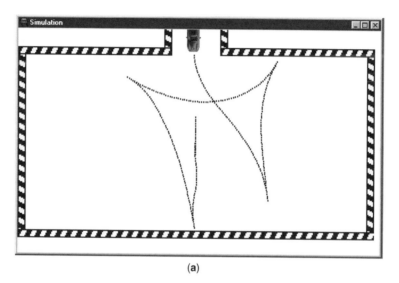

(a)

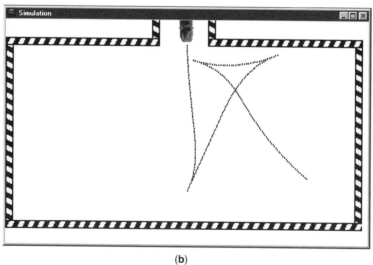

(b)

FIGURE 8.14 Routes after training: (**a**) position 2, (**b**) position 3.

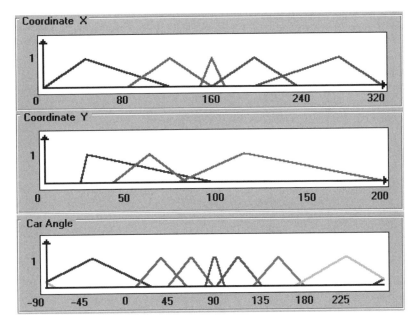

FIGURE 8.15 Membership functions after the genetic adjustment.

increase comes from the modifications made in the membership functions, which makes the vehicle change to a different route to reach the final position.

The genetic training modulus developed in this work added this program with an automatic technique for the adjustment of the membership functions parameters. This technique shows that the performance of a fuzzy control may be improved through the genetic algorithms, substituting for the trial-and-error method as used before by students for this purpose, with no good results.

The genetic algorithms provide distinctive advantages for the optimization of membership functions, resulting in a global survey, reducing the chances of ending into a local minimum, once it uses several sets of simultaneous solutions. The fuzzy logic supplied the evaluation function, a stage of the genetic algorithm where the adjustment is settled.

8.4 CONCLUSIONS

The use of fuzzy logic to solve control problems has been increasing considerably in the past years. The successfulness of fuzzy application depends on a number of parameters, such as fuzzy membership functions, which are usually decided upon subjectively. One way to improve the performance of the fuzzy reasoning model is the use of evolutionary algorithms.

The integration of fuzzy systems with one or more numeric or intelligent techniques has allowed computer systems to solve problems or find solutions that only

TABLE 8.5 Simulations Results

Position	x	y	Car angle	Iterations generated by fuzzy controls	
				Original	trained
1	1	126	182	450	329
2	6	46	132	167	154
3	8	41	190	1000	1000
4	10	187	228	453	328
5	15	70	−90	318	162
6	51	112	48	278	130
7	51	112	54	280	132
8	70	95	−40	275	261
9	74	69	190	164	164
10	76	193	232	605	363
11	88	46	44	283	305
12	115	120	0	182	280
13	120	90	45	182	156
14	131	140	−72	457	292
15	141	69	−28	342	314
16	154	166	−80	863	436
17	160	135	268	1101	545
18	161	191	178	315	286
19	173	140	−72	762	590
20	208	143	244	363	310
21	217	66	−50	684	325
22	228	194	−48	830	655
23	246	169	154	312	307
24	250	180	−40	739	800
25	265	170	−40	672	329
26	290	95	−40	280	190
27	300	124	258	317	306
28	305	156	−90	350	346
29	314	73	−46	235	355
30	314	194	−44	513	744
Total				13772	10894
Average				459.07	363.13

one of the techniques alone could not obtain. The use of techniques together allow limitations to be covered, always using the better characteristics of each technique.

REFERENCES

1. Zadeh LA. Fuzzy sets. Information Control 1965; 8:338–353.
2. Birkhoff GD. Lattice Theory. Vol. XXV, 2nd ed. Providence, RI: Colloquium Publications, 1948.

3. Goguen JA. L-fuzzy sets. J Math Anal Appl 1967; 18:145–174.

4. Zadeh LA. The concept of a linguistic variable and its applications to approximate reasoning. Information Sci 1975; 8:199–249, 301–357; 9:43–80.

5. Zadeh LA. Outline of a new approach to the analysis of complex systems and decision process. IEEE Trans 1973; SMC-3(1):28–44.

6. Tsukamoto Y. An approach to fuzzy reasoning method. In: Gupta MM, Ragade RK, Yager RR, eds. Advances in fuzzy sets theory and applications, 1st ed. New York: North-Holland; 1979. p. 137–149.

7. Lambert Torres G, Mukhedkar D. On-line fuzzy rule set generator. Advance Information Processing in Automatic Control Congress of IFAC, Nancy, France 1989.

8. Medsker LR. Hybrid intelligent systems. Boston: Kluwer Academic; 1995.

9. Ankenbrandt CA, Buckles BP, Petry FE. Scene recognition using genetic algorithms with semantic nets. Pattern Recognition Lett 1990; 11:285–293.

10. Karr CL. Genetic algorithms for fuzzy controllers. AI Expert 1991; 6(2):26–33.

11. Karr CL. Applying genetics to fuzzy logic. AI Expert 1991; 6(3):38–43.

12. Karr CL, Stanley DA. Fuzzy logic and genetic algorithms in time-varying control problems. In: Proc. NAFIPS-91. Proceedings of the North American Fuzzy Information Processing Society Workshop, Columbia University, Columbia, USA, May 15–17, 1991. p. 285–290.

13. Meredith DL, Kumar KK, Karr CL. The use of genetic algorithms in the design of fuzzy logic controllers. In: Proc. WNN-AIND 91. Proceedings of the Second Workshop on Neural Networks: Academic/Industrial/Nasa/Defense, Society of Photo Optimal, Auburn University, USA, ISBN-10:0819406422, Feb. 11–13, 1991. p. 695–702.

14. Bhandari D, Pal SK, Kundu MK. Image enhancement incorporating fuzzy fitness function in genetic algorithms. In: Proc. of the IEEE 2nd International Fuzzy Systems Conference, Vol. 2. Proceedings of IEEE 2nd International Fuzzy Systems Conference, IEEE Press, San Francisco, USA, ISBN:0-7803-0614-7, Mar. 28–Apr. 1, 1993. p. 1408–1413.

15. Feldman DS. Fuzzy network synthesis with genetic algorithm. In: Proc. of the Fifth International Conference on Genetic Algorithms. Proceedings of the Fifth International Conference on Genetic Algorithms, Morgan Kaufmann Publishers Inc., San Francisco, USA, ISBN:1-55860-299-2, July 17–22, 1993. p. 312–317.

16. Ishibuchi H, Nozaki K, Yamamoto N. Selecting fuzzy rules by genetic algorithm for classification problems. In: Proc. of the IEEE 2nd International Fuzzy System Conference, Vol. II. Proceedings of IEEE 2nd International Fuzzy Systems Conference, IEEE Press, San Francisco, USA, ISBN:0-7803-0614-7, Mar. 28–Apr. 1, 1993. p. 1119–1124.

17. Lee MA, Takagi H. Dynamic control of genetic algorithms using fuzzy logic techniques. Proceedings of the Fifth International Conference on Genetic Algorithms, Morgan Kaufmann Publishers Inc., San Francisco, USA, ISBN:1-55860-299-2, July 17–22, 1993. p. 76–83.

18. Park D, Kandel A, Langholz G. Genetic-based new fuzzy reasoning models with application to fuzzy control. IEEE Trans. SMC 1994; 24(1):39–47.

19. Ishibuchi H et al. Selecting fuzzy if-then rules by a genetic method. Electronics Commun Japan 1994; 77(2pt 3):94–104.

20. Barczak CL, Martin CA, Krambeck CP. Experiments in fuzzy control using genetic algorithms. In: Proc. of the IEEE International Symposium on Industrial Electronics. IEEE Press, Santiago, Chile, 1994. p. 426–428.

21. Brezdek JC, Hathaway RJ. Optimization of fuzzy clustering criteria using genetic algorithms. In: Proc. of the First IEEE Conference on Evolutionary Computation. IEEE Press, Orlando, USA, 1994. p. 589–594.

22. Xu HY, Vukovich G. Fuzzy evolutionary algorithms and automatic robot trajectory generation. In: Proc. of the First IEEE Conference on Evolutionary Computation. IEEE Press, Orlando, USA, 1994. p. 595–600.

23. Kim J, Moon Y, Zeigler BP. Designing fuzzy net controllers using GA optimization. In: Proc. of the IEEE Symposium on Computer-Aided Control Systems Design. IEEE Press, Tucson, USA, 1994. p. 83–88.

24. Karr CL, Gentry EJ. Cases in geno-fuzzy control. In: Ward TL, Ralston PAS, eds. Intelligent Control of Engineering Machines and Processes. New York: Dekker; 1995.

25. Hongesombut K, Mitani Y, Tsuji K. An automated approach to optimize power system damping controllers using hierarchical genetic algorithms. Proc. of IEEE PES Summer Meeting. 1999.

26. Hsiao Y-T. et al. Enhancement of restoration service in distribution system using combination fuzzy-GA method. IEEE Trans Power Systems 2000; 15:1394–1400.

27. Yalcinoz T, Altun H, Uzam M. Economic dispatch solution using a genetic algorithm based on arithmetic crossover. Proc. of the IEEE Porto Power Tech Conference. 2001.

28. Damousis IG et al. A fuzzy logic system for calculation of the interference of overhead transmission lines on buried pipelines. Electric Power Systems Res 2001; 57:105–113.

29. Lu J, Nehrir MH, Pierre DA. A fuzzy logic-based self tuning power system stabilizer optimized with a genetic algorithm. Electric Power Systems Res 2001; 60:77–83.

CHAPTER 9

Differential Evolution, an Alternative Approach to Evolutionary Algorithm

KIT PO WONG and ZHAOYANG DONG

9.1 INTRODUCTION

As a relatively new population-based optimization technique, differential evolution has been attracting increasing attention for a wide variety of engineering applications including power engineering. Unlike the conventional evolutionary algorithms that depend on predefined probability distribution function for mutation process, differential evolution uses the differences of randomly sampled pairs of objective vectors for its mutation process. Consequently, the object vectors' differences will pass the objective functions topographical information toward the optimization process and therefore provide more efficient global optimization capability. This chapter aims at providing an overview of differential evolution and presenting it as an alternative to evolutionary algorithms.

Differential evolution was first proposed over 1994–1996 by Storn and Price at Berkeley as a new evolutionary algorithm (EA) [1, 2]. Differential evolution (DE) is a stochastic direct search optimization method. It is generally considered as an accurate, reasonably fast, and robust optimization method. The main advantages of DE are its simplicity and therefore easy use in solving optimization problems requiring a minimization process with real-valued and multimodal (multiple local optima) objective functions. DE uses a nonuniform crossover that makes use of child vector parameters to guide through the minimization process. The mutation operation with DE is performed by arithmetical combinations of individuals rather than perturbing the genes in individuals with small probability compared with one of the most popular EAs, genetic algorithms (GAs). Another main characteristic of DE is its ability to search with floating point representation instead of binary representation as used in many basic EAs such as GAs. The characteristics together with other factors of DE make it a fast and robust algorithm as an alternative to EA, and it

Modern Heuristic Optimization Techniques. Edited by K. Y. Lee and M. A. El-Sharkawi

has found an increasing application in a number of engineering areas including power engineering.

This chapter aims at introducing differential evolution as an alternative approach to evolutionary algorithm. In order to learn the advances of differential evolution, it is necessary to first review the general evolutionary algorithms.

The chapter is organized as follows. After the introduction, we will review evolutionary algorithms in Section 9.2 The fundamentals of differential evolution are detailed in Sections 9.3 and 9.4. Conclusions are given in Section 9.6 at the end of the chapter.

9.2 EVOLUTIONARY ALGORITHMS

Evolutionary algorithm (EA; also known as evolutionary computation) includes stochastic search and optimization algorithms originated from natural evolution principles. EAs were proposed more than 40 years ago and include three mainstream algorithms, genetic algorithms (GAs), evolutionary programming (EP), and evolutionary strategies (ES) [3–9]. These algorithms are robust, adaptive, and have found their application in a wide variety of theoretical and practical problems involving search and optimization tasks. Different from traditional calculus-based optimization techniques, evolutionary algorithms are based on a population of encoded tentative solutions that are processed with some evolutionary operators to find a good acceptable solution if not the global optimum one. The search and optimization process follows the principle of the survival of the fittest to generate successively better results over generations to finally approximate the optimal solutions. These three well-established mainstream EAs are further discussed and compared for completeness in the sequel [10].

9.2.1 Basic EAs

In the implementation of GAs, a population of candidate solutions, which are referred to as chromosomes, evolves to satisfactory solutions, approximating the global solution in finite generations through genetic operators of reproduction, mutation, and crossover [7].

Initially, EP was proposed for evolution of the finite state machines for prediction purpose [5]. It differs from the fundamental GA mainly in two aspects. First, EP employs real values of parameters for evolution rather than the encoded parameters as the fundamental GA does; second, EP relies more on mutation and selection rather than crossover as GA does. However, with the advances in encoding techniques for GA, it has been increasingly flexible in handling parameter values to best suit the needs of the optimization problem. Consequently, the encoding difference between GA and EP is diminishing.

A simple ES is based on population size and a positive temperature that decreases to zero as in simulated annealing (SA) with the evolutionary process. The search capability of ES can be enhanced through controlling the dynamics of mutation and selection.

Among the many factors affecting the robustness and effectiveness of the evolutionary algorithms, population size and the total number of generations are among the most important ones. When the population size is increased more than its minimum size, the computation time for the evolution process may increase if the number of generations required for obtaining the optimal solutions cannot be reduced sufficiently. For GAs, the minimum population size can be reduced by forming high-quality chromosomes in a population [8]. A balance has to be made so as to explore the search space with the minimum number of individuals in order to save computational costs. The program flow chart of a typical GA is given in Fig. 9.1 [7].

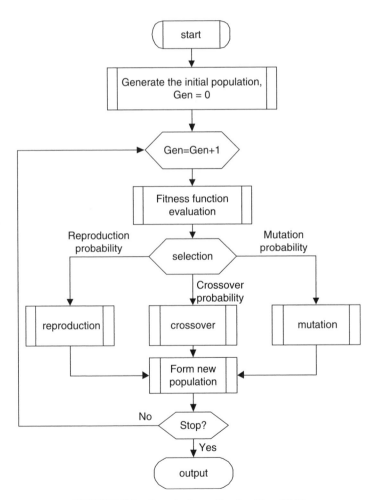

FIGURE 9.1 A typical genetic algorithm (GA).

Given the characteristics and advantages of evolutionary algorithms over the traditional optimization methods, they have become more and more popular in solving complex, nonlinear, nondifferentiable, nonconvex, and dynamically interactive engineering and science optimization problems. Over recent years, successful and attempted application of evolutionary algorithms include the areas of power systems engineering, such as system planning, security assessment, decision making, and electricity market management and control. Evolutionary algorithms have also been used to solve sophisticated control systems and communications engineering problems, such as system identification, linearization, and optimal and robust control. Other successful applications include very-large-scale integration (VLSI) design, planning and decision making, neural network, and other learning algorithms also employ evolutionary algorithms to enhance its learning and approximation capability.

It has been noted that recent advances in evolutionary algorithm research attempt to develop algorithms that can learn the structure of the solution space of an optimization problem and exploit this knowledge to search for a global optimum. Other advances include various techniques to enhance the search capability via controlling population size, mutation probability, and creating virtual populations [8]. The continuous advances will open up a wider horizon of areas where evolutionary computation will be applicable. Among these advances, the research work of Refs. 8 and 11 introduced new acceleration techniques to a basic GA. It is reviewed briefly because of its uniqueness.

9.2.2 Virtual Population-Based Acceleration Techniques

In Ref. 8, virtual population-based GA acceleration schemes were proposed to formulate high-quality chromosomes in a population. The virtual population consists of the current population of candidate chromosomes and a number of new populations derived from the current population. This enables sufficient diversity in the population to be formed in an early stage of the evolution process and effectively prevents premature convergence. This virtual population-based solution acceleration technique explores the acceleration techniques used for solving power system load flow problems with an iterative Gauss–Siedel method. Two solution acceleration techniques, numerical and analytical, can be used to form the virtual population.

The first scheme is through numerical solution acceleration as shown in (9.1).

$$\vec{\mathbf{V}}_{\text{acc}}^{G+1} = \vec{\mathbf{V}}_{\text{best}}^{G} + \lambda(\vec{\mathbf{V}}_{\text{best}}^{G} - \vec{\mathbf{V}}^{G}), \tag{9.1}$$

where $\vec{\mathbf{V}}_{\text{best}}^{G}$ is the best solution in current population G and is assumed closest to the global optimum, $\vec{\mathbf{V}}^{G}$ is a candidate solution to the current population, and $\vec{\mathbf{V}}_{\text{acc}}^{G+1}$ is the accelerated $\vec{\mathbf{V}}^{G}$. λ is the acceleration factor and $\lambda \in [0, 1]$. This process is shown in Fig. 9.2.

In addition to the above numerical acceleration technique, the analytical technique can also be used. The authors in Ref. [8] gave such an analytical technique as shown

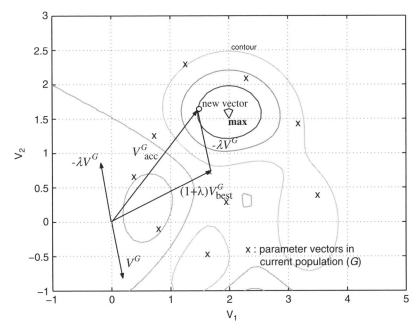

FIGURE 9.2 Numerical solution acceleration method for the virtual GA [8, 12].

in (9.2) for solving an unconstrained minimization problem of $\min f(v_1, v_2, \ldots, v_N)$,

$$\vec{\mathbf{V}}_{\text{acc}}^{G+1} = \vec{\mathbf{V}}^{G} + \mu\vec{\mathbf{U}}[(f(\vec{\mathbf{V}}_{\text{best}}^{G}) - f(\vec{\mathbf{V}}^{G})], \tag{9.2}$$

where $\vec{\mathbf{V}}^{G} = [v_1, v_2, \ldots, v_N]^{T}$, $\vec{\mathbf{U}} = [1, 1, \ldots, 1]^{T}$, $\vec{\mathbf{V}}_{\text{best}}^{G}$ is the best candidate solution in the current population G, and μ is a matrix of derivatives obtained by (9.3) provided that $f(v_1, v_2, \ldots, v_N)$ is differentiable at the point of evaluation

$$\mu = \begin{bmatrix} \left(\dfrac{\partial f}{\partial v_1}\right)^{-1} & 0 & \cdots & 0 \\ 0 & \left(\dfrac{\partial f}{\partial v_2}\right)^{-1} & \cdots & 0 \\ \vdots & \vdots & \ddots & \vdots \\ 0 & 0 & \cdots & \left(\dfrac{\partial f}{\partial v_N}\right)^{-1} \end{bmatrix}. \tag{9.3}$$

It can be seen that μ is similar to the inverse of the Jacobian matrix for solving $\min f(v_1, v_2, \ldots, v_N)$.

The virtual population-based acceleration method has proved to be able to effectively improve the convergence characteristics of GAs through benchmark test functions and power flow problems in Refs. 8 and 11.

Given all the characteristics of EAs, they have been used on many occasions as generic global optimization tools. However, various levels of performance of different EAs on different specific optimization problems often leads to argument on their suitability as a universal generic optimization tool. This is mainly because the evolutionary algorithms selected need to be problem specific in order to have an optimal performance. EAs as global optimizers need to face the challenges of peculiarities of various objective functions to avoid degrading performances. This generally may not be an easy task and introduces the advantages of differential evolution as an alternative to EAs.

9.3 DIFFERENTIAL EVOLUTION

The DE was presented as a heuristic optimization method that can be used to minimize nonlinear and nondifferentiable continuous space functions with real-valued parameters by Storn and Price in Ref. 1. The most important characteristic of DE is that it uses the differences of randomly sampled pairs of object vectors to guide the mutation operation instead of using probability distribution functions as other EAs. The distribution of the differences between randomly sampled object vectors is determined by the distribution of these object vectors. Because the distribution of the object vectors is mainly determined by the corresponding objective function's topography, the biases where DE tries to optimize the problem match those of the function to be optimized. This enables DE to function robustly and more as a generic global optimizer than other EAs.

According to Price [13], the main advantages of a DE include

- Fast and simple for application and modification
- Effective global optimization capability
- Parallel processing nature
- Operating on floating point format with high precision
- Efficient algorithm without sorting or matrix multiplication
- Self-referential mutation operation
- Effective on integer, discrete, and mixed parameter optimization
- Ability to handle nondifferentiable, noisy, and/or time-dependent objective functions
- Operates on flat surfaces
- Ability to provide multiple solutions in a single run and effective in nonlinear constraint optimization problems with penalty functions

In the following sections, we will look into the specific operators and mathematical algorithms within a DE.

9.3.1 Function Optimization Formulation

An EA is an efficient method to solve optimization problems. It is necessary to briefly discuss the problem formulation for an optimization problem. Because EA is basically used to solve minimization problems, only such problems are discussed here. Maximization problems can be easily converted into minimization ones.

A general constrained minimization problem can de described as

$$\text{Min } \mathbf{y} = \mathbf{f}(\mathbf{x}) \tag{9.4}$$

s.t.

$$\mathbf{g}(\mathbf{x}) \le 0, \tag{9.5}$$

where $\mathbf{x} = \{x_i\}$, $(i = 1, 2, \ldots, n)$, is the vector of variables where an optimal value is to be calculated to minimize $\mathbf{y}$; $\mathbf{f}(\mathbf{x}) = \{f_j(\mathbf{x})\}$, $(j = 1, 2, \ldots, m_1)$, is the vector of objective functions, and $\mathbf{g}(\mathbf{x}) = \{g_k(\mathbf{x})\}$, $(k = 1, 2, \ldots, m_2)$, is the vector of equality and inequality constraints. Let $m = m_1 + m_2$ and assume that $\mathbf{x}$ is represented as a vector of floating numbers.

This minimization problem can be further represented as

$$\text{Min } \mathbf{F}(\mathbf{x}), \tag{9.6}$$

where $\mathbf{F}(\mathbf{x})$ incorporates the objective function $\mathbf{f}(\mathbf{x})$ and the constraint function $\mathbf{g}(\mathbf{x})$. There are different ways to convert (9.4)–(9.5) into (9.6), among which weighted sum or minmax are two popular approaches as shown in (9.7)–(9.8) with the weighting factors $w_i > 0$

$$\mathbf{F}(\mathbf{x}) = \sum_{j=1}^{n} w_j f_j(\mathbf{x}) + \sum_{k=1}^{m} w_k g_k(\mathbf{x}) \tag{9.7}$$

$$\mathbf{F}(\mathbf{x}) = \max_{j=1:n, k=1:m} [w_j f_j(\mathbf{x}), w_k g_k(\mathbf{x})]. \tag{9.8}$$

For (9.6) and (9.7), the local and global minima can be calculated if the region of realizability of $\mathbf{x}$ is convex [1, 13, 14]. For (9.6) and (9.8), all local and multiple global minima (if any) can theoretically be located [1, 13]. For power system problems, the optimization problem may not always be convex and therefore requires heuristic algorithms such as DE to solve it.

Depending on the specific problem to be optimized, the weighting factor can have different practical meanings. For example, in risk management for power system planning, the weighting factor can represent the relative importance of each objective and constraint; it can also represent the probability of occurrence for future scenarios while considering their impact on the planning options.

9.3.2 DE Fundamentals

As a member of the EA family, DE also relies on the initial population generation, mutation, recombination, and selection through repeated generations until the termination criteria is met. The fundamentals of DE are introduced accordingly in the sequel.

9.3.2.1 Initial Population   DE is a parallel direct search method using a population of N parameter vectors for each generation. At generation G, the population P^G is composed of x_i^G, $i = 1, 2, \ldots, N$. The initial population P^{G0} can be chosen randomly under uniform probability distribution if there is nothing known about the problem to be optimized:

$$x_i^G = x_{i(L)} + \text{rand}_i[0, 1] \cdot (x_{i(H)} - x_{i(L)}), \tag{9.9}$$

where $x_{i(L)}$ and $x_{i(H)}$ are the lower and higher boundaries of d-dimensional vector $x_i = \{x_{j,i}\} = \{x_{1,i}, x_{2,i}, \ldots, x_{d,i}\}^T$. If some *a priori* knowledge is available about the problem, the preliminary solution can be included to the initial population by adding normally distributed random deviations to the nominal solution.

9.3.2.2 Mutation and Recombination to Create New Vectors   The key characteristic of a DE is the way it generates trial parameter vectors throughout the generations. A weighted difference vector between two individuals is added to a third individual to form a new parameter vector. The newly generated vector will be evaluated by the objective function. The value of the corresponding objective function will be compared with a predetermined individual. If the newly generated parameter vector has lower objective function value, it will replace the predetermined parameter vector. The best parameter vector is evaluated for every generation in order to track the progress made throughout the minimization process. The random deviations of DE are generated by the search distance and direction information from the population. Correspondingly, this adaptive approach is associated with the normally fast convergence properties of a DE.

For each parent parameter vector, DE generates a candidate child vector based on the distance of two other parameter vectors. For each dimension $j \in [1, d]$, this process is shown in (9.10) as is referred to as scheme DE 1 by Storn and Price [1]:

$$x' = x_{r3}^G + F \cdot (x_{r1}^G - x_{r2}^G), \tag{9.10}$$

where the random integers $r1 \neq r2 \neq r3 \neq i$ are used as indices to index the current parent object vector. As a result, the population size N must be greater than 3. F is a real constant positive scaling factor and normally $F \in (0, 1+)$. F controls the scale of the differential variation $(x_{r1}^G - x_{r2}^G)$ (Fig. 9.3) [1, 13].

Selection of this newly generated vector is based on comparison with another DE1 control variable, the crossover constant $CR \in [0, 1]$, to ensure the search diversity. Some of the newly generated vectors will be used as child vector for the next generation, and others will remain unchanged. The process of creating new candidates is described in the pseudo-code as shown in the following [1, 13, 15–17].

9.3.2.2.1 Pseudo-code of the Creation Procedure in DE (Mutation and Recombination)   Mutation and recombination:

```
For each individual j in the population
        Generate 3 random integers, r₁, r₂ and r₃ ∈ (1, N)
        and r₁ ≠ r₂ ≠ r₃ ≠ j
```

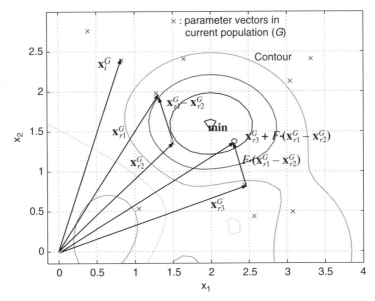

FIGURE 9.3 Vector representation of the child vector creation procedure with DE1 in (9.10) from vectors of current generation, where the dotted closed lines are the contour toward the minimal solution point.

```
Generate a random integer i_rand ∈ (1, N)
For each parameter i
    If rand(0,1)<CR or i=i_rand
        x'_i,j = x_i,r3+F· (x_i,r1 − x_i,r2)
    Else
        x'_i,j = x_i,j
    End If
End For i
End For j
```

The DE1 scheme in creating a candidate vector can be revised to include the impact from the best vector $\mathbf{x}_{\text{best}}^{G}$ of the current generation. An additional control variable λ is used in this so called DE2 scheme as shown in (9.11):

$$\mathbf{x}' = \mathbf{x}_i^G + \lambda \cdot (\mathbf{x}_{\text{best}}^G - \mathbf{x}_i^G) + F \cdot (\mathbf{x}_{r1}^G - \mathbf{x}_{r2}^G), \qquad (9.11)$$

where the current best vector through λ provides increased greediness of the search process. The contribution of the current best vector is shown in Fig. 9.4. This is useful for noncritical objective functions. The same procedure can be used to ensure the search diversity by comparing with the crossover probability CR as shown in the pseudo-code for creation procedure in DE (mutation and recombination).

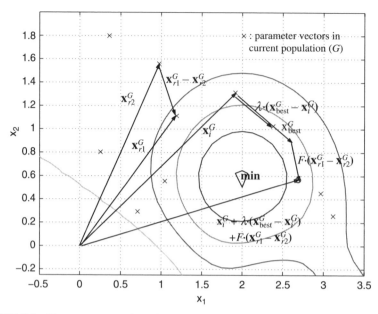

FIGURE 9.4   Vector representation of the child vector creation procedure with DE2 in (9.11) from vectors of current generation, where the dotted closed lines are the contour toward the minimal solution point.

9.3.2.3 *Selection and the Overall DE*   The selection process of DE follows the typical EA process. Each new vector $\mathbf{x}'$ is compared with $\mathbf{x}_i$. The new vector $\mathbf{x}'$ replaces $\mathbf{x}_i$ as a member of the next generation if it produces a better solution than $\mathbf{x}_i$. The procedure is given as below [1, 13, 15–17].

9.3.2.3.1 *Pseudo code of DE*   Generate and evaluate initial population:

```
Loop while the termination criteria is not met
(e.g., G < Gmax)
        for i = 1 to N
            mutation and recombination:
            Create candidate x' as in Fig. 3
            selection:
            Evaluate x'
            if (x'i is better than xiG, i.e., f(x'i) ≤ f(xiG))
                x'iG+1 = x'i
            else
                x'iG+1 = xiG
            end if
        end for
        (e.g., G=G+1)
end loop
```

9.4 KEY OPERATORS FOR DIFFERENTIAL EVOLUTION

As a member of the EA family, DE shares many common features with other EAs. In this section, some specific features of DE are analyzed in greater detail aiming at better understanding and application of the technique. These include encoding method, population, mutation, selection, and crossover operations.

9.4.1 Encoding

Although there are floating number encoding methods available, binary encoding scheme is still the most commonly used method for most EAs. Binary encoding is based on limited binary integers and therefore is disadvantaged in two major aspects: (i) limitation of ability to effectively represent variables possible of taking values of different magnitudes, where for large-valued parameters many bits may not be significant in final results; and (ii) difficulty in preserving the continuum's topology, where the scheme may not map consecutive binary integers to adjacent values of the original object variables. Efforts have been made to map the binary integers to alternative representation, which allows the adjacent objective variables to be transferred to each other by a single bit flit on the binary integers. Gary Codes approach is one of such techniques.

DE tackles the disadvantages on binary encoding by using floating point numbers to encode the parameter variables. The ordinary floating point arithmetic operations are used in DE mutation process as well (see Sections 9.4.2 and 9.4.3).

9.4.2 Mutation

The objective of mutation is to enable search diversity in the parameter space as well as to direct the existing object vectors with suitable amount of parameter variation in a way that will lead to better results at a suitable time. It keeps the search robust and explores new areas in the search domain. For real parameter optimization, mutation is the process of adding a randomly generated number to one or more parameters of an existing object vector. Zero mean probabilistic distribution should be used in generating mutation vectors. For real parameter optimization, a dynamically scalable zero mean probabilistic distribution should be used in generating the mutation vector for each parameter. This is usually achieved by self-adaptation of the mutation process in evolutionary strategies. In DE, the mutation vectors are generated by adaptively scaling and correlating the output of predefined, multivariate probability distribution. The DE mutation of a vector is achieved by adding the weighted difference of two randomly selected vectors as in (9.12):

$$\mathbf{x}_i'^{G+1} = \mathbf{x}_i^G + f_1 \cdot (\mathbf{x}_{r1}^G - \mathbf{x}_{r2}^G), \tag{9.12}$$

where G represents the Gth generation; $r1 \neq r2 \neq i$, and $r1, r2$ are randomly selected integers within the population size N, that is, $r1, r2 \in \{1, 2, \dots, N\}$. The condition that $r1$, $r2 \neq i$ is to avoid the mutation process becomes an arithmetic crossover process. Otherwise, without losing generality, if we assume $r2 = i$, we have (9.13) as

$$x_i'^{G+1} = x_i^G + f_2 \cdot (x_{r1}^G - x_i^G). \tag{9.13}$$

Clearly, (9.13) is a linear combination of two vectors. It is not a mutation but rather a arithmetic crossover. Both (9.12) and (9.13) can be used to guide the search process, however in different ways.

The mutation vector defined in (9.12) does not have any information of the original vector x_i^G. The difference defined by randomly selected $(x_{r1}^G - x_{r2}^G)$ has zero mean, and the scale factor f_1 only changes the scale without introducing bias into the search process. Population diversity is ensured with (9.12). In comparison, the crossover vector obtained in (9.13) contains the original vector x_i^G with only one addition vector x_{r1}^G. The constant scaling factor f_2 together with the vector $(x_{r1}^G - x_i^G)$ tends to move the newly generated vector $x_i'^{G+1}$ either toward or away from x_{r1}^G. As a result, a nonzero f_2 introduces a bias in searching process through (9.13).

In addition to the DE mutation defined in (9.10)–(9.11), which relies on two randomly chosen vectors, more objective vectors can be chosen to form the mutation vector as defined in the following equations [18]:

$$x_i'^{G+1} = x_{best}^G + F \cdot (x_{r1}^G - x_{r2}^G) \tag{9.14}$$

$$x_i'^{G+1} = x_i^G + F \cdot (x_{best}^G - x_i^G) + F \cdot (x_{r1}^G - x_{r2}^G) \tag{9.15}$$

$$x_i'^{G+1} = x_{best}^G + F \cdot (x_{r1}^G - x_{r2}^G) + F \cdot (x_{r3}^G - x_{r4}^G) \tag{9.16}$$

$$x_i'^{G+1} = x_1^G + F \cdot (x_{r2}^G - x_{r3}^G) + F \cdot (x_{r4}^G - x_{r5}^G) \tag{9.17}$$

$$x_i'^{G+1} = (x_{r1}^G + x_{r2}^G + x_{r3}^G)/3 + (p_2 - p_1) \cdot (x_{r1}^G - x_{r2}^G)$$
$$+ (p_3 - p_2) \cdot (x_{r2}^G - x_{r3}^G) + (p_1 - p_3) \cdot (x_{r3}^G - x_{r1}^G), \tag{9.18}$$

where F is the mutation constant, $r1 \neq r2 \neq r3 \neq r4 \neq r5 \neq i$ are the randomly chosen indices, and $r1$, $r2$, $r3$, $r4$, $r5 \in \{1, 2, \ldots, N\}$; Eq. (9.18) is the trigonometric mutation operator [19], and the values of p_i, $i = 1, 2, 3$ are obtained by

$$p_1 = \frac{|f(x_{r1}^G)|}{p'}, \quad p_2 = \frac{|f(x_{r2}^G)|}{p'}, \quad \text{and } p_3 = \frac{|f(x_{r3}^G)|}{p'},$$

with $p' = |f(x_{r1}^G)| + |f(x_{r2}^G)| + |f(x_{r3}^G)|$, where $f(x)$ is the function to be optimized. Eqs. (9.14)–(9.18) share the same principle of mutation as in (9.10)–(9.11). However, according to Ref. [19], the trigonometric mutation operation of (9.18) biases the new trial solution heavily in the direction of the best one of three individuals chosen for mutation. It is a local search operator.

According to Price, after comparison with other mutation operations, DE mutation can be considered as globally correlated and therefore effectively enhances the DE's global optimization capabilities.

9.4.3 Crossover

Crossover or recombination is the main operator for GAs and a complementary process for DE. Crossover aims at reinforcing prior successes by generating child individuals out of existing individuals or object vector parameters. The basic recombination process is a discrete recombination. The crossover constant CR is used to determine if the newly generated individual is to be recombined. Alternatively, the arithmetic crossover formula of (9.13) can be used to achieve the rotational invariance, which is otherwise difficult to achieve with the discrete recombination. The resulting expression of the mutation and crossover processes are given in (9.19) as a combination of (9.12)–(9.13),

$$\mathbf{x}_i'^{G+1} = \mathbf{x}_i^G + f_2 \cdot (\mathbf{x}_{r3}^G - \mathbf{x}_i^G) + F \cdot (\mathbf{x}_{r1}^G - \mathbf{x}_{r2}^G) \tag{9.19}$$

where the randomly generated integers $r1 \neq r2 \neq r3 \neq i$, F is the mutation constant, and f_2 controls the crossover constant. f_2 can take from $[0, 1]$ and remain constant throughout the evolution process. Generally, with a population size of $20d$ (d is the problem dimension), $F = 0.8$ and $f_2 = 0.5$ appear to be reasonably good value to start a DE process. It can be seen that $f_2 = 1$ is the discrete recombination model with $CR = 1$, and $f_2 = 0$ represents a mutation-only model. With the discrete crossover probability $CR \in [0, 1]$, starting from a lower limit of 0 for separable objective functions, a satisfactory range of CR appears to be within $0.8-1.0$.

9.4.4 Other Operators

Among other operators, population size and selection operation are the most important. DE usually employs fixed population size throughout the search process. Variable population size can also be an option with proper size control algorithms to suit different objectives of DE applications. Selection of the population size N involves conflicting objectives. In order to achieve fast computational speed, the population size should be as small as possible. However, too small N may lead to premature convergence or stagnation. For engineering problems, a population size of $20d$ will usually generate satisfactory results. Depending on the problem and available computational resources, the population size can be in a range as low as $2d$ to as high as $100d$.

DE adopts a very simple selection operator as shown in Figs. 9.3 and 9.4. A newly generated child individual $\mathbf{x}_i'$ is selected to replace the parent individual $\mathbf{x}_i^G$ only if $f(\mathbf{x}_i') \leq f(\mathbf{x}_i^G)$. This ensures that the cost of the individual and the overall population will not increase throughout the search process.

It is worth noting that most EAs closely carry out selection and crossover processes. They use either tournament selection or a deterministic selection scheme to determine which newly generated individuals are to enter the next population. Both selection processes require additional computational cost in sorting of the population at each generation.

Apart from the fundamental DE introduced so far, researchers have been continuously trying to enhance the performance of DE by incorporating different features. These advances include parallel differential algorithm [18], ant direction hybrid DE [20], DE for constrained optimization [17], Pareto-based multiobjective DE [21], and variable scaling hybrid DE [22].

Compared with other EAs, clearly the DE represented by (9.10)–(9.11) is simple. Considering other DE operators, the overall computational efficiency is high compared with most other EAs. However, it is worth noting that another recently proposed heuristic algorithm, particle swarm optimization (PSO), has a similar structure in its search mechanism [23, 24]. The PSO method can be described by the following two equations:

$$\mathbf{x}_i^{G+1} = \mathbf{x}_i^{G+1} + V_i^{G+1} \tag{9.20}$$

$$V_i^{G+1} = \omega V_i^G + c_1 r_1 \times (P_{\text{best}i}^G - \mathbf{x}_i^G) + c_2 r_2 \times (G_{\text{best}i}^G - \mathbf{x}_i^G), \tag{9.21}$$

where V_i^G is the velocity of individual i at iteration G; ω, c_1, and c_2 are weight parameters/factors; r_1 and r_2 are random numbers between 0 and 1; $\mathbf{x}_i^G$ is the position of individual i at iteration G; $P_{\text{best }i}^G$ is the best position of individual i until iteration k; and $G_{\text{best }i}^G$ is the best position of the group until iteration G. Eq. (9.20) defines the movement of each individual with the velocity defined in (9.21). Clearly, the PSO method has comparable computational efficiency with DE.

9.5 AN OPTIMIZATION EXAMPLE

An optimization example is given in this section to compare the performance of DE and GA in solving a minimization problem with respect to search diversity and solution efficiency under different parameter values. Without losing generosity, consider the following simple optimization problem,

$$\text{Min } y = x + \sin(3x) - \cos(5x), \text{ where } x \in [0, 3]. \tag{9.22}$$

A standard DE is used to solve the problem, and the performance is summarized in Table 9.1.

A GA with elitism is used to solve the same problem but converted to a maximization problem. The results of using GA are summarized in Table 9.2.

TABLE 9.1 **Solving the Optimization Problem Using a Basic DE**

CR	F	NP	No. of generations to converge	Best value	Standard deviation	Convergence time
0.4	0.9	10	35	−4.171806	0.1902	0.01
0.5	0.9	10	21	−4.171806	0.0645	0.02
0.6	0.9	10	24	−4.171806	0.0447	0.02
0.7	0.9	10	12	−4.171806	0.0854	0.01
0.8	0.9	10	27	−4.171806	0.0456	0.01
0.9	0.9	10	19	−4.171806	0.1035	0.02
0.4	0.8	10	54	−4.171806	0.0368	0.02
0.5	0.8	10	35	−4.171806	0.14	0.02
0.6	0.8	10	30	−4.171806	0.1584	0.02
0.7	0.8	10	20	−4.171806	0.2083	0.01
0.8	0.8	10	15	−4.171806	0.1577	0.02
0.9	0.8	10	20	−4.171806	0.0833	0.02
0.4	0.7	10	45	−4.171806	0.0284	0.02
0.5	0.7	10	36	−4.171806	0.0083	0.02
0.6	0.7	10	38	−4.171806	0.0444	0.02
0.7	0.7	10	18	−4.171806	0.0512	0.02
0.8	0.7	10	18	−4.171806	0.1344	0.02
0.9	0.7	10	24	−4.171806	0.1412	0.01

From Tables 9.1 and 9.2, clearly the basic DE has overall better performance than GA with elitism in the following aspects:

- Less number of generations to obtain the final solution.
- More search diversity as shown in the standard deviation of the last generation of individuals/solution vectors.

TABLE 9.2 **Solving the Optimization Problem Using a GA with Elitism**

P_c crossover probability	P_m mutation probability	NP population size	No. of generations to converge	Best value	Standard deviation	Convergence time
0	0	50	11	4.1695	0.0402	0.8112
0.1	0	50	52	3.9829	0	0.701
0.01	0	50	36	4.142	0	0.691
0.02	0	50	20	4.119	0	0.681
0.03	0	50	55	4.0916	0	0.6209
0.04	0	50	50	4.1665	0	0.691
0.05	0	50	62	4.1709	0	0.711
0.06	0	50	66	4.1154	0	0.7311
0.07	0	50	47	4.1024	0	0.701
0	0	10	24	4.1352	0	0.1903
0	0	20	47	4.14	0	0.3505

• Generally smaller size of population is required to obtain better result than GA of larger population size; this means significant savings of computational costs especially when solving large-scale optimization problems.

9.6 CONCLUSIONS

Differential evolution is an efficient heuristic algorithm for search and optimization. DE operates on floating point representation of variables to be optimized. Like other evolutionary algorithms, DE is capable of handling nonconvex, nondifferentiable complex optimization problems. The main advantages of a DE come from its simple but effective mutation process to ensure the search diversity as well as to enhance the search effectiveness with the information from the objective function directly. This chapter compared DE with other evolutionary algorithms to present DE as an alternative for EA. It should be noted that DE has been used in solving a variety of engineering optimization problems and is attracting more and more interest from scientists and engineers looking for an alternative optimization technique in many areas.

ACKNOWLEDGMENTS

This work was supported in part by the Department of Electrical Engineering, Hong Kong Polytechnic University, Hong Kong, and School of Information Technology and Electrical Engineering, The University of Queensland, St. Lucia, Australia.

REFERENCES

1. Storn R, Price K. Differential evolution—a simple and efficient adaptive scheme for global optimization over continuous spaces. Technical Report TR-95-012, March 1995. Available at ftp.ICSI.Berkeley.edu/pub/techreports/1995/tr-95-012.ps.Z.
2. Storn R, Price K. Minimizing the real functions of the ICEC'96 contest by differential evolution. Proc. of IEEE Int. Conf. on Evolutionary Computation, Nagoya, Japan, 1996.
3. Fraser AS. Simulation of genetic systems by automatic digital computers. Aust. J. Biol. Sci. 1957; 10:484–491.
4. Bremermann HJ. Optimisation through evolution and recombination. In: Yovits MC, Jacobi GF, Goldstine GD, eds. Self-organizing systems. Washington, DC: Spartan; 1962. p. 93–106.
5. Fogel LJ, Owens AJ, Walsh MJ. Artificial intelligence through simulated evolution. New York: Wiley; 1966.
6. Holland J. Adaptation in natural and artificial systems. Ann Arbor: University of Michigan Press; 1975.
7. Goldberg D. Genetic algorithms in search, optimisation and machine learning. Reading, MA: Addison-Wesley; 1989.

8. Wong KP, Li A. Virtual population and acceleration techniques for evolutionary power flow calculation in power systems. In: Sarker R, Mohammadian M, Yao X, eds. Evolutionary Optimisation. Boston: Kluwer Academic Publishers; 2002. p. 329–345.

9. Bäck T. Evolutionary algorithms in theory and practice: evolution strategies, evolutionary programming, genetic algorithms. Oxford: Oxford University Press; 1996.

10. Wong KP, Dong ZY. Special issue on evolutionary computation for systems and control applications. Int J Systems Sci 2004; 35(13–14):729–730.

11. Wong KP, Li A, Law MY. An advanced genetic-algorithm power flow method. IEE Proc. Genr. Transm. Distrib. 1999; 146(6):609–616.

12. Wong KP, Li A. A technique for improving the convergence characteristic of genetic algorithms and its application to a genetic-based load flow algorithm. In: Kim JH, Yao X, Furuhasi T, eds. Simulated evolution and learning. Lecture notes in artificial intelligence 1295. IEEE Press; 1997. p. 167–176.

13. Price K. An introduction to differential evolution. In: Corne D, Dorigo M, Glover F, eds. New ideas in optimization. New York: McGraw-Hill; 1999. p. 79–108.

14. Brayton H, Hachtel G, Sangiovanni-Vincentelli A. A survey of optimization techniques for integrated circuit design. IEEE Proc. 1981; 69:1334–1362.

15. Price K. Differential evolution: a fast and simple numerical optimizer. In: Smith M, Lee M, Keller J, Yen J, eds. 1996 Biennial Conference of the North American Fuzzy Information Processing Society. New York: IEEE Press; p. 524–527.

16. Price KV. Differential evolution vs. the functions of the 2nd ICEO. In: Proc. IEEE International Conference on Evolutionary Computation, IEEE Press; 13–16 April 1997, p. 153–157.

17. Becerra RL, Coello Coello CA. Culturizing differential evolution for constrained optimization. In: Proc. of the Fifth Mexican International Conference in Computer Science (ENC'04), IEEE Press; 2004. p. 304–311.

18. Tasoulis DK, Pavlidis NG, Plagianakos VP, Vrahatis MN. Parallel differential evolution. Proc. Congress on Evolutionary Computation (CEC2004), Vol. 2, IEEE Press; 19–23 June 2004, p. 2023–2029.

19. Fan HY, Lampinen J. A trigonometric mutation operation to differential evolution. J Global Optim. 2003; 27:105–129.

20. Chiou J-P, Chang C-F, Su C-T. Ant direct hybrid differential evolution for solving large capacitor placement problems. IEEE Trans. Power Systems 2004; 19(4):1794–1800.

21. Xue F, Sanderson AC, Graves RJ. Pareto-based multi-objective differential evolution. In: Proc. of the 2003 Congress on Evolutionary Computation (CEC'03), Vol. 2, 8–12 Dec., 2003, Canberra, Australia, IEEE Press; p. 862–869.

22. Chiou J-P, Chang C-F, Su C-T. Variable scaling hybrid differential evolution for solving network reconfiguration of distribution systems. IEEE Trans. Power Systems. 2005; 20(2): 668–674.

23. Park J-B, Lee K-S, Shin J-R, Lee KY. A particle swarm optimization for economic dispatch with nonsmooth cost functions. IEEE Trans. Power Systems, 2005; 20(1):34–42.

24. Kennedy J, Eberhart R. Particle swarm optimization. Proc. IEEE Int. Conf. Neural Networks (ICNN'95), Vol. IV, Perth, Australia, IEEE Press; 1995. p. 1942–1948.

CHAPTER 10

Pareto Multiobjective Optimization

PATRICK N. NGATCHOU, ANAHITA ZAREI, WARREN L.J. FOX, and
MOHAMED A. EL-SHARKAWI

10.1 INTRODUCTION

Compared with single-objective (SO) optimization problems, which have a unique solution, the solution to multiobjective (MO) problems consists of sets of trade-offs between objectives. The goal of multiobjective optimization (MOO) algorithms is to generate these trade-offs [1–4]. Exploring all these trade-offs is particularly important because it provides the system designer/operator with the ability to understand and weigh the different choices available to them.

Solving MO problems has traditionally consisted of converting all objectives into a SO function. The ultimate goal is to find the solution that minimizes or maximizes this single objective while maintaining the physical constraints of the system or process. The optimization solution results in a single value that reflects a compromise between all objectives. This simple optimization process is no longer acceptable for systems with multiple conflicting objectives: System engineers may desire to know all possible optimized solutions of all objectives simultaneously. In the business world, it is known as a trade-off analysis.

It is commonplace for real-world optimization problems in business, management, and engineering to feature multiple, generally conflicting objectives. For instance, many businesses face the problem of minimizing their operating cost while maintaining a stable work force. A common problem in the integrated circuit manufacturing industry is placing the largest number of functional blocks on a chip while minimizing the area and/or power dissipation.

The planning and operation of power systems inherently requires solving multiobjective optimization problems. For instance, in environmental/economic load dispatch, the problem consists of minimizing operation cost while minimizing fossil fuel emissions and system losses [4]. Also, when designing transmission networks, the

Modern Heuristic Optimization Techniques. Edited by K. Y. Lee and M. A. El-Sharkawi

optimization of reactive resources location and sizing of the transmission and distribution system is another example of a MO problem [6, 7].

This chapter focuses on heuristic multiobjective optimization, particularly with population-based stochastic algorithms such as evolutionary algorithms. The next section presents the basic principles behind MOO, notably introducing the Pareto optimality concepts and formulation. This is followed in Section 10.3 with a presentation of the different solution approaches. A distinction is made between classic and modern techniques. The former consist of converting the MO problem into an SO problem, whereas the latter take advantage of population-based meta-heuristics to explore the entire trade-off curve. Section 10.4 addresses the problem of performance evaluation of MO optimizers.

10.2 BASIC PRINCIPLES

For illustration purposes, consider the hypothetical problem of determining, given a choice of transportation means, the most efficient of them based on distance covered in a day and energy used in the process. In a MOO framework, the two objectives are maximization of distance (or equivalently, minimization of 1/distance) and minimization of energy. The transportation modes considered are walking, bicycle, cow, car, motorcycle, horse, airplane, rocket, balloon, boat, and scooter. Figure 10.1 shows a plot of 1/distance versus consumed energy for these transportation modes. Given the same amount of food, it would be generally expected that a cow would cover a shorter distance than a horse. Also, a rocket covers a longer distance than a plane

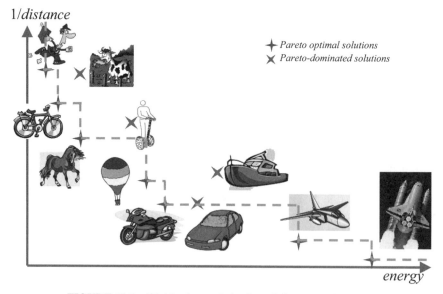

FIGURE 10.1 Biobjective optimization of distance and energy.

while consuming a higher amount of energy. The points on the dotted line represent a set of optimum solutions. Sections 10.2.1 and 10.2.2 will refer to this example.

10.2.1 Generic Formulation of MO Problems

The general MO problem requiring the optimization of N objectives may be formulated as follows:

$$\text{Minimize} \quad \vec{y} = \mathbf{F}(\vec{x}) = [f_1(\vec{x}), f_2(\vec{x}), \ldots, f_N(\vec{x})]^T$$
$$\text{subject to} \quad g_j(\vec{x}) \leq 0, j = 1, 2, \ldots, M, \tag{10.1}$$

where $\vec{x} \in \Omega$.

In Eq. (10.1), $\vec{y}$ is the objective vector, the g_j's represent the constraints, and $\vec{x}$ is the decision vector representing the decision variables within a parameter space Ω. The space spanned by the objective vectors is called the objective space. The subspace of the objective vectors that satisfies the constraints is called the feasible space.

The *utopian* solution is the solution that is optimal for all objectives.

$$\vec{x}_0{}^* \in \Omega : \forall \vec{x} \in \Omega, f_i(\vec{x}_0{}^*) \leq f_i(\vec{x})$$
$$\text{for } i \in \{1, 2, \ldots, N\}. \tag{10.2}$$

For $N = 1$, the MO problem is reduced to an SO problem. In that case, the utopian solution is simply the global optimum. It always exists, even if it cannot be found.

For the more general case where $N > 1$, the utopian solution does not generally exist because the individual objective functions are typically conflicting. Rather, there is a possibly uncountable set of solutions that represent different compromises or trade-offs between the objectives.

In the earlier example, $N = 2$ and the parameter space Ω is the set of transportation modes. The objective vector would then be:

$$\vec{y} = [f_1(x), f_2(x)] = [1/\text{distance}(x), \text{energy}(x)], x \in \Omega.$$

Maximization of distance and minimization of energy are two conflicting objectives in this example. Therefore, instead of a single solution, there exists a set of solutions that appear on the dotted line in Figure 10.1.

10.2.2 Pareto Optimality Concepts

To compare candidate solutions to the MO problems, the concepts of Pareto dominance and Pareto optimality are commonly used. These concepts were originally introduced by Francis Ysidro and then generalized by Vilfredo Pareto [3].

A solution belongs to the Pareto set if there is no other solution that can improve at least one of the objectives without degrading any other objective. Formally, a decision vector $\vec{u} \in \Omega$ is said to *Pareto-dominate* vector $\vec{v} \in \Omega$, in a minimization context, if and only if:

$$\forall i \in \{1, K, N\}, f_i(\vec{u}) \leq f_i(\vec{v}),$$
$$\text{and } \exists j \in \{1, K, N\}, f_i(\vec{u}) < f_i(\vec{v}). \tag{10.3}$$

In the context of MOO, Pareto dominance is used to compare and rank decision vectors: $\vec{u}$ dominating $\vec{v}$ in the Pareto sense means that $\vec{F}(\vec{u})$ is either better than or the same as $\vec{F}(\vec{v})$ for all objectives, and there is at least one objective function for which $\vec{F}(\vec{u})$ is strictly better than $\vec{F}(\vec{v})$. For instance, in Figure 10.1, a motorcycle Pareto dominates a car, because $f_1(\text{motorcycle}) = f_2(\text{car})$ (a motorcycle travels the same distance as a car), and $f_1(\text{motorcycle}) < f_2(\text{car})$ (a motorcycle consumes less energy than a car).

A solution $\vec{a}$ is said to be Pareto optimal if and only if there does not exist another solution that dominates it. In other words, solution $\vec{a}$ cannot be improved in one of the objectives without adversely affecting at least one other objective. The corresponding objective vector $\vec{F}(\vec{a})$ is called a Pareto dominant vector, or noninferior or nondominated vector. The set of all Pareto optimal solutions is called the Pareto optimal set. The corresponding objective vectors are said to be on the Pareto front. It is generally impossible to come up with an analytical expression of the Pareto front.

On Fig. 10.1, the Pareto optimal set consists of points on the dotted line. The boat, scooter, and cow do not belong to the Pareto optimal set. For instance, the scooter cannot belong to the Pareto optimal set due to the presence of another element, the horse, which improves both objectives by traveling a longer distance and consuming less energy. Similarly, the bicycle and horse both Pareto-dominate the cow, and the balloon, motorcycle, and car Pareto-dominate the boat.

In general, given a set of vectors in objective space, it is possible to determine the set of vectors corresponding with nondominated solutions by applying Pareto dominance comparison. Figure 10.2 depicts Pareto fronts for biobjective minimization and maximization problems.

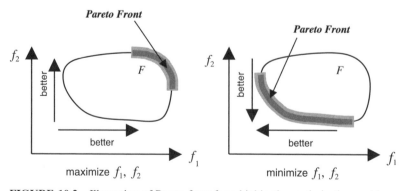

FIGURE 10.2   Illustration of Pareto front for a biobjective optimization problem.

10.2.3 Objectives of Multiobjective Optimization

MOO consists of determining all solutions to the MO problem that are optimal in the Pareto sense. In contrast with SO problems where the solution consists of a single element of the search space, the solution to MO problems consists of a set of elements of the search space. The objective of MOO is to determine the best approximation to this Pareto optimal set.

For stochastic optimizers in particular, solutions to MO problems are not always identical, and comparing them is often difficult because many criteria must be taken into account (Fig. 10.3). Solving MO problems is itself a multiobjective problem. The objectives are to

(a) Minimize the distance between the approximation set generated by the algorithm and the Pareto front;
(b) Ensure a good distribution of solutions along the approximation set (uniform if possible);
(c) Maximize the range covered by solutions along each of the objectives.

The above requirements guide the design of MOO algorithms (Section 10.3), the evaluation of their performance, and the quality of the results they generate (Section 10.4).

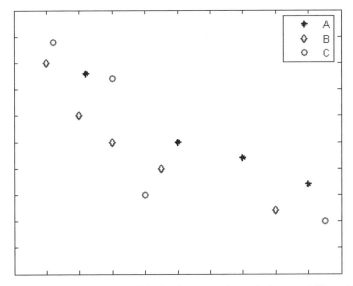

FIGURE 10.3 Illustration of the difficulty in comparing solutions to MO problems and motivating the objectives of MOO. In the objective space, we represent three hypothetical solutions of a biobjective minimization problem. The distribution of trade-off points in set A is fairly uniform, but the objective space coverage range is not good as for sets B and C. Set C covers the largest range but does not have a good distribution. Set B falls somewhere in between sets A and C. A good MOO Algorithm will try to generate an approximation set with a uniform distribution of trade-off points that at the same time covers the largest range in the objective space and is very close to the Pareto front.

10.3 SOLUTION APPROACHES

Solving MO problems has evolved over the years, and techniques or algorithms can be categorized along chronological lines. Classic approaches, which have roots in the operations research and optimization theory fields, essentially consist of converting the MO problem into a SO problem, which then can be solved using traditional scalar optimization techniques.

Classic approaches traditionally used mathematical programming (e.g., steepest descent, Newton–Raphson) to solve the SO problem. However these approaches are ill-suited for real-world problems where mixed-type decision variables, constraints, or nondifferentiable objective functions are all too common. It is only with the development of meta-heuristic algorithms and a parallel increase in computational power that these became incorporated into optimizers to tackle real-world problems.

The use of population-based meta-heuristics characterizes the second class of MO solvers, which is historically more recent. Indeed, population-based algorithms such as evolutionary algorithms, particle swarm optimization, or ant colony optimization allow direct generation of trade-off curves in a single run. Hence, the modern techniques are geared toward direct determination of the Pareto front by optimizing all the objectives concurrently.

10.3.1 Classic Methods

Classic methods were essentially techniques developed by the operations research community to address the problem of multicriteria decision making (MCDM). Given multiple objectives and preferential information about these objectives, the MO problem is converted into an SO problem by either aggregating the objective functions or optimizing the most important objective and treating the others as constraints. The SO problem can then be solved using traditional scalar-valued optimization techniques. These techniques are geared toward finding a single solution and are ideal for cases where preferential information about the objectives is known *a priori*. Using either ranking the objectives in order of importance, or target optimal values for each objective, the goal is to find the solution that best satisfies the criteria and additional information (preferences) provided by the Decision Maker (DM). In order to generate a trade-off curve, the solution procedure has to be reapplied after modifying the aggregation modalities or design criteria.

In the general case, and in order to generate an approximation to the nondominated front, all that is needed is to modify the aggregation parameters and solve the newly created SO problem. Below are some examples.

10.3.1.1 Weighted Aggregation   A simple and still very popular method is the weighted aggregation method. This is a special case of the utility function method, which converts the MO problem into an SO problem by applying a function operator to the objective vector. This function is designed by the DM to capture their preferences [2]. The utility function used in the case of weighted aggregation is a linear combination of the objectives:

$$\text{Minimize } Z = \sum_{j=1}^{N} w_j f_j(\vec{x}) \quad \text{with } w_j \geq 0 \text{ and } \sum_{j=1}^{N} w_j = 1, \qquad (10.4)$$

where the weights (w_j's) are chosen to reflect the relative importance the DM attaches to the objectives and must be specified for each of the N objectives *a priori*. Solving equation (10.4) yields a single solution that represents the best compromise given the chosen weights. To obtain an approximation set, the search has to be repeated with different values for the weights, which is computationally efficient.

Conventional weighted aggregation misses concave portions of the Pareto front. To alleviate this problem, a variant of this method called dynamic weighted aggregation (DWA) was developed. In DWA, the weights are incrementally and systematically modified using periodic functions [8, 9]. This method is unfortunately limited to biobjective problems.

10.3.1.2 *Goal Programming*

A closely related variation of the weighted aggregation technique is the goal programming or goal attainment technique. This method seeks to minimize deviation from prespecified goals. Equation (10.5) is a common formulation:

$$\text{Minimize } Z = \sum_{j=1}^{N} w_j |f_j(\vec{x_j}) - T_j|, \qquad (10.5)$$

where T_j represents the target or goal set by the DM for the jth objective function, and the w_j's now capture the priorities [2]. As in the weighted aggregation approach, the main drawback of this approach is the need for *a priori* information.

10.3.1.3 *ε-Constraint*

This is a method designed to discover Pareto optimal solutions based on optimization of one objective while treating the other objectives as constraints bound by some allowable range ε_i. The problem is repeatedly solved for different values of ε_i to generate the entire Pareto set:

Step 1: Minimize $f_k(\vec{x})$, $\vec{x} \in \Omega$ subject to $f_i(\vec{x}) \leq \varepsilon_i$ and $g_j(\vec{x}) \leq 0$; $i = 1$, K, N
($i \neq k$), and $j = 1$, K, M

Step 2: Repeat step 1 for different values of ε_i.

This is a relatively simple technique, yet it is computationally intensive. Furthermore, the solutions found are not necessarily globally nondominated [10].

10.3.1.4 *Discussion on Classic Methods*

Classic methods attempt to ease the decision-making process by incorporating *a priori* preferential information from the DM and are geared toward finding the single solution representing the best compromise solution. To examine all possible trade-offs requires systematic variation of the aggregation parameters before solving the problem, which makes the approach inefficient although simple to implement. There are further limitations: It is difficult to control the diversity of solutions in the approximation set, and more importantly, the techniques are sensitive to the shape of the global Pareto front [3, 10].

10.3.2 Intelligent Methods

10.3.2.1 Background   Meta-heuristic algorithms are increasingly used to tackle optimization problems. Meta-heuristic algorithms are successful SO solvers because they allow handling of real-world optimization problems for which, by virtue of their complexity (ill-defined function, nondifferentiability, etc.), it is impossible to find exact algorithms. For these types of problems, meta-heuristics are a practical way to generate acceptable solutions, even though they cannot guarantee optimality. Another advantage is the ability to incorporate problem-specific knowledge to improve the quality of the solutions. In the context of MOO, they offer greater flexibility for the DM, mainly in cases where no *a priori* information is available as is the case for most real-world MO problems. Hence, they were first hybridized with the mathematical programming techniques in the classic solvers. Several variants of meta-heuristic algorithms exist: single-point stochastic search algorithms such as simulated annealing, or population-based algorithms such as evolutionary algorithms of which genetic algorithms are a special class (Fig. 10.4).

The techniques presented here are mainly population-based algorithms, which are set apart by their ability to generate an approximation set in a single run. In population-based algorithms, several candidate solutions are evaluated in a single iteration. This is the characteristic that makes them suitable for MOO and mainly define the techniques presented in this section. When applied to MO problems, population-based algorithms can generate an approximation of the Pareto front in a single iteration [3, 11–13].

10.3.2.2 Structure of Population-Based MOO Solvers   The general structure of EA-based MO solvers is similar to the one used for SOO (Fig. 10.5). The different steps are now adapted so as to meet the objectives of MOO that were outlined in Section 10.2.3. *Fitness assignment* controls convergence (i.e., how to guide the population to nondominated solutions). To prevent premature convergence to a region of the front, *diversity* mechanisms such as niching are included in the determination of an individual's fitness. Diversity also allows one to control the spread of the approximation set. Finally, a form of *elitism* is applied to prevent the deterioration problem whereby nondominated solutions may disappear from one generation to the next. Once fitness of individuals is computed, selection and variation can be performed.

10.3.2.2.1 Fitness Assignment   There are three methods of fitness assignment: aggregation-based, criterion-based, and Pareto-based. Aggregation-based assignment

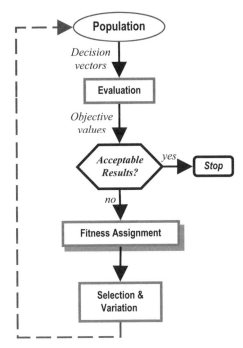

FIGURE 10.4 Generic structure of evolutionary algorithms. Evolutionary computing emulates the biological evolution process. A population of individuals representing different solutions is iteratively evolving to find the optimal solutions. At each iteration, the fittest individuals are chosen, and then variation operations (mutation and crossover) are applied, thus yielding a new generation (offspring).

consists in evaluating the fitness of each individual based on a weighted aggregation of the objectives. In this sense, it is an extension of the classic weighted aggregation method presented in Section 10.3.1. An early implementation of this technique is Haleja & Lin's genetic algorithm (HLGA). To explore the different parts of the Pareto front, they apply systematic variation of the aggregation weights [11].

An example of criterion-based assignment is Schaffer's vector-evaluated genetic algorithm (VEGA) [14], which is a non–Pareto-based technique that differs from the conventional genetic algorithm only in the way in which the selection step is performed. At each generation, the population is divided into as many equal-size subgroups as there are objectives, and the fittest individuals for each objective function are selected (Fig. 10.6).

The VEGA algorithm is easy to implement. However, it suffers from the speciation problem (i.e., evolution of species that excel in one of the objectives). This causes the algorithm to fail to generate compromise solutions (i.e., solutions that are not necessarily the best in one objective but are optimal in the Pareto sense). In addition, the algorithm is sensitive to the shape of the Pareto front.

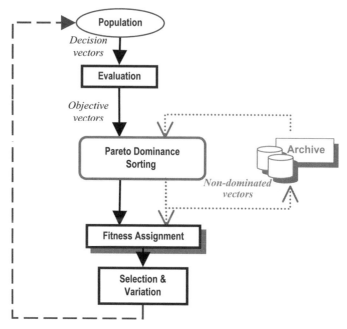

FIGURE 10.5 Pareto front generation using population-based techniques.

Pareto-based fitness assignment is the most popular and efficient technique. Here, Pareto-dominance is explicitly applied in order to determine the probability of replication of an individual. Given a population, a simple method is to find the set of nondominated individuals. These are assigned the highest rank and eliminated from further contention. The process is then repeated with the remaining individuals until the entire population is ranked and assigned a fitness value proportional to the

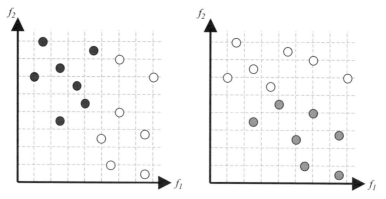

FIGURE 10.6 Illustration of VEGA approach for a biobjective minimization problem. A population of 14 candidate solutions is represented in object space. On the left, the individuals selected according to f_1; on the right, individuals selected according to f_2.

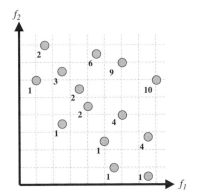

FIGURE 10.7 Illustration of fitness computation for MOGA in a biobjective minimization problem. The rank of a given individual corresponds with 1 plus the number of individuals by which it is dominated. Nondominated individuals have rank 1. The number by each individual is their rank.

ranks. The *multiobjective genetic algorithm* (MOGA) is an algorithm implementing Pareto-based fitness assignment [15]. Each individual is assigned a rank equal to 1 plus the number of individuals by which that individual is dominated. Hence, nondominated individuals have rank equal to 1. Fitness values are then assigned to the ranked population with the lowest rank having the highest fitness (Fig. 10.7).

10.3.2.2.2 Diversity In conjunction with fitness assignment mechanism, an appropriate niching mechanism is necessary to prevent the algorithm from converging to a single region of the Pareto front [10]. Niching methods are inspired from statistical sampling techniques [11]. A popular niching technique called sharing consists of regulating the density of solutions in the hyperspace spanned by either the objective vector or the decision vector. Sharing is often used in the computation of the fitness value. For example, in the MOGA algorithm discussed earlier, an objective space density-based fitness sharing is applied after population ranking. This allows differentiating individuals having identical ranks and favoring those that are in sparsely occupied regions of the objective space.

10.3.2.2.3 Elitism In EA-based solvers, an elitist strategy refers to a mechanism by which the fittest individuals found during the evolutionary search are always copied to the next generation. This increases exploitation, helps convergence, and allows the algorithm to concurrently search multiple local optima.

In MOO, elitism has been proved to be a requirement for convergence to the true Pareto front. The idea is that at each generation, the nondominated solutions found in the population are compared with an external set of nondominated solutions found throughout the entire search process. One of the early algorithms to explicitly incorporate elitism is the *strength pareto evolutionary algorithm* (SPEA) originally proposed in [16]. In SPEA, a repository or external archive is used to maintain nondominated solutions and is updated at each generation if better nondominated solutions are found. In modern MOO algorithms, the use of an external repository has become standard.

Elitism is intertwined with the diversity and fitness assignment mechanisms. If using a finite-size external repository, diversity criteria govern the insertion/deletion of nondominated solutions in the full archive. Also, elements from this archive may be used in fitness assignment and selection to create the next generation of individuals.

10.3.2.3 Common Population-Based MO Algorithms
This section presents some popular population-based MO algorithms. Most of them are based on EAs, but others are inspired from swarm intelligence. The list is by no means exhaustive and is not claimed to be the state-of-the-art MO algorithms. However, it showcases the use of different approaches for incorporating fitness assignment, diversity, and elitism.

The *nondominated sorting genetic algorithm* (NSGA) uses a layered classification technique [17]. All nondominated individuals are assigned the same fitness value and sharing is applied in the decision variable space. The process is repeated for the remainder of the population with a progressively lower fitness value assigned to the nondominated individuals. Because of the iterative ranking process, NSGA is computationally expensive. Furthermore, it does not have any elitism mechanism. NSGA-II, an improved version of NSGA, was introduced to address these issues. The elitist strategy employed in NSGA-II is peculiar in that it does not involve an external repository. Rather, the best offspring and best parents are combined. It uses nondominated ranking where the number of elements a particular solution dominates and is dominated by is taken into account, as well as a local neighborhood crowding for diversity [18] (Fig. 10.8).

In the *niched Pareto genetic algorithm* (NPGA) [19], instead of bilateral direct comparison, two individuals are compared with respect to a comparison set

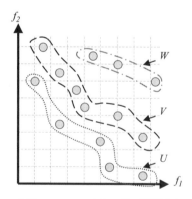

FIGURE 10.8 Illustration of fitness computation for NSGA in a biobjective minimization problem. A layered classification technique is used whereby the population is incrementally sorted using Pareto dominance. Individuals in set U have the same fitness value, which is higher than the fitness of individuals in set V, which in turn are superior to individuals in set W.

(usually 10% of the entire population). When one candidate is dominated by the set while the other is not, the latter is selected. If neither or both the candidates are dominated, fitness sharing is used to decide selection. NPGA introduces a new variable (size of the comparison set) but is computationally faster than the previous techniques as the selection step is applied only to a subset of the population.

As mentioned earlier, in SPEA, elitism is applied through an external archive that not only maintains the nondominated solutions found during the evolutionary search but also seeds the next generation [16]. Fitness values are assigned to both population and archive members. A MOGA-style fitness assignment is applied to archive members: the fitness, or strength, of each archive member is equal to the number of population members that are dominated by it, divided by the population size plus one. As for population members, their fitness is one added to the sum of the strength values of all archive members that dominate it. During selection, population and archive members are merged, and individuals and tournament selection is performed favoring individuals with *lowest* fitness to be selected. Hence, archive members have the highest chance of appearing in the next generation.

SPEA2 was developed to address some of the major drawbacks of SPEA [20]. First, in SPEA, the fitness assignment procedure often results in different individuals having the same fitness. Second, diversity is not explicitly taken into account outside of the archive management. This becomes especially problematic when there are more than two objectives. Finally, SPEA's archive management is such that the size of the archive varies during the optimization, and often solutions at the archive's boundaries are lost. SPEA2 uses a finer-grain fitness assignment procedure that takes into account the number of individuals a population member dominates in addition to the number of archive members that dominates it. Furthermore, a neighborhood distance-based density estimation term is added to the fitness computation. Finally, the archive update is performed so that the number of elements in the archive remains constant, and solutions at the boundaries of the space spanned by the approximation set are maintained.

The *Pareto archive evolution strategy* (PAES) also uses an external archive. The role of the archive is limited to storing the nondominated solutions and providing a source of comparison for ranking candidate solutions. However, archive members are not involved in the mutation and crossover procedures. The interesting feature of this algorithm is in the *adaptive grid* at the heart of its diversity and niching mechanisms. The objective space is divided into a recursive manner, thus creating a multidimensional coordinate system, or grid, over the objective space. A crowding-based fitness sharing mechanism is applied by first determining the location of solutions within this grid and estimating the density of solutions per cell. Individuals corresponding with solutions in the less crowded cells have higher fitness [21].

Other population-based approaches have also been applied to MO problems. Most notable is particle swarm optimization (PSO), a simple and robust optimizer motivated by social behavior such as bird flocking. In PSO, the individuals (called particles) fly around the search space. During their flight, each particle adjusts its position and velocity based on its own experience (personal best) and the group's experience (global best) [22, 23]. Several PSO-based MO solvers have been proposed. One example is the *multiobjective particle swarm optimizer* (MOPSO). In MOPSO, Pareto-dominance is used to update each particle's personal best. A new global best is selected at each PSO epoch from the set of nondominated solutions using a selection mechanism inspired by PAES's adaptive grid procedure. MOPSO is able to promote exploration of the less densely populated regions of the objective space while maintaining a uniform distribution [24].

10.3.2.4 Discussion on Modern Methods   The meta-heuristic techniques successfully address the limitations of the classic approaches. They can generate multiple solutions in a single run, and the optimization can be performed without *a priori* information about the objectives' relative importance. These techniques can handle problems with incommensurable or mixed-type objectives. They are not susceptible to the shape of the Pareto front. Their main drawback is performance degradation as the number of objectives increases, as there do not exist computationally efficient methods to perform Pareto ranking. Furthermore, they require additional parameters such as a sharing factor, or the number of Pareto samples.

10.4   PERFORMANCE ANALYSIS

10.4.1   Objective of Performance Assessment

Performance assessment is important in the development and application of meta-heuristic algorithms. Especially for stochastic optimization algorithms, it is essential to gauge the quality of the solutions generated by the optimizer and also the computational resources necessary to achieve the results. It is also important for tuning the multiple parameters all too common in this class of optimizers or even comparing the quality of two different optimizers for a given class of problems.

In the case of SO optimization, where the goal is to find the decision vector that satisfies a minimization or maximization objective, the typical performance analysis consists of Monte Carlo simulations, statistical characterization of computational intensity (CPU cycles, run time, and/or number of function evaluations), and quality of the solution, found by analyzing the distribution of the output of the solver both in the objective and decision spaces. If the global optima are known, some quantitative metrics are distance to nominal global optima, or average number of iterations necessary to get within a certain range of global optima. These metrics can then be used to compare different optimizers or measure the effect of the different algorithmic parameters.

In the case of MOO, however, a difficulty arises from the fact that the solution is not a point but rather a set of points. Assessing the intrinsic performance of the optimizer requires comparison operations on sets. In the early days of MOO, people relied on qualitative assessment by visual comparison of approximation sets. In recent years, because of the development of several MOO optimizers, it has become important to develop quantitative metrics and comparison methods, as well as reliable test functions, to evaluate the performance of these different optimizers [11, 25].

10.4.2   Comparison Methodologies

Comparison methodologies all involve transformation of the approximation sets generated by an algorithm into another representation. Statistical testing procedures may then be applied on the output representation to compare the algorithm. This section will present an outline of comparison methodologies popular in the literature. We assume we are comparing two optimizers, or the same optimizer with different tuning parameters. For each optimizer, we have a number of approximation sets and we wish to determine which optimizer is the best.

10.4.2.1   *Quality Indicators*   The most common approach consists of applying indicator functions on the approximation sets. These indicator functions or quality indicators convert each approximation set into a real number that is representative of the quality of an approximation set with respect to a specific property. Many indicators have been proposed, mainly to allow direct assessment of how good a solution is with respect to the objectives of MOO outlined in Section 10.2.3. Examples are

- *Cardinality* (i.e., number of solutions in the approximation set).
- *Generational distance*, which is a measure of the distance of an approximation set to the Pareto front (if known). It is the average distance between points on solution set and nearest point of the true Pareto front.
- *Spacing*, which quantifies the spread of solutions. It is the standard deviation of distance of points on solution front to nearest point on solution front.
- The *hypervolume indicator* measures the size of the dominated region bounded by some reference point [11].

The main attraction of these indicators is that statistical testing can be applied on the numbers they generate. However, they must be interpreted with care. Because they are measuring different aspects, it is frequently the case that two different indicators disagree on the superiority of different algorithms.

Some indicators can, even in nonpathological cases, judge an approximation set to be better than another when the latter dominates the former. The notion of indicator unreliability was introduced to formalize this property [25]. Formally, given two approximation sets A and B such that A dominates B, an unreliable indicator is an indicator that yields preference for the set B over A. For example, the cardinality indicator is unreliable. It can be shown that the spacing and generational distance indicators are unreliable as well.

Conversely, a reliable indicator *cannot* yield a preference for an approximation set D over another approximation set C when C dominates D. The hypervolume indicator is an example of a reliable indicator.

In addition to these unary indicators, binary quality indicators have been suggested. Their aim is to compare two approximation sets and give a quantitative measure of how much better set A is compared with set B. Binary indicator functions were shown to be more powerful than unary indicators.

10.4.2.2 Attainment Function Method This approach makes use of attainment surfaces. These are summary plots that delineate the region of the objective space that is dominated by the approximation set. It is possible to infer, from multiple runs of the optimizer, what is the probability that an objective vector is attained (i.e., weakly dominated by) an approximation set. The attainment function gives the probability that an objective vector is attained in one optimization run of the algorithm. By following this procedure for approximation sets generated from a large number of runs, one obtains an empirical attainment function, an estimate of the true attainment function. Statistical testing procedures can then be applied [25, 26] (Fig. 10.9).

The attainment function approach preserves more information than the quality indicators and the dominance ranking method. They are ideal for visual comparison and more sensitive to qualitative difference between optimizers than quality indicators. Their main drawback is that they are computationally expensive to generate and only practical for problems with two or three objective functions [27].

10.4.2.3 Dominance Ranking The dominance ranking approach was recently suggested in Ref. [25] as a preference-independent comparison method in an initial step to comparing MO optimizers. In this method, all approximation sets generated by the different optimizers are pooled, and then each approximation set is assigned a rank based on a ranking procedure. One such procedure could simply be the number of sets by which a specific approximation set is dominated. Once a rank is assigned to each set, the sets associated with each algorithm are transformed into a

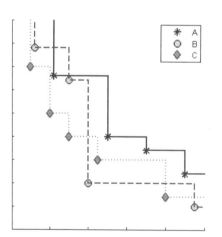
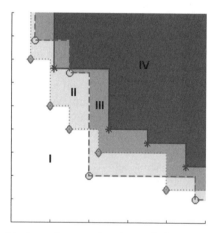

FIGURE 10.9 Attainment surfaces and empirical attainment functions. Left panel: Objective space representation of three approximation sets (*A*, *B*, and *C*), which are hypothetical outputs of a given MO optimizer for a biobjective minimization problem. The lines linking the objective vectors of each approximation set are the attainment surfaces. Right panel: Empirical attainment functions created from the three approximation sets, giving the probability that an objective vector is attained by the optimizer. In this case, objective vectors in regions I, II, III, and IV are respectively attained with probabilities 0, $1/3$, $2/3$, and 1.

set of values on which a statistical rank test can be applied to determine whether there is a significant statistical difference between the sets of ranks and by extension whether one algorithm outperforms another. If it is determined that an algorithm is significantly better than another, quality indicators and empirical attainment functions may be used to further characterize the differences in the optimizers, but are not necessary.

10.5 CONCLUSIONS

This chapter discussed to how to adapt population-based meta-heuristic algorithms, particularly EAs, to solving MO problems. As discussed, these problems are common in real-world optimization, especially in power systems. Thanks to the availability of cheap computational power, population-based techniques, which apply Pareto-optimality to concurrently optimize the different objectives, are practical, flexible turnkey approaches to generating approximation of Pareto fronts. They overcome the limitations of the classical methods such as aggregation-based techniques, which require *a priori* information about relative importance of the objectives and are sensitive to the shape of the Pareto front.

The main difficulty in MOO is that the solution consists of a set of trade-offs. As discussed in the chapter, Pareto-dominance can be incorporated in the basic operations of EAs (i.e., fitness evaluation, diversity, and elitism) to facilitate exploration and discovery of the Pareto front. The most popular algorithms in the literature to date were taken as illustrations. The subject of performance assessment, which is itself complicated by the need to compare sets of trade-offs, was briefly covered.

The hope is that this high-level overview of meta-heuristic MOO field has provided the reader with the basic tools to tackle MO problems. These algorithms are generic and flexible enough to be augmented and combined with application-specific knowledge to improve algorithmic performance and quality of the results.

ACKNOWLEDGMENTS

The authors would like to thank members of the University of Washington's Computational Intelligence Laboratory, especially Nissrine Krami and David Krout, for their input. Patrick Ngatchou was supported by the U.S. Office of Naval Research, contract no. N00014-01-G-0460.

REFERENCES

1. Stadler W. Multicriteria optimization in engineering and in the sciences. New York: Plenum Press; 1988.

2. Tabucanon MT. Multiple criteria decision making industry. Amsterdam: Elsevier Science; 1988.

3. Coello Coello CA, Van Veldhuizen DA, Lamont GB. Evolutionary algorithms for solving multi-objective problems. New York: Kluwer Academic Publishers; 2002.

4. Bagchi TP. Multiobjective scheduling by genetic algorithms. Boston: Kluwer Academic Publishers; 1999.

5. Abido MA. Environmental/economic power dispatch using multiobjective evolutionary algorithms: a comparative study. In: 2003 IEEE Power Engineering Society General Meeting, 13–17 July 2003, Toronto, Canada. IEEE: 2003.

6. Begovic M, Radibratovic B, Lambert F. On multiple Volt-VAR optimization in power systems. Proceedings of the 37th Hawaii International Conference on System Sciences, 2004.

7. Krami N, El-Sharkawi MA, Akherraz M. Multi-objective particle swarm optimization for reactive power planning. Accepted at the IEEE Swarm Intelligence Symposium, Indianapolis, 2006.

8. Jin Y, Olhofer M, Sendhoff B. Dynamic weighted aggregation for evolutionary multi-objective optimization: why does it work and how? Proc. Genetic and Evolutionary Computation Conf. (GECCO), 2001.

9. Parsopoulos KE, Vrahatis MN. Particle swarm optimization method in multiobjective problems. In: Proc. of the 2002 ACM Symposium on Applied Computing (SAC '02), Madrid, Spain. ACM Press: New York; 2002. p. 603–607.

10. Coello CAC. A comprehensive survey of evolutionary-based multiobjective optimization techniques. Knowledge and Information Systems 1999; 1(3):269–308.

11. Zitzler E, Deb K, Thiele L. Comparison of multiobjective evolutionary algorithms: empirical results. Evolutionary Computation 2000; 8(2):149–172.

12. Jones DF, Mirrazavi SK, Tamiz M. Multi-objective meta-heuristics: an overview of the current state-of-the-art. Eur J Operational Res 2002; 137(1):1–9.

13. Eiben AE, Smith JE. Introduction to evolutionary computing. Berlin: Springer; 2003.

14. Schaffer JD. Multiple objective optimization with vector evaluated genetic algorithms. In: Proceedings of the International Conference on Genetic Algorithms and Their Applications, Pittsburgh, IEEE; July 24–26, 1985. p. 93–100.

15. Fonseca C, Fleming PJ. Genetic algorithms for multiobjective optimization: formulation, discussion, generalization. In: Proceedings of the Fifth International Conference on Genetic Algorithms, San Mateo, CA, IEEE; 1993. p. 416–423.

16. Zitzler E, Thiele L. An evolutionary algorithm for multiobjective optimization: the strength Pareto approach. TIK Tech. Report No. 43, Swiss Federal Institute of Technology (ETH); 1998.

17. Srinivas N, Deb K. Multiobjective function optimization using nondominated sorting genetic algorithms. Evolutionary Computation 1994; 2(3):221–248.

18. Deb K, Pratap A, Agarwal S, Meyarivan T. A fast and elitist multiobjective genetic algorithm: NSGA–II. IEEE Trans Evol Computat 2002; 6(2):182–197.

19. Horn J, Nafpliotis N, Goldberg DE. A niched Pareto genetic algorithm for multiobjective optimization. In: Proceedings of the First IEEE Conference on Evolutionary Computation, IEEE World Congress on Computational Intelligence, Piscataway, NJ. Vol. 1. IEEE; 1994, p. 82–87.

20. Zitzler E, Laumanns M, Thiele L. SPEA2: improving the strength Pareto evolutionary algorithm. TIK Report No. 103, ETH Zurich; 2001.

21. Knowles J, Corne D. The Pareto archived evolution strategy: a new baseline algorithm for Pareto multiobjective optimisation. In: Proceedings of the 1999 Congress on Evolutionary Computation, IEEE; 1999, Vol. 1, p. 98–105.

22. Kennedy J, Eberhart RC. Swarm intelligence. Morgan Kaufmann; 2001.

23. Song M-P, Gu G-C. Research on particle swarm optimization: a review. In: Proceedings of the Third International Conference on Machine Learning and Cybernetics. IEEE; 2004; p. 2216–2241.

24. Coello Coello CA, Pulido GT, Lechuga MS. Handling multiple objectives with particle swarm optimization. IEEE Trans Evol Computat 2004; 8(3):256–279.

25. Knowles JD, Thiele L, Zitzler E. A tutorial on the performance assessment of stochastic multiobjective optimizers. TIK Report No. 214, ETH Zurich; July 2005.

26. da Fonseca VG, Fonseca CM, Hall AO. Inferential performance assessment of stochastic optimisers and the attainment function. Lecture Notes in Computer Science, Volume 1993, IEEE; Jan 2001, p. 213–225.

27. Knowles JD. A summary-attainment-surface plotting method for visualizing the performance of stochastic multiobjective optimizers. Proceedings 5th International Conference on Intelligent Systems Design and Applications, IEEE; 2005. p. 552–557.

CHAPTER 11

Trust-Tech Paradigm for Computing High-Quality Optimal Solutions: Method and Theory

HSIAO-DONG CHIANG and JAEWOOK LEE

11.1 INTRODUCTION

Optimization technology has practical applications in almost every branch of science, business, and technology. It deals with how to do things in the best possible manner for real-life problems. There are many diverse practical applications of the technology. Indeed, a large variety of the quantitative issues such as decision making, design, operation, planning, and scheduling arising in science, engineering, and economics can be perceived and modeled as (nonlinear) optimization problems [1–3].

The global minimum solution of a constrained optimization problem is the vector in the solution space that the objective function takes on its lowest value over the entire feasible solution space. On the other hand, a local minimum solution only guarantees that its objective function value is a minimum with respect to other points nearby in the solution space. Each local optimal solution has a different value of the objective function. The solution space generally contains only one global optimal solution and many local optimal solutions. From both the engineering viewpoint and the economic viewpoint, the reward and value of obtaining the global optimal solution is the highest when compared with local optimal solutions.

For most practical applications, the task of searching the solution space to find the global optimal solution is very challenging [4–7]. The primary challenges are that, in addition to the high-dimension solution space, there are many local optimal solutions in the solution space; the global optimal solution is just one of them, and yet both the global optimal solution and local optimal solutions all share the same local properties of characterization. Unfortunately, there is no practical criterion for verifying whether or not a local optimal solution is the global one. Indeed, searching for the

---

Modern Heuristic Optimization Techniques. Edited by K. Y. Lee and M. A. El-Sharkawi
Copyright © 2008 the Institute of Electrical and Electronics Engineers, Inc.

global optimal solution is a challenging and difficult task, especially for large-scale optimization problems.

The great majority of existing numerical methods/techniques, and heuristic for solving nonlinear optimization problems usually find a local optimal solution, but not the global one. They usually become trapped at every local optimal solution and are unable to escape from those solutions. This drawback has motivated the recent development of a number of more sophisticated numerical methods designed to find better solutions by introducing some mechanisms that allow the search process to escape from local optimal solutions. These sophisticated numerical methods include the simulated annealing algorithm, genetic algorithm, evolution programming, tabu search, and particle swarm operator. However, many research and numerical studies indicate that these sophisticated numerical methods may suffer from several difficulties, among others, requiring intensive computational efforts and yet still be unable to find the global optimal solution and yielding inconsistent solutions.

One reliable way to find the global optimal solution of a nonlinear optimization problem, in our viewpoint, is to first find all the local optimal solutions and then find, from them, the global optimal solution. This approach is reliable but may not be practical because the computational time required to find all of the local optimal solutions can be prohibitively high, especially for large-scale optimization problems. To this end, we have developed a transformation under stability-retained equilibrium characterization (Trust-Tech) method, which is deterministic in nature, attempting to compute all the local optimal solutions of constrained nonlinear optimization problems [8–10]. One distinguishing feature of the Trust-Tech method is that it systematically computes a set of (or all) the local optimal solutions of general constrained nonlinear optimization problems; furthermore, the set of local optimal solutions are deterministically obtained in a tier-by-tier manner. A set of high-quality local optimal solutions can thus be derived from the several sets of computed local optimal solutions. The global optimal solution may well be contained in the set of high-quality local optimal solutions. A theoretical basis of the Trust-Tech method will be presented. The Trust-Tech method has been applied to numerical examples arising from several disciplines with promising results.

From a theoretical viewpoint, the Trust-Tech paradigm is built on the optimization theory, nonlinear dynamical systems, theory of stability region, characterization of stability boundary, and advanced numerical methods. It translates the task of finding all the feasible components of general constrained optimization problems into the task of finding all the stable equilibrium manifolds of the so-called quotient gradient system whose vector field is associated with the constraint set of the optimization problems. In addition, it translates the task of finding all the local optimal solutions located in a feasible component of general constrained optimization problems into the task of finding all the stable equilibrium points of the so-called projected gradient system.

11.2 PROBLEM PRELIMINARIES

In this section, we review some concepts in nonlinear optimization problems and in nonlinear dynamical systems that are essential in the subsequent development in this paper. We consider a general constrained nonlinear optimization problem of the form:

$$\begin{array}{ll} \text{minimize} & f(x) \\ \text{subject to} & h_i(x) = 0,\ i \in I = \{1, \ldots, l\} \\ & g_j(x) \le 0,\ j \in J = \{1, \ldots, s\}, \end{array} \qquad (11.1)$$

where f, $h_i(x)$, $g_j(x)$ are assumed to be in $C^2(\mathfrak{R}^n, \mathfrak{R})$ for all $i \in I, j \in J$.

The *Lagrangian function* L: $\mathfrak{R}^{n+l+s} \to \mathfrak{R}$ of (1) is defined by

$$L(x, \lambda, \mu) = f(x) + \sum_{i \in I} \lambda_i h_i(x) + \sum_{j \in J} \mu_j g_j(x).$$

We call $\bar{x}$ a *critical point* of (1) (with Lagrange-multipliers $(\bar{\lambda}, \bar{\mu})$) if

$$\nabla_x L(\bar{x}, \bar{\lambda}, \bar{\mu}) = \nabla f(\bar{x}) + \sum_{i \in I} \bar{\lambda}_i \nabla h_i(\bar{x}) + \sum_{j \in J} \bar{\mu}_j \nabla g_j(\bar{x}) = 0$$

and

$$\begin{array}{ll} h_i(\bar{x}) = 0, & i \in I \\ g_j(\bar{x}) \le 0, & \bar{\mu}_j g_j(\bar{x}) = 0, \quad j \in J. \end{array}$$

The constraint set of (11.1) defines the following

$$S = \{x \in \mathfrak{R}^n \colon h_i(x) = 0,\ i \in I,\ g_j(x) \le 0, j \in J\},$$

which can be any closed subset of $\mathfrak{R}^n$ with very complex structure. The constraint set S is usually nonconvex and disconnected; i.e., it is composed of several (disjoint) connected feasible regions. The task of locating each connected feasible region of the set S is itself a difficult one. A reasonable local structure of the constraint set can be obtained under the following assumption, which is *generically* true [11]:

Assumption 11.1.

1. (Regularity) At each $x \in S$, the set of vectors

$$\{\nabla h_i(x),\ i \in I\} \cup \{\nabla g_j(x), j \in J_0(x)\}$$

   are linearly independent where $J_0(x) = \{j \in J \colon g_j(x) = 0\}$ is the index set for the *active* constraints.
2. (Nondegeneracy) At each critical point, $\bar{x} \in S$, $d^T \nabla^2_{xx} L(\bar{x}, \bar{\lambda}, \bar{\mu}) d \ne 0$ for all $d \ne 0$ satisfying $\nabla h_i(\bar{x})^T d = 0$, $\forall i \in I$ and $\nabla g_j(\bar{x})^T d = 0$, $\forall j \in J_0(\bar{x})$.

3. (Strict Complementarity) $\bar{\mu}_j \neq 0, j \in J_0(\bar{x})$.

4. (Finiteness and Separating Property) f has finitely many critical points in S at which it attains different values of f.

We hence consider, without loss of generality, the following optimization problem with equality constraints:

$$
\begin{aligned}
\text{minimize} \quad & f(x) \\
\text{subject to} \quad & h_i(x) = 0, i \in I = \{1, \dots, l\}.
\end{aligned}
\tag{11.2}
$$

We define the following equality vector $H(x) := (h_1(x), \dots, h_m(x))^T$. By applying the regularity condition in Assumption 11.1 and the implicit function theorem, it can be shown that the constraint set (or the feasible set or region)

$$
M = \{x \in \mathfrak{R}^n \colon H(x) := (h_1(x), \dots, h_m(x))^T = 0\}
$$

is a smooth manifold. In general, the constraint set M can be very complicated with several disjoint path-connected feasible components; in other words, the constraint set M can be decomposed into several disjoint path-connected, feasible components, say,

$$
M = \sum_k M_k
$$

where the M_k's are disjoint connected feasible components. Each path-connected component may contain several local optimal solutions of the optimization problem (11.2).

We next review some concepts in nonlinear dynamical systems that are relevant in this paper. We consider a general dynamical system described by the following:

$$
\dot{x} = F(x).
\tag{11.3}
$$

Several basic definitions and facts about the nonlinear dynamical system (11.3) are presented as follows. The solution of (11.3) starting from $x \in \mathfrak{R}^n$ at $t = 0$ is called a *trajectory*, denoted by $\phi(\cdot, x)\colon \mathfrak{R} \to \mathfrak{R}^n$. A state vector $x^* \in \mathfrak{R}^n$ is called an *equilibrium point* of system (11.3) if $F(x) = 0$. We say that an equilibrium point $x^* \in \mathfrak{R}^n$ of (11.3) is *hyperbolic* if the Jacobian matrix of $F(\cdot)$ at x^* has no eigenvalues with a zero real part. If the Jacobian of the hyperbolic equilibrium point has exactly k eigenvalues with positive real part, we call it a *type-k equilibrium point*. It can be shown that for a hyperbolic equilibrium point, it is a (asymptotically) *stable equilibrium point* if all the eigenvalues of its corresponding Jacobian have a negative real part and an *unstable equilibrium point* if all the eigenvalues of its corresponding Jacobian have a positive real part. For a type-k hyperbolic equilibrium point x^*, its *stable and unstable manifolds* $W^s(x^*)$, $W^u(x^*)$ are defined as follows:

$$W^s(x^*) = \left\{ x \in \Re^n: \lim_{t \to \infty} \phi(t, x) = x^* \right\}$$
$$W^u(x^*) = \left\{ x \in \Re^n: \lim_{t \to -\infty} \phi(t, x) = x^* \right\},$$

where the dimension of $W^u(x^*)$ and $W^s(x^*)$ is k and $n - k$, respectively. A set K in $\Re^n$ is called to be an *invariant set* of (11.3) if every trajectory of (11.3) starting in K remains in K for all $t \in \Re$.

The concept of stability region plays an essential role in the nonlinear system stability theory. It is also vital in the development of Trust-Tech method. It is defined as follows: the *stability region* (or *region of attraction*) of a stable equilibrium point x_s is the collection of all the points whose trajectories all converge to be defined as [13, 14]:

$$A(x_s) := \left\{ x \in \Re^n: \lim_{t \to \infty} \phi(t, x) = x_s \right\}.$$

The *quasi-stability region* $A_p(x_s)$ of a stable equilibrium point x_s is defined as [14] $A_p(x_s) = \text{int}(\overline{A(x_s)})$, where $\overline{A(x_s)}$ is the closure of the stability region $A(x_s)$, and $\text{int}(\overline{A(x_s)})$ is the interior of $\overline{A(x_s)}$. From a topological point of view, $A(x_s)$ and $A_p(x_s)$ are open, invariant, and path-connected sets; moreover, they are diffeomorphic to $\Re^n$. A comprehensive theory of quasi-stability boundary and quasi-stability region can be found in Ref. 14.

11.3 A TRUST-TECH PARADIGM

One reliable way to find the global optimal solution of the optimization problem (11.2) is to find first all the local optimal solutions and then find, from them, the global optimal solution. To this end, we develop the Trust-Tech method, which is deterministic in nature, attempting to compute all the local optimal solutions of constrained nonlinear optimization problems [8–10]. The Trust-Tech method is composed of two distinct phases: In phase I, starting from an arbitrary starting point, it systematically finds all the feasible components that satisfy the constraint set. In phase II, it computes all the local optimal solutions in each feasible component found in phase I.

11.3.1 Phase I

Phase I of the Trust-Tech method finds all the feasible components via exploring some trajectories of a particular nonlinear dynamical system. In order to visit each feasible component, phase I consists of two steps:

Step 1.1: Approach a (path-connected) feasible component of the optimization problem (11.2).

Step 1.2: Move away from the feasible component and approach another feasible component of the optimization problem (11.2).

We design a nonlinear dynamical system whose trajectories can be used to perform step 1.1 and step 1.2. The central idea in designing such a nonlinear dynamical system is that every path-connected feasible component corresponds with a stable equilibrium manifold (a generalized concept of stable equilibrium point) of the nonlinear dynamical system. In this way, the task of finding all the feasible components of the optimization problem (11.2) is equivalent to the task of finding all the stable equilibrium manifolds of the nonlinear dynamical system. To this end, we build the following so-called quotient gradient system whose vector field is associated with the constraint set of the optimization problem (11.2) characterized by the equality vector $H(x)$

$$\dot{x} = -DH(x)^T H(x),\tag{11.4}$$

where $DH(x)$ represents the Jacobian matrix of the vector $H(x)$.

It can be shown [16] that if a set is a feasible component of the optimization problem (11.2) then the set is a stable equilibrium manifold of the quotient gradient system (11.4). This analytical result provides a basis for our proposed method in searching for feasible components of the optimization problem (11.2). Hence, steps 1.1 and 1.2 are numerically implemented via the following two tasks:

Task 1.1: Approach a stable equilibrium manifold of the quotient gradient system (11.4).

Task 1.2: Move away from the stable equilibrium manifold and approach another stable equilibrium manifold of the quotient gradient system (11.4).

11.3.2 Phase II

Phase II of the Trust-Tech method finds all the local optimal solutions lying within each feasible component using some trajectories of a particular class of nonlinear dynamical systems. Phase II is conceptually composed of two steps:

Step 2.1: Start from a feasible point found in step 1.1, compute a local optimal solution located in the feasible component of the optimization problem (11.2).

Step 2.2: Move away from the local optimal solution and approach another local optimal solution lying in the same feasible component of the optimization problem (11.2).

We design a nonlinear dynamical system whose trajectories can be explored to perform step 2.1 and step 2.2. The central idea in designing such a nonlinear dynamical system is that all the local optimal solutions of the optimization problem (11.2) correspond with all the stable equilibrium points of the nonlinear dynamical system; in particular, every local optimal solution of the optimization problem (11.2) corresponds with a stable equilibrium point of the nonlinear dynamical system. To this end, we design the following projected gradient dynamical system

$$\dot{x}(t) = -P_H(x(t))\nabla f(x(t)), \tag{11.5}$$

where $x(0) = x_0 \in M$, and the projection matrix

$$P_H(x) = \left(I - DH(x)^T \left(DH(x)DH(x)^T\right)^{-1} DH(x)\right) \in \mathfrak{R}^{n \times n}$$

is a positive semidefinite matrix for every $x \in M$.

We will employ certain trajectories of the projected gradient dynamical system to find a set of local optimal solutions of the optimization problem (11.2). We note that $P_H(x)\nabla f(x)$ is the orthogonal projection of $\nabla f(x)$ to the tangent space $T_x M$, which means $P_H(x)\nabla f(x) \in T_x M$ for all $x \in M$; thus every trajectory of (11.5) starting from $x_0 \in M_k$ stays in M_k. In other words, M_k is an invariant set of (11.5). This implies that the trajectory of system (11.5) starting from a feasible point stays in the feasible component containing the point. Furthermore, it can be shown that the task of finding a local optimal solution of the optimization problem (11.2) is equivalent to the task of finding a stable equilibrium point of the projected gradient system (11.5).

One important issue in computing multiple optimal solutions or all of the local optimal solutions is how to effectively move (escape) from a local optimal solution and move on toward another local optimal solution. Significant efforts have been directed in the past attempting to address this issue, but without much success. Next, an effective method to overcome this issue will be presented and incorporated into phase II of the Trust-Tech method. The basis of the method is described as follows. We translate the problem of how to escape from a local optimal solution of the optimization problem (11.2) and to move on toward another local optimal solution (in the same path-connected feasible component) into the problem of how to escape from the stability region of the corresponding (asymptotically) stable equilibrium point (SEP) of the projected gradient system (11.5) and enter into the stability region of another SEP of the projected gradient system (11.5). Once a system trajectory lies inside the stability region of another SEP, the ensuing system trajectory will converge to the SEP; which is another local optimal solution of the optimization problem (11.2). Hence, another local optimal solution is found.

We next present a numerical method that moves from a local optimal solution of the optimization problem (11.2) and moves toward another local optimal solution. Let x_{opt} be a local optimal solution. We escape from x_{opt} and move to another local optimal solution via a decomposition point, which lies on the stability boundary of x_{opt}. Recall that an equilibrium point is termed a *decomposition point* if it is a type-one equilibrium point lying on the quasi-stability boundary of a stable equilibrium point [14]. Starting from a local optimal solution, say x_{opt}, we compute another local optimal solution via the following

> *Task 2.1*: Approach a stable equilibrium point of the projected gradient system (11.5).
>
> *Task 2.2*: Move from the stable equilibrium point to a decomposition point (in order to escape from the local optimal solution).
>
> *Task 2.3*: Approach another stable equilibrium point of the projected gradient system (11.5) (in the same path-connected feasible component) by moving along the unstable manifold of the decomposition point.

Task 2.1 can be implemented by following the trajectory of the projected gradient system (11.5) starting from any initial point located in a feasible component. Task 2.2 is rather challenging. A robust computational method for Task 2.2 will be presented in a later section. Task 2.3 can be implemented by following the trajectory starting from an initial point, which is close to the decomposition point but outside the stability region, until it approaches another stable equilibrium point of (11.5), which is another local optimal solution of the optimization problem (11.2).

Suppose there are six local optimal solutions in the search space of an optimization problem (Fig. 11.1). We define a projected gradient dynamical system described by the equation (11.5) that is completely stable and each stable equilibrium point corresponds with each local optimal solution of the optimization problem. The dashed line represents the quasi-stability boundary of each stable equilibrium point (i.e., each local optimal solution) (Fig. 11.2). Starting from an initial point, the corresponding system trajectory of the dynamical system (or a local search method) finds a stable equilibrium point (i.e., a local optimal solution) (Fig. 11.3). Starting from the found local optimal solution, the numerical method for computing tier-one local optimal solutions computes three initial points that lie inside the quasi-stability region of each stable equilibrium point (SEP). In addition, each initial point is close to the corresponding SEP (i.e., a tier-one local optimal solution) (Fig. 11.4). Starting from an initial point, the corresponding (dynamical) trajectory or a local

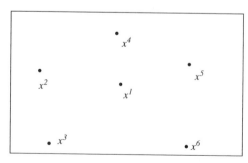

FIGURE 11.1 $x_s^1, x_s^2, \ldots, x_s^6$: stable equilibrium points.

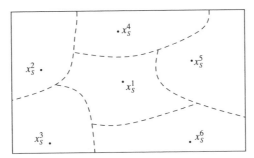

FIGURE 11.2 $x_s^1, x_s^2, \ldots, x_s^6$: stable equilibrium points; -----: stability boundary (i.e. boundary of stability region).

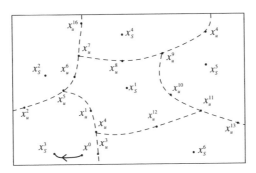

FIGURE 11.3 $x_s^1, x_s^2, \ldots, x_s^6$: stable equilibrium points; -----: stability boundary (i.e. boundary of stability region); $x_u^1, x_u^2, \ldots, x_u^{16}$: unstable equilibrium point; x^0: initial condition; ⟵: system trajectory.

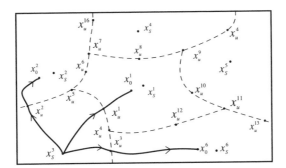

FIGURE 11.4 $x_s^1, x_s^2, \ldots, x_s^6$: stable equilibrium points; -----: stability boundary (i.e. boundary of stability region); $x_u^1, x_u^2, \ldots, x_u^{16}$: unstable equilibrium point; ╱ : search paths by Trust-Tech paradigm.

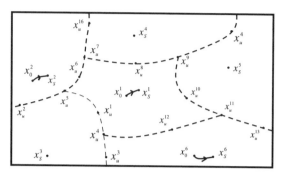

FIGURE 11.5 $x_s^1, x_s^2, \ldots, x_s^6$: stable equilibrium points; -----: stability boundary (i.e. boundary of stability region); $x_u^1, x_u^2, \ldots, x_u^{16}$: unstable equilibrium point; ⤳: search paths by Trust-Tech paradigm.

search method computes the corresponding stable equilibrium point (i.e., a local optimal solution) (Fig. 11.5).

We illustrate in Figs. 11.1 to 11.5 the search process of phase II. These figures serve to emphasize the capability of phase II in computing the surrounding local optimal solutions in a tier-by-tier manner. To find the local optimal solutions located in other path-connected feasible components, one needs to invoke phase I again to escape from one feasible component of the constraint set and approach another feasible component.

11.4   THEORETICAL ANALYSIS OF TRUST-TECH METHOD

The Trust-Tech method uses certain trajectories of the projected gradient system (11.5) and the quotient gradient system (11.4) to systematically locate the set of local optimal solutions in a deterministic manner. The global dynamical behaviors of the two constructed dynamical systems (11.4) and (11.5) play an important role in finding the set of local optimal solutions of the optimization problem (11.2). This section presents theoretical analysis of dynamical characterizations of these two dynamical systems.

Assumption 11.2.   For any bounded, closed interval $[a, b]$ and for any relatively closed subset K of $\left(f^{-1}([a,b])\backslash E_f\right)$ in M, we have

$$\inf\{\|P_H(x)\nabla f(x)\| : x \in K\} > 0,$$

where $E_f = \{x \in \Re^n : P_H(x)\nabla f(x) = 0\}$

Remark 11.2.   If each path-connected feasible component of M is compact, then Assumption 11.2 always holds. The constraint components of many practical optimization problems are compact, hence, the above assumption usually holds.

Assumption 11.3. The feasible vector $H \in C^2(\mathfrak{R}^n, \mathfrak{R}^m)$ satisfies one of the following:

1. $\|H(x)\|$ is a proper map (i.e., the pre-image of a compact set is compact), or
2. for any $\gamma > 0$ and for any closed subset K of

$$\{x \in \mathfrak{R}^n \colon \|H(x)\| \leq \gamma, \quad \|DH(x)^T H(x)\| \neq 0\}$$

we have

$$\inf\{\|DH(x)^T H(x)\| \colon x \in K\} > 0$$

The global behaviors of nonlinear dynamical systems can be very complicated. Their trajectories can behave unbounded or/and bounded but complicated such as almost periodic trajectory, or even chaotic motions. We next show that the global behaviors of the projected gradient system (11.5) and the quotient gradient system (11.4) are very simple: every trajectory converges to an equilibrium point. There is no complicated behaviors such closed orbit (i.e., limit cycle) and chaotic motion that can exist in these two systems. Note that a nonlinear dynamical system is said to be *completely stable* if every trajectory of the system converges to one of its equilibrium points.

Theorem 11.1: (Completely Stable). If the optimization problem (11.2) satisfies Assumptions 11.2 and 11.3, then the corresponding projected gradient system (11.5) is completely stable and the objective function $f(x)$ of optimization problem (11.2) is bounded from below on the constraint set M. *Proof.* It is easy to see that the objective function $f(x)$ satisfies the three conditions required for being an energy function of the projected gradient system (11.5). Hence, according to [13], the projected gradient system (11.5) is completely stable. This completes the proof. □

Theorem 11.1 implies that every trajectory of the projected gradient system (11.5) converges to one of the equilibrium points. The system does not admit other complex trajectories such as limit cycle, quasi-periodic, or chaotic trajectories. The theorem below gives a complete characterization of the quasi-stability region of the projected gradient system (11.5).

Theorem 11.2: [9, 14] (Characterization of Quasi-stability Boundary). Let $A_p(x_s)$ be the quasi-stability region of x_s of the projected gradient system (11.5) and let x_i ($i = 1, 2, \ldots$) be all the decomposition points of x_s. If the system satisfies the following conditions:

(i) The equilibrium points on $A_p(x_s)$ are hyperbolic and finite in number;
(ii) The stable and unstable manifolds of the equilibrium points on $A_p(x_s)$ satisfy the transversality condition;

then the quasi-stability boundary $\partial A_p(x_s)$ is the union of the closure of the stable manifold of x_i; i.e.,

$$\partial A_p(x_s) = \underset{x_i \in \partial A_p(x_s)}{Y} \overline{W^s(x_i)}$$

We next derive a dynamic relationship between a stable equilibrium point (SEP) and a decomposition point lying on the quasi-stability boundary of the SEP in the following theorem.

Theorem 11.3: [8] (Decomposition Points and the Stable Equilibrium Points). Let $A_p(x_s)$ be the quasi-stability region of x_s of the projected gradient system (11.5) that satisfies the following conditions

  (i) The equilibrium points on quasi-stability boundaries are hyperbolic and finite in number;
 (ii) The stable and unstable manifolds of the equilibrium points on quasi-stability boundaries satisfy the transversality condition.

If x_d is a decomposition point on the quasi-stability boundary $\partial A_p(x_s)$, then there exists another one and only one SEP, says $\bar{x}_s$, to which the unstable manifold of x_d converges. Conversely, if the set $\partial A_p(x_s) \cap \partial A_p(\bar{x}_s)$ is nonempty, then the set contains a decomposition point.

Theorem 11.3 reveals a relationship between SEPs and decomposition points. The unstable manifold of a dynamic decomposition point converges to two SEPs of the projected gradient system (11.5). In the context of optimization, Theorem 11.3 asserts that two neighboring local optimal solutions are connected by the unstable manifold of the corresponding dynamic decomposition point. Note that the two conditions stated in Theorem 11.3 are generic properties for general nonlinear dynamical systems.

Theorem 11.4: [9, 10] (Completely Stable). If the optimization problem (11.2) satisfies Assumption 11.2 and all the equilibrium manifolds are pseudo-hyperbolic and finite in number, then every trajectory of the quotient gradient system (11.4) is bounded and converges to one of the equilibrium manifolds.

Theorem 11.5: [9, 10] (Equilibrium Manifolds and Feasible Regions). If the optimization problem (11.2) satisfies Assumption 11.2 and all the equilibrium manifolds are pseudo-hyperbolic and finite in number, then each path-connected component of the constraint set M of the optimization problem (11.2) is a stable equilibrium manifold of the quotient gradient system (11.4). Conversely, if Σ is a stable equilibrium manifold of the quotient gradient system (11.4), then Σ is a nonisolated set of local minima of the following minimization problem

$$\min_{x \in R^n} \frac{1}{2} \|H(x)\|^2.$$

Theorem 11.5 provides a theoretical basis for phase I of the Trust-Tech method. They assert that each feasible component of the optimization problem (11.2) can be located via the stable equilibrium manifold of the quotient gradient system (11.4).

11.5   A NUMERICAL TRUST-TECH METHOD

We have shown that all the local optimal solutions of the optimization problem (11.2) can be found, at least conceptually, by alternating between phase I and phase II of the Trust-Tech method (hence by switching between the quotient gradient system (11.4) and the projected gradient system (11.5). In this section, we develop a numerical Trust-Tech method to systematically compute multiple high-quality local optimal solutions of general nonlinear programming problems with disconnected feasible regions. Specifically, a numerical Trust-Tech method for computing compute tier-one local optimal solution will be presented; followed by a numerical Trust-Tech method for computing compute tier-N local optimal solutions. Hence, starting from an initial point, the Trust-Tech method finds a set of surrounding local optimal solutions in a tier-by-tier manner. Hence, several sets of surrounding local optimal solutions can be computed. A set of high-quality local optimal solutions can thus be derived from the several sets of computed local optimal solutions. The global optimal solution may well be contained in the set of high-quality local optimal solutions.

Before presenting the Trust-Tech method, we present the definitions for the tier-one local optimal solutions and tier-N local optimal solutions with respect to a known local optimal solution in the continuous solution space of the optimization problem (11.2). A local optimal solution x_{s1} of the optimization problem (11.2) is said to be a *tier-one* local optimal solution to a known local optimal solution x_{s0}, if the closure of the stability region of x_{s1} of the associated nonlinear dynamical system, say (11.5) and that of x_{s0} intersect each other. We next extend the definition of tier-one local optimal solution to tier-N solution.

Definition 11.1 (Tier-N Local Optimal Solution). A local optimal solution x_{sn} of the optimization problem (11.2) is said to be a tier-N ($N > 1$) local optimal solution to a known local optimal solution x_{s0} if x_{sn} meets the follows two conditions:

(i) x_{sn} is neither a tier-$(N-2)$ nor tier-$(N-1)$ local optimal solution of x_{s0}.

(ii) x_{sn} is a tier-1 local optimal solution of one of the tier-$(N-1)$ local optimal solutions of x_{s0}.

11.5.1 Computing Another Local Optimal Solution

Theorem 11.3 in conjunction with Theorems 11.1 and 11.2 provide a theoretical basis to move away from a local optimal solution and approach another local optimal solution via the following two steps:

> *Step 1*: Starting from a local optimal solution, which is a SEP of the projected gradient system (11.5) of the optimization problem (11.2), move along a (given or desired) direction to find the corresponding dynamic decomposition point.
>
> *Step 2*: Starting from the dynamic decomposition point, move along the unstable manifold of the decomposition point, which will lead to another local optimal solution.

The existence of a dynamic decomposition point at step 1 is ensured by Theorem 11.2 and the existence of another local optimal solution of step 2 is ensured by Theorem 11.3. Theorem 11.3 provides a theoretical basis for phase II of the Trust-Tech method. They assert that each local optimal solution of the optimization problem (11.2) can be located via the unstable manifold of a dynamic decomposition point of the projected gradient system (11.5).

We present a numerical method that starts a local optimal solution and moves along a search direction to find another local optimal solution. For the purpose of illustration, we use the unconstrained optimization problem as an example. Given a local optimal solution of the unconstrained continuous optimization problem (11.2) (i.e., a stable equilibrium point (SEP) of the associated nonlinear dynamical system (11.5)) and a predefined search path starting from the SEP, we develop a method for computing the exit point of the nonlinear dynamic system (11.5) associated with an optimization problem (11.2).

11.5.1.1 *Method for Computing Exit Point*   The method is as follows: Starting from a known local optimal solution, say x_s, move along a predefined search path to compute said exit point, which is the first local maximum of the objective function of the optimization problem (11.2) along the predefined search path. Another method for computing exit point is as follows: move along the search path starting from x_s, and at each time-step, compute the inner-product of the derivative of the search path and the vector field of the nonlinear dynamic system (11.5). When the sign of the inner-product changes from positive to negative, say between the interval $[t_1, t_2]$, then either the point that is the search path at time t_1 or the point that is the search path at time t_2 can be used to approximate the exit point.

11.5.1.2 *Method for Computing Dynamic Decomposition Point (DDP)*   We present a method for computing the DDP of the associated nonlinear dynamical system (11.5) with respect to the local optimal solution x_s and with respect to the predefined search path as follows:

> *Step 1*: Starting from a known local optimal solution x_s, move along a predefined search path, say $\varphi_t(x_s)$, to compute the exit point of the

quasi-stability boundary of x_s of the system (11.5). Let the exit point be denoted as x_{ex}.

Step 2: Starting from x_{ex}, compute the minimal distance point (MDP) by the following steps:

   (i) Integrate the associated nonlinear dynamic system (11.5) for a few time-steps; letting the end point be denoted as the current exit point.

  (ii) Draw a ray connecting the current exit point and the local optimal solution, say x_s, and replace said current exit point with a corrected exit point, which is the first local maximal point of objective function along said ray, starting from x_s, and assign this corrected exit point to said exit point, namely x_{ex}.

 (iii) Repeat the steps (i) and (ii) until the norm of the vector field (11.5) at said current exit point obtained in step (ii) is smaller than a threshold value, i.e., $\|F^U(x_{\text{ex}})\| \leq \varepsilon$. Then said point is declared as a minimal distance point (MDP), say x_d^0.

Step 3: Using said MDP x_d^0 as an initial guess and solving the set of nonlinear algebraic equations of said vector field of system (11.5), wherein a solution is the DDP, say x_{ddp}, with respect to the x_s and with respect to said search path $\varphi_t(x_s)$.

We are now in a position to state the detailed procedure of the Trust-Tech method for finding another local optimal solution.

11.5.1.3 *Trust-Tech Method for Computing Another Local Optimal Solution*

Step 1: Starting from a known local optimal solution x_s, apply the method for computing dynamic decomposition point to system (11.5) to find a DDP, denoted as x_{ddp}.

Step 2: Set $x_0 = x_s + (1 + \varepsilon)(x_{\text{ddp}} - x_s)$ where ε is a small number.

Step 3: Starting from x_0, conduct a transitional search by integrating dynamic system (11.5) to obtain the corresponding system trajectory until it reaches an interface point. x_{inf}, at which an effective (hybrid) local search method outperforms the transitional search.

Step 4: Starting from the interface point x_{inf}, apply the effective (hybrid) local search method, chosen in step 3, to find the corresponding local optimal solution.

11.5.2 Computing Tier-One Local Optimal Solutions

We present a numerical Trust-Tech-based method, given a local optimal solution, say x_s^0 and a set of given search direction, for computing tier-one local optimal solutions of x_s^0. Each tier-one local optimal solution is computed via computing the corresponding DDPs and then following the unstable manifold of each DDP, and optionally, via a local method.

11.5.2.1 Trust-Tech Method for Computing Tier-One Local Optimal Solutions

Step 1: Initialize the set of decomposition points $V_{ddp} = \{\phi\}$ and set of tier-one local optimal solutions $V_s = \{\phi\}$.

Step 2: Define a set of search paths S_i for x_s^0, $i = 1, 2, \ldots, m_j$, and set $i = 1$.

Step 3: For each search path S_i starting from x_s^0, compute its corresponding local optimal solution using the following steps

   (i) Apply the method for computing dynamic decomposition point (DDP) to system (11.5) to find the corresponding DDP. If a DDP is found, proceed to the next step; otherwise, go to step (vi).

   (ii) Letting the found DDP be denoted as $x_{ddp,i}$, check if it belongs to the set V_{ddp}, i.e., $x_{ddp,i} \in V_{ddp}$. If it does, go to step (vi); otherwise, set $V_{ddp} = V_{ddp} \cup \{x_{ddp,i}\}$ and proceed to the next step.

   (iii) Set $x_{0,i} = x_s^0 + (1 + \varepsilon)(x_{ddp,i} - x_s^0)$, where ε is a small number.

   (iv) Starting from $x_{0,i}$, conduct a transitional search by integrating dynamic system (11.5) to obtain the corresponding system trajectory until it reaches an interface point, say $x_{inf,i}$, at which an effective local search method outperforms the transitional search.

   (v) Starting from the interface point $x_{inf,i}$, apply the effective local search method, chosen in step (iv), to find the corresponding local optimal solution with respect to $x_{ddp,i}$, denoted as x_s^i. And set $V_s = V_s \cup \{x_s^i\}$.

   (vi) Set $i = i + 1$.

   (vii) Repeat steps (i) through (vi) until $i > m_j$.

Step 4: Output the set of tier-one local optimal solutions V_s of x_s^0.

In order to enhance the efficiency of the method, we incorporate a local search method into the method. The basic idea for this incorporation is as follows. Once a decomposition point is found, approaching another local optimal solution from the decomposition point poses no particular challenge: it can be implemented by numerically integrating the associated projected gradient system in combination with a local search method. The purpose of incorporating a local search method is to find the local optimal solution in an effective way when the search process is near a local optimal solution. At Step 3(iv), we have incorporated a local search method into the solution algorithm in order to improve the speed of computing local optimal solutions.

11.5.3 Computing Tier-N Solutions

We next present a Trust-Tech solution algorithm for computing tier-one, tier-two, ... local optimal solutions. Starting from an initial point, say x_0, the solution algorithm computes a complete set of local optimal solutions as well as the global optimal solution of the continuous optimization problem (11.2).

Step 1: Starting from x_0, apply an effective local search method to find a local optimal solution, say x_s^0.

Step 2: Set $j = 0$. Initialize the set of local optimal solutions $V_s = \{x_s^0\}$, the set of tier-j local optimal solutions $V_{\text{new}}^j = \{x_s^0\}$, and the set of found decomposition points $V_{\text{ddp}} = \{\phi\}$.

Step 3: Initialize the set of tier-($j+1$) local optimal solutions $V_{\text{new}}^{j+1} = \{\phi\}$.

Step 4: For each local optimal solution in V_{new}^j, say, x_s^j, find its all tier-one local optimal solutions.

   (i) Define a set of search paths S_i^j for x_s^j, $i = 1, 2, \ldots, m_j$ and set $i = 1$.

   (ii) For each search path S_i^j starting from x_s^j, apply an effective method to nonlinear system (11.5) to find the corresponding decomposition point. If a DDP is found, proceed to the next step; otherwise, go to the step (viii).

   (iii) Letting the found DDP be denoted as $x_{\text{ddp}, i}^j$, check if it belongs to the set V_{ddp}, i.e., $x_{\text{ddp}, i}^j \in V_{\text{ddp}}$. If it does, go to step (viii); otherwise, set $V_{\text{ddp}} = V_{\text{ddp}} \cup \left\{ x_{\text{ddp}, i}^j \right\}$ and proceed to the next step.

   (iv) Set $x_{0,i}^j = x_s^j + (1 + \varepsilon)(x_{\text{ddp}, i}^j - x_s^j)$, where ε is a small number.

   (v) Starting from $x_{0,i}^j$, conduct a transitional search by integrating the nonlinear dynamic system (11.5) to obtain the corresponding system trajectory until it reaches an interface point. $x_{\text{inf},i}^j$, at which an effective hybrid local search method outperforms the transitional search.

   (vi) Starting from the interface point $x_{\text{inf},i}^j$, apply the effective hybrid local search method, chosen in step (v), to find the corresponding local optimal solution, denoted as $x_{s,j}^i$.

   (vii) Check if $x_{s,j}^i$ has been found before, i.e., $x_{s,j}^i \in V_s$. If it is a new local optimal solution, set $V_s = V_s \cup \left\{ x_{s,j}^i \right\}$, $V_{\text{new}}^{j+1} = V_{\text{new}}^{j+1} \cup \left\{ x_{s,j}^i \right\}$.

   (viii) Set $i = i + 1$.

   (ix) Repeat steps (ii) through (viii) until $i > m_j^k$.

Step 5: If the set of all newly computed local optimal solutions, V_{new}^{j+1}, is empty, continue with step 6; otherwise set $j = j + 1$ and go to step 3.

Step 6: Output the set of local optimal solutions V_s and identify the best optimal solution from the set of V_s by comparing their corresponding objective function values in set V_s and selecting the one with the smallest value. And output it as the global optimal solution.

11.6 HYBRID TRUST-TECH METHODS

The hybridization of two different optimization methods into one in order to merge their advantages and to reduce their disadvantages is a well-known technique. The question associated with this hybridization is what methods to merge and how to merge them. The task of finding each local optimal solution of an optimization problem itself is a challenging work. We believe the most effective method is a tailored "local" optimization method that is a blend of domain knowledge, problem structure exploration, rigor, and heuristics. The Trust-Tech method provides a platform where it is easy and straightforward to incorporate and use any tailored local optimal method to locate a set of local optimal solutions and find them effectively. In fact, the Trust-Tech method provides a seamless integration environment for local optimization methods to locate local optimal solutions. Hence, a simple hybrid Trust-Tech method is a combination of Trust-Tech method and a local optimization method.

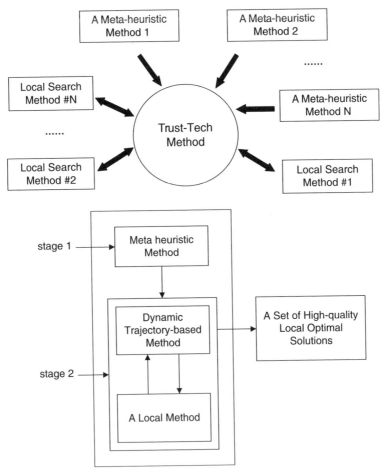

FIGURE 11.6 The structure of a hybrid Trust-Tech method is composed of two stages. Stage 1 is composed of a meta-heuristic method and stage 2 is composed of the Trust-Tech method, in combination with a local search method.

From a theoretical viewpoint, the Trust-Tech method can find all the local optimal solutions. However, from a computational viewpoint, it may well be due to the lack of computational methods that can compute the complete set of decomposition points; the Trust-Tech method can only compute multiple local optimal solutions (in each feasible component) for general constrained optimization problems. To overcome these difficulties, we develop a Trust-Tech method that combines the capability of the Trust-Tech method and that of a meta-heuristic method. The structure of the hybrid Trust-Tech method is composed of two stages, which takes advantages of both a meta-heuristic and the Trust-Tech method. Stage 1 is composed of a meta-heuristic method, such as genetic algorithm or simulated evolution method, and stage 2 is composed of the Trust-Tech method, in combination with a local search method (Fig. 11.6). The key function of stage 1 is to generate a set of "good" solution vectors located in several promising subregions in the search space. Starting from each solution vector generated at stage 1, the application of stage 2 finds a set of the corresponding tier-one local optimal solutions, tier-two local optimal solutions, and so on.

Depending on the structure of the search space, different local search methods can be selected for computing the local optimal solutions at the different parts of the search space. Regarding the local search method, effectiveness is the guideline of selection. Depending on the structure of the search space near the local optimal solution of interest, it can be a general-purpose one such as the steepest descent method, the Newton-type methods, conjugate gradient methods, or a tailored local method. Also, depending on the optimization problem under study, it may be desired to use an appropriate meta-heuristic to generate a set of good quality of solutions for stage 2. Hence, despite the general-purpose nature of the proposed Trust-Tech method, it is advantageous, depending on the underlying application, to adopt an appropriate meta-heuristic method, different local search methods, and even to design different nonlinear dynamical systems (11.4) and (11.5).

11.7 NUMERICAL SCHEMES

From a theoretical viewpoint, the Trust-Tech method can find all the local optimal solutions of the general constrained nonlinear programming (11.2) satisfying the required mild conditions. However, an effective numerical implementation of the method may represent a great challenge. We discuss some of these issues involved in this section.

Phase I of the Trust-Tech method can be numerically implemented by effective ordinary differential equation (ODE) solvers or difference equation solvers. One numerical difficulty that can arise in phase I is how to numerically differentiate two found equilibrium manifolds. This difficulty is due to the fact that equilibrium manifolds are in general collections of infinitely many points and, hence, it is difficult to differentiate two equilibrium manifolds. In addition, it may well be due to the lack of computational methods that can compute the complete set of unstable equilibrium manifolds, phase I of the Trust-Tech method can only compute multiple feasible components, instead of all of the feasible components (i.e., the feasible region) for general constrained optimization problems.

Another challenging task in the numerical implementation of phase II is the search of all the decomposition points that lie on the quasi-stability boundary of a given stable equilibrium point (i.e., a local optimal solution). In the past, significant efforts have been devoted to a similar topic in the fields of chemistry and physics [19–21], but without much success. We adopted, in this paper, a boundary search method proposed in Ref. 16 to compute decomposition points. The key idea of boundary search method is based on the following geometric observation: the "energy" landscape of the quasi-stability boundary resembles a valley in which a decomposition point geometrically corresponds with the local bottom of the valley. To this end, boundary search method first makes an approximation surface to the quasi-stability boundary where decomposition points become a stable equilibrium point of system (11.5) restricted to the approximation surface. Then by following the path of system (11.5) restricted to the approximation surface, it locates a decomposition point.

The task of selecting proper search direction vectors from a given SEP is also very challenging. Improper direction vectors can lead to the situation that several direction vectors all converge to the same decomposition point in which case computational resources are wasted and the computational efficiency of the Trust-Tech method is degraded. It appears that effective direction vectors in general have a close relationship with the structure of the objective function and feasible set. Hence, exploitation of the structure of the objective under study will prove fruitful in selecting direction vectors. (For the detailed numerical implementation of the boundary search method and other methods for this purpose, see Ref. 16.)

11.8 NUMERICAL STUDIES

In this section, we shall give two examples and a graphical explanation to illustrate the proposed Trust-Tech method. The first example emphasizes phase II of the method.

Example 11.1

Find all the local optimal solutions of the following minimization problem:

$$\text{minimize } f(x, y) := \sum_{i=0}^{20} (\sin((-2 + i/5)x + y) + 0.1(x^2 + y^2) - h_i)^2$$

$$\text{subject to } g(x, y) := -0.75 + 0.8838x + 3.9062x^2$$
$$-0.8838y + 4.6875y + 3.9062 \le 0,$$

where $10h_i$ is defined by the data -10, -5, -2, -1, 0, 2, 3, 6, 9, 10, 9, 6, 2, 2, 1, -2, -3, -5, -7, -9, -8. By transforming inequality constraint into equality constraint by adding a slack variable s^2, we obtain the following equivalent problem

$$\text{minimize } f(x, y) := \sum_{i=0}^{20} (\sin((-2 + i/5)x + y) + 0.1(x^2 + y^2) - h_i)^2$$

$$\text{subject to } g(x, y) := -0.75 + 0.8838x + 3.9062x^2$$

$$-0.8838y + 4.6875y + 3.9062 + s^2 = 0.$$

Phase I builds the following quotient gradient system

$$\begin{bmatrix} \dot{x} \\ \dot{y} \\ \dot{s} \end{bmatrix} = -h(x, y, s)\nabla h(x, y, s) = \begin{bmatrix} -h(x, y, s)\dfrac{\partial}{\partial x}h(x, y, s) \\ -h(x, y, s)\dfrac{\partial}{\partial y}h(x, y, s) \\ -h(x, y, s)\dfrac{\partial}{\partial s}h(x, y, s) \end{bmatrix} \tag{11.6}$$

and phase II builds the following projected gradient system

$$\begin{bmatrix} \dot{x} \\ \dot{y} \\ \dot{s} \end{bmatrix} = -\left(I - \frac{\nabla h(x, y, s)\nabla h(x, y, s)^T}{\|h(x, y, s)\|^2}\right)\nabla f(x, y, s). \tag{11.7}$$

First, we choose an initial guess $x_0 = (1, 0, 0)$ that is not a feasible point and set $\varepsilon = 0.001$. We next describe the search process of the proposed method in computing multiple local optimal solutions of this example.

Phase I:

(i) By integrating the quotient gradient system (11.4) starting from x_0, the system trajectory converges to the point $x_1 = (0.1, 0, 0.8)$, which is a feasible point located in a feasible component.

Phase II:

(i) By integrating the projected gradient system (11.5) starting from x_1, the corresponding trajectory converges to a local optimal solution $x_s^1 = (-0.0118, 0.0743)$ (with a non-negative S value).

(ii) Applying the algorithm for finding decomposition points proposed in the previous section with four direction vectors $v_1, -v_1, v_2, -v_2$, where v_1, v_2 are eigenvectors of the Hessian matrix of the objective function f restricted to the tangent space $T_{x_s}M$, we obtain the following three decomposition points:

$$x_d^1 = (-0.7201, 0.5472)$$
$$x_d^2 = (0.2269, 0.2712)$$
$$x_d^3 = (-0.1661, -0.2885).$$

TABLE 11.1 Local Minimal Solutions and Decomposition Points of Example 11.1

Local minimal solutions	Objective value $f(\cdot)$	Decomposition points
$x_s^1 = (-0.0118, 0.0743)$	7.1041	$x_d^1 = (-0.7201, 0.5472)$
$x_s^2 = (-0.8745, 0.8135)$	12.5141	$x_d^2 = (0.2269, 0.2712)$
		$x_d^3 = (-0.1661, -0.2885)$

(iii) By integrating the projected gradient system (11.5) starting from these decomposition points obtained in Phase II: Step (ii), the resulting trajectory converges to another local optimal solution, $x_s^2 = (-0.8745, 0.8135)$ (with a non-negative S value). Because no other new decomposition point of x_s can be found, the search process of phase II is completed.

We illustrate the search process of phase I and phase II in Table 11.1. We notice that there is only one connected constraint, and the numerical Trust-Tech method finds all the local optimal solutions. By comparing all the local optimal solutions, the global optimal solution is found at $x_s^1 = (-0.0118, 0.0743)$ with its value 7.1041.

Example 11.2: Applications to network partition problems.

We next apply the Trust-Tech method to optimal network partition of a very-large-scale integration (VLSI) network, which is a discrete optimization problem. Due to its complexity, a VLSI system is partitioned at several levels for analysis and design: (1) system level partitioning, in which a system is partitioned into a set of subsystems whereby each subsystem can be designed and fabricated independently on a single PCB; (2) board level partitioning, where a PCB is partitioned into VLSI chips; (3) chip level partitioning, where a chip is divided into smaller subcircuits. At each level, the constraints and objectives of the partitioning are usually different. A partition is also referred to as *a cut*.

The partition problem belongs to the category where no closed-form objective function is available. And it appears that no distance can be defined to fill the continuous space between two instances of partitions. The problem is thus a discrete optimization problem and our proposed Trust-Tech method cannot be directly applied.

We have developed a discrete version of our Trust-Tech method to solve the network partition problem. As to the local search method, the industry standard Fiduccia and Matheyses (FM) method is used. We have implemented the discrete version of our Trust-Tech method for solving the network partition problem and evaluated it on several benchmark problems. The test results shown in Table 11.2 are obtained using the minimum cut-sets of 50 runs from different initial points under 45% to 55% area balance criterion and real cell sized. The partition results obtained by using the standard FM method are also shown as a comparison.

TABLE 11.2   Partition Results on Benchmark Problems

Benchmark circuits	Number of cells	Number of cut-sets obtained by the FM method	Number of cut-sets obtained by the Trust-Tech method	Improvement by the Trust-Tech method
S38584	22451	199	54	268%
S38417	25589	405	120	238%
S13207	9445	91	78	17%
S15850	11071	144	79	82%
S35932	19880	120	100	20%
S5378	3225	98	76	29%

By incorporating the Trust-Tech method with the traditional FM method, the quality of the solution is significantly improved, which is shown in Table 11.2. The improvements range from 9.5% to 268%, where large circuits achieve a big improvement and small circuits achieve a small improvement. Overall, the Trust-Tech method is significantly better than the FM method. The other improvement of the Trust-Tech method is that the solution deviation is significantly smaller compared with the same indices of the FM method. This improvement means that the Trust-Tech method needs much fewer runs than the FM method to get a good solution.

11.9   CONCLUSIONS REMARKS

We have presented in this chapter a Trust-Tech method for systematically computing multiple high-quality local optimal solutions. One distinguished feature of the Trust-Tech method is that it systematically computes all the local optimal solutions of constrained (continuous) nonlinear optimization problems via a two-phase approach: phase I, starting from an arbitrary starting point, systematically finds all the feasible regions that satisfy nonlinear equality and inequality constraints. The starting point of phase I can be an infeasible point. Phase II then computes all the local optimal solutions in each (connected) feasible region found in phase I. To numerically implement phase I, we have introduced an induced gradient-like dynamical system whose trajectories can locate all the connected feasible regions. By following the system trajectory from an arbitrary initial point, a feasible point lying in a feasible region can be obtained. To account for the fact that the feasibility set is in general composed of several (disjoint) connected feasible regions, one can move away from the feasible region containing the feasible point and move toward another feasible point located in a different feasible region by following the system trajectories in reverse time. Once a feasible region is found, phase II employs the trajectories of a projected gradient system to visit all the local optimal solutions in the feasible region. The Trust-Tech paradigm translates the task of finding all the feasible components of general constrained optimization problems into the task of finding all the stable equilibrium manifolds of the so-called quotient gradient system whose vector field is associated with the constraint set of the optimization problems. In addition, it translates the task of finding all the local optimal solutions located in a feasible component of general constrained optimization problems into the task of finding all the stable equilibrium points of the so-called projected gradient system. The complexity of the two-phase method depends on the number of quasi-stability regions rather on the system dimension. The theoretical foundation of the two-phase method has been developed.

From a practical viewpoint, several numerical implementation issues of the Trust-Tech method still require further development. Further development of effective methods to compute the complete set of decomposition points and to compute the complete set of unstable equilibrium manifolds associated with the optimization problem under study are needed. A numerical implementation of the Trust-Tech method developed in this paper requires an ODE or differential and algebraic equation (DAE) solver. It was thought by many researchers that ODE or DAE solvers are too extensive to be of general use compared with, say, a conventional Newton-type method and are seen as some kind of occasional last resort for the case when the other methods experience difficulty. However, their performance is significantly affected by the choice of integration method and the way in which the step size is controlled. When suitably implemented, these methods deserve a place in the mainstream of optimization algorithm development [12].

REFERENCES

1. Fletcher P. Practical methods of optimization. New York: Wiley; 1987.

2. Floudas CA, Pardalos PM. Recent advances in global optimization. Princeton Series in Computer Science. Princeton, NJ: Princeton University Press; 1992.

3. Papalambros P, Wilde DJ. Principles of optimal design—modeling and computation. New York: Cambridge University Press; 1986.

4. Floudas CA, Pardalos PM, eds. State of the art in global optimization. Boston: Kluwer Academic Publishers; 1996.

5. Horst R, Tuy H. Global optimization: deterministsic approaches. Boston: Kluwer Academic Publishers; 1994.

6. Horst R, Pardalos PM, Thoai NV. Introduction to global optimization. Boston: Kluwer Academic Publishers; 1995.

7. Pardalos PM, Rosen JB. Constrained global optimization: algorithms and applications. Lecture Notes in Computer Science, Vol. 268. New York: Springer-Verlag, 1987.

8. Chiang HD, Chu C-C. A systematic search method for obtaining multiple local optimal solutions of nonlinear programming problems. IEEE Trans Circuits Systems, 1996; 43:99–106.

9. Lee J, Chiang H-D. A dynamical trajectory-based methodology for systematically computing multiple optimal solutions of nonlinear programming problems. IEEE Trans Automatic Control, 2004; 49(6):888–899.

10. Lee J. Trajectory-based methods for global optimization. Ph.D. dissertation, Cornell Univ., Ithaca, NY; 1999.

11. Jongen HTh, Jonker P, Twilt F. Nonlinear optimization in R_n. Frankfurt: Peter Lang Verlag; 1986.

12. Brown AA, Bartholomew-Biggs MC. Some effective methods for unconstrained optimization based on the solution of systems of ordinary differential equations. J Optim Theory Appl. 1989; 62:211–224.

13. Chiang HD, Hirsch MW, Wu FF. Stability region of nonlinear autonomous dynamical systems, IEEE Trans Automatic Control, 1988; 33:16–27.

14. Chiang HD, Fekih-Ahmed L. Quasi-stability regions of nonlinear dynamical systems: Theory. IEEE Trans Circuits Systems, 1996; 43(8):627–635.

15. Guckenheimer J, Homes P. Nonlinear oscillations, dynamical systems, and bifurcations of vector fields. Applied Mathematical Sciences, Vol. 42. New York: Springer; 1986.

16. Lee J, Chiang HD. Theory of stability regions for a class of nonhyperbolic dynamical systems and its application to constraint satisfaction problems. IEEE Trans Circuits Systems I: Fundam Theory Appl. 2002; 49(2):196–209.

17. Liotard DA. Algorithmic tools in the study of semiempirical potential surfaces. Int J Quantum Chem 1992; 44:723–741.

18. Mousseau N, Barkerma GT. Traveling through potential energy landscapes of disordered materials: The activation-relaxation technique. Phys Rev E 1998; 57(2):2419–2424.

19. Sinclair JE, Fletcher R. A new method of saddle-point location for the calculation of defect migration energies. J. Phys Solid State Phys 1974; 7:864–870.

20. Liotard DA. Algorithmic tools in the study of semiempirical potential surfaces. Int J Quantum Chem 1992; 44:723–741.

21. Mousseau N, Barkerma GT. Traveling through potential energy landscapes of disordered materials: The activation-relaxation technique. Phys Rev E 1998; 57(2):2419–2424.

SELECTED APPLICATIONS OF MODERN HEURISTIC OPTIMIZATION IN POWER SYSTEMS

CHAPTER 12

Overview of Applications in Power Systems

ALEXANDRE P. ALVES da SILVA, DJALMA M. FALCÃO, and KWANG Y. LEE

12.1 INTRODUCTION

This survey covers the broad area of evolutionary computation applications to optimization, model identification, and control in power systems. Almost all reviewed papers have been published in the *IEEE Transactions* and the *IEE Proceedings*. A total of 166 articles are listed in this survey, and these sources are grouped by topical area in the References section. This survey shows the development of this broad area and identifies the current trends. The following techniques are considered under the scope of evolutionary computation: evolutionary algorithms (e.g., genetic algorithms, evolution strategies, evolutionary programming, and genetic programming), simulated annealing, tabu search, and particle swarm optimization.

12.2 OPTIMIZATION

Optimization is the basic concept behind the application of evolutionary computation (EC) to any problem in power systems [1, 2]. Besides the problems in which optimization itself is the final goal [3–89], it is also a means for modeling/forecasting [109–119], control [120–135], and simulation [145, 146]. Optimization models can be roughly divided in two classes: continuous (involving real variables only) and discrete (with at least one discrete variable). The objective function(s) (single or multiple) and the constraints of the problem can be linear or nonlinear, convex or concave. Optimization techniques have been applied to several problems in power systems. Thermal unit commitment/hydrothermal coordination and economic dispatch/optimal power flow, maintenance scheduling, reactive sources allocation, and expansion planning are among the most important applications.

Modern heuristic search techniques such as evolutionary algorithms are still not competitive for continuous optimization problems such as economic dispatch and optimal power flow. Successive linear programming, interior point methods, projected augmented Lagrangian, generalized reduced gradient, augmented Lagrangian methods, and sequential quadratic programming have all a long history of successful applications to this type of problem. However, the trade-off between modeling precision and optimality has to be taken into account.

One good example is when the input–output characteristics of thermal generators are highly nonlinear (non-monotonically increasing) due to effects such as "valve points" [32, 40, 42]. In this situation, their incremental fuel cost curves cannot be reasonably approximated by quadratic or piecewise quadratic (or linear) functions. Therefore, traditional optimization techniques, although achieving *mathematical optimality*, have to sacrifice modeling accuracy, providing suboptimal solutions in a practical sense. Evolutionary computation algorithms allow precise modeling of the optimization problem, although usually not providing mathematically optimal solutions, but near optimal ones. Another advantage of using EC for solving optimization problems is that the objective function does not have to be differentiable.

For discrete optimization problems, classic mathematical programming techniques have collapsed for large-scale problems. Optimization algorithms such as "branch and bound" and dynamic programming, which seek for the best solution, have no chance in dealing with the above-mentioned problems, unless significant simplifications are assumed (e.g., besides the curse of dimensionality, dynamic programming has difficulty in dealing with time-dependent constraints). Problems such as generation scheduling have the typical features of large-scale combinatorial optimization problems (i.e., non-deterministic polynomial time (NP)-complete problems cannot be solved for the optimal solution in a reasonable amount of time). For this class of problems, general-purpose heuristic search techniques (problem independent), such as EC, have been very efficient for finding near optimal solutions in reasonable time.

12.3 POWER SYSTEM APPLICATIONS

Generation scheduling is one of the most popular applications of EC to power systems [3–31]. The pioneer work of Zhuang and Galiana [3] inspired subsequent papers on the application of general-purpose heuristic methods to unit commitment and hydrothermal coordination. Realistic modeling is possible when solving these problems with such methods. The general problem of generation scheduling is subject to constraints such as power balance, minimum spinning reserve, energy constraints, minimum and maximum allowable generations for each unit, minimum up- and down-times of thermal generation units, ramp rates limits for thermal units, and level constraints of storage reservoirs. Incorporation of crew constraints, take-or-pay fuel contract [7], water balance constraints caused by hydraulic coupling [9], rigorous environmental standards [13], multiple-fuel-constrained generation scheduling [16], and dispersed generation and energy storage [21] are possible with EC techniques.

A current trend is the utilization of EC only for dealing with the discrete optimiz-ation problem of deciding the on/off status of the generation units. Mathematical pro-gramming techniques are employed to perform the economic dispatch while meeting all plant and system constraints. Expansion planning is another area that has been extensively studied with the EC approach [55–89]. The transmission expansion plan-ning for the North-Northeastern Brazilian network, for which optimal solution is unknown, has been evaluated using genetic algorithms (GAs). The estimated cost is about 8.8% less than the best solution obtained by conventional optimization [78]. Economic load dispatch/optimal power flow [32–50] and maintenance schedul-ing [51–54] have also been solved by EC methods. Another interesting application is the simulation of energy markets [145, 146].

12.4 MODEL IDENTIFICATION

System identification methods can be applied to estimate mathematical models based on measurements. Parametric and non-parametric are the two main classes in system identification methods. The parametric methods assume a known model structure with unknown parameters. Their performances depend on a good guess of the model order, which usually requires previous knowledge of the system character-istics. System identification can be used for modeling a plant or a problem solution (e.g., pattern recognition [112, 118]). The following sections show a few examples of successful applications of EC to identification.

12.4.1 Dynamic Load Modeling

The fundamental importance of power system components modeling has been shown in the literature. Regardless of the study to be performed, accurate models for trans-mission lines, transformers, generators, regulators, and compensators have already been proposed. However, the same has not occurred for loads. Although the import-ance of load modeling is well-known, especially for transient and dynamic stability studies, the random nature of a load composition makes its representation very difficult.

Two approaches have been used for load modeling. In the first one, based on the knowledge of the individual components, the load model is obtained through the aggregation of the load components models. The second approach does not require the knowledge of the load physical characteristics. Based on measurements related to the load responses to disturbances, the model is estimated using system identifi-cation methods.

The composition approach requires information that is not generally available, which consists in a disadvantage of this method. This approach does not seem to be appropriate because the determination of an average (and precise) compos-ition for each load bus of interest is virtually impossible. The second approach does not suffer from this drawback, because the load to be modeled can be assumed a *black box*. However, a significant amount of data related to staged tests and natural disturbances affecting the system need to be collected.

Considering the shortcomings of the two approaches, and the fact that data acquisition and processing are becoming very cheap, it seems that the system identification approach is more in accordance with current technology. This approach allows real-time load monitoring and modeling, which are necessary for on-line stability analysis. As the dynamic characteristics of the loads are highly nonstationary, structural adaptation of the corresponding mathematical models is necessary. Evolutionary computation can be used in this adaptive process, searching for new model structures and parameters. Examples of this possibility are described in Refs. 113 and 115.

12.4.2   Short-Term Load Forecasting

The importance of short-term load forecasting has increased lately. With deregulation and competition, energy price forecasting has become a big business. Load bus forecasting is essential for feeding the analytical methods used for determining energy prices. The variability and nonstationarity of loads are getting worse due to the dynamics of energy tariffs. Besides, the number of nodal loads to be predicted does not allow frequent interventions from load forecasting specialists. More autonomous load predictors are needed in the new competitive scenario.

With power systems growth and the increase in their complexity, many factors have become influential in electric power generation and consumption (load management, energy exchange, spot pricing, independent power producers, nonconventional energy, etc.). Therefore, the forecasting process has become even more complex, and more accurate forecasts are needed. The relationship between the load and its exogenous factors is complex and nonlinear, making it quite difficult to model through conventional techniques, such as time series linear models and linear regression analysis. Besides not giving the required precision, most of the traditional forecasting techniques are not robust enough. They fail to give accurate predictions when quick weather changes occur. Other problems include noise immunity, portability, and maintenance.

Linear methods interpret all regular structure in a data set, such as a dominant frequency, as linear correlations. Therefore, linear models are useful if and only if the power spectrum is a useful characterization of the relevant features of a time series. Linear models can only represent an exponentially growing or a periodically oscillating behavior. Therefore, all irregular behavior of a system has to be attributed to a random external input to the system. Chaos theory has shown that random input is not the only possible source of irregularity in a system's output.

The goal in creating an autoregressive moving average (ARMA) model is to have the residual as white noise [111]. This is equivalent to producing a flat power spectrum for the residual. However, in practice, this goal cannot be perfectly achieved. Suspicious anomalies in the power spectrum are very common (i.e., the residual's power spectrum is not really flat). Consequently, it is difficult to say if the residual corresponds with white noise or if there is still some useful information to be extracted from the time series. Neural networks can find predictable patterns that cannot be detected by classic statistical tests such as auto(cross)correlation coefficients and power spectrum.

Besides, many observed load series exhibit periods during which they are less predictable, depending on the past history of the series. This dependence on the past of the series cannot be represented by a linear model [119]. Linear models fail to consider the fact that certain past histories may permit more accurate forecasting than others. Therefore, differently from nonlinear models, they cannot identify the circumstances under which more accurate forecasts can be expected.

The neural networks (NNs) ability in mapping complex nonlinear relationships is responsible for the growing number of their applications to load forecasting. Several electric utilities, all over the world, have been applying NNs to short-term load forecasting in an experimental or operational basis. Despite their success, there are still some technical issues that surround the application of NNs to load forecasting, particularly with regard to parameterization. The main issue in the application of feedforward NNs to time series forecasting is the question of how to achieve good generalization. The NN ability to generalize is extremely sensitive to the choice of the network's architecture, preprocessing of data, choice of activation functions, number of training cycles, size of training sets, learning algorithm, and the validation procedure.

The greatest challenges in NN training are related to the issues raised in the previous paragraph. The huge number of possible combinations of all NN training parameters makes its application not very reliable. This is especially true when a nonstationary system has to be tracked (i.e., adaptation is necessary), as it is the case in load forecasting. Nonparametric NN models have been proposed in the literature [114]. With nonparametric modeling methods, the underlying model is not known, and it is estimated using a large number of candidate models to describe available data. Application of this kind of model to short-term load forecasting has been neglected in the literature. Although the very first attempt to apply this idea to short-term load forecasting dates back to 1975 [109], it is still one of the few investigations on this subject, despite the tremendous increase in computational resources.

12.4.3 Neural Network Training

The main motivation for developing nonparametric NNs is the creation of fully data-driven models (i.e., automatic selection of the candidate model of the right complexity to describe the training data). The idea is to leave for the designer only the data-gathering task. Obviously, the state of the art in this area has not reached that far. Every so-called nonparametric model still has some dependence on a few preset training parameters. A very useful by-product of the automatic estimation of the model structure is the selection of the most significant input variables for synthesizing a desired mapping. Input variable selection for NNs has been performed using the same techniques applied to linear models. However, it has been shown that the best input variables for linear models are not among good input variables for nonlinear ones.

12.4.3.1 *Pruning Versus Growing* Nonparametric NN training uses two basic mechanisms for finding the most appropriate architecture: pruning and

growing. Pruning methods assume that the initial architecture contains the *optimal* structure. It is common practice to start the search using an oversized network. The excessive connections and/or neurons have to be removed during training while adjusting the remaining parts. The pruning methods have the following drawbacks:

- There is no mathematically sound initialization for the neural network architecture, therefore initial guesses usually use very large structures; and
- Because of the previous argument, a lot of computational effort is wasted.

As the growing methods operate in the opposite direction of the pruning methods, the shortcomings mentioned before are overcome. However, the incorporation of one element has to be evaluated independently of other elements that could be added later. Therefore, pruning should be applied as a complementary procedure to growing methods, in order to remove parts of the model that become unnecessary during the constructive process.

12.4.3.2 Types of Approximation Functions The greatest concern when applying nonlinear NNs is to avoid unnecessary complex models to not overfit the training patterns. The *ideal* model is the one that matches the complexity of the available data. However, it is desirable to work with a general model that could provide any required degree of nonlinearity. Among the models that can be classified as universal approximators (i.e., the ones that can approximate any continuous function with arbitrary precision), the following types are the most important:

- Multilayer networks
- Local basis function networks
- Trigonometric polynomials
- Algebraic polynomials

The universal approximators above can be linear, although nonlinear in the inputs, or nonlinear in the parameters. Regularization criteria, analytic (e.g., Akaike's information criterion, minimum description length, etc.) or based on resampling (e.g., cross-validation), have been proposed. In practice, model regularization considering nonlinearity in the parameters is very difficult. An advantage of using universal approximators that are linear in the parameters is the possibility of decoupling the exploration of architecture space from the weight space search. Methods for selecting models with nonlinearity in the parameters attempt to explore both spaces simultaneously, which is an extremely hard nonconvex optimization problem.

12.5 CONTROL

Another important application of EC in power systems is the parameter estimation and tuning of controllers. Complex systems cannot be efficiently controlled by standard feedback, because the effects caused by plant parameter variations are not eliminated. Adaptive control can be applied in this case (i.e., when the plant is time invariant with partially unknown parameters or the plant is time variant).

In practical applications, it is difficult to express the real plant dynamics in mathematical equations. Adaptive control schemes can adjust the controller according to process characteristics, providing a high performance level. The adaptive control problem is concerned with the dynamic adjustment of the controller parameters, such that the plant output follows the reference signal.

However, conventional adaptive control has some drawbacks. Existent adaptive control algorithms work for specific problems. They do not work as well for a wide range of problems. Every application must be analyzed individually (i.e., a specific problem is solved using a specific algorithm). Besides, a compatibility study between the model and the adaptive algorithm has to be performed.

The benefits of coordination between voltage regulation and damping enhancements in power systems are well-known. This problem has been aggravated as power networks operational margins decrease. GAs can be used to simultaneously tune the parameters of automatic voltage regulators with the parameters of power system stabilizers and terminal voltage limiters of synchronous generators [120–135].

One of the objectives is to maximize closed-loop system damping, which is basically achieved by tuning the stabilizers parameters. The second objective is to minimize terminal voltage regulating error, which is mainly accomplished by tuning the automatic voltage regulators and terminal voltage limiters parameters. Minimum closed-loop damping and maximum allowable overshoot are constraints that can be incorporated into the optimization problem. The design of the control system takes into account system stabilization over a prespecified set of operating conditions (nominal and contingencies).

12.5.1 Examples

Examples of the application of GAs to the coordinated tuning of power system damping controllers are described in [123, 130]. In the first work, a GA is used to tune the controllers of two flexible alternating current transmission system (FACTs) devices (static VAR compensator (SVC) and thyristor-controlled series capacitor (TCSC)) used to damp oscillations in a hypothetical 3 areas and 6 machines power system. In the second one, the same method is used to simultaneously tune 22 power system stabilizers in a configuration of the Brazilian power system with 57 synchronous machines and 1762 buses.

In both applications, the tuning of the power system controllers is formulated as an optimization problem as follows:

$$\max F = \sum_{i=1}^{m} \left[\sum_{j=1}^{n} (\zeta_j) \right]_i, \tag{12.1}$$

subject to

$$
\begin{aligned}
K_{ii_{min}} &\leq K_{ii} \leq K_{ii_{max}} \\
\alpha_{i_{min}} &\leq \alpha_i \leq \alpha_{i_{max}} \\
\omega_{i_{min}} &\leq \omega_i \leq \omega_{i_{max}} \\
\zeta_{min} &\leq (\zeta_j)_i,
\end{aligned}
\tag{12.2}
$$

where n is the system order, m is the number of operating conditions, ζ is the closed-loop system eigenvalue damping ratio, and K, α, and ω are controllers parameters. The problem defined in (12.1) and (12.2) is a complex optimization problem with an implicit objective function, which depends on the evaluation of the eigenvalues of a large matrix. This problem is very difficult to solve using conventional methods.

The fitness function f used by the GA is defined as

$$
f = \begin{cases}
0 \rightarrow if\,any(\zeta_j)_i \leq \zeta_0 \\
\beta_0 \rightarrow if\,all(\zeta_j)_i > \zeta_0\,and\,if\,any(\zeta_j)_i \leq 0 \\
\beta_1 \rightarrow if\,all(\zeta_j)_i > 0\,and\,if\,any(\zeta_j)_i \leq \zeta_1 \\
\beta_2 \rightarrow if\,all(\zeta_j)_i > \zeta_1\,and\,if\,any(\zeta_j)_i \leq \zeta_2 \\
\quad \vdots \\
\beta_K \rightarrow if\,all(\zeta_j)_i > \zeta_{K-1}\,and\,if\,any(\zeta_j)_i \leq \zeta_{min} \\
F \rightarrow if\,all(\zeta_j)_i \geq \zeta_{min}
\end{cases}
\tag{12.3}
$$

where $\beta_0 < \beta_1 < \beta_2 < \cdots < \beta_K < F$ are positive scalars specified to discriminate the stable solutions with performance constraint not satisfied, first from the unstable solutions, and second from the stable solutions with all constraints satisfied,

$$
\zeta_0 < 0 < \zeta_1 < \zeta_2 < \cdots < \zeta_{K-1} < \zeta_{min}.
$$

In the tuning of FACT devices [123], the GA uses a binary encoding, proportional selection, and crossover and mutation probabilities set to 0.6 and 0.001, respectively. Considering three operating conditions, the GA found a set of controller parameters that produce damping ration for all oscillation modes considerably higher than the ones obtained with a conventional design technique. For a large-scale model of the Brazilian system, three operating conditions were also considered. In this case, the GA uses real encoding, tournament selection, elitism, and a special technique to vary the search space during execution. The GA approach was able to find a solution with at least 5% damping ratio, which is a performance similar to the controllers actually implemented in the system. Nonlinear simulation tests with the controllers obtained with the GA approach indicated a performance slightly better than the ones set up by the experts.

The result of the work reported above showed that in a multimachine power system, fixed-structure damping controllers can be tuned to provide satisfactory damping performance over a prespecified set of operating conditions. The GA-based tuning process has shown robustness in achieving controller satisfying the design criteria in a large-scale realistic power system.

12.6   DISTRIBUTION SYSTEM APPLICATIONS

Distribution system operational and expansion planning offer a great opportunity for the application of EC and other intelligent system techniques [147]. Several problems in this area require modeling of uncertainties not easily modeled by probabilistic methods like load forecasting and estimation. Others require classification and clustering techniques. Several problems, like the ones described in this section, are of the combinatorial optimization type.

12.6.1 Network Reconfiguration for Loss Reduction

Distribution network reconfiguration, performed by changing the state of a set of switches normally open/closed, may be required for several purposes like loss reduction, load balancing, service restoration, and so forth. Although different in purpose, all reconfiguration problems have in common the combinatorial optimization characteristic and nonlinear mathematical models. In all cases, the reconfiguration method must take into consideration some constraints like network connectivity, radial configuration, limits on voltage drop and power flow in branches, and so forth.

Reconfiguration for loss reduction, the main purpose of the work described in this section, is usually performed considering the network in its normal operating state with the objective of achieving energy efficiency. The modeling of distributed generation (DG) directly connected to the distribution system is a feature that is becoming important nowadays. As a by-product of the optimization process, usually a better voltage regulation is also achieved.

The GA used to generate the potential solutions to the reconfiguration problem uses a binary representation in which each bit represents the state of a switch (open/closed). In order to reduce the dimensionality of the network to be analyzed, the concept of load block is used (i.e., all load points inside a region of the network isolated by switches are treated as a unique load point).

The fitness function is defined as the total active loss of the network evaluated using a load flow algorithm. The result of the load flow calculations is also used to check the voltage drop constraints. The constraints on the radial and connect characteristics of the network are dealt with using two approaches:

- *Rejection*: Any solution given by the GA not satisfying at least one constraint is attributed a zero fitness value (i.e., it will be certainly rejected in the next selection cycle); or
- *Repair*: Some solutions not satisfying the constraints go through a process of slight modification (repair) in order to make it feasible.

The method described in this section was tested using both a small test system available in the literature and an actual distribution network [148]. The first one is composed of 5 distribution feeders arranged in two substations, with a total of 26 switching devices. The second one is composed of 7 feeders with a total of 121 switching devices used for reconfiguration purposes. The influence of distributed generation was modeled in both tests. The GA used in the tests uses the computational platform described in Ref. [147], with tournament selection, a 4 individuals elitist strategy, and with crossover probability and mutation probabilities set to 0.95 and 0.001, respectively. It was observed that a key point for the efficiency of the method is to start with a population of feasible solutions. The results of the GA solutions were compared with a basic configuration established by an expert. For the first system, the GA found configurations that provided reductions of 28% and 48% for the cases without and with DG, respectively. For the second system, the corresponding percentages in the active loss reduction were 11.4% and 11.7%.

12.6.2 Optimal Protection and Switching Devices Placement

Most of the customer interruptions occur due to failures in the distribution network. One way of reducing the number and duration of these interruptions is by installing protection and switching devices in certain locations of the network in order to reduce the amount of customers affected by a fault. Device type and placement should be determined taking into consideration investment and operational costs and the improvement in the network reliability measured by several indexes.

The problem of selecting places for installation of protection devices in a large radial distribution network is a very complex combinatorial optimization problem. The traditional criteria used to decide which type and place to install these devices are based on the protection philosophy adopted in the company, including selectivity and coordination analysis. More recently, criteria based on the benefits of these devices to system reliability have been increasingly used in this type of analysis.

The approach reported in this section [148] is composed of two parts:

- A procedure to compute the impact of new switching and protection devices installation on the system reliability indexes based on analytical techniques. This procedure can be used to evaluate the solutions proposed by an expert or as a component of the methodology described below.
- An automatic methodology that uses a GA as an optimization tool and the analytical procedure described above for the assessment of the potential solutions. The GA replaces the expert in the task of generating potential solutions to the problem. Considerations regarding the protection system selectivity and coordination are included in the methodology as weights in the fitness function used in the GA, as will be explained later.

The solution found by the methodology using GA is a compromise between the costs incurred by the installation and operation of the switching and protection devices and the reliability indexes while avoiding configurations that violate practical experience rules of distribution protection engineering. The fitness function used to drive the GA to the solution is the ratio between the benefit achieved by the installation of switching and protection devices, measured by the improvements in the expected outage cost to customers (ECOST), and the investment and operational costs incurred by these device installations. Formally, this function is defined as

$$\mathbf{F} = K \frac{\text{Benefit}}{\text{Total cost}}, \tag{12.4}$$

where Benefit = ECOST (before) − ECOST (after); Total cost = $C_{inv} + C_{inst} + C_{man}$; C_{inv} is investment costs; C_{int} is installation costs; C_{inv} is maintenance costs; and $K < 1$: constant enforcing penalties on potential solutions that do not satisfy fully the constraints defined in the next section.

Experts on distribution network protection impose restrictions on configuration of some switching and protection devices. The information from practical experience is introduced in the GA formulation through the constant K. In this work, the potential solutions violating the following constraints are penalized for:

- More than three fuse devices in series owing to coordination difficulties;
- Fuse devices on main feeders;
- Fuse devices located upstream of sectionalizing switches; and
- System average interruption duration index (SAIDI) and system average interruption frequency index (SAIFI) below predefined thresholds.

Besides the technical constraints introduced above, potential solutions with total cost above predefined values are also penalized. The GA used in this work is based on the simple or canonical GA described in Ref. [149] with a few changes to improve its performance. The main characteristics of the GA are

- *Solution encoding*: The possible solution to the problem is encoding using 2 bits for each gene as follows:
    - 00: no device
    - 01: sectionalizing switch
    - 10: fuse
    - 11: recloser
- *Genetic operators*: The GA uses a proportional selection scheme based on the roulette wheel method and uniform crossover; the mutation ratio decreases exponentially with the number of generations.
- *Stopping rule*: The rule adopted is a fixed number of generations.

The approach to protection devices placement presented above has been preliminarily tested with examples reported in the literature as well with actual distribution networks of Brazilian utilities. Part of the results achieved is reported in Ref. 149. These results show a considerable improvement in the results compared with the ones based only in practical expertise.

12.6.3 Prioritizing Investments in Distribution Networks

Privatization and deregulation have imposed severe challenges for distribution companies, both in terms of investments and charges for not achieving the degree of quality in the energy supply imposed by the regulatory agencies. In such a scenario,

TABLE 12.1   Example of the Prioritizing

6/5/8	4/1	3/7	2/10
Int. #1	Int. #2	Int. #3	Int. #4

the optimization of the investments in the distribution system becomes a priority for the utilities.

The usual procedure concerning the planning of reinforcements in distribution networks (renewal or replacement of components) starts with the establishment of a long-term plan (several years), based on costs, rough load growth forecasting, and technical aspects of the problem. This long-term plan is materialized as sequence of projects in time in which each project corresponds with the renewal or replacements of a network component. The decision whether to actually implement or not the long-term plan has to be taken looking to a shorter-term horizon (1 year, for instance) and taking into consideration a cash flow analysis and technical evaluation. Several factors have to be taken into consideration in this analysis:

- Investment costs
- Operation and maintenance costs
- Minimum reliability requirements
- Utility public image
- Sales opportunities
- Et cetera

The distribution company policy is, usually, to postpone the investments in the network as much as possible but to ensure the adequacy of the supply. In the prioritizing investment problem formulation described in this section, the planning horizon is divided in intervals (quarter year, year, etc.) and the executions of the projects are allocated to each interval. For example, suppose that there are 10 projects (numbered 1 to 10) planned to be accomplished in the period considered, which has been divided in four intervals. Then, a possible allocation of projects can be as shown in Table 12.1. It should be noted that project 9 does not appear in this representation, which means that it has been postponed until the next period.

The net present value[1] (NPV) of each project is evaluated taking into consideration the corresponding cash flow and a given interest rate. The NPV varies, of course, according to the implementation date of each project. Also performed for each interval is a reliability evaluation and load flow calculations considering the projects implemented in this interval and the corresponding load level.

Using the above model, the prioritizing investment problem can be formulated as a combinatorial optimization problem in which the variables are the position of the

---

[1]The NPV is defined as the difference between the present value (corrected by the discount rate) of the receipts and the present value of the expenditures.

project execution in the considered interval; the objective is to maximize the sum of the projects NPV, and there are constraints on the minimum value of certain reliability indexes (SAIDI, SAIFI, etc.), voltage drop, total losses, and so forth.

The fitness function is defined as

$$F = \sum_{i=1}^{NP} \omega_i \times NPV_i + \sum_{i=1}^{NP} \frac{w_i}{L_i} + \sum_{i=1}^{NP} \frac{\psi_i}{\Delta V_i} + \sum_{i=1}^{NP} \frac{\alpha_i}{SAIDI_i} + \sum_{i=1}^{NP} \frac{\beta_i}{SAIFI_i}, \quad (12.5)$$

where NPV_i is net present value of project j; L_i is active losses related to project j; ΔV_i is voltage deviation related to project j; $SAIDI_i$ is system average interruption duration index related to project j; $SAIFI_i$ is system average interruption frequency index related to project j; ω_i is weight to take into consideration subjective factors like the utility public image and so forth; and ψ_i, α_i, β_i are weights.

The GA utilized in this application uses a chromosome in which the number of genes equal to the number of projects and each gene is an integer representing the interval in which the project implemented. Special crossover and mutation operators were developed to operate with such encoding method. The experiments to test the effectiveness of the method described in this section are still in a preliminary stage. However, the results obtained so far indicate that the method may be adequate for the use in practical applications.

12.7 CONCLUSIONS

This survey has covered papers on evolutionary computation applications to power systems. System expansion planning and generation scheduling have been the most carefully investigated problems. Great progress, particularly for the transmission system expansion planning, has been reported. Distribution systems have also attracted significant attention.

Regarding different EC techniques, GAs are by far the most popular. The application of GAs to large-scale problems is still in progress. Only a few truly large power systems have been used for testing the EC approach. Evolutionary programming, although less popular than GAs, has shown great potential for tackling complex practical applications.

After recognizing the limitations of the initial ideas, the EC research community has merged many ideas that had been independently proposed. In fact, algorithms based on evolution theory are becoming pretty much similar. Simulated annealing, although the oldest heuristic search technique applied to power systems, has shown limited capacity for dealing with large-scale problems. Tabu search and particle swarm still need more empirical evidence of their potential for solving such problems.

There is a clear trend toward the combination of EC methods among themselves and with classic mathematical programming techniques (e.g., [5, 7, 8, 16, 24, 28]). The challenge for power engineers is to incorporate domain-specific knowledge into the heuristic search process, without deteriorating the EC exploration capability. A good start has been made, but the real challenges still lie ahead. Recent developments can be found in Refs. 150 to 166.

REFERENCES

Surveys

1. Miranda V, Srinivasan D, Proença LM. Evolutionary computation in power systems. In: 12th Power Systems Computation Conference, Dresden, Germany, Published by Power Systems Computation Conference, August 1996, Vol. 1, p. 25–40.
2. Nara K. State of the arts of the modern heuristics application to power systems. IEEE PES Winter Meeting, Singapore, Published by IEEE Power Engineering Society, January 2000, Vol. 2, p. 1279–1283.

Generation Scheduling

3. Zhuang F, Galiana F. Unit commitment by simulated annealing. IEEE Trans Power Systems 1990; 5(1):311–317.
4. Dasgupta D, McGregor DR. Thermal unit commitment using genetic algorithms. IEE Proc Generation Transmission Distribution 1994; 141(5):459–465.
5. Wong KP, Wong YW. Thermal generator scheduling using hybrid genetic/simulated-annealing approach. IEE Proc Generation Transmission Distribution 1995; 142(4):372–380.
6. Kazarlis SA, Bakirtzis AG, Petridis V. A genetic algorithm solution to the unit commitment problem. IEEE Trans Power Systems 1996; 11(1):83–92.
7. Wong KP, Suzannah YWW. Combined genetic algorithm/simulated annealing/fuzzy set approach to short-term generation scheduling with take-or-pay fuel contract. IEEE Trans Power Systems 1996; 11(1):128–136.
8. Bai X, Shahidehpour SM. Hydro-thermal scheduling by tabu search and decomposition method. IEEE Trans Power Systems 1996; 11(2):968–974.
9. Chen P-H, Chang H-C. Genetic aided scheduling of hydraulically coupled plants in hydro-thermal coordination. IEEE Trans Power Systems 1996; 11(2):975–981.
10. Yang P-C, Yang H-T, Huang C-L. Scheduling short-term hydrothermal generation using evolutionary programming techniques. IEE Proc Generation Transmission Distribution 1996; 143(4):371–376.
11. Maifeld TT, Sheblé GB. Genetic-based unit commitment algorithm. IEEE Trans Power Systems 1996; 11(3):1359–1370.
12. Srinivasan D, Tettamanzi AGB. Heuristics-guided evolutionary approach to multiobjective generation scheduling. IEE Proc Generation Transmission Distribution 1996; 143(6):553–559.
13. Srinivasan D, Tettamanzi AGB. An evolutionary algorithm for evaluation of emission compliance options in view of the clean air act amendments. IEEE Trans Power Systems 1997; 12(1):336–341.
14. Huang S-J, Huang C-L. Application of genetic based neural networks to thermal unit commitment. IEEE Trans Power Systems 1997; 12(2):654–660.

15. Yang H-T, Yang P-C, Huang C-L. A parallel genetic algorithm to solving the unit commitment problem: implementation on the transputer networks. IEEE Trans Power Systems 1997; 12(2):661–668.

16. Wong KP, Suzannah YWW. Hybrid genetic/simulated annealing approach to short-term multiple-fuel constraint generation scheduling. IEEE Trans Power Systems 1997; 12(2):776–784.

17. Mantawy AH, Abdel-Magid YL, Selim SZ. Unit commitment by tabu search. IEE Proc Generation Transmission Distribution 1998; 145(1):56–64.

18. Mantawy AH, Abdel-Magid YL, Selim SZ. A simulated annealing algorithm for unit commitment. IEEE Trans Power Systems 1998; 13(1):197–204.

19. Orero SO, Irving MR. A genetic algorithm modelling framework and solution technique for short term optimal hydrothermal scheduling. IEEE Trans Power Systems 1998; 13(2):501–518.

20. Chang H-C, Chen P-H. Hydrothermal generation scheduling package: a genetic based approach. IEE Proc Generation Transmission Distribution 1998; 145(4):451–457.

21. MacGill IF, Kaye RJ. Decentralised coordination of power system operation using dual evolutionary programming. IEEE Trans Power Systems 1999; 14(1):112–119.

22. Huse ES, Wangensteen I, Faanes HH. Thermal power generation scheduling by simulated competition. IEEE Trans Power Systems 1999; 14(2):472–477.

23. Huang S-J. Application of genetic based fuzzy systems to hydroelectric generation scheduling. IEEE Trans Energy Conversion 1999; 14(3):724–730.

24. Mantawy AH, Abdel-Magid YL, Selim SZ. Integrating genetic algorithms, tabu search, and simulated annealing for the unit commitment problem. IEEE Trans Power Systems 1999; 14(3):829–836.

25. Werner TG, Verstege JF. An evolution strategy for short-term operation planning of hydrothermal power systems. IEEE Trans Power Systems 1999; 14(4):1362–1368.

26. Juste KA, Kita H, Tanaka E, Hasegawa J. An evolutionary programming solution to the unit commitment problem. IEEE Trans Power Systems 1999; 14(4):1452–1459.

27. Rudolf A, Bayrleithner R. A genetic algorithm for solving the unit commitment problem of a hydro-thermal power system. IEEE Trans Power Systems 1999; 14(4):1460–1468.

28. Cheng C-P, Liu C-W, Liu C-C. Unit commitment by Lagrangian relaxation and genetic algorithms. IEEE Trans Power Systems 2000; 15(2):707–714.

29. Richter CW, Sheblé GB. A profit-based unit commitment GA for the competitive environment. IEEE Trans Power Systems 2000; 15(2):715–721.

30. Liang R-H, Kang F-C. Thermal generating unit commitment using an extended mean field annealing neural network. IEE Proc Generation Transmission Distribution 2000; 147(3):164–170.

31. Wu Y-G, Ho C-Y, Wang D-Y. A diploid genetic approach to short-term scheduling of hydro-thermal system. IEEE Trans Power Systems 2000; 15(4):1268–1274.

Economic/Reactive Dispatch and Optimal Power Flow

32. Walters DC, Sheblé GB. Genetic algorithm solution of economic dispatch with valve point loading. IEEE Trans Power Systems 1993; 8(3):1325–1332.

33. Wong KP, Fung CC. Simulated annealing based economic dispatch algorithm. IEE Proc Generation Transmission Distribution 1993; 140(6):509–515.

34. Bakirtzis A, Petridis V, Kazarlis S. Genetic algorithm solution to the economic dispatch problem. IEE Proc Generation Transmission Distribution 1994; 141(4):377–382.

35. Wong KP, Wong YW. Genetic and genetic/simulated-annealing approaches to economic dispatch. IEE Proc Generation Transmission Distribution 1994; 141(5):507–513.

36. Sheblé GB, Brittig K. Refined genetic algorithm-economic dispatch example. IEEE Trans Power Systems 1995; 10(1):117–124.

37. Wu QH, Ma JT. Power system optimal reactive power dispatch using evolutionary programming. IEEE Trans Power Systems 1995; 10(3):1243–1249.

38. Wong KP, Fan B, Chang CS, Liew AC. Multi-objective generation dispatch using bi-criterion global optimization. IEEE Trans Power Systems 1995; 10(4):1813–1819.

39. Chen P-H, Chang H-C. Large-scale economic dispatch by genetic algorithm. IEEE Trans Power Systems 1995; 10(4):1919–1926.

40. Yang H-T, Yang P-C, Huang C-L. Evolutionary programming based economic dispatch for units with non-smooth fuel cost functions. IEEE Trans Power Systems 1996; 11(1):112–118.

41. Xu JX, Chang CS, Wang XW. Constrained multiobjective global optimisation of longitudinal interconnected power system by genetic algorithm. IEE Proc Generation Transmission Distribution 1996; 143(5):435–446.

42. Orero SO, Irving MR. Economic dispatch of generators with prohibited operating zones: a genetic algorithm approach. IEE Proc Generation Transmission Distribution 1996; 143(6):529–534.

43. Song YH, Wang GS, Wang PY, Johns AT. Environmental/economic dispatch using fuzzy logic controlled genetic algorithms. IEE Proc Generation Transmission Distribution 1997; 144(4):377–382.

44. Wong KP, Yuryevich J. Evolutionary-programming-based algorithm for environmentally-constrained economic dispatch. IEEE Trans Power Systems 1998; 13(2):301–306.

45. Chang CS, Fu W. Stochastic multiobjective generation dispatch of combined heat and power systems. IEE Proc Generation Transmission Distribution 1998; 145(5):583–591.

46. Das DB, Patvardhan C. New multi-objective stochastic search technique for economic load dispatch. IEE Proc Generation Transmission Distribution 1998; 145(6):747–752.

47. Yuryevich J, Wong KP. Evolutionary programming based optimal power flow algorithm. IEEE Trans Power Systems 1999; 14(4):1245–1250.

48. Gomes JR, Saavedra OR. Optimal reactive power dispatch using evolutionary computation: extended algorithms. IEE Proc Generation Transmission Distribution 1999; 146(6):586–592.

49. Li N, Xu Y, Chen H. FACTS-based power flow control in interconnected power system. IEEE Trans Power Systems 2000; 15(1):257–262.

50. Yoshida H, Kawata K, Fukuyama Y, Takayama S, Nakanishi Y. A particle swarm optimization for reactive power and voltage control considering voltage security assessment. IEEE Trans Power Systems 2000; 15(4):1232–1239.

Maintenance Scheduling

51. Satoh T, Nara K. Maintenance scheduling by using simulated annealing method (for power plants). IEEE Trans Power Systems 1991; 6(2):850–857.

52. Kim H, Hayashi Y, Nara K. An algorithm for thermal unit maintenance scheduling through combined use of GA, SA and TS. IEEE Trans Power Systems 1997; 12(1):329–335.

53. Bretthauer G, Gamaleja T, Handschin E, Neumann U, Koffmann W. Integrated maintenance scheduling system for electrical energy systems. IEEE Trans Power Delivery 1998; (13)2:655–660.

54. Burke EK, Smith AJ. Hybrid evolutionary techniques for the maintenance scheduling problem. IEEE Trans Power Systems 2000; 15(1):122–128.

Generation, Transmission, and VAr Planning

55. Hsiao Y-T, Liu C-C, Chiang H-D, Chen Y-L. A new approach for optimal VAr sources planning in large scale electric power systems. IEEE Trans Power Systems 1993; 8(3):988–996.

56. Hsiao Y-T, Chiang H-D, Liu C-C, Chen Y-L. A computer package for optimal multi-objective VAr planning in large scale power systems. IEEE Trans Power Systems 1994; 9(2):668–676.

57. Iba K. Reactive power optimization by genetic algorithm. IEEE Trans Power Systems 1994; 9(2):685–692.

58. Chen Y-L, Liu C-C. Multiobjective VAr planning using the goal-attainment method. IEE Proc Generation Transmission Distribution 1994; 141(3):227–232.

59. Wong KP, Wong YW. Short-term hydrothermal scheduling. Part I: Simulated annealing approach. IEE Proc Generation Transmission Distribution 1994; 141(5):497–501.

60. Wong KP, Wong YW. Short-term hydrothermal scheduling. Part II: Parallel simulated annealing approach. IEE Proc Generation Transmission and Distribution 1994; 141(5):502–506.

61. Chen YL, Liu C-C. Interactive fuzzy satisfying method for optimal multi-objective VAr planning in power systems. IEE Proc Generation Transmission Distribution 1994; 141(6):554–560.

62. Chen Y-L, Liu C-C. Optimal multi-objective VAr planning using an interactive satisfying method. IEEE Trans Power Systems 1995; 10(2):664–670.

63. Jwo W-S, Liu C-W, Liu C-C, Hsiao Y-T. Hybrid expert system and simulated annealing approach to optimal reactive power planning. IEE Proc Generation Transmission Distribution 1995; 142(4):381–385.

64. Lee KY, Bai X, Park Y-M. Optimization method for reactive power planning by using a modified simple genetic algorithm. IEEE Trans Power Systems 1995; 10(4):1843–1850.

65. Romero R, Gallego RA, Monticelli A. Transmission system expansion planning by simulated annealing. IEEE Trans Power Systems 1996; 11(1):364–369.

66. Fukuyama Y, Chiang H-D. A parallel genetic algorithm for generation expansion planning. IEEE Trans Power Systems 1996; 11(2):955–961.

67. Ma JT, Lai LL. Evolutionary programming approach to reactive power planning. IEE Proc Generation Transmission Distribution 1996; 143(4):365–370.

68. Rudnick H, Palma R, Cura E, Silva C. Economically adapted transmission systems in open access schemes—application of genetic algorithms. IEEE Trans Power Systems 1996; 11(3):1427–1440.

69. Chen Y-L. Weak bus oriented reactive power planning for system security. IEE Proc Generation Transmission Distribution 1996; 143(6):541–545.

70. Chen Y-L. Weak bus-oriented optimal multi-objective VAr planning. IEEE Trans Power Systems 1996; 11(4):1885–1890.

71. Gallego RA, Alves AB, Monticelli A, Romero R. Parallel simulated annealing applied to long term transmission network expansion planning. IEEE Trans Power Systems 1997; 12(1):181–188.

72. Lai LL, Ma JT. Application of evolutionary programming to reactive power planning—comparison with nonlinear programming approach. IEEE Trans Power Systems 1997; 12(1):198–206.

73. Liu C-W, Jwo W-S, Liu C-C, Hsiao Y-T. A fast global optimization approach to VAr planning for the large scale electric power systems. IEEE Trans Power Systems 1997; 12(1):437–443.

74. Zhu J, Chow M-Y. A review of emerging techniques on generation expansion planning. IEEE Trans Power Systems 1997; 12(4):1722–1728.

75. Lee KY, Yang FF. Optimal reactive power planning using evolutionary algorithms: a comparative study for evolutionary programming, evolutionary strategy, genetic algorithm, and linear programming. IEEE Trans Power Systems 1998; 13(1):101–108.

76. Chang CS, Huang JS. Optimal multiobjective SVC planning for voltage stability enhancement. IEE Proc Generation Transmission Distribution 1998; 145(2):203–209.

77. Chen Y-L. Weighted-norm approach for multiobjective VAr planning. IEE Proc Generation Transmission Distribution 1998; 145(4):369–374.

78. Gallego RA, Monticelli A, Romero R. Comparative studies on non-convex optimization methods for transmission network expansion planning. IEEE Trans Power Systems 1998; 13(3):822–828.

79. Lai LL, Ma JT. Practical application of evolutionary computing to reactive power planning. IEE Proc Generation Transmission Distribution 1998; 145(6):753–758.

80. Paterni P, Vitet S, Bena M, Yokohama A. Optimal location of phase shifters in the French network by genetic algorithm. IEEE Trans Power Systems 1999; 14(1):37–42.

81. Park Y-M, Won J-R, Park J-B, Kim D-G. Generation expansion planning based on an advanced evolutionary programming. IEEE Trans Power Systems 1999; 14(1):299–305.

82. Urdaneta AJ, Gómez JF, Sorrentino E, Flores L, Diaz R. A hybrid genetic algorithm for optimal reactive power planning based upon successive linear programming. IEEE Trans Power Systems 1999; 14(4):1292–1298.

83. Chou C-J, Liu C-W, Lee J-Y, Lee K-D. Optimal planning of large passive-harmonic-filters set at high voltage level. IEEE Trans Power Systems 2000; 15(1):433–441.

84. Gallego RA, Romero R, Monticelli AJ. Tabu search algorithm for network synthesis. IEEE Trans Power Systems 2000; 15(2):490–495.

85. Park J-B, Park Y-M, Won J-R, Lee KY. An improved genetic algorithm for generation expansion planning. IEEE Trans Power Systems 2000; 15(3):916–922.

86. Delfanti M, Granelli GP, Marannino P, Montagna M. Optimal capacitor placement using deterministic and genetic algorithms. IEEE Trans Power Systems 2000; 15(3):1041–1046.

87. da Silva EL, Gil HA, Areiza JM. Transmission network expansion planning under an improved genetic algorithm. IEEE Trans Power Systems 2000; 15(3):1168–1174.

88. da Silva EL, Areiza Ortiz JM, de Oliveira GC, Binato S. Transmission network expansion planning under a tabu search approach. IEEE Trans Power Systems 2001; 16(1):62–68.

89. Ramirez-Rosado IJ, Bernal-Agustin JL. Reliability and costs optimization for distribution networks expansion using an evolutionary algorithm. IEEE Trans Power Systems 2001; 16(1):111–118.

Distribution Systems Planning and Operation

90. Hasselfield CW, Wilson P, Penner L, Lau M, Gole AM. An automated method for least cost distribution planning. IEEE Trans Power Delivery 1990; 5(2):1188–1194.

91. Nara K, Shiose A, Kitagawa M, Ishihara T. Implementation of genetic algorithm for distribution systems loss minimum reconfiguration. IEEE Trans Power Systems 1992; 7(3):1044–1051.

92. Richards GG, Yang H. Distribution system harmonic worst case design using a genetic algorithm. IEEE Trans Power Delivery 1993; 8(3):1484–1491.

93. Chu RF, Wang J-C, Chiang H-D. Strategic planning of LC compensators in nonsinusoidal distribution systems. IEEE Trans Power Delivery 1994; 9(3):1558–1563.

94. Sundhararajan S, Pahwa A. Optimal selection of capacitors for radial distribution systems using a genetic algorithm. IEEE Trans Power Systems 1994; 9(3):1499–1507.

95. Miranda V, Ranito JV, Proença LM. Genetic algorithms in optimal multistage distribution network planning. IEEE Trans Power Systems 1994; 9(4):1927–1933.

96. Chiang H-D, Wang J-C, Tong J, Darling G. Optimal capacitor placement, replacement and control in large-scale unbalanced distribution systems: modeling and a new formulation. IEEE Trans Power Systems 1995; 10(1):356–362.

97. Chiang H-D, Wang J-C, Tong J, Darling G. Optimal capacitor placement, replacement and control in large-scale unbalanced distribution systems: system solution algorithms and numerical studies. IEEE Trans Power Systems 1995; 10(1):363–369.

98. Yeh E-C, Venkata SS, Sumic Z. Improved distribution system planning using computational evolution. IEEE Trans Power Systems 1996; 11(2):668–674.

99. Jiang D, Baldick R. Optimal electric distribution system switch reconfiguration and capacitor control. IEEE Trans Power Systems 1996; 11(2):890–897.

100. Billinton R, Jonnavithula S. Optimal switching device placement in radial distribution systems. IEEE Trans Power Delivery 1996; 11(3):1646–1651.

101. Jonnavithula S, Billinton R. Minimum cost analysis of feeder routing in distribution system planning. IEEE Trans Power Delivery 1996; 11(4):1935–1940.

102. Huang Y-C, Yang H-T, Huang C-L. Solving the capacitor placement problem in a radial distribution system using tabu search approach. IEEE Trans Power Systems 1996; 11(4):1868–1873.

103. Miu KN, Chiang H-D, Darling G. Capacitor placement, replacement and control in large-scale distribution systems by a GA-based two-stage algorithm. IEEE Trans Power Systems 1997; 12(3):1160–1166.

104. Ramírez-Rosado IJ, Bernal-Agustín JL. Genetic algorithms applied to the design of large power distribution systems. IEEE Trans Power Systems 1998; 13(2):696–703.

105. Zhu J, Bilbro G, Chow M-Y. Phase balancing using simulated annealing. IEEE Trans Power Systems 1999; 14(4):1508–1513.

106. Chuang AS, Wu F. An extensible genetic algorithm framework for problem solving in a common environment. IEEE Trans Power Systems 2000; 15(1):269–275.

107. Chen T-H, Cherng J-T. Optimal phase arrangement of distribution transformers connected to a primary feeder for system unbalance improvement and loss reduction using a genetic algorithm. IEEE Trans Power Systems 2000; 15(3):994–1000.

108. Hsiao Y-T, Chien C-Y. Enhancement of restoration service in distribution systems using a combination fuzzy-GA method. IEEE Trans Power Systems 2000; 15(4):1394–1400.

Load Forecasting/Management and Model Identification

109. Dillon TS, Morsztyn K, Phua K. Short term load forecasting using adaptive pattern recognition and self-organizing techniques. 5th Power System Computation Conference, Cambridge, England, September, 1975.

110. Mori H, Kobayashi H. Optimal fuzzy inference for short-term load forecasting. IEEE Trans Power Systems 1996; 11(1):390–396.

111. Yang H-T, Huang C-M, Huang C-L. Identification of ARMAX model for short term load forecasting: an evolutionary programming approach. IEEE Trans Power Systems 1996; 11(1):403–408.

112. Souza JCS, Leite da Silva AM, Alves da Silva AP. Data debugging for real-time power system monitoring based on pattern analysis. IEEE Trans Power Systems 1996; 11(3):1592–1599.

113. Ju P, Handschin E, Karlsson D. Nonlinear dynamic load modelling: model and parameter estimation. IEEE Trans Power Systems 1996; 11(4):1689–1697.

114. Kwok T-Y, Yeung D-Y. Constructive algorithms for structure learning in feedforward neural networks for regression problems. IEEE Trans Neural Networks 1997; 8(3):630–645.

115. Alves da Silva AP, Ferreira C, Zambroni de Souza AC, Lambert Torres G. A new constructive ANN and its application to electric load representation. IEEE Trans Power Systems 1997; 12(4):1569–1575.

116. Gaul AJ, Handschin E, Hoffmann W, Lehmkoster C. Establishing a rule base for a hybrid ES/XPS approach to load management. IEEE Trans Power Systems 1998; 13(1):86–93.

117. Kung C-H, Devaney MJ, Huang C-M, Kung C-M. Fuzzy-based adaptive digital power metering using a genetic algorithm. IEEE Trans Instrumentation and Measurement 1998; 47(1):183–188.

118. Souza JCS, Leite da Silva AM, Alves da Silva AP. On-line topology determination and bad data supression in power system operation using artificial neural networks. IEEE Trans Power Systems 1998; 13(3):796–803.

119. Alves da Silva AP, Moulin LS. Confidence intervals for neural network based short-term load forecasting. IEEE Trans Power Systems 2000; 15(4):1191–1196.

Control

120. Asgharian R, Tavakoli SA. A systematic approach to performance weights selection in design of robust H/sup/spl infin// PSS using genetic algorithms. IEEE Trans Energy Conversion 1996; 11(1):111–117.

121. Ju P, Handschin E, Reyer F. Genetic algorithm aided controller design with application to SVC. IEE Proc Generation Transmission Distribution 1996; 143(3):258–262.

122. Abdel-Magid YL, Bettayeb M, Dawoud MM. Simultaneous stabilisation of power systems using genetic algorithms. IEE Proc Generation Transmission Distribution 1997; 144(1):39–44.

123. Taranto GM, Falcão DM. Robust decentralised control design using genetic algorithms in power system damping control. IEE Proc Generation Transmission Distribution 1998; 145(1):1–6.

124. Reformat M, Kuffel E, Woodford D, Pedrycz W. Application of genetic algorithms for control design in power systems. IEE Proc Generation Transmission Distribution 1998; 145(4):345–354.

125. Wen J, Cheng S, Malik OP. A synchronous generator fuzzy excitation controller optimally designed with a genetic algorithm. IEEE Trans Power Systems 1998; 13(3):884–889.

126. Chen XR, Pahalawaththa NC, Annakkage UD, Kumble CS. Design of decentralised output feedback TCSC damping controllers by using simulated annealing. IEE Proc Generation Transmission Distribution 1998; 145(5):553–558.

127. Abido MA, Abdel-Magid YL. Hybridizing rule-based power system stabilizers with genetic algorithms. IEEE Trans Power Systems 1999; 14(2):600–607.

128. Abdel-Magid YL, Abido MA, Al-Baiyat S, Mantawy AH. Simultaneous stabilization of multimachine power systems via genetic algorithms. IEEE Trans Power Systems 1999; 14(4):1428–1439.

129. Welsh M, Mehta P, Darwish MK. Genetic algorithm and extended analysis optimisation techniques for switched capacitor active filters-comparative study. IEE Proc Electric Power Appl 2000; 147(1):21–26.

130. do Bomfim ALB, Taranto GN, Falcão DM. Simultaneous tuning of power system damping controllers using genetic algorithms. IEEE Trans Power Systems 2000; 15(1):163–169.

131. Abdel-Magid YL, Abido MA, Mantaway AH. Robust tuning of power system stabilizers in multimachine power systems. IEEE Trans Power Systems 2000; 15(2):735–740.

132. Zhang P, Coonick AH. Coordinated synthesis of PSS parameters in multi-machine power systems using the method of inequalities applied to genetic algorithms. IEEE Trans Power Systems 2000; 15(2):811–816.

133. Abido MA. Robust design of multimachine power system stabilizers using simulated annealing. IEEE Trans Energy Conversion 2000; 15(3):297–304.

134. Abido MA, Abdel-Magid YL. Robust design of multimachine power system stabilisers using tabu search algorithm. IEE Proc Generation Transmission Distribution 2000; 147(6):387–394.

135. Saleh RAF, Bolton HR. Genetic algorithm-aided design of a fuzzy logic stabilizer for a superconducting generator. IEEE Trans Power Systems 2000; 15(4):1329–1335.

Alarm Processing, Fault Diagnosis, and Protection

136. Wen ES, Chang CS. Tabu search approach to alarm processing in power systems. IEE Proc Generation Transmission Distribution 1997; 144(1):31–38.
137. Wen FS, Chang CS. Probabilistic approach for fault-section estimation in power systems based on a refined genetic algorithm. IEE Proc Generation Transmission Distribution 1997; 144(2):160–168.
138. Lai LL, Sichanie AG, Gwyn BJ. Comparison between evolutionary programming and a genetic algorithm for fault-section estimation. IEE Proc Generation Transmission Distribution 1998; 145(5):616–620.
139. Wen FS, Chang CS. Possibilistic-diagnosis theory for fault-section estimation and state identification of unobserved protective relays using tabu-search method. IEE Proc Generation Transmission Distribution 1998; 145(6):722–730.
140. So CW, Li KK. Time coordination method for power system protection by evolutionary algorithm. IEEE Trans Industry Appl 2000; 36(5):1235–1240.

State Estimation and Analysis

141. Irving MR, Sterling MJH. Optimal network tearing using simulated annealing. IEE Proc Generation Transmission Distribution 1990; 137(1):69–72.
142. Baldwin TL, Mili L, Boisen MB Jr, Adapa R. Power system observability with minimal phasor measurement placement. IEEE Trans Power Systems 1993; 8(2):707–715.
143. Mori H, Takeda K. Parallel simulated annealing for power system decomposition. IEEE Trans Power Systems 1994; 9(2):789–795.
144. Wong KP, Li A, Law MY. Development of constrained-genetic-algorithm load-flow method. IEE Proc Generation Transmission Distribution 1997; 144(2):91–99.

Energy Markets

145. Richter CW Jr, Sheblé GB. Genetic algorithm evolution of utility bidding strategies for the competitive marketplace. IEEE Trans Power Systems 1998; 13(1):256–261.
146. Richter CW Jr, Sheblé GB, Ashlock D. Comprehensive bidding strategies with genetic programming/finite state automata. IEEE Trans Power Systems 1999; 14(4):1207–1212.

Recent Papers

147. Falcão DM. Genetic algorithms applications in electrical distribution systems. 2002 IEEE World Congress on Computational Intelligence, Honolulu, Hawaii, May, 2002.
148. Borges CLT, Manzoni A, Viveros EC, Falcão DM. A parallel genetic algorithm based methodology for network reconfiguration in the presence of dispersed generation. 17th International Conference on Electricity Distribution (CIRED), Barcelona, Spain, May, 2003.

149. Velasquez RMG, Falcão DM. Optimal switching and protection devices placement in distribution networks for reliability improvement using genetic algorithms. ISAP2001 Conference, Budapest, Hungary, June, 2001.

150. Toune S, Fudo H, Genji T, Fukuyama Y, Nakanishi Y. Comparative study of modern heuristic algorithms to service restoration in distribution systems. IEEE Trans Power Delivery 2002; 17(1):173–181.

151. Lin WM, Cheng FS, Tsay MT. An improved tabu search for economic dispatch with multiple minima. IEEE Trans Power Systems 2002; 17(1):108–112.

152. Ghezelayagh H, Lee KY. Intelligent predictive control of a power plant with evolutionary programming optimizer and neuro-fuzzy identifier. 2002 World Congress on Computational Intelligence, Honolulu, Hawaii, May, 2002.

153. Park J-B, Kim J-H, Lee KY. Generation expansion planning in a competitive environment using a genetic algorithm. IEEE Power Engineering Society Summer Meeting, Chicago, July, 2002.

154. Kassabalidis IN, El-Sharkawi MA, Marks RJ, Moulin LS, Alves da Silva AP. Dynamic security border identification using enhanced particle swarm optimization. IEEE Trans Power Systems 2002; 17(3):723–729.

155. Abido MA. Optimal design of power-system stabilizers using particle swarm optimization. IEEE Trans Energy Conversion 2002; 17(3):406–413.

156. Jeon Y-J, Kim J-C, Kim J-O, Shin J-R, Lee KY. An efficient simulated annealing algorithm for network reconfiguration in large-scale distribution systems. IEEE Trans Power Delivery 2002; 17(4):1070–1078.

157. Sinha N, Chakrabarti R, Chattopadhyay PK. Evolutionary programming techniques for economic load dispatch. IEEE Trans Evol Computat 2003; 7(1):83–94.

158. Naka S, Genji T, Yura T, Fukuyama Y. A hybrid particle swarm optimization for distribution state estimation. IEEE Trans Power Systems 2003; 18(1):60–68.

159. El-Sharkh MY, El-Keib AA. Maintenance scheduling of generation and transmission systems using fuzzy evolutionary programming. IEEE Trans Power Systems 2003; 18(2):862–866.

160. Jeon Y-J, Kim J-C, Yun S-Y, Lee KY. Application of ant colony algorithm for network reconfiguration in distribution systems. IFAC Symposium for Power Plants and Power Systems Control, Seoul, Korea, June, 2003.

161. Park J-B, Lee K-S, Shin J-R, Lee KY. Economic load dispatch for non-smooth cost functions using particle swarm optimization. IEEE Power Engineering Society General Meeting, Toronto, Canada, July, 2003.

162. Gaing Z-L. Particle swarm optimization to solving the economic dispatch considering the generator constraints. IEEE Trans Power Systems 2003; 18(3):1187–1195.

163. Park J-B, Lee K-S, Shin J-R, Lee KY. Economic load dispatch based on a hybrid particle swarm optimization. ISAP2003 Conference, Lemnos, Greece, August, 2003.

164. Park J-B, Lee K-S, Shin J-R, Lee KY. A particle swarm optimization for economic dispatch with non-smooth cost functions. IEEE Trans Power Systems 2005; 20(1):34–42.

165. Zhao B, Guo CX, Cao YJ. A multiagent-based particle swarm optimization approach for optimal reactive power dispatch. IEEE Trans Power Systems 2005; 20(2):1070–1078.

166. Huang C-M, Huang C-J, Wang M-L. A particle swarm optimization to identifying the ARMAX model for short-term load forecasting. IEEE Trans Power Systems 2005; 20(2):1126–1133.

CHAPTER 13

Application of Evolutionary Technique to Power System Vulnerability Assessment

MINGOO KIM, MOHAMED A. EL-SHARKAWI, ROBERT J. MARKS, and IOANNIS N. KASSABALIDIS

13.1 INTRODUCTION

The trend toward deregulation is altering the manner in which electric power systems are operated. In a deregulated environment, increased interconnections and unforeseen changes in the system topology and load can cause system instability, which needs to be addressed in real time. In the past, monopolistic environment, power companies could afford increased security margins; it is no longer probable. For these reasons and because of the limited construction of new power plants, the system is required to operate closer to its vulnerability boundary. This, in turn, requires the industry to develop better methods of quantifying the real-time vulnerability status of their systems, including vulnerability border tracking. However, the high-dimensional nature of power systems' operating space makes this task extremely difficult without using approximations, more efficient computational techniques, and more powerful computers.

Among the effective computational methods that can be effectively applied to the vulnerability assessment problem are the *computational intelligent techniques*. This chapter investigates the use of these paradigms to assess the vulnerability of a power system by addressing the various computational challenges as well as providing a visualization tool to describe the vulnerability status of the system.

The purpose of vulnerability assessment is to determine the ability of the power system to continue providing service in case of an unforeseen, but probable, catastrophic contingency. A power system can become vulnerable for various reasons, including major component failures, communication interruptions, human errors, unfavorable weather conditions, and even sabotage. A power system is invulnerable

Modern Heuristic Optimization Techniques. Edited by K. Y. Lee and M. A. El-Sharkawi
Copyright © 2008 the Institute of Electrical and Electronics Engineers, Inc.

if it withstands all postulated credible contingencies without violating any of the system constraints. If there is at least one contingency (or one sequence of events) for which the system constraints are violated, the system is said to be vulnerable, or insecure. A vulnerable system could experience a catastrophic failure of system components leading to blackouts that often affect large portions of the power network and typically millions of customers.

The failure of a major component may not necessarily lead to extensive blackout under all operating conditions. The blackout is often a result of failed components, operating states, environmental conditions, network topology, status of other equipment, and operators' actions. Most catastrophic events occur as a sequence of events. For example, a tripped generator may cause unbalance in the system's energy, thus leading to tripping of loads and other generators. This is known as *cascading outage*, which often leads to loss of service to a large number of customers.

Cascaded outages can be triggered by a number of natural or man-made events. The list is long, but the typical events are

- Faults in transmission lines could lead to excessive currents in the system. The faulted section of the network is isolated quickly to prevent any thermal damage to system components such as transformers or transmission lines. This is done by tripping (opening) the circuit breakers on both ends of the faulted line. The loss of major and heavily loaded transmission lines often results in outages.
- Natural calamities such as lightning, earthquake, strong wind, and heavy frost can damage major power system equipment. When lightning hits a power line, the line insulators can be damaged leading to short-circuits (faults). Major earthquakes could damage substations, thus interrupting power to the areas served by the substations. Heavy winds may cause trees to fall on power lines creating faults that trip transmission lines.
- When major power system equipment such as generators and transformers fails, it may lead to outages.
- The protection and control devices may not operate properly to isolate faulty components. This is known as *hidden failure* and can cause the fault to affect a wider region in the system.
- The breaks in communication links between control centers in the power system may lead to wrong information being processed, thus causing control centers to operate asynchronously and probably negating each other.
- Human errors can lead to tripping of essential equipment.

A good example of a cascading outage is the blackout that occurred in North American on August 14, 2003, which is the worst in history, so far. About 50 million people were left without electric energy for days. The blackout occurred on a hot summer day when the system was heavily loaded. All bulk power transmission lines were in service until a 230-kV transmission line was tripped due to a fault. A small but increasing power shift out to Ontario, Canada, was observed. Then, a sudden power surge entered New York and moved westward into the Ontario system. Within 6 seconds, critical tie lines were tripped and the system collapsed. The black-out affected customers in the northeastern and midwestern United States as well as in Ontario. These cascaded events resulted in the tripping of 21 power plants including 7 nuclear plants.

Another major blackout is the one that occurred on August 10, 1996, in the western states of the United States. The event was triggered by an outage of a 500-kV line between Seattle, Washington, and Portland, Oregon. This outage led to overloading lines in the Portland area and a voltage depression in the lower Columbia River area. These eventually led to the tripping of 13 generators at the McNary power plant. As a result, tripping of several transmission lines broke the interconnection into four electrical islands, bringing about service interruptions to 7.5 million customers. This serious event could have been avoided if a mere 0.4% of the load had been shed for 30 minutes [1].

13.2   VULNERABILITY ASSESSMENT AND CONTROL

Vulnerability assessment (VA) and vulnerability control (VC) are two different but related processes. With VA, the power system is analyzed assuming credible system contingencies or sequence of events had occurred. If the analysis indicates that a system is vulnerable, the VC should provide preventive strategies by changing system operating conditions to a more viable status, thus forestalling the possibility of cascading outages. A power system is said to be invulnerable if it can withstand all credible contingencies without violating any of the system constraints. If there is at least one contingency, or sequence of probable events, which violates the system con-straints, the system is judged to be vulnerable. Therefore, the goal of vulnerability assessment is to determine when disruptions of service are likely to occur, and the goal of the vulnerability control is to take steps to reduce the risk of blackouts.

The most common approach to VA is the *time-domain* simulation to calculate the generators' behaviors relative to a given contingency. By examining the generators' activities, one can determine if stability has been maintained. The time-domain approach is directly applicable to any level of detail for power system models and can be very accurate. For a typical large power system, however, thousands of non-linear differential and algebraic equations must be solved before stability is assessed, which requires intensive computation time. Therefore, the usefulness of this method is limited to a few off-line studies of small systems.

An alternative approach is the *direct method*, which employs energy functions. This method was originally proposed in the late 1940s and was further developed in the 1950s and 1960s [2–6]. In contrast with the time-domain approach, the direct method determines system stability by comparing the system energy level to a critical energy value. Direct methods are very attractive because they avoid the time-consuming computation and also provide qualitative measures of the degree of system stability without a need for step-by-step calculation. Although direct methods have a long history of development [2–17], many have thought of them as impractical for detailed large-scale power systems analysis because of intrinsic modeling limitations [6].

In order to overcome such weaknesses, computational intelligence (CI) techniques, such as neural networks, particle swarm optimization, and evolutionary computation, have been proposed for VA [18–27]; especially *neural networks* for its pattern classification capability. As a pattern classifier, neural networks need extensive off-line simulation so as to acquire a large-enough set of training data to represent the different operating conditions of typical power systems. Once trained, neural networks not only provide extremely fast solutions but also have the ability to accommodate new patterns or new operating conditions by generalizing the training data. Moreover, such techniques do not entail the sacrifice of system modeling details and/or operational constraints.

13.3 VULNERABILITY ASSESSMENT CHALLENGES

The key challenges associated with vulnerability assessment are five:

1. The large numbers of contingencies and sequence of events that are typically needed to provide accurate VA. In addition, the wide range of operating conditions and topology of the power system makes the operating space very complex.
2. The speed by which the vulnerability can be assessed on-line.
3. The large number of measurements available in the power system, and the lack of methods to enhance the correlations between measurements and VA.
4. The lack of effective vulnerability index.
5. The lack of identifiable border of vulnerability.

13.3.1 Complexity of Power System

The complexity of a power system due to its immense size, wide operating range, and its changing topology make the VA a rather challenging task given today's computational technology. Approximations and rules are often used instead of elaborate simulations. Among these methods is the dynamic equivalent technique where the power system is divided into local and external subsystems; the local subsystem (the system under study) is modeled in detail and the external system is replaced by a simplified model that can emulate the dynamic interaction effects [28].

Because of the wide range operating condition of the power system, VA cannot be performed on-line using the current computational power. Therefore, the vulnerability assessment is performed in two steps: off-line simulations and on-line assessment. The off-line simulation is done when the computational power is available. In this step, data on power system states along with corresponding vulnerability status are collected. Once the data constitutes sufficient representation of the different operating conditions of the power system, the VA model can be developed to recognize the VA status of any operating patterns. This requires the data to be of sufficient size and highly correlated with the VA status.

13.3.2 VA On-line Speed

For on-line assessment, the VA model must execute very fast in real time. This is the area where CI can provide the computational edge needed. Neural network trained by off-line data has proved to be a fast VA [25].

13.3.3 Feature Selection

Thus far, the work on intelligent systems for VA, including neural networks, is based on at least one of the following processes:

- Reducing the operating space to a subspace of a manageable size. This is done by ignoring large portions of a system's operation.
- Use of extensive off-line simulations to cover as much area in the operating space as possible.

The first process cannot be implemented without severely sacrificing the accuracy of the assessment technique beyond the training subspace. The second is also an unrealistic process because the size and operating space of the power system is immense. For example, the VA of a power system requires extensive data on:

- Predisturbance network topology.
- Predisturbance load/generation conditions.
- Dynamic machine data.
- Postdisturbance control actions.
- Type and duration of disturbance.
- Operating conditions

Clearly, using this list, a plethora of data can exist for a given power system. Moreover, some variables in the data vector may likely have a weak correlation with the VA status. The curse of dimensionality states that, as a rule of thumb, the required cardinality of the training set for accurate training increases exponentially with the input dimension. Attempts to reduce the dimensionality of the data vector by identifying correlations [24] require approximation to the system dynamics under restrictive conditions. If the correlation among elements in the input vector and the VA status is weak, the classifier will try to force a mapping that is unlikely to result in worthwhile information. In fact, such weak correlation can bias the classifier to an undesirable region with high testing errors. Therefore, the key to the success of the classifier is to use "effective" features. The term *feature* is sometimes synonymous with the term *measurement*. Measurements are simply observations, whereas feature is a single or combined data that has a good correlation with the vulnerability assessment.

For a given power system, the amount of features are immense and cannot be all used in the assessment process using the computational power we have today. Therefore, a feature reduction method must be employed. Feature reduction refers to the process of reducing the dimensions of the feature vector while preserving the needed information. There are two basic approaches to feature reduction: feature selection and feature extraction. Feature selection and extraction are techniques used to massage data into a form more conducive to training. Feature selection, important when cost is associated with the acquisition of features, has, as its goal, choosing the best subset of features able to effectively perform accurate classification or regression. Reducing the cardinality of the feature set by combining two or more features into a single composite feature is the goal of feature extraction. The goal is to synthesize an augmented feature space conducive to classification or regression.

Feature selection involves reducing the dimensions of the input vector by selecting only features that are highly correlated with the desired analysis. It can be considered as a problem of global combinatorial optimization; searching out the most suitable features among all possible combinations. A commonly used criterion that can be used for feature selection is Fisher discriminant proposed in 1936 [29]. This method seeks to find the optimal linear separation for two classes of data. Recently, in order to solve the high combinatorial problem of feature selection, evolutionary computation techniques have been applied [30].

On the other hand, feature extraction is a transformation of the original data set into a new space of lower dimensions. By developing a function, $f(\cdot)$, that maps a given input vector in a high-dimension space into a lower-dimension extracted-feature space, feature dimensionality can be effectively reduced. The well-known feature extraction techniques include the principal-components algorithm and the Karhuren–Loève expansion [31–34].

Feature extraction has advantages over feature-selection methods in that all of the features are used to form a new set of composite features and therefore, in some cases, may preserve more of the original information [32–35].

The simplest way of evaluating the candidate features is to treat each of them separately and check their effect on the desired analysis. However, a feature in and of itself could be important, yet when added to other important features forms a less effective combination.

13.3.3.1 *Fisher's Linear Discriminant: Selection Criteria* The classic method of linear discrimination was described by Fisher [29] for two data classes and extended to higher dimension [36]. The Fisher approach is based on the projection of D-dimensional data onto a line. The hope is that such projections onto a line will be well separated by class. Thus, the line is oriented to maximize this class separation [37].

For a two-class example, a given training set

$$H = \{x_1, x_2, \ldots, x_n\} = \{H_1 \cup H_2\}$$

is partitioned into an $n_1 \leq n$ training vector in subset H_1, corresponding with class w_1, and an $n_2 = n - n_1$ training vector in set H_2, corresponding with class w_2. The task is to find a linear mapping $y = \mathbf{w}^T \mathbf{x}$ such as to maximize

$$F(\mathbf{w}) = \frac{|m_1 - m_2|^2}{\sigma_1^2 + \sigma_2^2},$$

where m_i is the mean of class H_i and σ_i^2 is the variance of H_i.

The criterion function F can be rewritten as an explicit function of $\mathbf{w}$ as

$$F(\mathbf{w}) = \frac{\mathbf{w}^T S_B \mathbf{w}}{\mathbf{w}^T S_W \mathbf{w}},$$

where S_B is referred to as the *between-class scatter* matrix and S_W is the *within-class scatter* matrix. The within-class scatter S_W is defined as

$$S_W = S_1 + S_2,$$

where S_1 and S_2 are

$$S_i = \sum_{\mathbf{x} \in C_i} (\mathbf{x} - m_i)(\mathbf{x} - m_i)^T$$

and

$$m_i = \frac{1}{n_i} \sum_{\mathbf{x} \in H_i} \mathbf{x}.$$

The between-class scatter S_B is

$$S_B = (m_1 - m_2)(m_1 - m_2)^T.$$

Finally, the solution for $\mathbf{w}$ to maximize F can be written as

$$\mathbf{w} = S_W^{-1}(m_1 - m_2).$$

One of the main problems of any feature selection algorithm based on the Fisher approach is how to deal with the computational complexity. If we have a total of D possible features, then, because each feature can be either present or absent, there are, in all, 2^D possible feature subsets to be considered. For a relatively small number of features, we might consider simply searching through all possible combinations and calculate their respective Fisher values. For a large number of features, however, such a method would become prohibitively expensive. If we want to select d features, the number of checks is given by

$$N = \frac{D!}{(D-d)!d!}$$

This number may be significantly smaller than 2^D but still be impracticably large in many applications.

13.3.3.2 Neural Network Feature-Extraction (NNFE)

A neural network can be used effectively for nonlinear dimensionality reduction, thereby overcoming some of the limitations of linear principal-component analysis. Figure 13.1 shows a schematic of neural network (NN) structure that can be used to achieve the reduction.

The neural network has d input neurons, d output neurons and m hidden neurons with $m < d$. The input vector constituted by a training data set, represented by d-dimensional vector **X**, is presented sequentially as the input and output of the neural network. In other words, the neural network is designed and trained to reproduce its own input vector. This is just attempting to map each input vector onto itself. The optimization goal is to minimize a sum-of-squares error of the form

$$E = \frac{1}{2} \sum_{n=1}^{N} \sum_{k=1}^{d} \left\{ y_k(x^n) - x_k^n \right\}^2$$

where d is the number of features and N is the number of training data sets.

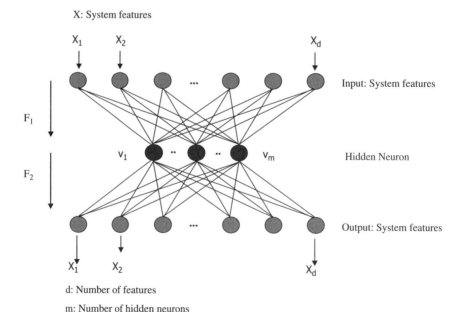

d: Number of features

m: Number of hidden neurons

FIGURE 13.1 Neural network structure to extract features (autoencoder).

After the neural network is trained, the vector $\mathbf{v}$ of the hidden variables represents the extracted features of the system. This network can be understood as a two-stage successive mapping operation F_1 and F_2. F_1 is a projection from the large d-dimensional input onto a small m-dimensional subspace S. This is done through the input to the hidden layer of the neural network. F_2 is a mapping from that subspace back to the full d-dimensional space. This process is represented by the hidden to output layer of the neural network.

Figure 13.2 is a simple geometric interpretation for the case $d = 3$ (three-dimensional space) and $m = 2$ (two-dimensional space). The mapping F_1 defines a projection of points from the original three-dimensional space onto the two-dimensional subspace S. Points in $S(F_2)$ are then mapped through F_2 back to the three-dimensional space of the original data.

Because learning in this network is nonlinear, there is the risk of finding a suboptimal local minimum for the error function. Also, one must take care to choose an appropriate number of m-neurons. Assuming that the two-stage mapping has been successively achieved, the middle layer of this network represents the extracted features and is a valid representation of the original system.

Such a network is said to be an *autoassociator* or *autoencoder*. This feature-reduction system works with any kind of activation function. If the activation functions are set to be linear, this procedure is similar to the conventional principal-component approach. On the contrary, if the hidden neurons have nonlinear activation functions, such a network effectively performs a nonlinear principal-component analysis and has the advantage of not being limited to linear transformations.

Figure 13.3 shows the implementation of the NN for vulnerability assessment using the encoder of Fig. 13.1. After the NN of Fig. 13.1 is trained, the outputs of its hidden layer are used as inputs to train the vulnerability-assessment network as shown.

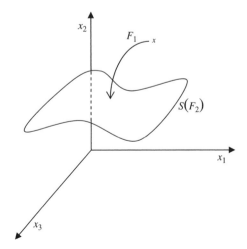

FIGURE 13.2 Geometric interpretation of the mapping.

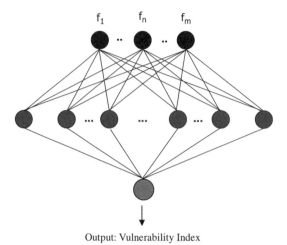

Output: Vulnerability Index

FIGURE 13.3 Neural network structure to calculate the vulnerability index.

By this two-step method, the computational burden is placed on the feature-extraction network, which is used off-line and only once per system. In the second step, the NN is rapidly trained to identify the vulnerability index on-line.

13.3.3.3 Support Vector Machine Feature-Extraction *Support vector machines* (SVMs) [38] project feature data into a higher-dimensional space where linear classification is possible. SVMs perform nonlinear transformations of the input space into a higher-dimensional feature space in contrast with the feature selection and extraction techniques discussed earlier. In the new feature space, optimal linear separating hyperplanes are computed. Classes that cannot be separated in the original space are often separable in higher-dimensional spaces. With the use of geometric properties, optimal separating hyperplanes can be directly calculated from the data. The SVM parameterization provides a meaningful characterization of the data complexity via a number of support vectors (i.e., the data points at the margin of the classes) that are independent of the problem's dimensionality. The generalization ability of the optimal separating hyperplane can be directly related to the number of support vectors. The bigger the number of support vectors, the larger the expected error rate.

13.3.4 Vulnerability Border

Knowledge of the vulnerability border can provide an operator with valuable guidance for steering the power system away from vulnerable operating regions. Correct identification of the vulnerability border can provide the operator with easily understood visual information. Moreover, the distance from the border provides a direct assessment of the degree of vulnerability. The vulnerability boundary, however, cannot be determined analytically for a large-scale power system, requiring, rather, extensive computation by numerical methods.

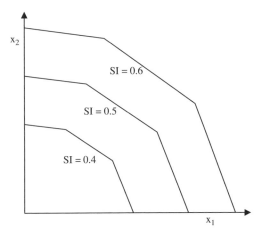

FIGURE 13.4 Nomogram for two parameters showing three security levels.

Until now, utility companies in North America have used a *nomogram* for characterizing security boundaries [39]. Figure 13.4 is an example of a typical nomogram, showing three levels of the vulnerability index (VI). In developing a nomogram, two critical parameters are chosen and are denoted, respectively, by x_1 and x_2 in Fig. 13.4. All other critical parameters are fixed at some appropriate values. Points on the nomogram curve are plotted by repeated computer simulations. Although two parameters are used, this process may require a great deal of computer simulation. Hence, usually a few points on the boundary, often only the corner points, are calculated. The remaining segments of the curve are obtained using a straight line through the calculated points. This approximation can result in significant inaccuracies. The number of critical parameters is usually limited [39].

A better approach to VA is to analyze all key features of the system for all credible contingencies. If the analysis indicates that a system is vulnerable, there should be preventive strategies such as changing the system operating conditions to a more viable status, thus forestalling the possibility of cascading outages.

An even a better approach is to identify the vulnerability border, where the operating point is located on either side of the border, and a margin can be obtained to assess the degree of vulnerability. However, due to the complexity of power systems, it is not possible to determine the vulnerability borders analytically. Instead, numerical solutions such as the boundary tracking techniques can be used in conjunction with fast search algorithm to identify sections of the border near the current operating points. Ideally, the process should be able to answer questions such as:

- What is the nearest unstable operating point?
- What is the most stable operating configuration?
- Given recent trajectory, when is instability likely to occur?

Border tracking is based on searching the functional input–output relationship of the system model (e.g., NN). Therefore, the accuracy of the model is critical to its performance. It is especially important that the model be accurate in operating regions of interest such as near the vulnerability boundary. Several methods can be used to track the vulnerability border, among them are the gradient method, evolutionary computation, and multiagent randomized search techniques (e.g., *particle swarm optimization*).

13.3.4.1 Gradient Method

The gradient method is used to track along specific contours in order to locate new areas of interest [40]. Figure 13.5 shows the concept. Given a search point, x_0, boundary tracking can be used to find the point $\hat{x}_0$ that is closest to x_0 and lies on the surface $f(x) = c$, where c is the value of the VA boundary. This point must also satisfy the operational constraint of the system. The process begins by generating random points around the search point and projecting them onto the surface, $f(x) = c$, by the model of the system such as an inverted neural network. By identifying the boundary point, a VA margin is computed. Information about the location of the nearest invulnerable operating state is valuable in that it outlines operating strategies that should be avoided. In the event of vulnerability, it defines the nearest stable operating state.

13.3.4.2 Evolutionary Computation Method

The *evolutionary computation* (EC) method is based on using a search technique to place as many points on the border as needed to achieve the desired interpolation accuracy among them. This process requires the use of a fast VA technique and a rapid optimization search algorithm.

A query-based evolutionary algorithm for boundary marking using neural network was originally proposed in Ref. 41. By this method, an evolutionary algorithm is used to spread points evenly on a contour of interest. These points are then verified via simulations thus quantifying the accuracy of the vulnerability boundary. Areas of inaccuracy can then be improved by augmenting the training database and retraining the neural network.

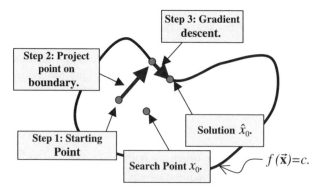

FIGURE 13.5 Illustration of boundary tracking wherein training data is dynamically generated in regions only of interest in the classification process.

Instead of locating and then querying individual points, EC can be used to achieve the following:

- Evaluation of a population of solutions, thus offering the ability to query entire areas of interest.
- Evaluation of evenly distributing the points across the area. Evenly distributing the points is important because a global view of the vulnerability boundary in multiple dimensions is provided thus allowing the entire boundary to be queried and potentially improved.

After the points are spread, they are simulated and their true vulnerability index is determined. If all the points are within tolerance, the algorithm stops. Otherwise, the points with unacceptably large errors are added to the training database and the neural network is retrained.

In the evolutionary boundary algorithm, all reproduction is *asexual* (i.e., no mating or crossover takes place). Offspring are produced as perturbations of single parents. This concentrates the search in the area close to the vulnerability boundary and speeds convergence. The algorithm seeks to minimize a fitness function, F, of the following form:

$$F = |f(\mathbf{x}) - S| + \frac{1}{D_{\text{avg}}},$$

where f is the neural network function, $\mathbf{x}$ is the current point, S is the vulnerability boundary, and D_{avg} is the average distance to the nearest neighbors.

The evolutionary algorithm is randomly initialized with N points and then proceeds as follows.

- The population is sorted based on fitness, F.
- The M points with the lowest fitness scores are deleted.
- Replacements are generated for each deleted point:
- M parents are selected proportional to fitness from the remaining points.
- New offspring are created as perturbations of the selected parents,

$$\mathbf{x}_{\text{new}} = \mathbf{x}_{\text{parent}} + \mathbf{n}, \text{ where } \mathbf{n} \sim N(0, \sigma).$$

- Feasibility constraints are enforced on the new offspring via the solution of a standard power flow.
- Repeat until convergence.

By successively deleting points with poor fitness values and replacing them with perturbations of points with high fitness, the population tends to spread evenly across the solution contour. Typical values used in this chapter are $N = 100$, $M = 20$, $m = 3$ and $\sigma = 0.05$.

13.3.4.3 *Enhanced Particle Swarm Optimization Method* *Particle swarm optimization* (PSO) is a novel optimization method developed in Refs. 42–45. It is a multiagent search technique that traces its origin to the emergent motion of a flock of birds searching for food. It uses a number of agents (particles) that constitute a swarm. Each agent (particle) traverses the search space looking for the global minimum (or maximum). PSO has been recently proposed for power systems applications such as those reported in Refs. 46 and 47. The goal of the PSO is to find operating points that lie on the border under selected conditions.

While the agents in the PSO algorithm are searching the space, each agent remembers two positions. The first is the position of the best point the agent has found (SelfBest). The second one is the position of the best point found among all agents (GroupBest). The equations that govern the motion of each agent are

$$\vec{s}(k + 1) = \vec{s}(k) + \vec{v}(k)$$
$$\vec{v}(k + 1) = w.\vec{v}(k) + a_1.r(0, 1).[\vec{s}_{\text{SelfBest}}(k) - \vec{s}(k)]$$
$$+ a_2.r(0, 1).[\vec{s}_{\text{GroupBest}}(k) - \vec{s}(k)],$$

where $\vec{s}$ is a solution vector of a single particle, $\vec{v}$ is the velocity of this particle, a_1 and a_2 are two scalar parameters of the algorithm, w is an inertia weight, $r(0,1)$ is a uniform random number between 0 and 1, GroupBest is the best solution of all particles, and SelfBest is the best solution observed for the current particle. A maximum velocity, v_{max}, that cannot be exceeded may also be imposed.

It has been found through experimentation [43–46] that the design parameter values that often work best are $a_1 = a_2 = 2$, and w between 0.9 and 1.2.

The goal of border tracking is to identify as many points as possible on the border. Therefore, a separate swarm for each point to be placed on the border should always be initiated. Each point is sequentially identified by running one swarm after the other. This method also provides the ability to impose constraints on the distribution of the points placed on the border. For this reason, we propose the following three different approaches to the border identification problem:

Case 1. Cover the entire vulnerability border with a predetermined number of points, without restrictions on the distribution of the points on the border.

Case 2. Place points uniformly on the entire border.

Case 3. Uniformly cover the portion of the border that is closest to the operating state.

In the first approach, the goal is to find points as accurately on the border as possible. The PSO variation requires an objective function to perform this task. In VA application, we seek to minimize the objective function:

$$f_1(\vec{x}) = |NN(\vec{x}) - c|,$$

where NN is the vulnerability index produced by the model (neural network), $\vec{x}$ is the current position of the particle (operating point of the power system), and c is the vulnerability index of the border and is set by the user.

In the second approach, proximity to the closest neighbor is penalized, where neighbors are defined as the points found by previous swarms. The objective function in this case has a penalty term for the closeness on border point as follows:

$$f_2(\vec{x}) = f_1(\vec{x}) - w_c d_c,$$

where $\vec{x}$ is the current position of the particle, d_c is the distance to the closest border point, and w_c is the weighting factor for d_c.

Covering the entire border with points in a high-dimensional power system, however, is very time-consuming. This is because the number of border points, as a rule of thumb, increases exponentially with the increase in the feature space dimensionality.

The third approach addresses this issue. In order to accomplish this, we modify the fitness function to reward for proximity of the border point to the current operating state. In addition, the penalty for closeness of border points is maintained. These are two contradicting demands and must be balanced through appropriate weighting in the objective function.

$$f_3(\vec{x}) = f_1(\vec{x}) - w_c d_c + w_o d_o,$$

where d_o is the distance from the operating point and w_o is the weighting factor for d_o.

These weighting factors affect the objective function in a very crucial manner and their choice should be such that the desired effect is accomplished. In our case, the desired effect is to be able to find points on the border while imposing restrictions on their distance from each other, which is done through w_c, or their proximity to the current operating state, which is done through w_o. These restrictions can be conflicting with the goal of finding points close to the border, as points already on the border can, in effect, push the others away from the current operating state, while the current operating state can attract the points toward it and thus away from the border. Thus we choose w_c and w_o so that the maximum of the products $w_c d_c$ and $w_o d_o$ does not exceed, say, 1% of the desired border value. Also, if w_o is greater than w_c, the points will tend to be close to the operating state but not as uniformly distributed, while if w_c is greater than w_o, the points will tend to be more uniformly distributed but not as close to the current operating state.

To determine if enough points are on the border, a technique utilizing the midpoints between neighboring border points can be used. The midpoint is defined as the point derived by linear interpolation between a pair of closest neighbors. Each point on the border has one closest neighbor, thus the total number of midpoints is at least equal to half the number of points, but not greater than the total number of points. Figure 13.6 shows the midpoint of A and B. By definition, no other points

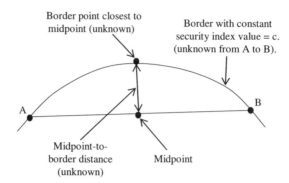

FIGURE 13.6 Midpoint and its proximity to the border in the input space.

exist on the segment AB. The proximity of the midpoint to the border can only be determined in the output space, because the segment AB is unknown.

13.3.5 Selection of Vulnerability Index

The VA should provide the operator with a *vulnerability index* (VI). The VI should reflect the level of system strength or weakness relative to the occurrence of an undesired event. The vulnerability of power system will increase if:

- The operating state changes in such a way that the number of contingencies that would lead to blackout is increased.
- Environmental conditions change so as to increase the likelihood of a blackout.
- System equipment status changes such as to increase the likelihood of a blackout in case of cascaded events.

The most common VI techniques are the *critical clearing time* (CCT) and the *energy margin* (EM). The CCT [48] represents the maximum elapsed time from the initiation of a fault until the fault is isolated and the power system remains transiently stable. A number of time domain simulations can be conducted to ascertain the CCT, which is generally accepted as the best vulnerability analysis tool with regard to accuracy, reliability, and modeling capability [49]. This approach, however, requires intensive computation time, thereby limiting its usefulness to a few off-line studies.

To overcome this weakness, direct methods have been developed [6] to determine system stability from energy functions by comparing system energy levels to a critical energy value. The difference is called *energy margin* (EM) and is used as a vulnerability index. Direct methods have a long history of development [2–17], but, until recently, many researchers have thought of them as impracticable for large-scale systems because of its intrinsic modeling limitations. Moreover, it is less accurate than the time domain approach [6].

These methods, nevertheless, have received considerable attention because of their ability to calculate the sensitivity at the margin to power system parameters [6]. The operator needs to know not only whether a system is invulnerable at any given time, but also which parameters are critical for vulnerability. In Ref. 50, the energy margin ΔE is used as an indicator of the level of security and its sensitivity $\partial \Delta E / \partial p$ to a changing system parameter p as an indicator of its trend. A low value of ΔV with a high value of $\partial \Delta E / \partial p$ means the system is vulnerable to changes in the parameter. However, it is relatively difficult for the operator to use sensitivity as a basis for estimating the true vulnerability of a power system.

13.3.5.1 *Vulnerability Index Based on Distance from a Border* Knowing the vulnerability border can provide an operator with valuable guidance for steering the power system away from vulnerable operating regions. By contrast with the sensitivity analysis mentioned above, correct identification of the vulnerability border can provide easily understood visual information. Moreover, the distance from the border provides a direct assessment of the degree of vulnerability, which is, in itself, a new kind of vulnerability index (VI) as shown in Fig. 13.7. The vulnerability boundary, however, cannot be determined analytically for a large-scale power system, requiring, rather, extensive computation by numerical methods.

If a given operating point lies inside the border, the operating state of the system is said to be invulnerable and the distance from the closest point on the border would be the margin of safety. On the other hand, if an operating point lies outside the border, the operating state is said to be vulnerable and the shortest distance to the border will be the degree of vulnerability. By this approach, the VI can be changed from output to input space.

The algorithm starts by training a neural network to predict the vulnerability status of a known power system. The NN module provides a VI in the output space. This process can be modeled by mapping $VI = f_{NN}(\vec{x})$, where $\vec{x}$ comprises the power system features chosen to represent the operating state, VI is the vulnerability

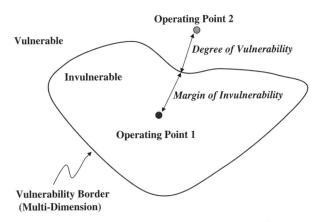

FIGURE 13.7 Vulnerability border and margin.

index, and $f_{NN}(\ )$ is the neural network model. The output of the neural network module is the VI or any other possible index such as the critical clearing time or the energy margin. The desired VI of the border is set by the operator. The PSO algorithm seeks to find the closest border point by minimizing an objective function F, of the following form

$$F = |f_{NN}(\vec{\mathbf{x}}) - VI_{\text{desired}}| + w \times \|\vec{\mathbf{x}} - \vec{\mathbf{x}}_{\text{Operating}}\|,$$

where f_{NN} is the neural network function, $\vec{\mathbf{x}}$ is the current position of a given particle, $\vec{\mathbf{x}}_{\text{Operating}}$ is the position of the operating point, $\|\ \|$ is the Euclidean distance between vectors, and w is a weighting factor. As explained earlier, the PSO method can be used to find the border points as well as the minimum distance from the operating point.

In addition to just computing the Euclidean distance between the current operating point and the nearest border point, the VA can be visualized. The visualization can provide the operator with an illustration of system vulnerability. To be effective, the visualization should be developed for the input space, where the variables are explicit (voltages, powers, etc.). Figures 13.8 and 13.9 show such a visualization of the VI before and after an event, respectively. The superimposed bars in the figures represent the magnitude of the variables in the operating space (dark shading) and the value of the variables at the nearest feasible border point (light shading). The vulnerability index as defined by the Euclidean distance is also shown at the far right in the figures. A negative vulnerability index represents an invulnerable operation. Its magnitude is the vulnerability margin. A positive magnitude represents a vulnerable operation, and its magnitude is an indication for the degree of system vulnerability. As seen in Fig. 13.8, all variables are less or equal to these at the nearest border point. This is invulnerable operation, and the vulnerability index is about 20%.

After an event (tripping of a line), the system VI is shown in Figure 13.9. As seen in the figure, most variables clearly exceed the values at the nearest border point, and the system is vulnerable. The vulnerability index is positive and its numerical value (50%) indicates how far the system is from the border.

13.3.5.2 Vulnerability Index Based on Anticipated Loss of Load
Normally, power systems are operated under equilibrium conditions, where total load consumption and system losses equal the total generation. System frequency is governed by this equilibrium, and consequently, any unbalance in loads could result in frequency excursions that may lead to loss of synchronism. Excess of load will result in a system frequency drop, and load shedding has to be employed in order to rapidly balance demand and generation.

Because the purpose of vulnerability assessments is to make it possible to avoid catastrophic power outages, the index should reflect how much load might be lost at such times. Hence, using the anticipated loss of load as an alterative index makes great sense. This concept is simple but requires a great deal of computation. In the case of small systems, one could examine all possible combinations of load

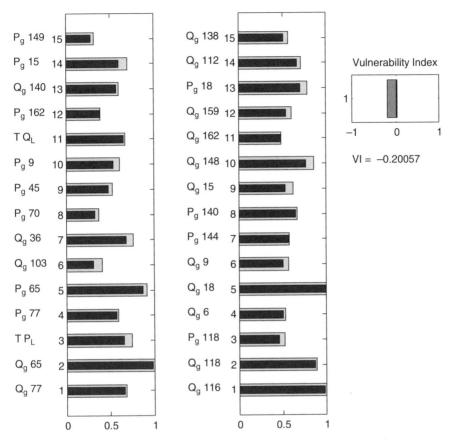

FIGURE 13.8 Current operating point and nearest border point after the first event.

reductions to find an appropriate minimum. However, for systems of realistic size, an exhaustive search is impracticable. Suppose that we shed each load from 0% to 100% in increments of 1%. If a system has N loads, there would be 100^N possible combinations of load reductions. Such a number is well beyond the limits of current computations. An alternative method is to use the particle swarm optimization method to find the best combination of load reductions on key buses while maintaining system stability. The amount of load shed is the vulnerability index.

Load shedding is often accomplished using frequency-sensitive relays that detect the onset of decay in power system frequency where both frequency and rate of frequency decline are measured. Load shedding is usually implemented in stages; each is triggered at different frequency level or at specified rate of frequency decline [51–54].

Load shedding is a combinatorial optimization problem that lends itself to the PSO technique. The first step in the PSO load shedding algorithm is to determining the load buses that are candidates for load shedding. These buses are often known to the operator as they have large impacts on system stability. Alternatively, sensitivity

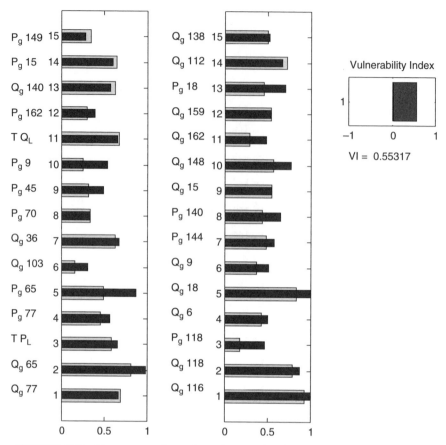

FIGURE 13.9 Current operating point and nearest border point after the second event.

analysis can be employed. Each PSO agent (particle) traverses the search space looking for the global minimum combination of load shedding using the initial values obtained in the previous step. These best points are verified using the power system model.

The fitness values to maximize by the PSO should be the inverse of the summation of all load reductions.

$$F = \frac{1}{\sum MVA}, \quad \text{if system is stable}$$

$$F = -\frac{1}{\sum MVA}, \quad \text{if system is unstable.}$$

This fitness function will have a maximum value when all loads are maintained. If the system is unstable, the fitness function has a negative sign.

After each iteration, the fitness value of the global best point is compared with the fitness at the previous iteration. If the difference is smaller than a set value and is maintained small for several iterations, a solution is achieved and the amount of load shedding is the vulnerability index.

13.4 CONCLUSIONS

It is evident that computational intelligence can substantially advance the vulnerability assessment analysis of complex power systems. CI can be used in every step of the vulnerability assessment tasks.

When feature-extraction techniques are used, the training and testing times are substantially reduced. This should facilitate the task of on-line assessment.

The anticipated loss of load can be used as a vulnerability index instead of the conventional energy function or critical clearing time. Such an index would be fully applicable in the case of cascading events. Moreover, any relevant control actions could be incorporated into the ongoing process of calculation. The CI can effectively address the computational problems associated with the large number of possible combinations of load reductions and at the same time provide the minimum load shedding. This is a form of on-line or preventive control.

The ultimate goal of vulnerability assessment is not only to estimate the vulnerability of a power system but also to provide guidance to an operator for avoiding potentially vulnerable states. This can be achieved by identifying the border of vulnerability and by providing the operator with a visualization of the operating space with data on the vulnerability margin.

REFERENCES

1. Western Systems Coordinating Council. Disturbance report for the power system outage that occurred on the Western interconnection. August 10, 1996.

2. Magnusson PC. The transient-energy method of calculating stability. Trans AIEE 1947; 66:747–755.

3. Aylett PD. The energy integral criterion of transient stability limits of power systems. Proc IEE 1958; 77:527–536.

4. Gless GE. Direct method of Lyapunov applied to transient power system stability. IEEE Trans Power Apparatus Systems 1966; PAS-85:159–168.

5. El-Abiad AH, Nagappan K. Transient stability regions for multi-machine power systems. IEEE Trans Power Apparatus Systems 1996; PAS-85:169–179.

6. Chiang H, Chu C, Cauley G. Direct stability analysis of electric power systems using energy functions: theory, applications, and perspective. Proc IEEE 1995; November: 1497–1529.

7. Athay T, Podmore R, Virmani S. A practical method for the direct analysis of transient stability. IEEE Trans Power Apparatus Systems 1979; PAS-98(2):573–584.

8. Fouad AA, Stanton S. Transient stability of a multi-machine power system. Part I: Investigation of system trajectories. Part II: Critical transient energy. IEEE Trans Power Apparatus Systems 1981; PAS-100(7):3408–3424.

9. Bose A. Application of direct methods to transient stability analysis of power systems. IEEE Trans Power Apparatus Systems 1984; PAS-103(7):1629–1636.

10. Chiang HD, Wu FF, Varaiya PP. Foundations of the potential energy boundary surface method for power system transient stability analysis. IEEE Trans Circuit Systems 1988; 35:712–728.

11. El-Kady M, et al. Dynamic security assessment utilizing the transient energy function method. IEEE Trans Power Systems 1988; PWRS-1(3):284–291.

12. Chiang H. A theory-based controlling UEP method for direct analysis of power system transient stability. In: Proceedings of the International Symposium on Circuits and Systems (IEEE ISCAS '89), Portland, Oregon, IEEE Press, May, 1989. p. 1980–1983.

13. Pai MA. Energy function analysis for power system stability. Dodrecht: Kluwer Academic Publishers; 1989.

14. Fouad AA, Vittal V. Power system transient stability analysis using the transient energy function method. Englewood Cliffs, NJ: Prentice-Hall; 1992.

15. Chiang H, Wu F, Varaiya P. A BCU method for direct analysis of power system transient stability. IEEE Trans Power Systems 1994; 9(3):1194–1208.

16. Llamas A, et al. Clarifications of the BCU method for transient stability analysis. IEEE Trans Power Systems 1995; 10:210–219.

17. Vaahedi E, et al. Enhanced "second kick" methods for on-line dynamic security assessment. IEEE Trans Power Systems 1996; 11:1976–1982.

18. Sobajic D, Pao Y. Artificial neural-net based dynamic security assessment for electric power systems. IEEE Trans Power Systems 1989; 4(1):220–228.

19. El-Sharkawi M, et al. Dynamic security assessment of power systems using back error propagation artificial neural networks. Second Symposium on Expert System Applications to Power Systems, Seattle, WA, 1989.

20. El-Sharkawi MA, Marks RJ II, eds. Applications of neural networks to power systems. Piscataway, NJ: IEEE Press; 1991.

21. El-Sharkawi MA, Huang SS. Query-based learning neural network approach to power system dynamic security assessment. International Symposium on Nonlinear Theory and Its Applications, Waikiki, Hawaii, 5–10 December, 1993.

22. Jeyasurya B. Application of artificial neural networks to power system transient energy margin evaluation. Elsevier, Electric Power Systems Res 1993; 26:71–78.

23. Sing LB, Po WK. Transient stability assessment: an artificial neural network approach. In: Proceedings IEEE International Conference on Neural Networks, Vol. 2, IEEE Press, 1995. p. 702–707.

24. El-Sharkawi MA, Niebur D, eds. Application of artificial neural networks to power systems. Piscataway, NJ: IEEE Press, 1996.

25. Mansour Y, Vaahedi E, El-Sharkawi M. Dynamic security contingency screening and ranking using neural networks. IEEE Trans Neural Networks 1997; 8(4):942–950.

26. Jensen CA, El-Sharkawi MA, Marks RJ. Power system security boundary enhancement using evolutionary-based query learning. Invited paper, International Journal of Engineering Intelligent Systems, CRL Publishing, December 1999, p. 215–218.

27. Jensen CA, El-Sharkawi M, Marks RJ. Power system security assessment using neural networks: feature selection using Fisher discrimination. IEEE Trans Power Systems 2001; 16:757–763.

28. El-Sharkawi MA. Choice of model and topology for external equivalent systems. IEEE Trans Power Apparatus Systems 1983; 102:3761–3768.

29. Fisher RA. The use of multiple measurements in taxonomic problems. Ann Eugenics 1936; 7(pt II):179–188.

30. Jensen C. Application of computational intellgence to power system security assessment. PhD Dissertation, University of Washington; 1999.

31. Weerasooriya S, El-Sharkawi MA. Use of Karhunen-Loève expansion in training neural networks for static security assessment. In: Proceedings of the First International Forum on Applications of Neural Networks to Power Systems, Seattle, WA, University of Washington, July 1991. p. 59–64.

32. Schalkoff RJ. Pattern recognition: statistical, structural and neural approaches. New York: Clemson University and John Wiley & Sons; 1992.

33. Bishop CM. Neural networks for pattern recognition. Oxford: Oxford University Press; 1995.

34. Ripley BD. Pattern recognition and neural networks. Cambridge: Cambridge University Press; 1996.

35. Zayan MB, El-Sharkawi MA, Prasad NR. Comparative study of feature extraction techniques for neural network classifier. In: Proceeding of the International Conference on Intelligent Systems Applications to Power Systems, 28 Jan.–2 Feb. 1996. p. 400–404.

36. Rao CR. The utilization of multiple measurements in problems of biological classfication. J Roy Statistical Soc Ser 1948; 159–203.

37. Duda RO, Hart PE. Pattern classification and scene analysis. New York: John Wiley & Sons; 1973.

38. Vapnik VN. Statistical learning theory. New York: John Wiley & Sons; 1998.

39. McCalley J, et al. Security boundary visualization for systems operation. IEEE Trans Power Systems 1997; 12:940–947.

40. Bonissone P, El-Sharkawi MA, Fogel D, Freer S. Evolutionary computation and applications. IEEE CD-ROM # EC128; 2001.

41. Reed RD, Marks RJ II. An evolutionary algorithm for function inversion and boundary marking. IEEE International Conference on Evolutionary Computation, Perth, West Australia, December, 1995.

42. Eberhart R. Particle swarm optimization. In: Proceedings IEEE International Conference on Neural Networks, Perth, Australia. Piscataway, NJ: IEEE Service Center, 1995; p. 1942–1948.

43. El-Sharkawi MA, Marks RJ, Eberhart R. Evolutionary techniques and fuzzy logic in power systems. IEEE Tutorial 2000. TP-142.

44. Shi Y, Eberhart R. A modified particle swarm optimizer. In: IEEE World Congress on Computational Intelligence. IEEE Press, 1998. p. 69–73.

45. Eberhart R, Kennedy J. A new optimizer using particle swarm theory. In: Proceedings of the Sixth International Symposium on Micro Machine Human Sci 1995. p. 39–43.

46. Yoshida H, Fukuyama Y, Takayama S, Nakanishi Y. A particle swarm optimization for reactive power and voltage control in electric power systems considering voltage security

assessment. In: IEEE International Conference on Systems, Man, and Cybernetics, SMC '99. IEEE Press, 1999. p. 497–502.

47. Kassabalidis IN, El-Sharkawi MA, Marks RJ II, Moulin LS, Alves da Silva AP. Dynamic security border identification using enhanced particle swarm optimization. IEEE Trans Power Systems 2002; 17:723–729.

48. Kundar P. Power system stability and control. New York: McGraw-Hill; 1994.

49. Moechtar M, Cheng TC, Hiu L. Transient stability of power system—a survey. WESCON '95. Conference on "Microelectronics Communications Technology Producing Quality Products Mobile and Portable Power Emerging Technologies". 7–9 Nov. 1995. p. 166–171.

50. Fouad AA, Zhou Q, Vittal V. System vulnerability as a concept to assess power system dynamic security. IEEE Trans Power Systems 1994; 9:1009–1015.

51. Eborn S, Ait-Kheddache A. Optimal load shedding methodologies in power systems. Southeastcon '88. IEEE Conference Proceedings. IEEE Press, 1988. p. 269–272.

52. Grewal S, Konowalec J, Hakim M. Optimization of a load shedding scheme. IEEE Industry Applications Magazine 1998; 4:25–30.

53. Anderson R. Power System Protection. Piscataway, NJ: IEEE Press; 1999.

54. Lindahl S, Runvik G, Stranne G. Operational experience of load shedding and new requirements on frequency relays. Developments in Power System Protection 1997; March: 262–265.

■■■■■■ CHAPTER 14

Applications to System Planning

EDUARDO NOBUHIRO ASADA, YOUNGJAE JEON, KWANG Y. LEE,
VLADIMIRO MIRANDA, ALCIR J. MONTICELLI, KOICHI NARA,
JONG-BAE PARK, RUBÉN ROMERO, and YONG-HUA SONG

14.1 INTRODUCTION

It is well-known that most of the optimization problems in power system planning are complex, large-scale, hard, nonlinear combinatorial problems of mixed integer nature. This means that the number of solutions to be evaluated grows exponentially with the system size, that is, the problem is a non-polynomial time (NP-complete) with a large number of local optimal solutions, which makes the solution space a potentially high multimodal landscape.

The very nature of the planning exercise has led to the subdivision of the problem into subproblems that have gained autonomy on their own. We may therefore refer for instance to the following problems with such a complex nature:

(a) Long-term generation expansion planning, where a capacity addition plan must be produced.

(b) Network expansion planning, either at transmission or at distribution level, typically adopting a DC model to evaluate power flows, and where a plan for network additions and reinforcement must be produced.

(c) Reactive power planning, where an investment plan for new reactive sources at selected load buses must be produced, which nowadays is useful both at transmission and at distribution level, due to the growing importance of distributed generation

For all these problems, early attempts to solve them by applying mathematical analytical models had only a limited success and in some cases with no real-world application. In practice, the use of heuristics allowed a more realistic representation

Modern Heuristic Optimization Techniques. Edited by K. Y. Lee and M. A. El-Sharkawi
Copyright © 2008 the Institute of Electrical and Electronics Engineers, Inc.

of system characteristics, constraints, and objectives of the decision makers, even if potentially sacrificing optimality.

The emergence of meta-heuristics has given robustness to nonanalytical methods, because of the rationale behind them. Besides, evolutionary algorithms have provided a high degree of confidence in a stochastic convergence to the optimum and have supported this confidence with a mathematical background explaining not only how they achieve convergence but also how to improve the convergence rate.

In the following sections, we will refer to a number of models proposed to solve planning problems in power systems, some of them having achieved practical application. We will especially focus on the coding strategies adopted, because they are crucial to the efficiency of every meta-heuristic, and will try to point out the advantages brought by hybrid approaches to those old yet largely unsolved problems. Mind that this chapter is not meant to discuss all of the state-of-the-art models about power system planning but rather to provide good examples of how the use of meta-heuristics has improved modeling, has allowed us to obtain better results than with alternative models, or even has allowed us to obtain results that were not previously at hand with conventional models.

14.2 GENERATION EXPANSION

Generation expansion planning (GEP) is one of the most important decision-making activities associated with power supply guarantee. In classic vertically organized power systems, this was a function of electric utilities, but in unbundled market-organized systems, GEP is still important as a means of developing reference plans for system growth, even if the construction of new power plants depends on the initiative of private investors.

The least-cost GEP objective is to determine the minimum-cost capacity addition plan (i.e., the type and number of candidate plants) that meets forecasted demand within a prespecified reliability criterion over a planning horizon. This is a problem well suited to be solved by evolutionary algorithms, and this section is devoted to show that a classic genetic algorithm, enhanced with features dealing with the coding scheme and especially with the crossover operator (allowing a better search and avoiding the destruction of promising substrings), may outperform other methods including some commonly accepted in practice.

To solve the GEP problem, a number of methods have been successfully applied during the past decades. Masse and Gilbrat [1] applied a linear programming approach that necessitates the linear approximation of an objective function and constraints. Bloom [2] applied a mathematical programming technique using a decomposition method and solved it in a continuous space. Park *et al.* [3] applied Pontryagin's maximum principle whose solution also lies in a continuous space. Although the above-mentioned mathematical programming methods have their own advantages, they possess one or both of the following drawbacks in solving a GEP problem. That is, they treat decision variables in a continuous space, and there is no guarantee to get the global optimum because the problem is not mathematically convex.

Dynamic programming (DP) based framework is one of the most widely used algorithms in GEP [4–8]. However, the so-called curse of dimensionality has interrupted direct application of the conventional full DP in practical GEP problems. For this reason, WASP [4] and EGEAS [5] use a heuristic tunneling technique in the DP optimization routine where users prespecify states and successively modify tunnels to arrive at a local optimum. David and Zhao developed a heuristic-based DP [7] and applied the fuzzy set theory [8] to reduce the number of states. Recently, Fukuyama and Chiang [9] and Park et al. [10] applied a genetic algorithm (GA) to solve sample GEP problems and showed promising results. However, an efficient method for a practical GEP problem that can overcome a local optimal trap and the dimensionality problem simultaneously has not been developed yet.

Mathematically, solving a least-cost GEP problem is equivalent to finding a set of optimal decision vectors over a planning horizon that minimizes an objective function under several constraints. The GEP problem to be considered is formulated as follows [11]:

$$\underset{U_1, \ldots, U_T}{\text{Min}} \sum_{t=1}^{T} \{f_t^1(U_t) + f_t^2(X_t) - f_t^3(U_t)\} \tag{14.1}$$

$$\text{s.t.} \quad X_t = X_{t-1} + U_t \quad (t = 1, \ldots, T) \tag{14.2}$$

$$\text{LOLP}(X_t) < \varepsilon \quad (t = 1, \ldots, T) \tag{14.3}$$

$$\underline{R} \leq R(X_t) \leq \overline{R} \quad (t = 1, \ldots, T) \tag{14.4}$$

$$\underline{M_t^j} \leq \sum_{i \in \Omega_j} x_t^i \leq \overline{M_t^j} \quad (t = 1, \ldots, T \text{ and } j = 1, \ldots, J) \tag{14.5}$$

$$0 \leq U_t \leq \overline{U_t} \quad (t = 1, \ldots, T), \tag{14.6}$$

where T is number of periods (years) in a planning horizon, J is number of fuel types, Ω_j is index set for jth fuel type plant, X_t is cumulative capacity (MW) vector of plant types in year t, x_t^i is cumulative capacity (MW) of ith plant type in year t, U_t is capacity addition (MW) vector by plant types in year t, $\overline{U_t}$ is maximum construction capacity (MW) vector by plant types in year t, u_t^i is capacity addition (MW) of ith plant in year t, $\text{LOLP}(X_t)$ is loss of load probability (LOLP) with X_t in year t, $R(X_t)$ is reserve margin with X_t in year t, ε is reliability criterion expressed in LOLP, $\overline{R}$, $\underline{R}$ are upper and lower bounds of reserve margin, $\overline{M_t^j}$, $\underline{M_t^j}$, are upper and lower bounds of jth fuel type in year t, $f_t^1(U_t)$ is discounted construction costs (\$) associated with capacity addition U_t in year t, $f_t^2(X_t)$ is discounted fuel and operating and maintenance (O&M) costs (\$) associated with capacity X_t in year t, and $f_t^3(U_t)$ is discounted salvage value (\$) associated with capacity addition U_t in year t.

The objective function is the sum of tripartite discounted costs over a planning horizon. It is composed of discounted investment costs, expected fuel and O&M costs, and salvage value. To consider investments with longer lifetimes than a planning horizon, the linear depreciation option is utilized [4]. In this paper, five types of constraints are considered. Equation (14.2) implies state equation for dynamic planning problem [11]. Equations (14.3) and (14.4) are related to the LOLP reliability criteria and the reserve margin bands, respectively. The capacity mixes by fuel types are considered in (14.5). Plant types give another physical constraint in (14.6), which reflects the yearly construction capabilities.

14.2.1 A Coding Strategy for an Improved GA for the Least-Cost GEP

Although the state vector, X_t, and the decision vector, U_t, have dimensions of MW, we can easily convert those into vectors that have information on the number of units in each plant type. This mapping strategy is very useful for GA implementation of a GEP problem such as encoding and treatment of inequality (14.6) and is illustrated in the following equations:

$$X_t = (x_t^1, \ldots, x_t^N)^T \quad \rightarrow \quad X_t' = (x_t'^1, \ldots, x_t'^N)^T \tag{14.7}$$

$$U_t = (u_t^1, \ldots, u_t^N)^T \quad \rightarrow \quad U_t' = (u_t'^1, \ldots, u_t'^N)^T, \tag{14.8}$$

where N is number of plant types including both existing and candidate plants, X_t' is cumulative number of units by plant types in year t, U_t' is addition number of units by plant types in year t, $x_t'^i$ is ith plant type's cumulative number of units in year t, and $u_t'^i$ is ith plant type's addition number of units in year t.

Because it is convenient to use integer values for GA implementation of a GEP problem, the reordered structure of (14.8) by plant types covering a planning horizon is used for encoding of a string as shown in (14.9). Here, each element of a string (i.e., $\hat{U}'^n = (u_1'^n, u_2'^n, \ldots, u_T'^n)^T$ for $n = 1, \ldots, N$) corresponds with a substring, and its structure is depicted in Fig. 14.1:

$$\hat{U}' = (u_1'^1, u_2'^1, \ldots, u_T'^1, \ldots, u_1'^n, u_2'^n, \ldots, u_T'^n, \ldots, u_1'^N, u_2'^N, \ldots, u_T'^N)^T$$
$$= (\hat{U}'^1, \ldots, \hat{U}'^n, \ldots, \hat{U}'^N)^T \tag{14.9}$$

14.2.2 Fitness Function

The objective function or cost of a candidate plan is calculated through the probabilistic production costing and the direct investment costs calculation [4, 5]. The fitness value of a string is calculated by

$$f'(i) = \frac{f(i) - f_{min}}{f_{max} - f_{min}}, \tag{14.10}$$

where f_{max}, f_{min} is maximum and minimum fitness value in a generation, $f'(i)$ is modified fitness value of string i, $f(i)$ is basic fitness value of string i using $f = \alpha/1 + J$, where α is a constant, and J is objective function of (14.1).

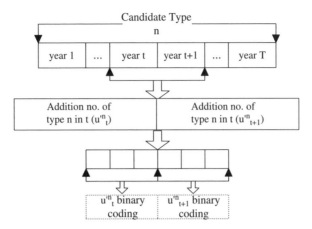

FIGURE 14.1   A substring structure.

This scheme has proved to avoid premature convergence when a roulette procedure is used as a selection operator.

14.2.3   Creation of an Artificial Initial Population

It is important to create an initial population of strings spread out throughout the whole solution space, especially in a large-scale problem. One alternative method could be to increase the population size, which yields a high computational burden. Instead, one adopted a new artificial initial population (AIP) scheme, which also takes the random creation scheme of the conventional GA into account. The procedures are illustrated in the following and in Table 14.1.

Step 1. Generate all possible binary seeds of each plant type considering (14.6). For example, if ith plant type has an upper limit of 3 units per year, then generate four possible binary seeds (i.e., 00, 01, 10, 11).

Step 2. Find the least common multiple (LCM) m from the numbers of the binary seeds of all types, and fill m binary seeds in a look-up table for all plant types and planning years. For example, if three plant types have upper limits of 3, 3, and 5 units per year, respectively, then the numbers of binary seeds are 4, 4, and 6, and m becomes 12.

Step 3. Select an integer within $[1, m]$ at random for each element $u_t'^n$ of a string in (14.9). Fill the string with the corresponding binary digits, and delete it from the look-up table. Repeat until m different strings are generated.

Step 4. Check the constraints of (14.3), (14.4) and (14.5). If a string satisfies these constraints for all years, then it becomes a member of an initial population. Otherwise, the only parts of the string that violate the constraints in year t are generated at random until they satisfy the constraints. Go to step 3 n times for $n \cdot m$ less than P, where P is the number of strings in a population and n is an arbitrary positive integer.

TABLE 14.1 Example of Look-Up Table with 3 Plant Types for 3 Planning Years

m	Type 1 (Upper limit: 3 units/year)			Type 2 (Upper limit: 3 units/year)			Type 3 (Upper limit: 5 units/year)		
	Year 1	Year 2	Year 3	Year 1	Year 2	Year 3	Year 1	Year 2	Year 3
1	00	00	00	00	00	00	000	000	000
2	01	01	01	01	01	01	001	001	001
3	10	10	10	10	10	10	010	010	010
4	11	11	11	11	11	11	011	011	011
5	00	00	00	00	00	00	100	100	100
6	01	01	01	01	01	01	101	101	101
7	10	10	10	10	10	10	000	000	000
8	11	11	11	11	11	11	001	001	001
9	00	00	00	00	00	00	010	010	010
10	01	01	01	01	01	01	011	011	011
11	10	10	10	10	10	10	100	100	100
12	11	11	11	11	11	11	101	101	101

Generated string 1: 011100101011000101010

Step 5. The remaining $P - n \cdot m$ strings are created using uniform random variables with binary number $\{0,1\}$. Go to step 4 to check the constraints and regenerate them if necessary. This process is repeated until all strings, which satisfy the constraints, are generated.

This AIP is based on both artificial and random selection schemes, which allows all possible string structures to be included in an initial population.

14.2.4 Stochastic Crossover, Elitism, and Mutation

The model includes two different schemes for genetic operation: a stochastic crossover technique and the application of elitism. The stochastic crossover scheme uses three different crossover methods; 1-point crossover, 2-point crossover, and 1-point substring crossover as illustrated in Fig. 14.2. Each crossover method has its own merits. The 1-point substring crossover can provide diverse bit structures to search solution space, however it easily destroys the string structure that may have partial information on the optimal structure.

Although the 1- and 2-point crossovers can not explore solution space as widely as the above crossover, the probability of destroying an already-found partial optimal structure is very low. The stochastic crossover strategy is similar to the process of stochastic selection of reproduction candidates from a mating pool. That is, one of the three different crossover methods is selected from a biased roulette wheel, where each crossover method has a roulette wheel slot sized according to its performance. The weight for each crossover method has been determined.

The second feature lies in the application of elitism [12]. The roulette wheel selection scheme, as illustrated in Fig. 14.3, gives a reproduction opportunity to a set of recessive members and might not give the set of dominant strings (i.e., an elite group) a chance to reproduce. Furthermore, the application of genetic operations changes the string structures of the fittest solutions. Thus, the best solutions in the current generation might not appear in the next generation. To circumvent these problems, an elite group is directly copied into the next generation.

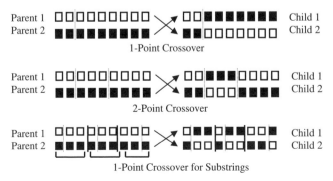

FIGURE 14.2 Three different crossover methods used.

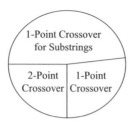

FIGURE 14.3 Roulette wheel for stochastic selection of crossover method.

We have applied the conventional mutation scheme where it performs bit-by-bit for the strings that have undergone the stochastic crossover operator. However, the mutation procedure is not applied to the set of dominant strings to preserve the elitism.

After genetic operations, we check all strings whether they satisfy the constraints of (14.X) to (14.X) or not. If any string violates the constraints of (14.4) to (14.6), only the parts of the string that violate the constraints in year t are generated at random until they satisfy the constraints as described in the AIP scheme.

14.2.5 Numerical Examples

The performance of the GA scheme was tested by comparing it with other methods. For this purpose, the IGA (improved GA), SGA (simple GA, based on the simple crossover–mutation–roulette scheme), tunnel-constrained dynamic programming (TCDP) employed in WASP, and full dynamic programming (DP) were implemented in FORTRAN77 on a PC.

The IGA, SGA, TCDP, and DP methods have been applied in two test systems: case 1 for a power system with 15 existing power plants, 5 types of candidate options, and a 14-year study period; and case 2 for a real-scale system with a 24-year study period. The planning horizons of 14 and 24 years are divided into 7 and 12 stages (2-year intervals), respectively. The forecasted peak demand over the study period is given in Table 14.2.

Tables 14.3 and 14.4 show the technical and economic data of the existing plants and candidate plant types for future additions, respectively.

TABLE 14.2 Forecasted Peak Demand

Stage (Year)	0 (1996)	1 (1998)	2 (2000)	3 (2002)	4 (2004)	5 (2006)	6 (2008)
Peak (MW)	5000	7000	9000	10,000	12,000	13,000	14,000
Stage (Year)	—	7 (2010)	8 (2012)	9 (2014)	10 (2016)	11 (2018)	12 (2020)
Peak (MW)	—	15,000	17,000	18,000	20,000	22,000	24,000

TABLE 14.3 Technical and Economic Data of Existing Plants

Name (Fuel type)	No. of units	Unit capacity (MW)	FOR (%)	Operating cost ($/kWh)	Fixed O&M cost ($/kW-Month)
Oil #1 (heavy oil)	1	200	7.0	0.024	2.25
Oil #2 (heavy oil)	1	200	6.8	0.027	2.25
Oil #3 (heavy oil)	1	150	6.0	0.030	2.13
LNG G/T #1 (LNG)	3	50	3.0	0.043	4.52
LNG C/C #1 (LNG)	1	400	10.0	0.038	1.63
LNG C/C #2 (LNG)	1	400	10.0	0.040	1.63
LNG C/C #3 (LNG)	1	450	11.0	0.035	2.00
Coal #1 (anthracite)	2	250	15.0	0.023	6.65
Coal #2 (bituminous)	1	500	9.0	0.019	2.81
Coal #3 (bituminous)	1	500	8.5	0.015	2.81
Nuclear #1 (PWR)	1	1000	9.0	0.005	4.94
Nuclear #2 (PWR)	1	1000	8.8	0.005	4.63

FOR = forced outage rate; O&M = operation and maintenance; PWR = pressurized water reactor.

We used 8.5% as a discount rate, 0.01 as LOLP criterion, and 15% and 60% as the lower and upper bounds for reserve margin. The lower and upper bounds of capacity mix were 0% and 30% for oil-fired power plants, 0% and 40% for LNG-fired, 20% and 60% for coal-fired, and 30% and 60% for nuclear generation.

14.2.6 Parameters for GEP and IGA

Parameters for the IGA were selected through experiments (Table 14.5). Especially, the dominant parameters such as crossover probabilities and weights for crossover techniques were determined empirically from a test system with a 6-year planning horizon with other data being the same as cases 1 and 2.

To decide the weight of each crossover method in a biased roulette wheel for stochastic crossover, nine experiments were performed by changing the probability of crossover from 0.6 to 0.8, and the results were compared with the optimal solution obtained by the full DP. Among the three crossover methods, the 1-point substring crossover showed the best performance in every case. Thus, we set the 1-point substring crossover with the biggest weight and others with an equal smaller weight.

To determine the weight of each crossover method in a biased roulette wheel, 18 simulations were performed with different weights and crossover probabilities. Among 18 simulations, we have found the optimal solution 7 times and the second-best solution 4 times. Furthermore, the optimal or the second-best solution was found by applying the stochastic crossover technique when the probability of crossover is 0.6. Also, when the weight of 1-point substring crossover is 0.7 and weights for others are 0.15, it always found optimal or the second-best optimal solution. Therefore, we have set the weights in the stochastic crossover technique as 0.15:0.15:0.70 among the three crossover methods.

TABLE 14.4 Technical and Economic Data of Candidate Plants

Candidate type	Construction upper limit	Capacity (MW)	FOR (%)	Operating cost ($/kWh)	Fixed O&M cost	Capital cost ($/kW)	Life-time (years)
Oil	5	200	7.0	0.021	2.20	812.5	25
LNG C/C	4	450	0	0.035	0.90	500.0	20
Coal (bituminous)	3	500	9.5	0.014	2.75	1062.5	25
Nuclear (PWR)	3	1000	9.0	0.004	4.60	1625.0	25
Nuclear (PHWR)	3	700	7.0	0.003	5.50	1750.0	25

FOR = forced outage rate; O&M = operation and maintenance; LNG = liquid natural gas; C/C = combined cycle; PWR = pressurized water reactor; PHWR = pressurized heavy water reactor.

TABLE 14.5 Parameters for IGA Implementation

Parameters	Values
• Population size	300
• Maximum generation	300
• Probabilities of crossover and mutation	0.6, 0.01
• Number of elite strings	3 (1%)
• Weights of 1-point, 2-point, and 1-point	0.15:0.15:0.70
• Crossover for substrings in a biased roulette wheel	

14.2.7 Numerical Results

The IGA was applied to two test systems and compared with the results of DP, TCDP, and SGA. Throughout the tests, the solution of the conventional DP provided the global optimum whereas TCDP could only guarantee a local optimum. Both the global and a local solution can be obtained in case 1; however, in case 2 the "curse of dimensionality" prevents the use of the conventional DP. Figure 14.4 illustrates the convergence characteristics of various GA-based methods in case 1. It also shows the improvement of IGA over SGA. The IGA employing the stochastic crossover scheme (IGA2) has shown better performance than the IGA using the artificial initial population scheme (IGA1). By considering both schemes simultaneously (IGA3), the performance is significantly enhanced.

Table 14.6 summarizes costs of the best solution obtained by each solution method. In case 1, the solution obtained by IGA3 is within 0.18% of the global solution costs while the solutions by SGA and TCDP are within 1.3% and 0.4%, respectively. In case 1 and case 2, IGA3 has achieved a 0.21% and 0.61% improvement of

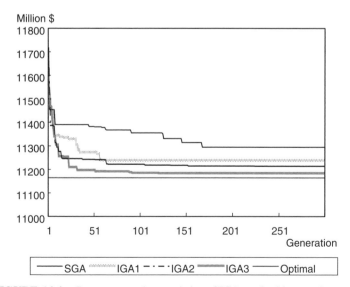

FIGURE 14.4 Convergence characteristics of IGA method in case 1 system.

TABLE 14.6   Summary of the Best Results Obtained by Each Solution Method

		Cumulative discounted cost (10^6 \$)	
Solution method		Case 1 (14-year study period)	Case 2 (24-year study period)
DP		11,164.2	Unknown
TCDP		11,207.7	16,746.7
SGA		11,310.5	16,765.9
IGA	IGA1	11,238.3	16,759.2
	IGA2	11,214.1	16,739.2
	IGA3	11,184.2	16,644.7

costs over TCDP, respectively. Although SGA and IGAs have failed in finding the global solution, all IGAs have provided better solution than SGA. Furthermore, solutions of IGA3 are better than that of TCDP in both cases, which implies that it can overcome a local optimal trap in a practical long-term GEP. Table 14.7 summarizes generation expansion plans of case 1 and case 2 obtained by IGA3.

The execution time of GA-based methods is much longer than that of TCDP. That is, IGA3 requires approximately 3.7 and 6 times of execution time in case 1 and case 2, respectively. However, it is much shorter than the conventional DP. Fig. 14.5 shows the observed execution time on a PC/Pentium (166 MHz) of IGA3 and DP as the stages are expanded. Execution time of IGA3 is almost linearly proportional to the number of stages while that of DP exponentially increases. In the system with 11 stages, it takes more than 9 days for DP and requires about 1.2 millions of array memories to obtain the optimal solution whereas it takes only 11 hours by IGA3 to get the near optimum.

TABLE 14.7   Cumulative Number of Newly Introduced Plants in Case 1 and Case 2 by IGA3

Type year	Oil (200 MW)	LNG C/C (450 MW)	Coal (500 MW)	PWR (1000 MW)	PHWR (700 MW)
1998	3 (5)*	2 (1)	2 (3)	0 (1)	2 (0)
2000	5 (6)	3 (1)	5 (6)	0 (1)	4 (1)
2002	5 (7)	3 (1)	5 (6)	0 (2)	4 (1)
2004	8 (10)	7 (3)	6 (7)	0 (2)	4 (1)
2006	10 (12)	10 (3)	6 (7)	0 (2)	6 (2)
2008	10 (13)	10 (3)	6 (9)	0 (2)	6 (2)
2010	10 (13)	10 (3)	6 (9)	0 (2)	6 (4)
2012	14	11	8	1	7
2014	17	14	8	1	7
2016	19	15	10	1	9
2018	19	17	10	3	9
2020	20	18	12	3	9

LNG = liquid natural gas; C/C = combined cycle; PWR = pressurized water reactor; PHWR = pressurized heavy water reactor.
*The figures within parentheses denote the results of IGA3 in case 1.

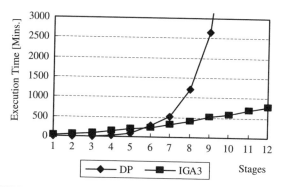

FIGURE 14.5 Observed execution time for the number of stages.

This method definitely provides quasi-optimums in a long-term GEP within a reasonable computation time. Also, the results of the IGA method are better than those of TCDP employed in the WASP, which is viewed as a very powerful and computationally feasible model for a practical long-term GEP problem. Because a long-range GEP problem deals with a large amount of investment, a slight improvement by the IGA method can result in substantial cost savings for electric utilities.

14.3 TRANSMISSION NETWORK EXPANSION

The objective of a network expansion planning consists of determining an (optimal) expansion plan of the electric network, with circuits that satisfy the operational conditions for a forecasted demand growth under a particular generation expansion plan.

Network expansion, both at distribution and at transmission level, may be studied by *static* or by *dynamic* models. A static model tries to discover an optimal network structure (*where* and *which* type of new equipment should be installed in an optimal way that minimizes the installation and operational costs) for a given scenario of generation and load, typically in a long-term context. A dynamic model is more complex and it aims at, besides answering the questions of *where* and *which*, defining *when* to install the network additions, creating therefore a plan of investment along successive periods of time.

In this section, we will address static transmission expansion planning, as a good example of the application of meta-heuristics to network expansion problems. Three solution approaches will be discussed: tabu search, simulated annealing, and genetic algorithm. The dynamic expansion problem will be addressed in a following section devoted to distribution expansion planning.

14.3.1 Overview of Static Transmission Network Planning

For a long-term study some assumptions are made, for example, the reactive power allocation is not considered in the first moment, being executed in the operational planning studies. The main concern in this stage is to identify the principal power corridors that probably will become part of the expanded system. Basically, the DC power flow model is the most employed one and it is considered a reference because in general, networks synthesized by this model satisfy the basic conditions stated by operation planning studies. The topologies found in this phase will be further analyzed by operation planning tools such as AC power flow and other security-related tools such as short-circuit analysis and transient and dynamic stability analysis.

When the DC power flow model represents the power grid, the transmission network expansion planning problem can be formulated as follows:

$$\min v = \sum_{(i,j)\in\Omega} c_{ij} n_{ij} \tag{14.11}$$

subject to

$$\mathbf{Sf} + \mathbf{g} = \mathbf{d} \tag{14.12}$$

$$f_{ij} - \gamma_{ij}(n_{ij}^0 + n_{ij})(\theta_i - \theta_j) = 0 \tag{14.13}$$

$$|f_{ij}| \le (n_{ij}^0 + n_{ij})\bar{f}_{ij} \tag{14.14}$$

$$\mathbf{0} \le \mathbf{g} \le \bar{\mathbf{g}}$$

$$0 \le n_{ij} \le \bar{n}_{ij}$$

n_{ij} integer; f_{ij} and θ_j unbounded

$$(i, j) \in \Omega,$$

where (c_{ij}) is the cost of a circuit that can be added to right-of-way $i - j$, (γ_{ij}) is the susceptance of the circuit, (n_{ij}) is the number of circuits added in right-of-way $i - j$, (n_{ij}^0) is the number of circuits in the base case, (f_{ij}) is the power flow in the right-of-way, $(\bar{f}_{ij})$ is the maximum power flow of a circuit in $i - j$, v is the investment cost, $\mathbf{S}$ is the branch-node transposed incidence matrix, $\mathbf{f}$ is a vector with elements f_{ij}, $\mathbf{g}$ is a vector with elements g_k (generation in bus k) whose maximum value is $\bar{g}$, $\bar{n}_{ij}$ is the maximum number of circuits that can be added in right-of-way, $\mathbf{d}$ is the vector of net demand, and Ω is the set of all right-of-ways.

Constraint (14.12) represents the power conservation in each node, which represents Kirchhoff's current law (KCL) in the equivalent DC network. Constraint (14.13) represents Kirchhoff's voltage law (KVL) and the constraints are nonlinear. The transmission network expansion problem as formulated above is a nonlinear mixed integer problem (NLMIP). It is a hard combinatorial problem, which can lead to a combinatorial explosion on the number of alternatives. However, if the integrality constraints of variables n_{ij} are relaxed with $n_{ij} \ge 0$, the DC model becomes a nonlinear problem (NLP).

The DC model is widely used in transmission planning problems. However, algorithms capable of providing optimal solutions for large and complex systems using this model have never been reported. Considering the difficulties to deal with this problem, relaxed versions of DC models have been employed with the available optimization techniques. The most used relaxed models are the transportation model and the hybrid models.

The transportation model was originally proposed by Garver [13] and from a mathematical point of view it is considered a relaxed DC model. Therefore, in the transportation model, the constraints related to Kirchhoff's second law are omitted [constraints (14.13)]. The transportation model becomes as follows:

$$\min v = \sum_{(i,j) \in \Omega} c_{ij} n_{ij} \tag{14.15}$$

subject to

$$\mathbf{Sf} + \mathbf{g} = \mathbf{d} \tag{14.16}$$

$$|f_{ij}| \le (n_{ij}^0 + n_{ij})\bar{f}_{ij} \tag{14.17}$$

$$\mathbf{0} \le \mathbf{g} \le \bar{\mathbf{g}}$$

$$0 \le n_{ij} \le \bar{n}_{ij}$$

n_{ij} integer; f_{ij} unbounded

$(i,j) \in \Omega$.

The transportation model can be represented by a mixed integer linear programming problem. Garver proposed a constructive heuristic algorithm for the transportation model. Usually, solutions found with the transportation model are not feasible for the DC formulation. Moreover, if the integrality constraint (n_{ij}) is relaxed, the problem becomes a linear programming problem (LP). The hybrid linear model (HLM) is obtained by relaxing constraints (14.17) in the DC model for all new circuits added to the system. It must be observed that constraints (14.17) for circuits belonging to the base topology are linear. The hybrid model becomes as follows:

$$\min v = \sum_{(i,j) \in \Omega} c_{ij} n_{ij}$$

subject to

$$\mathbf{Sf} + \mathbf{S}^0 \mathbf{f}^0 + \mathbf{g} = \mathbf{d}$$

$$f_{ij}^0 - \gamma_{ij} n_{ij}^0 (\theta_i - \theta_j) = 0 \quad \forall (i,j) \in \Omega_0$$

$$|f_{ij}^0| \le n_{ij}^0 \bar{f}_{ij} \quad \forall (i,j) \in \Omega_0 \tag{14.18}$$

$$|f_{ij}| \le n_{ij} \bar{f}_{ij} \quad \forall (i,j) \in \Omega$$

$$\mathbf{0} \le \mathbf{g} \le \bar{\mathbf{g}}$$

$$0 \le n_{ij} \le \bar{n}_{ij}$$

n_{ij} integer; f_{ij} and θ_j unbounded,

where $\mathbf{S}^0$ is the transposed branch-node incidence matrix formed by existing circuits in the base topology, $\mathbf{f}^0$ is the vector with power flow of the circuits in the base topology with elements f_{ij}^0, $\mathbf{S}$ is the complete transposed branch-node incidence matrix, $\mathbf{f}$ is the vector with power flows of the new circuits, $\mathbf{d}$ is the demand vector, and Ω_0 is the set of the circuits belonging to the base topology. It must be observed that in the optimal solution, the new circuits are not obliged to comply with the KVL. Hence, new circuits (which present similar characteristics to the existing one) may carry a different power flow in a parallel connection with the circuits of the base topology (the new circuit must obey only Kirchhoff's first law while the circuit of the base topology obeys both laws). Villasana and Garver proposed the hybrid model [14], and they adapted their model for a constructive heuristic algorithm that provided good solutions for the DC model.

It is also possible to formulate a nonlinear approach to the hybrid model. The linear disjunctive model [15] represents an option that has been considered in the transmission expansion problem. More details about these models can be found in Refs. 15 and 16.

14.3.2 Solution Techniques for the Transmission Expansion Planning Problem

The first relevant work with transmission network planning appeared in the early 1960s with the application of heuristic methods for optimization. Some important considerations must be made regarding its application. The heuristic methods in general are not difficult to program, and they are robust and very fast. However, the major drawback is the degeneration of the solution quality when the system complexity increases. In the 1970s and 1980s, the heuristic techniques achieved significant improvement, and new applications to conventionally intractable problems have been proposed. The heuristics methods have also been used as generators of high-quality initial topologies to the optimization methods such as the meta-heuristic techniques and branch-and-bound algorithms [16–19].

Garver presented the transportation model and the solution algorithm based on a constructive heuristic algorithm [13], and the most relevant contribution was the application of a constructive heuristic method that employs a sensitivity index for guiding the search. Since then, every new transmission planning constructive heuristic algorithm differed from Garver's model just in three aspects: (1) the sensitivity index, (2) the use of different mathematical models, and (3) the use of a local optimization method.

In Garver's algorithm, a linear program (LP) is solved for the current iteration [the integrality constraint for n_{ij} is relaxed (i.e., $n_{ij} \geq 0$)]. Basically, after solving the LP, the algorithm identifies the circuit with $n_{ij} \neq 0$ that carries the largest power flow and adds a similar circuit in the path $i-j$. The process stops when the LP solution indicates that all of $n_{ij} = 0$ (i.e., no additional circuit is needed).

In order to achieve best results, Villasana et al. [14] proposed an algorithm based on a hybrid model. After a LP solution indicates the best circuit, a circuit is added to the model. This circuit will comply with both KCL and KVL in the next LP subroutine call. As a consequence of complying with both Kirchhoff laws, the resulting topology will present better quality when evaluated by DC model.

In the 1980s, several constructive algorithms employed the DC model [20–22]. These algorithms explore the information related to the power system performance indices as the sensitivity indicators. The least effort criterion algorithm [21] calculates an approximate performance index that identifies the most attractive circuit. When the circuit chosen with the proposed index is added to the system, it provides a best power flow distribution, and consequently a decreasing in the system overload. The least effort criterion algorithm allows working with overloaded circuits, and the optimization process stops when the overload is no longer detected. It can be observed that this approach does not explicitly take into account the circuit costs. An improvement consists of dividing the performance index by the circuit cost. In Ref. 21, the concept of fictitious circuit was employed in order to make a nonconnected circuit artificially connected. It allowed the calculation of the performance index for the entire system.

The proposal presented in Ref. 22, called minimum load shedding algorithm, presents the same structure of the previous method. In this case, the aim of the algorithm was to expand the system avoiding the overloads and the operational problems were converted to a load shedding equivalent representation. For this reason, the sensitivity index is an approximate mathematical function that identifies the circuit that when added to the system should produce the least load shedding. There are many interesting heuristic algorithms proposed in the specialized literature and some of them adopt a hybrid approach by using concepts from classic optimization. Examples of hybrid approach can be found in Refs. 23 and 24.

In the transmission expansion planning problem, classic optimization has also been employed, such as Benders decomposition and branch-and-bound algorithms. The Benders decomposition was the most employed one and some of its application can be found in Refs. 15, 25 and 26. The use of Benders decomposition represents a very interesting approach from the theoretical point of view, however for large realistic systems its performance may not result as expected. One reason is the nonlinear characteristic of the problem. When Benders is employed in a hierarchical structure, the results have showed best performance [26]. The Benders approach in a hierarchical structure provided the optimal solution of the Garver's 6-bus system with generation rescheduling and the optimal solution of the Brazilian Southern System (46 bus and 79 circuit addition paths) with and without generation rescheduling. The same results were achieved with a different Benders model [15]. Another very employed algorithm is based on the branch-and-bound algorithm, which uses the transportation model. The branch-and-bound algorithm proposed in Ref. 7 converges rapidly for problems with medium size and medium complexity; however for large size and complex systems, the computational time becomes prohibitively high, even when using specialized strategies. Although it cannot find the global optimal topology of the Brazilian North-Northeastern system, it provided the best solution found for the North-Northeastern P1 Plan.

In the 1990s, many algorithms based on meta-heuristic techniques have been proposed. These algorithms have many interesting features such as (1) they easily find optimal solutions for small and medium systems; (2) they find suboptimal solutions for very complex systems; (3) they are robust (i.e., always find a feasible solution); (4) they are easy to program.

The simulated annealing was employed in the transmission expansion planning in Ref. 28 and in Ref. 29. The genetic algorithm was employed in Refs. 30 and 31 and the tabu search meta-heuristic was proposed in Refs. 18, 32, and 33. Hybrid versions of these algorithms have also been developed and a comparative analysis was presented in Ref. 17. New algorithms based on GRASP (greedy randomized adaptive search procedure) were presented in Ref. 16. All modern heuristic techniques are still in development, and in the next years many improvements and new alternatives are expected.

14.3.3 Coding, Problem Representation, and Test Systems

The meta-heuristic techniques presented in this chapter employ the coding proposed in Ref. 28: only integer variables are coded, that is, only the investment variables n_{ij}, in a chromosome of integer values. The operation variables such as the power flows, node voltage angles, generation level, and so forth, are determined with a linear programming algorithm.

Once the investment proposal is defined (i.e., the n_{ij} variables are known), the feasibility of the new topology must be analyzed and eventual load shedding calculated. The load shedding is detected by solving the LP problem whose n_{ij} variables are fixed in the desired solution. Hence, the objective function of an investment proposal specified by n_{ij}^k assumes the following formulation:

$$ v = \sum c_{ij} n_{ij}^k + \alpha w^k. \tag{14.19} $$

Additionally, new artificial variables are inserted in the mathematical formulation presented in (14.11) in order to facilitate its resolution. These variables represent artificial generation at each load bus (when artificial generation is active, there is load shedding in the bus). Therefore, for each investment proposal identified by $n_{ij}^k \in \mathbf{n}^k$, the following LP problem must be solved:

$$ \min w = \sum_{s \in \Gamma} r_s $$

subject to

$$ \mathbf{Sf} + \mathbf{g} + \mathbf{r} = \mathbf{d} $$

$$ f_{ij}^0 - \gamma_{ij}(n_{ij}^0 + n_{ij}^k)(\theta_i - \theta_j) = 0 $$

$$ |f_{ij}^0| \le (n_{ij}^0 + n_{ij}^k)\bar{f}_{ij} \tag{14.20} $$

$$ \mathbf{0} \le \mathbf{g} \le \bar{\mathbf{g}} $$

$$ \mathbf{0} \le \mathbf{r} \le \mathbf{d} $$

$$ f_{ij} \text{ and } \theta_j \text{ unbounded} $$

$$ (i,j) \in \Omega, $$

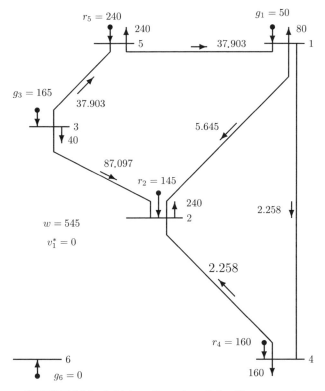

FIGURE 14.6 Initial configuration of the 6-bus network.

where w represents the total load shedding of the system to the investment proposal n_{ij}^k, $\mathbf{r}$ is the vector of artificial generators, and Γ represents the set of load buses. As can be observed, once n_{ij}^k variables are specified, the problem becomes a linear programming problem whose solution may indicate whether the proposal allows the adequate system operation ($w = 0$) or not (with load shedding corresponding with $w \neq 0$).

To illustrate the coding employed in this problem, we will present the 6-bus Garver system with 15 candidate circuits [13], whose initial topology is shown in Fig. 14.6. It can be noticed that the initial topology is infeasible with a load shedding of $w = 545$ MW.

The optimal solutions of Garver's 6-bus system are

1. With generation rescheduling (see Fig. 14.7):
   - $v = 110.0$ with additions of $n_{3-5} = 1$; $n_{4-6} = 3$.

2. Without rescheduling (see Fig. 14.8)
   - $v = 200$ with additions of $n_{2-6} = 4$; $n_{3-5} = 1$; $n_{4-6} = 2$.

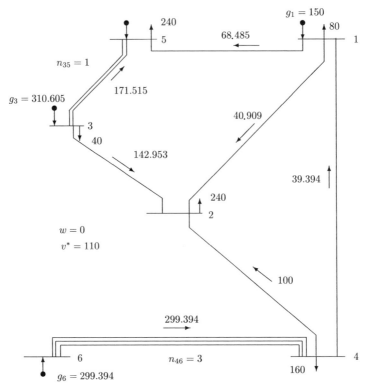

FIGURE 14.7 Optimal solution for 6-bus system with generation rescheduling.

A significant difference is observed for the two cases above. In case 1, the number of circuits needed is 4, and for the second case, 7 circuits are needed. The main difference in the topology is noted with the missing connection in path 2–6 (Fig. 14.7), and when the investment cost is compared, the difference is almost twice between them.

The coding adopted represents only the *added* circuits. Therefore, the topologies presented in Fig. 14.7 and Fig. 14.8 can be represented by two vectors $\mathbf{p}^0$, $\mathbf{p}^{opt}$ shown in Fig. 14.9.

A much more challenging system is the Brazilian North-Northeastern system (see Fig. 14.10) with 87 buses and 183 connection paths. The system data presented in Ref. 35 are for generation rescheduling. Its global optimal solution is still unknown, even when modeled by relaxed DC models. Therefore, it is considered as a benchmark in the transmission network expansion planning problem.

14.3.4 Complexity of the Test Systems

Garver's system presents 6 buses and 15 circuits; it is a small system widely used as a test system in network planning [13]. However, when generation rescheduling is considered, the global optimal point becomes very difficult to determine, and there is no constructive algorithm that could find the optimal point. After a long time since its proposition in 1970 [13], the global optimal solution was found by using a

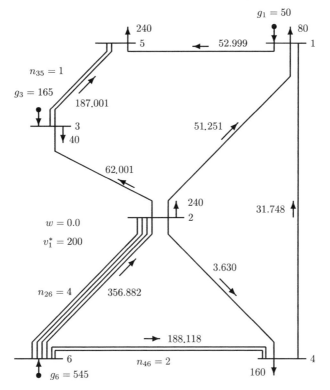

FIGURE 14.8 Optimal solution for the 6-bus system without rescheduling.

hierarchical Benders decomposition approach [26]. On the other hand, the case without generation rescheduling is very simple, and practically every constructive algorithm finds its global optimal solution.

The North-Northeastern Brazilian system corresponds with the actual network in the north-northeast of Brazil. The results presented are for Plan-2 without generation rescheduling. It is a medium-size network but with high complexity, which makes it a benchmark in network planning problem. Due to its highly disconnected nature, it

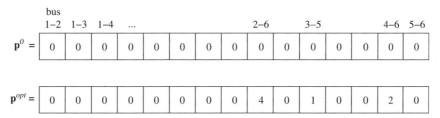

FIGURE 14.9 Initial configuration ($\mathbf{p}^0$) and optimal solution ($\mathbf{p}^{\mathrm{opt}}$) to the Garver's system (without generation rescheduling).

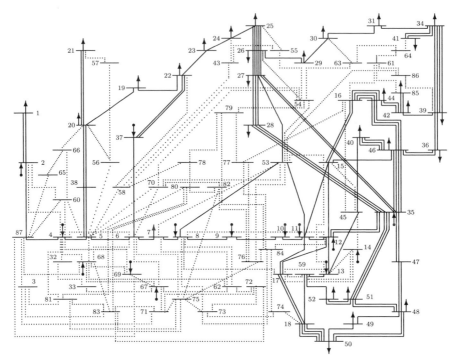

FIGURE 14.10 The Brazilian North-Northeastern network (solid lines, existing circuits; dotted lines, possible circuit paths).

involves the reinforcement of existing network in addition to the necessity to build from scratch many parts where new loads and generations are required. In these cases, the addition of a single branch, represented by a transmission line or transformer, may not be enough to guarantee network connectivity. In a highly disconnected system, entire paths may have to be built. Due to these facts, addition of 100 circuits to achieve suboptimal solutions are not uncommon. Considering a horizon of 10 to 15 years, the estimated costs are above US$2.0 billion.

The search space in the North-Northeastern system is also very large. For example, if 11 circuits could be introduced on each circuit path, it means $12^{183}-2^{656}$ possibilities! Even considering a reduction in the circuit number to 6, the search space remains very large. However, what really makes this system complex is the large number of good quality solutions (local optimal solutions) is distributed in a large multimodal landscape.

14.3.5 Simulated Annealing

The SA algorithm finds the optimal solutions for the 6-bus system after solving about 600 to 1300 LPs according to the parameters used for the adjustment of SA. The typical parameters are $\rho = 1.2$, $\alpha = 10$, $\beta = 0.7$, $\mu = 2$, and $T_0 \in [200, 400]$.

For the Brazilian North-Northeastern system, the solution given by a SA algorithm [10] was

- $v = $ US\$2,676,704,000 with load shedding $w = 0$ MW.
- The circuit additions are as follows:

$n_{1-2} = 1$ $n_{2-4}; = 1$ $n_{4-5} = 5$ $n_{4-6} = 3$ $n_{4-68} = 1$ $n_{5-56} = 1$ $n_{5-58} = 5$ $n_{6-7} = 2$
$n_{7-8} = 2$ $n_{8-17} = 2$ $n_{12-15} = 1$ $n_{12-17} = 1$ $n_{13-14} = 1$ $n_{13-15} = 4$ $n_{14-59} = 1$ $n_{15-16} = 3$
$n_{15-45} = 1$ $n_{15-46} = 1$ $n_{16-44} = 6$ $n_{17-18} = 2$ $n_{18-50} = 11$ $n_{18-74} = 3$ $n_{21-57} = 2$ $n_{22-23} = 1$
$n_{22-58} = 2$ $n_{24-43} = 1$ $n_{25-55} = 3$ $n_{26-54} = 3$ $n_{30-31} = 4$ $n_{30-63} = 3$ $n_{31-34} = 2$ $n_{35-51} = 1$
$n_{36-46} = 2$ $n_{40-45} = 2$ $n_{41-64} = 2$ $n_{42-44} = 2$ $n_{42-85} = 1$ $n_{43-55} = 2$ $n_{43-58} = 2$ $n_{48-49} = 1$
$n_{49-50} = 4$ $n_{52-59} = 1$ $n_{53-86} = 1$ $n_{54-58} = 2$ $n_{54-63} = 1$ $n_{56-57} = 1$ $n_{61-64} = 1$ $n_{61-85} = 2$
$n_{61-86} = 1$ $n_{67-68} = 1$ $n_{67-69} = 1$ $n_{67-71} = 3$ $n_{71-72} = 1$ $n_{72-73} = 1$ $n_{73-74} = 1$

In total, 116 circuits have been added in 55 different paths. Moreover, the investment was significantly reduced compared with the results provided by constructive heuristic algorithms. Improved versions of SA algorithms and other meta-heuristics have found better solutions. The computational effort required for finding similar solutions as above is proportional to the resolution of 200,000 to 300,000 LP problems. Typical parameter values of the SA algorithm are $\rho = 1.2$, $\alpha = 1000$, $\beta = 0.7$, $\mu = 5$, and $T_0 \in [50,000, 80,000]$.

The implementation of a parallel SA algorithm in transmission planning was introduced in Ref. 29 and consists of dividing the effort in generating the Markov chains over np processors. Two algorithms have been proposed: division algorithm and a hybrid division/clustering algorithm. The first version of parallel SA algorithm provided a topology with $v = $ US\$2,630,290,000 and a load shedding of 0 MW.

The resulting topology is significantly different from the topology obtained by the sequential SA algorithm, as observed with other serial implementations that present close investment values. Improved SA algorithms and parallel versions have reached a barrier of US\$2,600,000,000, which is hard to overcome. Fortunately, new evolutionary programs, such as genetic algorithms, have broken the barrier, as we will show in the next paragraphs.

14.3.6 Genetic Algorithms in Transmission Network Expansion Planning

The first models applying GA to network expansion planning appeared in 1993 and dealt with distribution systems [30] while the first proposals for the transmission planning problem appeared in 1998 [31] and were followed by an improved model in 2000 [32]. The following paragraphs focus on the basic model [31] as a good example of application of GA to static transmission network expansion planning (so far, no solid proposals are known about dynamic transmission expansion planning with application of meta-heuristics).

Basically, the same coding with integer numbers and objective function evaluation as above was used. Tests have shown that a chromosome with integer values proved to be more efficient, faster, and required less memory; in particular, binary coding caused harmful disruptions in chromosome structures from the application of cross-over and mutation operators, generating a chaotic behavior of the algorithm.

The selection operator with best performance was tournament selection. Recombination has been implemented by swapping parts between two configurations in order to generate two new offspring. Mutation has been implemented as an addition or removal of one circuit directly in the chromosome. In Ref. 31, a variable mutation rate controlled by simulated annealing was proposed. This feature helped avoid local minima and led to improved performance of the genetic algorithm.

The research done has shown that high-quality initial configurations decrease the overall computation time and also increase the solution quality [31]. Good results have also been obtained from constructive heuristic algorithms [13, 14, 21, 22], such as Garver's algorithm that identifies feasible solutions for the transportation model and additionally shows which circuit should be part of the optimal solution.

The parameters such as population size, recombination and mutation rates have all been tuned by trial and error (i.e., after several trials testing for a range of values). The stopping criterion was a classic one: the process stops whenever the incumbent solution (least-cost configuration) does not improve after a given number of iterations.

The application of GA to the Brazilian North-Northeastern system presents convergence problems when the initial population is randomly generated. The reason is the high number of islanded buses in the initial topology. However, when constructive heuristic is used for generating the initial population, a suboptimal solution is achieved after solving about 180,000 to 250,000 LP problems. The typical parameters are $n_{pop} = [160, 200]$, $\rho_c = 0.9$, $\rho_m = [0.01, 0.03]$, $\alpha = 1000$, $K = [5,500,000, 3,500,000]$, and stopping the process after 80 iterations without improvement in the incumbent solution. The best solution found without generation rescheduling and using a modified Garver's algorithm for generating the initial population proposes investments costs of US$2,600,593,000, and the proposal is as follows:

$$n_{1-2} = 1 \quad n_{2-4} = 1 \quad n_{4-5} = 4 \quad n_{4-81} = 3 \quad n_{5-56} = 1 \quad n_{5-58} = 4 \quad n_{6-75} = 1 \quad n_{12-15} = 1$$
$$n_{13-15} = 4 \quad n_{14-59} = 1 \quad n_{15-16} = 3 \quad n_{15-46} = 2 \quad n_{16-44} = 6 \quad n_{18-50} = 11 \quad n_{18-74} = 5 \quad n_{21-57} = 2$$
$$n_{22-58} = 2 \quad n_{24-43} = 1 \quad n_{25-55} = 4 \quad n_{30-31} = 1 \quad n_{30-63} = 2 \quad n_{35-51} = 2 \quad n_{36-46} = 2 \quad n_{39-86} = 1$$
$$n_{40-45} = 1 \quad n_{40-46} = 1 \quad n_{41-64} = 2 \quad n_{42-44} = 1 \quad n_{43-55} = 2 \quad n_{43-58} = 3 \quad n_{48-49} = 1 \quad n_{49-50} = 4$$
$$n_{52-59} = 1 \quad n_{53-86} = 1 \quad n_{54-58} = 1 \quad n_{54-63} = 1 \quad n_{56-57} = 1 \quad n_{61-85} = 3 \quad n_{61-86} = 1 \quad n_{63-64} = 1$$
$$n_{67-68} = 1 \quad n_{67-69} = 1 \quad n_{67-71} = 3 \quad n_{69-87} = 1 \quad n_{71-72} = 1 \quad n_{72-73} = 1 \quad n_{73-74} = 2 \quad n_{73-75} = 1$$
$$n_{75-81} = 1$$

Tests with modified versions of the algorithm provided many topologies with investment costs varying in the range US$2,600,000,000 to US$2,585,000,000. The important characteristic that must be observed is that the obtained investment costs are practically equivalent, nevertheless the topology may vary significantly. This feature indicates the presence of many suboptimal solutions in the Brazilian North-Northeastern system, which justifies having it considered as a benchmark for transmission expansion planning algorithms. In Ref. 32, a different version of GA approach to the transmission planning problem is presented.

14.3.7 Tabu Search in Transmission Network Expansion Planning

A tabu search modeling has also been experimented with in this problem using the same basic model and solution coding [18, 33, 34]. The neighborhood structure used in Ref. 18 comprises three parts:

1. One for the basic TS algorithm based on short-term memory, where infeasible transitions are allowed (constructive heuristics presented in [2, 3, 5], are used in this part);
2. Another for the intensification phase, where only feasible solutions are allowed (a greedy search is carried out); and
3. Yet another for the diversification phase, where the strategic oscillation takes the current solution back to the infeasible region (this is achieved by simply removing one or more circuits of the current configuration).

For the three cases, the size of neighborhoods are kept at minimum presenting only high-quality topologies. The search at each reduced neighborhood is made by a limited tabu search (LTS), which comprises a tabu list with an aspiration criterion, short-term memory, reduced neighborhood, strategic oscillation, intensification for local search, and limited diversification.

In Ref. 16, a three-step strategic oscillation is performed, consisting of (1) a basic TS for the infeasible region, (2) a TS strategy for the feasible region, and (3) another TS strategy for leaving the feasible region. The overall method is formed by the repeated application of the above three-step procedure plus diversification. For a comparative evaluation of the various iterative approximation methods applied to the transmission network expansion problem, see Ref. 17.

The TS method presents three basic steps: (1) determination of the initial population and elite configurations, (2) execution of LTS, and (3) the diversification process. Garver's algorithm is used for generating the initial configuration, and it is updated whenever a new feasible solution that is better than the worst configuration already stored is found. The elite configurations are also updated whenever a new configuration presents (1) better quality than the worst configuration in the elite list; (2) it is a topology that differs in a specified number of circuits when compared with each elite configuration. This selection results in an elite list that comprises quality and diversity attributes. The diversification process consists of finding new configurations to reinitialize the process, which can be achieved by realizing small perturbations in the population or applying path relinking with the elite list.

A critical issue in TS is the definition and generation of the neighborhood. To avoid a computational burden in the complete evaluation of a neighborhood for large systems, the model in Ref. 18 presented a technique for the reduction of neighborhood with the use of sensitivity indicators from constructive heuristics. A neighborhood is considered as a reduced list of circuit additions or swaps. The reduced neighborhood is obtained by using the ordering provided by the sensitivity indices of three approximation methods: (1) Garver's method based on transportation model

[22], (2) least-effort criterion [21], and (3) minimum load shedding criterion [22]. It must be observed that three LPs are involved in the neighborhood generation, and the list is completed by adding circuits or sets of circuits adjacent to buses with significant lack of generation or load. Finally, the neighborhood generation is completed with the random addition of circuits in order to create different solutions.

The neighborhood generation seen above has the compromise between quality (by greed search) and diversity (random search). The list may also include simultaneous additions of circuit set if the network has a high degree of disconnection.

Also, rather than using a single configuration, as happens in conventional TS algorithms, the use of a set of configurations allows a more comprehensive search on the solution space. The first configuration is generated with Garver's method in the same way as GA and SA algorithms and then the next one by applying tabu to some essential operations (e.g., addition of a component) on the first generated configuration. This allows the exploitation of new search regions and may reveal the main circuits that would eventually be found in an optimal solution. Garver's algorithm has proved to be more effective regarding computation time and quality.

In the application of TS to the Brazilian North-Northeastern systems, the algorithm was initiated with 20 topologies generated by a modified Garver's algorithm. During the process, 70 topologies are stored for the optimization routines and 6 of them are considered as elite topologies. On average, 170,000 to 200,000 LP calls are required depending on the selected parameters.

The best topology found by TS [18] presents investment cost of US\$2,574,745,000 (this solution represents the best known solution to the static generation/load case):

$$
\begin{array}{llllllll}
n_{1-2}=1 & n_{2-4}=1 & n_{4-5}=4 & n_{4-81}=3 & n_{5-58}=4 & n_{12-15}=1 & n_{13-15}=4 & n_{14-45}=1 \\
n_{15-16}=4 & n_{15-46}=1 & n_{16-44}=6 & n_{16-61}=2 & n_{18-50}=11 & n_{18-74}=6 & n_{19-22}=1 & n_{20-21}=3 \\
n_{20-38}=2 & n_{22-37}=1 & n_{22-58}=2 & n_{24-43}=1 & n_{25-55}=3 & n_{26-54}=1 & n_{27-53}=1 & n_{30-31}=2 \\
n_{30-63}=2 & n_{35-51}=2 & n_{36-39}=1 & n_{36-46}=3 & n_{40-45}=2 & n_{41-64}=2 & n_{43-55}=2 & n_{43-58}=2 \\
n_{48-49}=1 & n_{49-50}=4 & n_{52-59}=1 & n_{54-58}=1 & n_{54-63}=3 & n_{61-64}=1 & n_{61-85}=3 & n_{67-69}=2 \\
n_{67-71}=3 & n_{68-69}=1 & n_{69-87}=1 & n_{71-72}=1 & n_{72-73}=1 & n_{73-74}=2 & n_{73-75}=1 & n_{75-81}=1
\end{array}
$$

14.3.8 Hybrid TS/GA/SA Algorithm in Transmission Network Expansion Planning

Based on the best characteristics of simulated annealing, tabu search and genetic algorithms, a hybrid approach was proposed in Ref. 17. The base structure was formed by TS algorithm, and new components related to intensification and diversification were provided by GA and SA algorithms. For example, when the neighborhood is composed only of low-quality topologies, the intensification phase is executed based on SA acceptance criterion and the diversification phase; new population (as genetic algorithms) is generated with application of recombination operators. The best solution obtained from the hybrid approach proposes investments of US\$2,573,941,000 with 4 MW of load shedding, which is negligible when compared with the total demand (0.01% of total demand) as presented in following:

$$n_{1-2} = 1 \quad n_{2-87} = 1 \quad n_{4-5} = 4 \quad n_{4-81} = 2 \quad n_{5-56} = 1 \quad n_{5-58} = 3 \quad n_{6-37} = 1 \quad n_{12-15} = 1$$
$$n_{13-15} = 4 \quad n_{14-59} = 1 \quad n_{15-16} = 4 \quad n_{15-46} = 1 \quad n_{16-44} = 6 \quad n_{16-61} = 1 \quad n_{18-50} = 11 \quad n_{18-74} = 6$$
$$n_{21-57} = 2 \quad n_{22-37} = 1 \quad n_{24-43} = 2 \quad n_{25-55} = 4 \quad n_{30-31} = 1 \quad n_{30-63} = 2 \quad n_{35-51} = 1 \quad n_{36-46} = 1$$
$$n_{39-42} = 1 \quad n_{39-86} = 3 \quad n_{40-45} = 1 \quad n_{40-46} = 2 \quad n_{41-64} = 2 \quad n_{42-44} = 2 \quad n_{43-55} = 3 \quad n_{43-58} = 3$$
$$n_{48-49} = 1 \quad n_{49-50} = 4 \quad n_{52-59} = 1 \quad n_{53-86} = 1 \quad n_{54-55} = 1 \quad n_{54-63} = 1 \quad n_{56-57} = 1 \quad n_{61-64} = 1$$
$$n_{61-85} = 2 \quad n_{67-69} = 2 \quad n_{67-71} = 3 \quad n_{68-83} = 1 \quad n_{69-87} = 1 \quad n_{71-72} = 1 \quad n_{72-73} = 2 \quad n_{72-83} = 1$$
$$n_{73-74} = 2 \quad n_{81-83} = 1$$

It is worth observing that the configuration above presents an investment cost difference of US$840,000 when compared with the best solution obtained by tabu search. Moreover, in spite of apparent small differences in the investment value, both topologies are significantly different.

14.3.9 Comments on the Performance of Meta-heuristic Methods in Transmission Network Expansion Planning

It is difficult to conclude which algorithm presents the best performance to the network planning, as these algorithms are not deterministic and in many cases the efficiency depends on several points, for instance, the tuning of control parameters, the problem formulation, and, mainly, the way that the problem can be tailored in the selected meta-heuristic approach. However, our experience indicated that tabu search and its hybrid versions have a slight advantage over the other meta-heuristics due to their flexible characteristic to incorporate new intelligent strategies.

The three methods have shown ability to avoid getting entrapped into local optima, and despite large solution space, only a small fraction of alternatives are analyzed and the search is guided to better solutions. The results confirm the superiority of the meta-heuristic approach in dealing with large size, complex, nonlinear, and combinatorial problems whose alternatives increase exponentially with network dimension. Finally, new approaches have been proposed, which include the creation of hybrid algorithms that merge the best qualities of each meta-heuristic, such as strategies to avoid local optimal solutions by diversification methods.

14.4 DISTRIBUTION NETWORK EXPANSION

Distribution systems present many problems that may be tackled by meta-heuristics; for instance, network reconfiguration in a fault state, loss minimum reconfiguration, capacitor installation planning or network expansion. Because it is difficult to find the global optimum solution to these problems because of the number of combinations, many approximation algorithms have been developed to find an acceptable near optimum solution in the past several decades. In the 1970s, mostly only mathematically heuristic approaches were (unsuccessfully) tried. Also, algorithms such as branch and bound were suggested for static optimization, either with single cost objective [36] or coupling cost and reliability criteria in a single objective function [37]. However, there is no notice of an industrial application of the methods suggested.

In the early 1980s, expert systems came as a new approach to solve the problem. However, the result was not so encouraging for complex problems. In the early 1990s, finally, meta-heuristic methods came to light, especially in network reconfiguration and distribution planning problems.

Table 14.8 presents the result of a survey of papers published in IEEE PES transactions (proceedings of the regular meeting in 1999) between 1990 and 2004 referring to simulated annealing (SA), genetic and evolutionary algorithms (GA, ES, EP), and tabu search (TS) applications (and their combined use) to distribution systems. In the table, the names of the first author and the corresponding article numbers in the references are shown. The survey did not include expert systems or artificial neural network because their applications have been independently surveyed many times elsewhere. Also not included is the emergent particle swarm optimization method.

14.4.1 Dynamic Planning of Distribution System Expansion: A Complete GA Model

The problem of dynamic planning of a network is highly complex and large scale. It involves deriving a strategy for system expansion and not only arriving at a specific network design. This strategy develops along a series of time steps and is spatially dependent on load growth forecasts. It involves not only an optimization of the network design at a given moment but also taking into account the cross influence of decisions already taken and decisions to be taken in the future. A classic optimal solution will be, therefore, a sequence of network designs that optimize some objective function consisting of a lumped sum of costs over time.

The problem is further complicated because it must take into account multiple objectives and not only cost: reliability is a major concern, but also voltage quality and loss minimization may be of importance. Furthermore, uncertainty cannot be swept under the carpet. In long-term planning, decisions must take into account this factor and be robust, in the sense that they are acceptable in a wide range of scenarios without excessive regret. In some cases, one reads in the literature about "flexible planning," with the same idea: decisions that may still be good in several scenarios. In a framework of uncertainty, one gives up the concept of optimization, understanding that this is possible only if the future is known with absolute certainty.

The first heavy proposal for a dynamic expansion planning model using meta-heuristics was a GA model [30]. Later, models exploring several ways of representing uncertainty in spatial load forecasting were presented, such as in Ref. 62. The process involves a binary coding described in Ref. 30, where a network design for each time step is coded in a long chromosome. It also calls for the definition of several objectives to rank solutions (evolution of network designs):

- Topology feasibility: a connected network is required at all stages, and the transition between stages must come only from additions or deletions of

TABLE 14.8 Modern Heuristics Applications to Distribution System in IEEE PES Transactions (1990–2004)

Application	Years					
	1900–1994	1995–1996	1997–1998	1999–2000	2001–2002	2003–2004
SA	Chiang [38] Chu [39]	Chiang [40] Chiang [41] Billinton [42] Jiang [43] Jonnavithula [44]		Jeon [45]	Jeon [46]	
GA, EP, ES	Nara [47] Miranda [30]	Yeh [48]	Ramirez-Rosado [49]	Cavalho [50] Chuang [51] Huang [52] Chen [53] Hsiao [54] Delfanti [55]		Masoum [56]
TS				Nara [57] Ramirez-Rosado [58]	Gallego [59]	Hayashi [60] Rosado [61]

elements from the network of the precedent stage; furthermore, the network must be radial at each step (open loops are allowed).

- Load flow feasibility: a radial system able to carry load to all nodes without violating line/transformer flow limits.
- Voltage level feasibility: a radial system where voltage drops would not originate nodes with voltages over or under a specified admissible band.
- Investment: minimizing cost of all network additions.
- Losses: minimizing estimated cost of power losses as a sum of all stages.
- Voltage quality: minimizing an index of voltage level quality (values closer to nominal voltage).
- Reliability: minimizing energy or power not supplied due to failures in the system. To assess such value, the method includes an algorithm that estimates the influence of adding switches to the radial networks and takes into account the existence of open loops, so that a realistic value is calculated even in the planning stage.

The representation of uncertainties uses three tools, according to their nature:

- Events related to the life cycle of components or systems (such as reliability) are represented by probabilistic models;
- Events or possibilities perceived as discrete are modeled into a tree of futures; for instance, different economic background scenarios that lead to clear different levels of demand, not only global-but also in a regional basis;
- Other continuously perceived uncertainties may be modeled as fuzzy numbers, such as cost of equipment, load forecasts, or failure rates within a scenario in the tree of futures.

Uncertainties dealt with may relate to loads, importance and location of dispersed generation, line or substation capacity, costs of equipment and operation, delays in construction, costs of energy not supplied, reliability indices such as failure rates and repair times, and others.

All uncertainties related to future developments are organized in a "tree of fuzzy futures." This concept is illustrated in Fig. 14.11, where it is applied to load growth. A path along the tree of futures (such as $A-B-C-D_1-E_2$) defines a set of ordered, forecasted load values: the problem is that one does not know which path will actually be followed.

The tree of fuzzy futures is a way of introducing granularity in the information we have about the possible futures—the paths form a discrete set, but each path is in fact represented as "blurred," contaminated with uncertainty, and the way to represent this continuous uncertainty is with intervals or with fuzzy sets.

When fuzzy futures are considered, two new criteria may be added to the evaluation of an expansion plan: robustness and inadequacy.

Robustness is a single index related to the minimum level of load that the system still is able to carry without load disconnections.

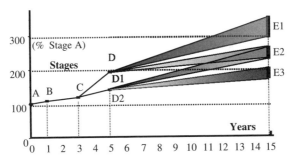

FIGURE 14.11 A tree of futures. Four paths for load growth are defined, in five stages A–B–C—D–E at years 0, 1, 3, 5, and 15. From stage D to E, the uncertainty in each path becomes fuzzy. The load forecast at year 15 is no longer a single value but an interval for each scenario.

Inadequacy is a fuzzy description of the sum of possible excessive line flows in the entire network because the load growth may be excessive to branch capacity. This criterion is useful to assess the possibility of occurrence of cases of repressed demand, because if the system is not planned to the actual demand growth, new consumers will not be allowed connection to the system and existing consumers will be limited in their consumption, having a negative effect in the economic activity of the region.

The core of the methodology can be described as follows:

1. Previous step: define all the data required, including the tree of fuzzy futures.
2. For each path in the tree of fuzzy futures, determine the (conditional) ideal plan in terms of the attribute values, as if one would know for certain that such path would occur—using a first formulation of a GA.
3. Considering all the paths and all the conditional ideals, determine the strategy that minimizes the regret (felt for the decisions taken), no matter which future happens—using a second formulation of a GA.

How is the regret measured? Among many possibilities, it can be assessed through the difference between what the actual decision (or plan) implies and what would be required if one would perfectly know the future—and therefore would have no fear in implementing an optimal plan.

A solution is taken as a dynamic sequence of network topologies through time. In stage 2 of the methodology, the fitness fuzzy value f of a solution x, within a trajectory in the tree of futures, is obtained from the fuzzy equation

$$f(x) = M - c_1(\text{IC} + \text{PL}) - c_2\text{VQ} - c_3\text{RB} - c_4(1 - \text{RO}) - c_5\text{IN},$$

where M is large (enough) constant value, c_i is constants externally fixed, IC is fuzzy investment, PL is fuzzy power losses, VQ is fuzzy voltage quality, RB is fuzzy reliability, RO is robustness (crisp), and IN is fuzzy inadequacy.

Varying constants c_i allows one to obtain a picture of a nondominated region of the domain of feasible solutions.

When dealing with the tree of futures as a whole, in stage 3, fitness is now evaluated as follows: if, in each criterion

$f_{\text{opt};ik}$: Value of attribute i in future k, for the ideal

$f_{x;ik}$: Value of attribute i in future k, for alternative x

then the general objective will be

$$\min[\text{Max}\{(f_{x;ik} - f_{\text{opt};ik}), \quad k = 1, \ldots, \text{number of futures}\}],$$

which means minimizing the overall regret in the decisions to be taken [63].

In stage 3, chromosomes have the same representation as in stage 2, but the fitness function is different; it now evaluates a path in the tree of futures relative to all optimal trajectories in search of a compromise solution.

This is also a multicriteria problem (each criterion represents the possible regret over one objective), and it is dealt with accordingly: weights are associated with each criterion in the fitness function, and then their values undergo perturbations along the succession of the generations, in order to ensure that the genetic process pushes evolution of the alternatives along the nondominated region of the solutions.

It has been commented by the authors that an application inspired in this model has been supplied to at least one utility.

14.4.2 Dynamic Planning of Distribution System Expansion: An Efficient GA Application

This section is about the application of meta-heuristics to distribution systems and not a full treaty on distribution planning. One cannot, however, change topic without mentioning two other proposals for dynamic planning: (1) An evolutionary programming (EP) proposal reported in Ref. 64, where mutations are achieved by line swapping and branch deletion and no recombination is used; (2) A GA model that shares some characteristics with the model described above and has been referred to as having had a successful practical application in the distribution utility of Portugal. It is first referred to Ref. 50 and it has, as distinctive characteristics, a different coding process and definition of the objective function that allows one to consider different scenarios ensuring a fast convergence to a compromise solution.

Its original coding idea represents radial structures instead of branches of the system. The problem with binary chromosome coding, in distribution planning, is the possibility of generation of a huge number of unfeasible topologies, either under the application of mutation or crossover operators. Any coding process that allows the generation of radial structures and generates viable offspring constitutes a major advantage in producing an efficient GA.

Also, the algorithm implements in a clever way a hedging strategy to reach a compromise among several scenarios. Therefore, it deals with spatial uncertainty as well as with time uncertainty.

14.4.3 Application of TS to the Design of Distribution Networks in FRIENDS

A new type of power delivery system named FRIENDS (flexible, reliable and intelligent energy delivery system) has been proposed for some time [57, 65, 66] as an answer to the question of what kind of system is desirable or suitable for power distribution where many distributed generators (DGs) and distributed energy storage systems (DESSs) exist. The FRIENDS concept allows reliable and energy conservation oriented operating strategies of the power system, taking into consideration ways of enhancing service to consumers through DGs and DESSs. FRIENDS tries to attain the following functions:

- Flexibility in reconfiguration of the system.
- Reliability in power supply.
- Multimenu services or customized power quality services to allow consumers to select the quality of electrical power and the supplier.
- Load leveling and energy conservation.
- Enhancement of information services to customers.
- Efficient demand side management.

To realize the above, various forms of framework can be developed. For urban areas, the development has followed the integrated concept shown in Fig. 14.12.

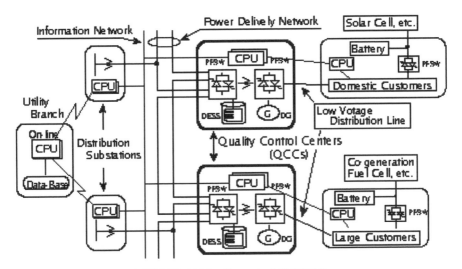

FIGURE 14.12 Concept of FRIENDS.

An important point is the so-called "Power Quality Control Center (QCC)" introduced to realize multiple power quality services named "customized power quality services" for each consumer. QCC supplies power to the area almost equivalent to a section of the current distribution system.

For realizing FRIENDS, one of the important problems is to determine power-line configuration and DGs installations in QCC, which minimizes the weighted sum of total facility installation cost and distribution loss cost under the constraint that the estimated amount of line overload is less than the predetermined value. The problem can be mathematically formulated as shown below under the assumption that the location of each QCC is known.

[Objective function]

$$\text{Min } \alpha \left(\sum_{n=1}^{ND} (aX_n + bYN_n) + \sum_{m=1}^{BR} c_m YL_m \right) + \beta \sum_{t=1}^{T} Oloss^t \tag{14.21}$$

[Constraints]
(DG's maximum capacity)

$$X_n \in \left\{ x^1n, x^2n, \ldots, x^i n, \ldots, x^L n \right\} \ (n = 1, \ldots, ND) \tag{14.22}$$

(Expected line overloads)

$$\sum_{t=1}^{T} \sum_{r=1}^{FLT} \frac{1}{T} p_r Oover^{rt} \le \varepsilon \tag{14.23}$$

(Power supply to all the loads)

$$\exists R_{sn}^r \tag{14.24}$$

where ND is total number of QCCs, BR is total number of potential lines, X_n is discrete variable for DG's capacity in nth QCC, x_n^i is ith available capacity of DG for nth QCC (X_n must be selected from x_n^i), $YN_n = 1$ if X_i is not 0; $=0$ otherwise, YL_m is 0–1 variable for installation of line m ($=1$ if line m is installed; $=0$ otherwise), FLT is number of fault cases, a is variable cost for capacity of DG installed at nth QCC, b is fixed cost of newly installed DG at nth QCC, c_m is installation cost of line m, $\exists R_{sn}^r$ is power supply route from every node n to power source (substation) must exist in any fault case r, $Oloss^t$ is distribution loss at load pattern t, T is number of load patterns under consideration, ε is limit of expected amount of line overload, α, β are weighting coefficients, and $Oover^{rt}$ is the sum of overloads of lines in fault case r at load pattern t.

To solve the above problem, the minimum cost facilities necessary to reduce overloads to satisfy Eq. (14.23) for all the predetermined operating states are found and installed. To apply TS (tabu search), the complex constraint Eq. (14.23) is relaxed.

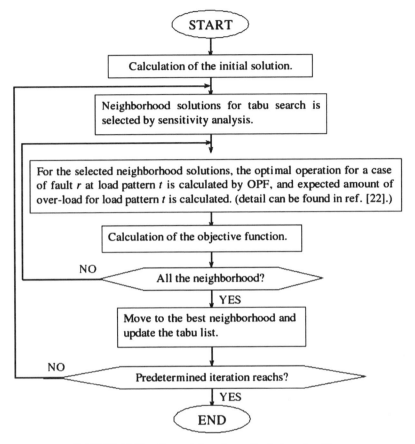

FIGURE 14.13 General flowchart of solution algorithm.

Other constraints such as Eqs. (14.22), (14.24), and load flow equations can be rigidly satisfied in the solution algorithm. Therefore, the objective function is changed into the following equation:
[Revised objective function]
Minimize

$$
\alpha \left(\sum_{n=1}^{ND} (aX_n + bYN_n) + \sum_{m=1}^{BR} c_m YL_m \right) + \beta \sum_{t=1}^{T} Oloss^t
$$
$$
+ \delta \left(\sum_{t=1}^{T} \sum_{r=1}^{FLT} \frac{1}{T} p_r Oover^{rt} - \varepsilon \right),
$$
(14.25)

where δ is a penalty factor for the constraint violation.

Sensitivity analysis also is introduced to reduce the computational burden, namely the most sensitive facilities are found to eliminate the overload. The general flowchart of the solution algorithm is shown in Fig. 14.13. In one iteration of the tabu search, the most effective distribution line candidate or distributed generators are selected and installed (eliminated) to reduce the total overload of distribution lines. This search operation is continued until a predetermined iteration number. Details of the solution algorithm are shown in Ref. 28.

14.5 REACTIVE POWER PLANNING AT GENERATION–TRANSMISSION LEVEL

The reactive power or VAR planning problem at the generation–transmission level is a nonlinear optimization problem. Its main object is to find the most economic investment plan for new reactive sources at selected load buses that will guarantee proper voltage profile and the satisfaction of operational constraints. Usually, the planning problem, is divided into operational and investment planning subproblems. In the operational planning problem, the available shunt reactive sources and transformer tap-settings are optimally dispatched at minimal operation cost. In the investment planning problem, new reactive sources are optimally allocated over a planning horizon at a minimal total cost (operational and investment).

During the past decades, there has been a growing concern in power systems about reactive power operation and planning [68–75]. Recent approaches to the VAR planning problem are becoming very sophisticated in minimizing installation cost and for the efficient use of VAR sources to improve system performance. Various mathematical optimization formulations and algorithms have been developed, which, in most cases, use nonlinear [76], linear [77], or mixed integer programming [78], and decomposition methods [79–82]. With the help of powerful computers, it is now possible to do a large amount of computation in order to achieve a global optimal instead of a local optimal solution.

Hsiao et al. [82] provided an approach using simulated annealing with a modified fast decoupled load flow. However, only the new configuration (VAR installation) is checked with the load flow, and existing resources such as generators and regulating transformers are not fully exploited.

This section presents an improved method (MSGA; modified simple genetic algorithm) of operational and investment planning by using a simple genetic algorithm (SGA) combined with the successive linear programming method. Benders cuts are constructed during the SGA procedure to enhance the robustness and reliability of the algorithm. The method takes advantage of both the robustness of the SGA and the accuracy of conventional optimization method.

The proposed VAR planning approach is in the form of a two-level hierarchy. In the first level, the SGA is used to select the location and the amount of reactive power sources to be installed in the system. This selection is passed on to the operation optimization subproblem in the second level in order to solve the operational planning problem. It is a common practice to use a successive linear programming (LP) formulation to improve the computation speed and to enhance the computation accuracy; the LP method is fast and robust. The operational planning problem is decoupled into coupled real (P) and reactive (Q) power optimization modules; and the successive linearized formulation of the P–Q optimization modules speeds up computation and allows the LP to be used in finding the solution of the nonlinear problem [83]. The dual variables in the LP are transferred from the P–Q optimization modules to the SGA module in the first level to set up the Benders cut for investment planning. This hierarchical optimization approach allows the SGA to obtain correct VAR installations and at the same time satisfies all the operational constraints and the requirement of minimum operation cost.

14.5.1 Benders Decomposition of the Reactive Power Planning Problem

The reactive power planning problem is to determine the optimal investment of VAR sources over a planning horizon. The cost function to be minimized is the sum of the operation cost and the investment cost. The investment cost is the cost to install new shunt reactive power compensation devices for the system. The operation cost considered is the fuel cost for generation.

The problem can be written in the following form:

$$\min_{Y,U} f(Y,U) = L_o(Y) + L_u(U) \tag{14.26a}$$

subject to

$$G_1(Y,U) \leq 0 \tag{14.26b}$$

$$G_2(U) \leq 0, \tag{14.26c}$$

where $Y = [P^T, V^T, N^T]$ is vector of operational variables, P is vector of real power generations, V is vector of bus voltage magnitudes, N is vector of tap-settings, U is vector of investment variables, $L_o(Y)$ is operation cost, $L_u(U)$ is investment cost, $G_1(Y,U)$ is constraint involving both Y and U, and $G_2(U)$ is constraint involving U only.

Equation (14.26a) consists of investment and operation cost. Equation (14.26b) is coupled constraints for operation and investment variables. It includes load flow balance and other important operational constraints. Equation (14.26c) includes constraints relative to only investment variables.

The operation cost is nonlinear, and the investment cost can be assumed to be linear with respect to the amount of newly added reactive power compensation. According to this assumption, the minimization problem (14.26) can be expressed in a nonlinear programming formulation:

$$\min_{Y,U} C^T U + f(Y) \tag{14.27a}$$

subject to

$$H(Y) + BU \leq b_1 \qquad (14.27b)$$

$$DU \leq b_2, \qquad (14.27c)$$

where Y is vector of operation variables, U is vector of investment variables, C is vector of cost coefficients, B, D are matrices of constraints, f, H are cost and constraint functions, respectively, and b_1, b_2 are vector of constraints.

Because of the structure of the constraints, it is quite natural to consider two-level hierarchical approach to solve the problem. That is, use the SGA to select the device and amount, and use an optimization method to obtain optimal results under the given installation.

In this chapter, the generalized Benders decomposition (GBD) method [84] is used in the SGA module in setting up a Benders cut in order to improve the convergence characteristics. The procedure is as follows:

(i) Assuming a feasible investment U, the feasible decision Y is obtained by solving the Y (operation) subproblem:

$$\min_{Y} f(Y) \qquad (14.28a)$$

subject to

$$H(Y) \leq b_1 - BU. \qquad (14.28b)$$

(ii) Having found optimal Y from the first stage, the decision for the feasible investment U is obtained by solving the U (investment) subproblem:

$$\min_{Y,U} Z = C^T U + \sigma \qquad (14.29a)$$

subject to

$$DU \leq b_2 \qquad (14.29b)$$

$$W(U) \leq \sigma, \qquad (14.29c)$$

where σ is low limit varable, and $W(U)$ is called the Benders cut and is a function that supplies information concerning the capacity decision U in terms of the operation feasibility. Then the problem would determine a solution (U, Y) that would minimize the global function (14.27a).

The Benders decomposition method builds the function $W(U)$ based on the solution of the Y subproblem. In nonlinear optimization, $W(U)$ can be determined if we observe that the simplex multiplier vector associated with the first stage (Y subproblem) is the basic feasible solution for the dual problem. Therefore,

$$W(U) = v(U^k + \lambda^k(BU^k - BU), \qquad (14.30)$$

where $v(\cdot)$ is the optimal operation cost with the installation of U^k.

The dual solution λ^k is the simplex multiplier associated with the constraint in the operation subproblem, where k is the iteration number. Because the revised simplex method is used for solving the operation subproblem, λ^k is obtained as a by-product, and new constraints, each corresponding with a different investment installation, are established.

From equations (14.29a) to (14.29c), it can be seen that λ_i^k presents the change of the cost caused by the unit change in investment for unit i. If $\lambda_i^k > 0$, then $U_i > U_i^k$ is helpful in generating a new member of population, which may decrease the total cost Z. If only one constraint is considered, a decreasing direction similar to the steepest descent method can be found. The Benders cut is viewed as a coordinator between investment and operation subproblems, and the GBD method iterates between the two. At each iteration, a new constraint is added into $W(U)$ to form a new constraint set.

14.5.2 Solution Algorithm

The problem is decomposed into investment and operation subproblems and solved iteratively until convergence [80].

The operation subproblem is again decomposed into economic real (P) and reactive (Q) power dispatch problems to minimize the fuel cost function [83, 85]. In the P module, optimal values of real power generation, and in the Q module, the optimal values of bus voltage magnitudes and transformer tap-settings, are obtained. In addition, the optimal values of reactive power dispatched by the generators and compensators are also obtained.

In each population, total operating and investment costs are calculated for each investment. The fitness is simply the inverse of this total cost. The ratio of the average fitness and the maximum fitness of the population is computed and generation is repeated until:

$$\frac{\text{Average fitness}}{\text{Maximum fitness}} \geq \text{AP}$$

where AP is a given number that represents the degree of satisfaction. If the convergence has been reached at given accuracy, then optimal values for investment are found. Other criteria, such as the difference between the maximum and minimum fitnesses and the rate of increase in maximum fitness can also be used as the stopping criteria. One other possibility is to stop the algorithm at some fintess number of generations and designate the result as the best fit from the population.

The iterative process is as follows:

Step 1. Initial population generation: compute the fitness of each member according to operation sub-problem results.

Step 2. Generate new population: typical SGA methods, reproduction crossover and mutation, are used. The Benders cut is used on a subset of strings to obtain one new and better member of the population.

Step 3. Compute the fitness of the new generation.

Step 4. If convergence condition is satisfied, stop computation. Otherwise, return to step 2, and begin a new generation.

The most important step is step 2. A new population is generated according to the fitness of the old population through the simulated spin of a weighted wheel in the SGA [1]. Some modifications are made to the SGA for our planning problem, resulting in a modified SGA (MSGA):

1. In the GBD, the iteration procedure is an alternate computation between investment and operation until the convergence is reached. The Benders cuts are selected and constructed from old population. It is used to obtain a new member of the population. The number of cuts can be adjusted as a part of the procedure. Some better fitted strings and some worse fitted strings are selected to construct the cuts. The Benders cut helps in narrowing down the space of possible solutions and thus speeds up the convergence.

2. An abandoning rate is considered in giving up some poor alternatives by assorting the fitness of the alternatives.

3. Different crossovers are also considered, that is, the tail–tail crossover and head–tail crossover, and the crossover position is selected randomly. The head–tail can also be used in producing new strings from two identical parents.

In the original SGA, only the fitness value resulting from the operation subproblem is used to generate a new generation. However, the new population generated only by its fitness is random and blind. By using the Benders cut, which makes use of both the dual variable information and the cost function, a new and better string can be found. If this new string is a good one (it may be the best one), that is, it has a higher fitness value, it will survive to the next generation. Otherwise, it will likely die afterwards. In this method, the robust characteristics of the SGA can still be maintained; at same time, it increases the chance to find the optimal result earlier. The Benders cut can be set up without difficulty because all variables are made available when the operation optimization subproblem is solved.

14.5.3 Results for the IEEE 30-Bus System

For the IEEE 30-bus system [83], there are six generator buses. Seven buses are selected to add capacitors. Each candidate bus has 3 bits for parameter resolution, and it can represent eight different values for installation. The length of chromosome string is 21 bits, and the population size is 25.

An initial optimization is run for the operational variables. The result shows that the system can maintain all operation constraints without any new capacitor installed, but at higher cost. In order to test the effectiveness of the program, high unit installation cost is used. It was anticipated that the SGA method should find an optimal result after certain generations and in which case the additional installation should be zero.

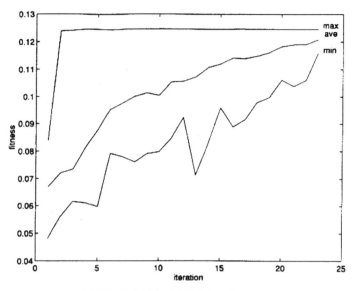

FIGURE 14.14   MSGA iteration result.

Figure 14.14 shows the iteration result for the test case using the MSGA with Benders cut added. There were the total of 264 crossovers and 104 discards for some string with bed fitness values. Figure 14.15 shows the iteration result for the test case using only the SGA, where there were 325 crossovers and 141 discards.

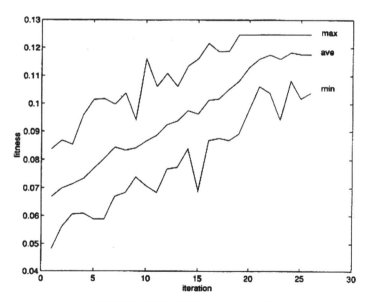

FIGURE 14.15   SGA iteration result.

It can be seen that when the Benders cuts are added by the MSGA, only two generations are needed to find the optimal result. After that the optimal results are still maintained during later iterations. As indicated in Fig. 14.15, the SGA method needs 18 generations to find the final result. Due to random search, the optimal result can only be reached after a considerable number of iterations. The convergence procedure is slower than the MSGA method.

14.6 REACTIVE POWER PLANNING AT DISTRIBUTION LEVEL

In this section, we refer to an evolutionary model that is distinct from the previous one in two aspects: it takes into account the shape of the load curve at different buses, and it is based on an evolutionary algorithm that gets contributions from sensitivities in repairing chromosomes. Sensitivities are an indication of the gradient and obtained from a mathematical model; therefore, we sometimes refer to these algorithms as hybrid instead of pure evolutionary.

Sensitivities had already been used in Ref. 86. However, this approach was organized in a two-phase model, where the GA acts in a first stage and a sensitivity-based heuristic performs a sort of postoptimization. In Ref. 87, an evolutionary algorithm instead uses the information about the gradient of the objective function (e.g., minimizing losses) to repair chromosomes and improve solutions, giving a push in the right direction to the evolution procedure.

The method as published in Ref. 87 is especially suitable for distribution networks, because it deals with the location of capacitors in P—Q buses and it pays attention to the load profiles instead of assuming a constant load. However, it has the potential to be extended to consider other sources of reactive injection.

14.6.1 Modeling Chromosome Repair Using an Analytical Model

Although the method has been published as a genetic algorithm, it is more close to an evolution strategy or evolutionary programming approach (EP/ES). Each individual is represented by a chromosome composed of integer values. An integer number indicates the number of capacitor modules of a predetermined value to be added to a given node, at a given level of the load curve.

To represent load curves, load profiles are defined for all the buses of the network. Typically, one has three time periods: peak, normal load, and low load. A chromosome is then composed of three sections, one for each time period, and within a section, we have a string of integer numbers denoting how many capacitor units are added to the system for that period.

The objective function adds investment cost with cost of losses and a penalty factor for voltage quality. In the calculation of the investment, the capacitor cost is calculated taking into account if one refers to fixed capacity or switching capacitors—at a location, a constant requirement for reactive compensation may be achieved with a fixed capacity and a variable need may be tracked by a switched capacitor bank, and one calculates the fixed capacitor requirement from the minimum value of the number of capacitor units installed at each location by examining the solution proposal for all time periods.

One interesting feature of this model is the chromosome repair technique. Changing of system losses P_{loss} can be identified as active power P_G changing at the reference bus. The inverse of the Jacobian matrix in the Newton–Raphson method is readily available because the method is used to evaluate losses.

Then, using the sensitivity coefficients $\partial P_G / \partial \delta$ of the reference bus derived from the nodal power injection equations and the elements $\partial \delta / \partial Q$ and $\partial \delta / \partial V$ of the inverse of the Jacobian matrix, the gradient vector of losses with respect to reactive power changing Q_i at bus i, applying the chain rule, and introducing variables δ for bus voltage angles and V for bus voltage values, is given as:

$$\frac{\Delta P_{\text{loss}}}{\Delta Q_i} = \sum_{i=1}^{n} \frac{\partial P_G}{\partial \delta_i} \frac{\partial \delta_i}{\partial Q_i} + \sum_{i=1}^{n} \frac{\partial P_G}{\partial V_i} \frac{\partial V_i}{\partial Q_i}. \tag{14.31}$$

The gradient vector of losses with respect to reactive power changing is used to find out a possible better reactive power setting in the current system status; the following equation is used:

$$Q_i^{\text{new}} = Q_i^{\text{old}} - \alpha \frac{\Delta P_{\text{loss}}}{\Delta Q_i}$$

where Q_i^{old} is reactive power at bus i, Q_i^{new} is new reactive power setting at bus i after a gradient vector move, and α is positive scale factor.

During the procedure, the gradient search is included when necessary to increase the diversity, to overcome a possible local optimum, and to move solutions in the right direction toward the global optimum solution without disturbing the evolutionary nature of the algorithm. The gradient search is only activated when the fitness of the current individuals is worse than a specified value: the best individual fitness of a previous generation (say, of generation -10 from the current one) is used as a threshold to decide which individuals of the current generation are to be repaired.

A full utilization of the gradient method to find the new reactive power setting is not however advisable:

(a) Rounding off the gradient vector is necessary because of the discrete nature of the reactive power setting; and

(b) Such a procedure risks getting trapped in local optima.

14.6.2 Evolutionary Programming/Evolution Strategies Under Test

The evolutionary process relies on an overlapping population with elitism, where only some of the elements of the population are replaced by new offspring. So, a percentage of parents is kept for the next generation.

The algorithm uses uniform crossover and Gaussian mutation in all genes. In this case, additive mutations were used where random numbers were rounded to the closest integer; the mutation rate σ was fixed with a value allowing effective small jumps between integer numbers.

In this type of model where the load curve is split among several periods at every bus, a problem of coherency arises. In fact, the model is built almost as if there were

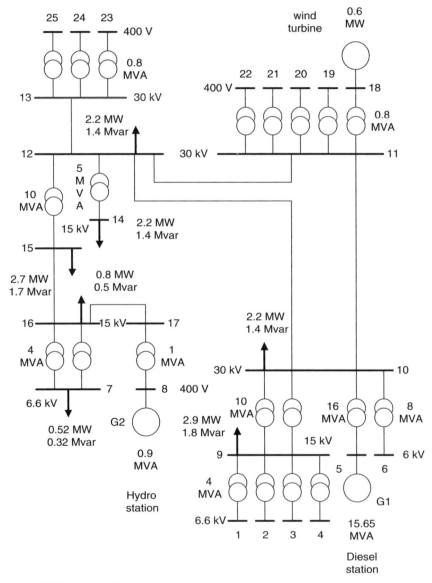

FIGURE 14.16 Single-line diagram of the simplified Azores Islands test system.

several independent problems of reactive power planning at the same time, but, in fact, there must be some kind of coordination among the several solutions for each bus.

The characteristics of stochastic convergence of the evolutionary algorithms are well-known and their robustness is one property that is very much appreciated, when observed; meaning that in several runs, the algorithm will offer basically the same solution. Therefore, one important test for these evolutionary models is to observe their behavior when the load is the same for all periods: if the number of capacitor additions becomes equal in all periods, then we may have more confidence in the quality of the solution for a general problem with distinct load levels.

Such a test was conducted for the 25-bus distribution system in one island of the Azores, from which a simplified diagram is shown in Fig. 14.16.

Table 14.9 compares the results of the algorithm, in such a flat load scenario for three load periods, with a simple genetic algorithm (SGA) with binary coding and no chromosome repair. Notice that the EP/ES solution is robust and gives the same result for all periods. Such result could not be achieved by a GA in the classic simple formulation.

TABLE 14.9   **Control Setting of Capacitor Banks Based on Best Run of Two Algorithms**

Bus no.	SGA			EP/ES		
	Low	Normal	Peak	Low	Normal	Peak
1	1	2	1	2	2	2
2	0	0	2	1	1	1
3	0	2	1	1	1	1
4	1	1	1	0	0	0
6	3	3	3	0	0	0
7	3	3	2	4	4	4
9	8	7	8	6	6	6
10	2	0	4	0	0	0
11	3	4	3	0	0	0
12	3	3	5	9	9	9
13	6	5	5	4	4	4
14	2	1	0	9	9	9
15	3	1	3	0	0	0
16	15	14	14	13	13	13
17	15	15	14	15	15	15
18	11	13	12	9	9	9
19	0	0	0	0	0	0
20	2	1	2	0	0	0
21	0	0	0	0	0	0
22	0	0	0	0	0	0
23	0	0	0	0	0	0
24	0	0	0	0	0	0
25	0	1	0	0	0	0

The results presented are for modules of 100-kV Ar capacitors and confirmed in 10 independent runs. Each generation was composed of 30 individuals and results evaluated after 300 generations. The study confirmed the advantage of using a chromosome repair technique based on an analytical method and on using an evolution strategies or evolutionary programming approach over a genetic algorithm in this problem.

14.7 CONCLUSIONS

The application of meta-heuristics (especially evolutionary algorithms) in planning is tempting because of the complexity and nonlinearities of the problems, which typically involve an evaluation of investment and an assessment of operation costs. The dimension of the problems is always very large, because in many cases planning implies an evolution in the time domain. Besides, many of the options available are of discrete, nature and the problem becomes mixed-integer with a combinatorial characteristic.

We have visited in this chapter models that have application in the generation, the transmission, and the distribution systems. Their authors claim reasonable amounts of success and also leave open the avenue for further improvement.

We have witnessed reports stating that simulated annealing does not seem able to compete with more recent and sophisticated meta-heuristics. In this respect, it also seems that simple genetic algorithms are not the answer. Almost all authors would agree today that real-valued or integer-valued chromosomes (such as in EP/ES models) are better suited to guarantee convergence and robustness to the models. Furthermore, almost all of them would also agree that hybrid models, where the convergence power of a meta-heuristic is enhanced with knowledge from mathematical analytic models, perform much better. They lack generality in the tool used but this comes at the benefit of quality of results.

As a general comment, we may state that dynamic expansion planning models are still a challenge, whereas for static expansion planning quite satisfactory experiences have been reported. The objective of the chapter was not to be exhaustive in the description of all the models proposed in the literature—certainly some other proposals with their own merits have come to light—but rather to provide enough evidence that the meta-heuristic approach works and should be seriously considered in practical planning applications.

REFERENCES

1. Masse P, Gilbrat R. Application of linear programming to investments in the electric power industry. Management Sci., 1957; 3(2):149–166.
2. Bloom JA. Long-range generation planning using decomposition and probabilistic simulation. IEEE Trans PAS. 1982; 101(4):797–802.

3. Park YM, Lee KY, Youn LTO. New analytical approach for long-term generation expansion planning based maximum principle and Gaussian distribution function. IEEE Trans PAS 1985; 104:390–397.

4. Jenkins ST, Joy DS. Wien Automatic System Planning Package (WASP)—an electric utility optimal generation expansion planning computer code. Oak Ridge, TN: Oak Ridge National Laboratory; ORNL-4945. 1974.

5. Electric Power Research Institute (EPRI). Electric generation expansion analysis system (EGEAS). EPRI EL-2561. Palo Alto, CA: EPRI; 1982.

6. Nakamura S. A review of electric production simulation and capacity expansion planning programs. Energy Res, 1984; 8:231–240.

7. David AK, Zhao R. Integrating expert systems with dynamic programming in generation expansion planning. IEEE Trans PWRS 1989; 4(3):1095–1101.

8. David K, Zhao R. An expert system with fuzzy sets for optimal planning. IEEE Trans. PWRS 1991; 6(1):59–65.

9. Fukuyama Y, Chiang H. A parallel genetic algorithm for generation expansion planning. IEEE Trans PWRS 1996; 11(2):955–961.

10. Park YM, Park JB, Won JR. A genetic algorithms approach for generation expansion planning optimization. In: Control of Power Plants and Power Systems, Proceedings of the IFAC Symposium Sipower' 95, pp. 257–262, IFAC Proceedings Volumes Series, Pergamon Pr, 1996.

11. Iba K. Reactive power optimization by genetic algorithm. IEEE Trans PWRS 1994; 9(2): 685–692.

12. Sheble GB, Maifeld TT, Brittig K, Fahd G. Unit commitment by genetic algorithm with penalty methods and a comparison of Lagrangian search and genetic algorithm-economic dispatch algorithm. Int. Electric Power Energy Systems 1996; 18(6):339–346.

13. Garver LL. Transmission network estimation using linear programming. IEEE Trans Power Apparatus Systems 1970; PAS-89:1688–1697.

14. Villasana R, Garver LL, Salon SJ. Transmission network planning using linear programming. IEEE Trans Power Apparatus Systems 1985; PAS-104(2):349–356.

15. Binato S, Pereira MVF, Granville S. A new benders decomposition approach to solve power system network design problems. IEEE Trans Power Systems 2001; 16(2):247–253.

16. Binato S, De Oliveira GC, De Araujo JL. A greedy randomized adaptive search procedure for transmission expansion planning. IEEE Trans Power Systems 2000; 16(2):247–253.

17. Gallego RA, Monticelli A, Romero R. Comparative studies of non-convex optimization methods for transmission network expansion planning. IEEE Trans Power Systems 1998; 13(3):822–828.

18. Gallego RA, Monticelli A, Romero R. Taboo search algorithm for network synthesis. IEEE Trans Power Systems 2000; 15(2):490–495.

19. Bahiense L, Oliveira GC, Pereira M, Granville S. A mixed integer disjunctive model for transmission network expansion. IEEE Trans Power Systems 2001; 16(3):560–565.

20. Dechamps C, Jamoulle A. Interactive computer programming for planning the expansion of meshed transmission networks. Int J Electrical Power Energy Systems 1980; 2(2):103–108.

21. Monticelli A, Santos A Jr, Pereira MVF, Cunha SH, Parker BJ, Praça JCG. Interactive transmission network planning using a least-effort criterion. IEEE Trans Power Apparatus Systems 1982; PAS-101(10):3909–3925.

22. Pereira MVF, Pinto LMVG. Application of sensitivity analysis of load supplying capability to interactive transmission expansion planning. IEEE Trans Power Apparatus Systems 1985; PAS-104(2):381–389.

23. Latorre-Bayona G, Peres-Arriaga JI. CHOPIN, a heuristic model for long term transmission expansion planning. IEEE Trans Power Systems 1994; 10(4):1828–1834.

24. Oliveira GC, Costa APC, Binato S. Large scale transmission network planning using optimization and heuristic techniques. IEEE Trans Power Systems 1995; 10(4): 1828–1834.

25. Granville S, Pereira MVF. Analysis of the linearized power flow model in Benders decomposition. EPRI-Report RP 2473–6. Stanford, CA: Stanford University, 1985.

26. Romero R, Monticelli A. A hierarchical decomposition approach for transmission network expansion planning. IEEE Trans Power Systems 1994; 9(1):373–380.

27. Haffner S, Monticelli A, Garcia A, Romero R. Specialized branch and bound algorithm for transmission network expansion planning. IEE Proc Generation Transmission Distribution 2001; 148(5):482–488.

28. Romero R, Gallego RA, Monticelli A. Transmission system expansion planning by simulated annealing. IEEE Trans Power Systems 1996; 11(1):364–369.

29. Gallego RA, Alves AB, Monticelli A, Romero R. Parallel simulated annealing applied to long term transmission network expansion planning. IEEE Trans Power Systems 1997; 12(1):181–188.

30. Miranda V, Ranito JV, Proença LM. Genetic algorithms in optimal multistage distribution network planning. IEEE PES Winter Meeting 1993, New York, Jan, 1993. IEEE Trans Power Systems 1994; 9(4).

31. Gallego RA, Monticelli A, Romero R. Transmission system expansion planning by extended genetic algorithm. IEE Proc Generation Transmission Distribution 1998; 145(3):329–335.

32. Da Silva EL, Gil HA, Areiza JM. Transmission network expansion planning under an improved genetic algorithm. IEEE Trans Power Systems 2000; 15(3):1168–1175.

33. Wen F, Chang CS. Transmission network optimal planning using the taboo search method. Electric Power System Res 1997; 42:153–163.

34. Da Silva EL, Gil HA, Areiza JM, De Oliveira GC, Binato S. Transmission network expansion planning under a taboo search approach. IEEE Trans Power Systems 2001; 16(1):62–68.

35. Romero R, Monticelli A, Garcia A, Haffner S. Test systems and mathematical models for transmission network expansion planning. IEE Proc Generation Transmission Distribution, 2002; 149(1):27–36.

36. Hindi KS, Brameller A. Design of low voltage distribution networks: a mathematical programming method. Proc IEE, Part C, 1977; 124(1):54–58.

37. de Oliveira MF, Miranda V. Optimization studies in distribution networks including reliability calculations. Proceedings of CIRED – International Conference on Electric Distribution, Session 6, AI.M./IEE ed, Liège, Belgium, May 1979.

38. Chiang HD, Jean-Jumeau R. Optimal network reconfigurations in distribution systems: part 1: a new formulation and a solution methodology. IEEE Trans PWRD 1990; 5:1902–1908.

39. Chu FR, Wang JC, Chiang HD. Strategic planning of LC compensators in nonsinusoidal distribution systems. IEEE Trans PWRD 1994; 9:1558–1563.

40. Chiang HD et al. Optimal capacitor placement, replacement and control in large-scale unbalanced distribution systems: system modeling and a new formulation. IEEE Trans PWRS 1995; 10:356–362.

41. Chiang HD et al. Optimal capacitor placement, replacement and control in large-scale unbalanced distribution systems: system solution algorithms and numerical studies. IEEE Trans PWRS 1995; 10:363–369.

42. Billinton R, Jonnavithula S. Optimal switching device placement in radial distribution system. IEEE Trans PWRD 1996; 11:1646–1651.

43. Jiang D, Baldick R. Optimal electric distribution system switch reconfiguration and capacitor control. IEEE Trans PWRS 1996; 11:890–897.

44. Jonnavithula S, Billinton R. Minimum cost analysis of feeder routing in distribution system planning. IEEE Trans PWRD 1996; 11(4):1935–1940.

45. Jeon YJ, Kim JC. An efficient algorithm for network reconfiguration in large scale distribution systems. In: Proc. of IEEE PES SM'99. IEEE Press; 1999. p. 243–247.

46. Jeon YJ, Kim JC, Kim JO, Shin JR, Lee KY. An efficient simulated annealing algorithm for network reconfiguration in large-scale distribution systems. IEEE Trans PWRD 2002; 17(4):1070–1078.

47. Nara K et al. Implementation of genetic algorithm for distribution systems loss minimum re-configuration. IEEE Trans PWRS 1992; 7(3):1044–1051.

48. Yeh EC, Venkata SS, Sumic Z. Improved distribution system planning using computational evolution. IEEE Trans PWRS 1996; 11(2):668–674.

49. Ramirez-Rosado IJ, Jose LBA. Genetic algorithms applied to the design of large power distribution systems. IEEE Trans PWRS 1998; 13(2):696–703.

50. Cavalho MS, Ferreira LAFM et al. Distribution network expansion planning under uncertainty: a hedging algorithm in an evolutionary approach. IEEE Trans PWRD 2000; 15(1):412–416.

51. Chuang AS, Wu F. An extensible genetic algorithm framework for problem solving in a common environment. IEEE Trans PWRS 2000; 15(1):269–275.

52. Huang SJ. An immune-based optimization method to capacitor placement in a radial distribution system. IEEE Trans PWRD 2000; 15(2):744–749.

53. Chen TH, Cherng JT. Optimal phase arrangement of distribution transformers connected to a primary feeder for system unbalance improvement and loss reduction using a genetic algorithm. IEEE Trans PWRS 2000; 15(3):994–1000.

54. Hsiao YT, Chien CY. Enhancement of restoration service in distribution systems using a combination fuzzy-GA method. IEEE Trans PWRS 2000; 15(4):1394–1400.

55. Delfanti M, Granelli GP, Marannino P, Montagna M. Optimal capacitor placement using deterministic and genetic algorithms. IEEE Trans PWRS 2000; 15(3):1041–1046.

56. Masoum MAS, Ladjevardi M, Jafarian A, Fuchs EF. Optimal placement, replacement and sizing of capacitor banks in distorted distribution networks by genetic algorithms. IEEE Trans PWRD 2004; 19(4):1794–1801.

57. Nara K, Hasegawa J. Configuration of new power delivery system for reliable power supply. In: Proc. of IEEE PES SM'99. IEEE Press; 1999. p. 248–253.

58. Ramirez-Rosado IJ, Dominguez-Navarro JA, Yusta-Loyo JM. A new model for optimal electricity distribution planning based on fuzzy set techniques. Proc. of IEEE PES SM'99. 1999. IEEE Press; p. 1048–1054.

59. Gallego RA, Monticelli AJ, Romero R. Optimal capacitor placement in radial distribution networks. IEEE Trans PWRS 2001; 16(4):630–637.

60. Hayashi Y, Matsuki J. Loss minimum configuration of distribution system considering N-1 security of dispersed generators. IEEE Trans PWRS 2004; 19(1):636–642.

61. Rosado IJR, Navarro JAD. Possibilistic model based on fuzzy sets for the multiobjective optimal planning of electric power distribution networks. IEEE Trans PWRS 2004; 19(4):1801–1810.

62. Miranda V, Proença LM, Oliveira L. Dynamic planning of distribution networks for minimum regret strategies. Proceedings of CIRED Argentina, Buenos Aires, Dec. 1997.

63. Miranda V, Proença LM. Why risk analysis outperforms probabilistic choice as the effective decision support paradigm for power system planning. IEEE Trans PWRS 1998; 13(2):643–648.

64. Miranda V, Srinivasan D, Proença L. Evolutionary computation in power systems. Int J Electrical Power Energy Systems 1998; 20(2):89–98.

65. Nara K et al. FRIENDS—forwarding to future power delivery system. Proc. of International Conference on Harmonics and Quality of Power 2000. Orlando (FL), USA, 2000.

66. Nara K, Hasegawa J. A new flexible, reliable and intelligent electrical energy delivery system. Electrical Eng Japan 1997; 121(1):26–34.

67. Nara K et al. Optimal configuration of new power delivery system for customized power supply services. Proc. of PSCC'02, Sevilla, Spain, 2002.

68. Happ HH. Optimal power dispatch: a comprehensive survey IEEE Trans Power Apparatus Systems 1977; PAS-96:841–854.

69. Alsac O, Bright J, Prais M, Stott B. Further development in LP-based optimal power flow IEEE Trans Power Systems 1990; 5(3):697–711.

70. Hobson E. Network constrained reactive power control using linear programming. IEEE Trans Power Apparatus and Systems 1980; PAS-99(4):1040–1047.

71. Fernandes R, Lange, Burchettr F, Happ H, Wirgau K. Large scale reactive power planning. IEEE Trans Power Apparatus and Systems 1983; PAS-102(5):1083–1088.

72. Happ HH, Wirgau KA. Static and dynamic var compensation in system planning. IEEE Trans Power Apparatus Systems 1978; PAS-97(5):1564–1578.

73. Hughes A, Jee G, Hsiang P, Shoults RR, Chen MS. Optimal power planning. IEEE Trans Power Apparatus and Systems 1981; PAS-100:2189–2196.

74. Shoults RR, Chen MS. Reactive power control by least square minimization. IEEE Trans Power Apparatus and Systems 1976; PAS-95:397–405.

75. Lebow WM, Mehra RK, Nadira R, Rouhani R, Usoro PB. Optimization of reactive volt-ampere (VAR) sources in system planning, EPRI Report, EL-3729, Project 2109–1, Vol.1, Nov. 1984.

76. Billington R, Sachdev SS. Optimum network VAR planning by nonlinear programming. IEEE Trans Power Apparatus and Systems 1973; PAS-92:1217–1225.

77. Hegdt GT, Grady WM. A matrix method for optimal var siting. IEEE Trans Power Apparatus and Systems 1975; PAS-94(4):1214–1222.

78. Aoki K, Fan M, Nishikori A. Optimal var planning by approximation method for recursive mixed integer linear planning. IEEE Trans Power Systems. 1988; PWRS-3(4): 1741–1747.

79. Lee KY, Ortiz JL, Park YM, Pond LG. An optimization technique for reactive power planning of subtransmission network under normal operation. IEEE Trans Power Systems, 1986; PWRS-1:153–159.

80. Mangoli MK, Lee KY, Park YM. Optimal long-term reactive power planning using decomposition technique. Electric Power Systems Res 1993; 26:41–52.

81. Nadira R, Lebow W, Usoro P. A decomposition approach to preventive planning of reactive volt-ampere (VAR) source expansion. IEEE Trans Power Systems 1987; PWRS-2(1): 72–77.

82. Hsiao Y-T, Liu C-C, Chiang H-D, Chen Y-L. A new approach for optimal VAR sources planning in large scale electric power systems. IEEE Trans Power Systems 1993; 8(3):988–996.

83. Mangoli MK, Lee KY, Park YM. Optimal real and reactive power control using linear programming. Electric Power System Res 1993; 26:1–10.

84. Geoffrion M. Generalized Benders decomposition. J Optim Theory Appl 1972; 10(4):237–261.

85. Lee KY, Park YM, Ortiz JL. United approach to optimal real and reactive power dispatch. IEEE Trans Power Apparatus and Systems 1985; PAS-104:1147–1153.

86. Mui KN, Chiang H-D, Darling G. Capacitor placement, replacement and control in large-scale distribution systems by a GA-based two-stage algorithm. IEEE Trans PWRS 1997; 12(3):1160–1166.

87. Miranda V, Oo NW, Fidalgo JN. Experimenting in the optimal capacitor placement and control problem with hybrid mathematical-genetic algorithms. Proceedings of the Intelligent Systems Application to Power Systems Conference ISAP 2001, Budapest, Hungary, June 2001.

■■■■■ CHAPTER 15

Applications to Power System Scheduling

KOAY CHIN AIK, LOI LEI LAI, KWANG Y. LEE, HAIYAN LU,
JONG-BAE PARK, YONG-HUA SONG, DIPTI SRINIVASAN,
JOHN G. VLACHOGIANNIS, and I.K. YU

15.1 INTRODUCTION

In electricity supply systems, there exist a wide range of problems involving the optimization process. Among these, power system scheduling is one of the most important problems in the operation and management of the system. In the past, various optimization methods have been applied, and some of them have been implemented into practice. In this chapter, some of the modern heuristic optimization techniques have been applied or enhanced, including genetic algorithms (GA), particle swarm optimization (PSO), evolutionary strategy (ES), and colony search. The power system scheduling problems addressed here range from economic dispatch, maintenance scheduling, cogeneration scheduling, short-term generation scheduling of thermal units, to constrained load flow problems.

15.2 ECONOMIC DISPATCH

15.2.1 Economic Dispatch Problem

The objective of conventional economic dispatch (ED) problems is to find the optimal combination of power generation that minimizes total generation costs while satisfying an equality constraint and several inequality constraints [1, 2].

The most simplified type of objective function in the ED problem can be expressed as a summation of all generating units' operating cost in the shape of a smooth function:

$$\text{Minimize} \sum_{i \in I} F_i(P_i) \tag{15.1}$$

Modern Heuristic Optimization Techniques. Edited by K. Y. Lee and M. A. El-Sharkawi
Copyright © 2008 the Institute of Electrical and Electronics Engineers, Inc.

$$F_i(P_i) = \alpha_i + \beta_i P_i + \gamma_i P_i^2,$$

where α_i, β_i, and γ_i represent cost coefficients of generating unit i, P_i the electrical output of generating unit i, and I indicates the set for all generating units.

While minimizing the total generation cost, the following constraints should be satisfied. For energy balance, the following equality constraint should be satisfied as the following equation:

$$\sum_{i \in I} P_i = D + P_{\text{loss}}, \tag{15.2}$$

where D implies the total system demand, and P_{loss} means the total network losses. However, the transmission losses are not considered in this study.

Also, generation of power from each unit should be between its maximum and minimum limits:

$$P_{i,\min} \le P_i \le P_{i,\max} \quad \forall i \in I, \tag{15.3}$$

where $P_{i,\min}$ corresponds with the minimum output of unit i and $P_{i,\max}$ corresponds with the maximum output of unit i.

To consider a more realistic and accurate representation of the objective function, nonsmooth cost functions with a few shapes have been applied to ED problem [3–5]. That is, the objective function of an ED problem has discontinuous and nondifferentiable points according to valve loading, change of fuels, and prohibited zones. Therefore, it is more realistic to treat the cost function as a set of piecewise quadratic functions as illustrated in Fig. 15.1 when considering multifuel problems, which are defined as follows [4]:

$$F_i(P_i) = \begin{cases} \alpha_{i1} + \beta_{i1} P_i + \gamma_{i1} P_i^2 & \text{if } P_i^{\min} \le P_i \le P_{i1} \\ \alpha_{i2} + \beta_{i2} P_i + \gamma_{i2} P_i^2 & \text{if } P_{i1} \le P_i \le P_{i2} \\ \quad \vdots & \quad \vdots \\ \alpha_{im} + \beta_{im} P_i + \gamma_{im} P_i^2 & \text{if } P_{im-1} \le P_i \le P_i^{\max} \end{cases}, \tag{15.4}$$

where α_{ij}, β_{ij}, and γ_{ij} correspond with the cost coefficients of generating unit i for the jth power level, respectively.

Nonsmooth cost function considering multiple valve effects is shown in Fig. 15.2. The function has a quadratic input–output curve and nonlinear sinusoid function approximating valve-point characteristics as follows:

$$\hat{F}(P_i) = F(P_i) + |e_i \sin(f_i(\underline{P}_i - P_i))|, \tag{15.5}$$

where $F_i(P_i)$ is the quadratic cost curve, and e_i and f_i are cost coefficients with a valve point curve.

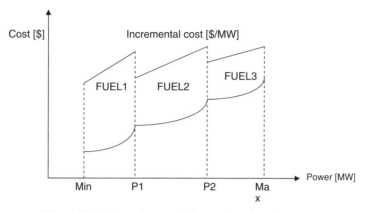

FIGURE 15.1 Piecewise quadratic cost function of a generator.

Finally, there is the cost function with prohibited operation zones to apply a more realistic and accurate objective function. A unit with prohibited operating zones has a discontinuous input-output power generation characteristic as illustrated in Fig. 15.3 Classic quadratic cost curves were used and, for units with prohibited operating zones, additional operating range constraints were imposed.

15.2.2 GA Implementation to ED

Until now, many studies have been carried out on the application of GA to the ED problem. In this section, a brief comparison of the studies [6–11] is made in terms

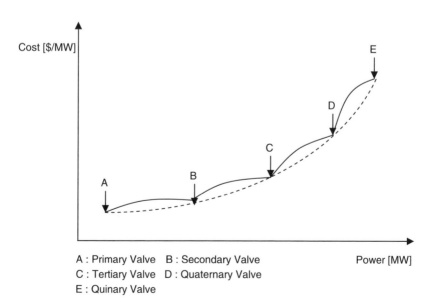

A : Primary Valve   B : Secondary Valve
C : Tertiary Valve   D : Quaternary Valve
E : Quinary Valve

FIGURE 15.2 Cost function with five valves.

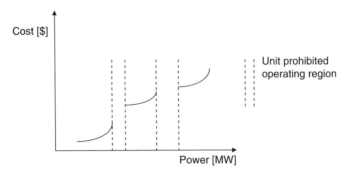

FIGURE 15.3   A unit with prohibited operating zone.

of the encoding method, constraints handling, genetic operation, fitness function, and so on.

15.2.2.1 Encoding Method   Two types of encoding schemes, that is, series encoding and embedded encoding, have been explored in Refs. 6 and 8. Both utilize the binary number $\{0, 1\}$. Illustrations of these encoding methods are shown in Fig. 15.4. The first encoding simply stacks each unit's output value in megawatt (MW) structure in cascade with every other string. The second method uses the

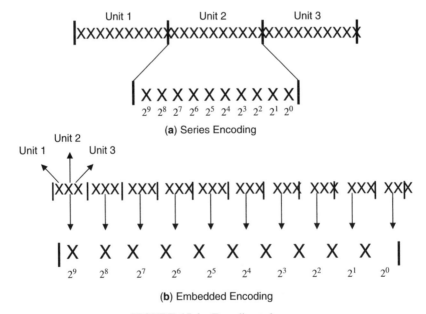

FIGURE 15.4   Encoding schemes.

same scheme for representation and decoding as the first, except that the assigned gene structures are embedded within each other through the string.

Bakirtzis, Petridis, and Kazarlis [9] have developed a GA version that produces better results. This is based on the observation that when N units with valve points are dispatched, at least $N - 1$ units operate at a valve point in the optimum solution. That is, at most one reference unit operates on a free point, between valve points, in order to satisfy the equality power balance constraint. Therefore, $N - 1$ units are described by a much smaller bit, say 4 bits, allowing only the valve point value of each unit to be encoded in it.

In Refs. 10 and 11, for the implementation of the dispatch problem, a binary series encoding scheme was used as it had been reported that series encoding provided a better ED solution to that of an embedded one [8].

15.2.2.2 *Constraints Handling*   To implement GA in the ED problem, electrical outputs covering all generating units are selected as encoding variables. By applying a linear mapping strategy in the random creation of a population, a generating unit's inequality constraint (15.3) can be easily satisfied. However, the strings that satisfy the inequality constraints do not always produce offspring satisfying the inequality constraints due to genetic operations. In these cases, offspring not satisfying the inequality constraints are discarded and genetic operations are repeated until they produce offspring that satisfy the inequality constraints.

However, it is not easy to treat the equality constraint (15.2). Therefore, the penalty function approach has been widely applied to include the equality constraint in the fitness function.

Park et al. [7] proposed a projection method, described in (15.6), to create chromosomes that always satisfy the equality constraint:

$$X = X_0 - \left\{ (\bar{a}^T X_0 - b) \cdot (\bar{a}^T \bar{a})^{-1} \right\} \bar{a} = prof(X_0). \tag{15.6}$$

By applying the projection method, any generated string X_0 that does not satisfy the supply/demand equality constraint can be transformed into the string X satisfying the constraint. Therefore, the fitness function does not require the penalty function to be included any more. Because the fuel cost function f should be minimized, the fitness function (F) for GA implementation has the following inverse form:

$$F = \frac{\alpha}{1 + f}, \tag{15.7}$$

where α is a positive scaling factor and f is the total generation cost.

Walters and Sheble [6] and Sheble and Brittig [8] implemented an objective function and constraints of the ED problem within the fitness function. Unit limits are automatically satisfied by the normalization process in the chosen encoding method, which only allows solutions in the operating range of the units. Equality power balance constraint was incorporated in the augmented objective function as a form of penalty term and was handled by the fitness function evaluation through the weighting factor.

Bakirtzis, Petridis, and Kazarlis [9] handled the equality constraint based on the observation that when N units with valve points are dispatched, at least $N - 1$ units operate at a valve point in the optimum solution. That is, at most one reference unit operates on a free point, between valve points, in order to satisfy the equality power balance constraint. Using the LaGrange function method where constraints are implemented by imposing penalties in individuals, Song and Chou [10] included the constraints of the ED problem in the fitness function formulation. Both equality and inequality constraints are taken into consideration during the optimization process. The revised power balance equation and the redistributed power output of each unit are as follows:

$$lerr = \left| \sum_{i \in I} P_i - D - P_{\text{loss}} \right| + \sum_{i \in I} \rho_i P_i \tag{15.8}$$

$$P_{i,\text{new}} = \frac{P_i}{\sum_{i \in I} P_i} (D + P_{\text{loss}}), \tag{15.9}$$

where $lerr$ is the revised power balance equation and ρ_i is a power balance penalty factor for unit i.

Orero and Irving [11] took care of the unit minimum and maximum loading limits in the problem encoding, and they used a penalty function approach to treat the constraints of the unit prohibited operating zones and the power balance equation.

15.2.2.3 Genetic Operations

In Ref. 6, a one-point crossover and mutation operation was used. The authors used both series and embedded encoding schemes for the economic dispatch problem solution and compared the two encoding methods in terms of generation cost and performance. In Ref. 8, several different techniques to enhance program efficiency and accuracy, such as mutation and crossover prediction, elitism, and interval approximation, were proposed. When crossover is needed, a binary string that is the same length as the population string is used to cue the two parent strings as to whether the child will get its bit value from the first parent or the second one. A pseudocode for this new crossover technique is illustrated in Fig. 15.5.

In addition, elitism was used to improve the performance capability of conventional GA such that early solutions are saved by ensuring the survival of the most fit strings in each population.

To accelerate the local gradient descend, the so-called phenotype mutation that is based on perturbation of real power vector (not binary vector) was introduced in Ref. 9. This new operator was designed only for the best genotype of every generation, so that it was not expected to load the algorithm with many more fitness evaluations.

```
Create a binary string, c, randomly that is the same length
As the population string

i=1
DO WHILE (i ≤ String Length)
    Choose two parents, p1 and, p2, randomly

    For the offspring c1,
    IF [c[i] = 1] THEN
        c1[i] = p1 [i]
    ELSE IF [c[i] = 0] THEN
        c1[i] = p2[i]

    For the offspring c2,
    IF [c[i] = 1] THEN
        c2[i] = p2[i]
    ELSE IF [c[i] = 0] THEN
        c2[i] = p1[i]
    i=i+1
END DO
```

FIGURE 15.5   Pseudocode for the new crossover technique.

An advanced version of GA was also proposed in Ref. 9, where $N - 1$ units can be described by a small number of bits, say 4 bits, allowing only the valve point value of each unit to be encoded. Therefore, every unit except the reference is allowed to produce a small number of discrete power output values, and these discrete values are chosen from the set of the unit's valve points.

An advanced version of GA was proposed in Ref. 10 that showed a more accurate and less time-consuming properties. This algorithm combines a strategy involving local search algorithm and GAs. Moreover, several techniques, such as elite policy, adaptive mutation prediction, nonlinear fitness mapping, and different crossover techniques, were explored to enhance program efficiency and accuracy.

To save early solution by ensuring the survival of the most fit strings in the current population to be carried over to the next generation, the elitism technique was adopted. Comparing the results of the most recent population to the elite population, elitism combines the two populations and determines the best results from both populations. Various crossover techniques including one-point, two-point, and uniform crossover were explored and compared. As a result, two-point crossover was reported to yield a better result in ED applications. By using the geometric mean probability theory to work out the probability of mutation, adaptive mutation prediction was explored, and much of the computation effort can be saved without sacrificing accuracy.

```
DO WHILE (GEN. ≤ MAX. GENERATION)
  Reshuffle population
  i=1
  DO WHILE(i ≤ population size)
    Choose two parents, p1 and p2, randomly, without replacement
    Perform cross over and mutation to produce offspring c1 and c2
    Evaluate fitness, f, of parents and offspring

    IF Distance (p1,c1)+Distance(p2,c2) ≤
                            Distance(p2,c1)+Distance(p1,c2)

    THEN
      IF (f(c1)>f(p1) replace p1 with c1
      IF (f(c2)>f(p2) replace p2 with c2
    ELSE
      IF (f(c2)>f(p1) replace p1 with c2
      IF (f(c1)>f(p2) replace p2 with c1
    i=i+2
  END DO
END DO
```

FIGURE 15.6   Pseudocode for the deterministic crowding GA.

In Ref. 11, a deterministic crowding GA was presented to maintain population diversity as the algorithm proceeds. A pseudocode for the deterministic crowding GA is shown in Fig. 15.6.

15.2.2.4 Fitness Function   Walters and Sheble [6] defined the fitness function in the following form:

$$Fit = sf_1[(1 - \%obj)^{sp_1}] + sf_2[(1 - \%l_{err})^{sp_2}] \tag{15.10}$$

$$obj = \sum_{i \in I} \alpha_i + \beta_i P_i + \gamma_i P_i^2 + \left| e_i \sin \left(f_i (\underline{P}_i - P_i) \right) \right| \tag{15.11}$$

$$l_{err} = \left| \sum_{i \in I} P_i - D - P_{loss} \right|, \tag{15.12}$$

where *sf* is a scaling factor for emphasis within the function itself, *sp* is a scaling power factor for emphasis over the entire population, *%obj* is a percentage value of the objective function, and *%l*$_{err}$ is a percentage value of the string's constraint equation error.

Sheble and Brittig [8] also explored the same fitness function of Walters and Sheble selected as in (15.10), except that the objective function of the ED problem was defined as a smooth function of the quadratic cost curve.

In Ref. 10, Song and Chou designed a fitness function based on the percentage rating. The lowest-cost operating solution is the point at which the minimum incremental cost rate (λ) is equal and simultaneously satisfies the specified demand. An error term was introduced and a measure of this error was calculated.

$$\lambda_{err} = \sum_{i \in I} \left| \lambda_{avg} - \lambda_i \right|, \tag{15.13}$$

where λ_{err} is an incremental cost error term, λ_i is an incremental cost for unit i, and λ_{avg} is an average incremental cost of individual string. By combining the two errors resulting from (15.8) and (15.12), the fitness function becomes:

$$Fit = sf_1[(1 - \%\lambda_{err})^{sp_1}] + sf_2[(1 - \%l_{err})^{sp_2}], \tag{15.14}$$

where $\%\lambda_{err}$ is the percentage of incremental cost error, $\%l_{err}$ is the percentage of power balance error, sf is a scaling factor for emphasis within function, and sp is a scaling power factor for emphasis over the entire population.

Orero and Irving [11] explored the use of a GA for the solution of an ED problem where some of the units had prohibited operating zones. The unit minimum and maximum loading limits were embedded in the problem encoding process. Other constraints to be considered were the power balance equation and the unit prohibited operating zones. These two constraints were handled using a penalty function approach. The problem objective function is formulated as follows:

$$\min \sum_{i \in I} F_i(P_i) + \Psi \left[\sum_{i \in I} P_i - (D + P_{loss}) \right] + \Phi \left[\sum_{k=1}^{n_i} P_i(zone_k) \right], \tag{15.15}$$

where Ψ is the penalty function for not satisfying the load demand, Φ represents the penalty function for a unit loading that has fallen within a prohibited operating zone, and n_i is the number of prohibited zones for unit i.

15.2.2.5 Multistage Method and Directional Crossover

Genetic algorithms simulate an analogous biological mechanism. Each chromosome has its adequate number of bits, and each bit has the corresponding genetic information. In continuous variable optimization problems, numerous bits are needed to obtain precise optimal solutions. As a kind of local search method, the multistage method was proposed by Park et al. [7], which divides solution space into N segments and finds the best solution point from generation to generation in the current stage. Some advantages are observed when this method is applied. First, it requires smaller bits than the traditional encoding schemes. Second, as the stage is increased, solution space is subdivided further and a more accurate solution can be obtained.

In addition, a directional crossover was also presented [7] to overcome the limitation of the traditional GA mechanism, which cannot guarantee the convergence to the optimal point and sometimes causes the loss of critical patterns [12, 13]. The directional operator helps to produce more competitive offsprings when combined with conventional operators.

15.2.3 PSO Implementation to ED

The practical ED problems with valve-point and multifuel effects are represented as a nonsmooth optimization problem with equality and inequality constraints, and this makes the problem of finding the global optimum difficult. To solve this problem, many salient methods have been proposed such as a mathematical approach [4], dynamic programming [1], evolutionary programming [3, 14, 15], tabu search [16], neural network approaches [5, 17], and genetic algorithm [18].

In this section, an alternative approach is proposed to the nonsmooth ED problem using a modified PSO (MPSO) [19], which focuses on the treatment of the equality and inequality constraints when modifying each individual's search. Additionally, to accelerate the convergence speed, a dynamic search-space reduction strategy is devised based on the distance between the best position of the group and the inequality boundaries.

15.2.3.1 Constraints Handling   In this section, a new approach to implement the PSO algorithm will be described in solving the ED problems. Especially, a suggestion will be given on how to deal with the equality and inequality constraints of the ED problems when modifying each individual's search point in the PSO algorithm. The process of the MPSO algorithm can be summarized as follows:

Step 1. Initialization of a group at random while satisfying constraints.

Step 2. Velocity and position updates while satisfying constraints.

Step 3. Update of *Pbest* and *Gbest*.

Step 4. Activation of space reduction strategy.

Step 5. Go to step 2 until satisfying stopping criteria.

In Ref. 19, Park et al. proposed the simple heuristic method for handling the inequality/equality constraints keeping the randomness of each individual. The position of each individual is modified by (4.3) in PSO. The resulting position of an individual is not always guaranteed to satisfy the inequality constraints because of over- under-velocity. If any element of an individual violates its inequality constraint because of over- under-speed, then the position of the individual is fixed to its maximum/minimum operating point as illustrated in Fig. 15.7, and this can be formulated as:

$$P_{ij}^{k+1} = \begin{cases} P_{ij}^{k+1} + v_{ij}^{k+1} & \text{if} \quad P_{ij,\min} \leq P_{ij}^{k+1} + v_{ij}^{k+1} \leq P_{ij,\max} \\ P_{ij,\min} & \text{if} \quad P_{ij}^{k+1} + v_{ij}^{k+1} < P_{ij,\min} \\ P_{ij,\max} & \text{if} \quad P_{ij}^{k+1} + v_{ij}^{k+1} > P_{ij,\max}. \end{cases} \tag{15.16}$$

Although the aforementioned method always produces the position of each individual satisfying the inequality constraints (15.3), the problem of equality constraint

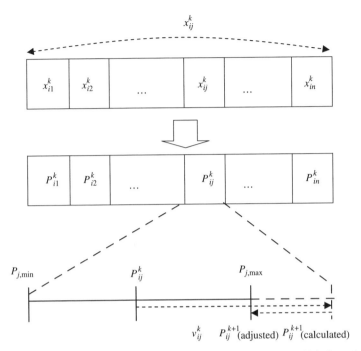

FIGURE 15.7 Adjustment strategy for an individual's position within boundary.

still remains to be resolved. Therefore, it is necessary to develop a new strategy such that the summation of all elements in an individual (i.e., $\sum_{j=1}^{n} P_{ij}^k$) is equal to the total system demand. To resolve the equality constraint problem without intervening the dynamic process inherent in the PSO algorithm, we propose the following heuristic procedures:

Step 1. Set $j = 1$. Let the present iteration be k.

Step 2. Select an element (i.e., generator) of individual i at random and store in an index array $A(n)$.

Step 3. Modify the value of element j using (4.1), (4.3), and (15.16).

Step 4. If $j = n-1$, then go to Step 5; otherwise $j = j + 1$ and go to Step 2.

Step 5. The value of the last element of individual i is determined by subtracting $\sum_{j=1}^{n-1} P_{ij}^k$ from D. If the value is not within its boundary, then adjust the value using (15.16) and go to step 6; otherwise go to step 8.

Step 6. Set $l = 1$.

Step 7. Readjust the value of element l in the index array $A(n)$ to the value satisfying the equality condition (i.e., $D - \sum_{j=1, j \neq l}^{n} P_{ij}^k$). If the value is within its boundary, then go to step 8; otherwise, change the value of element l using (15.16). Set $l = l + 1$, and go to step 7. If $l = n + 1$, go to step 1.

Step 8. Stop the modification procedure.

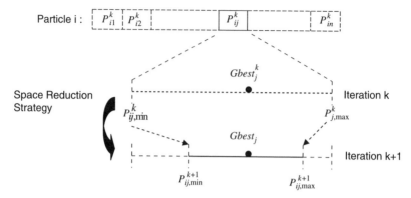

FIGURE 15.8   Schematic of dynamic space reduction strategy.

15.2.3.2   *Dynamic Space Reduction Strategy*

To accelerate the convergence speed to the solutions in PSO algorithms, Park et al. proposed the dynamic space reduction strategy [19]. This strategy is activated in the case when the performance is not increased during a prespecified iteration period. In this case, the search space is dynamically adjusted (i.e., reduced) based on the "distance" between the *Gbest* and the minimum and maximum output of generator j. To determine the adjusted minimum/maximum output of generator j at iteration k, the distance is multiplied by the predetermined step-size Δ and subtracted (added) from the maximum (minimum) output at iteration k as described in (15.17) and the mechanism of dynamic space reduction is illustrated as in Fig. 15.8:

$$
\begin{aligned}
P^{k+1}_{j,\max} &= P^k_{j,\max} - (P^k_{j,\max} - Gbest^k_j) \times \Delta \\
P^{k+1}_{j,\min} &= P^k_{j,\min} + (P^k_{j,\min} - Gbest^k_j) \times \Delta.
\end{aligned}
\tag{15.17}
$$

15.2.4   Numerical Example

15.2.4.1   *GA Implementation to ED with Smooth Cost Function*

In this section, we will briefly present the results from the genetic algorithms using the multistage method and directional crossover and compare those from the simple genetic algorithms [12] and the conventional lambda-iteration method [1].

For numerical tests, we have considered the ED problem with six generating units whose input data are given in Table 15.1.

The solutions obtained from the GAs using multistage and directional crossover are compared with those of simple GA and the conventional lambda-iteration method. Numerical tests are performed without considering the transmission loss for the total demand of 1680 MW. The three cases assumed are given in Table 15.2: case I, Goldberg's simple GA (SGA); case II, applying the directional

TABLE 15.1   Cost Coefficient of the Test System

	α_i	β_i	γ_i	$P_{i,\text{max}}$ (MW)	$P_{i,\text{min}}$ (MW)
U1	561.0	7.92	0.00156	600.0	150.0
U2	310.0	7.85	0.00194	400.0	100.0
U3	78.0	7.97	0.00482	200.0	50.0
U4	102.0	5.27	0.00269	500.0	100.0
U5	51.0	9.90	0.00172	350.0	40.0
U6	178.0	8.26	0.00693	280.0	100.0

TABLE 15.2   Basic Characteristics and Parameters in Each Case

Case	I (SGA)	II (SGA + DC)	III (SGA + DC + MS)
Crossover probability	0.9	0.9	0.9
Mutation probability	0.05	0.05	0.05
Encoding	6-bit (decimal)	6-bit (decimal)	2-bit (decimal)
Average convergence iteration number	400	400	210
DC probability	—	0.15	0.15

crossover (DC) strategy to the SGA structure; and case III , applying both the multi-stage (MS) method and the DC strategy to the SGA Based on empirical studies, various GA parameters are set and the resulting average iteration number in each case is given in Table 15.2.

The projection method to cope with the equality constraint [7] is applied for all the cases. In order to see the performance results, the population size is gradually increased and the results are summarized in Table 15.3. From Table 15.3, we can see that the results of cases II and III provide a better solution than the SGA does except in one place with a population size of 380.

Regarding the convergence, case III with an average of 210 iterations shows the best performance, and cases I and II show similar performance with an average of 400 iterations. Therefore, one can improve the performance of GAs by applying the various strategies or a combination of the strategies introduced in this section.

15.2.4.2 PSO Implementation to ED with Smooth/Nonsmooth Cost Function

In this section, MPSO [19] has been applied to ED problems where the objective functions can be either smooth or nonsmooth. The results obtained from the MPSO are compared with those of other methods: the numerical lambda-iteration method (NM) [1], the hierarchical numerical method (HM) [4], the GA [19], the tabu search (TS) [16], the evolutionary programming (EP) [14, 15], the modified Hopfield neural network (MHNN) [5], and the adaptive Hopfield neural network (AHNN) [17].

The MPSO is applied to an ED problem with three generators with the quadratic cost functions. Table 15.4 shows the input data of the system, and the system demand

TABLE 15.3 Objective Function Values in Each Case

Population Size	Case I	Case II	Case III
40	14924.9	14916.8	14901.0
60	14922.2	14906.4	14902.1
80	14906.2	14900.9	14911.2
100	14920.3	14902.1	14953.2
120	14909.6	14905.3	14897.7
140	14921.7	14906.4	14908.8
160	14907.3	14917.2	14901.6
180	14926.5	14900.4	14899.0
200	14924.4	14899.4	14902.1
220	14967.7	14897.9	14896.9
240	14925.7	14898.1	14911.6
260	14921.2	14899.5	14902.9
280	14911.3	14908.2	14903.1
300	14907.0	14904.9	14899.2
320	14909.2	14900.6	14896.6
340	14904.4	14902.8	14900.0
360	14924.7	14898.2	14897.4
380	14901.9	14905.1	14901.9
400	14936.9	14896.5	14904.8

The good performance is indicated in shadings.

considered is 850 MW. Table 15.5 shows the comparison of the results from MPSO, NM [1], improved evalutionary programming (IEP) [3], and MHNN [5], where the values of parameters are set as $\omega_{max} = 0.5$, $\omega_{min} = 0.1$, $c_1 = c_2 = 2$, and $\Delta = 0.31$.

As seen in Table 15.5, the MPSO has provided the global solution (the same result of the lambda-iteration method), exactly satisfying the equality and inequality constraints. In addition, the MPSO is applied to two ED problems with three generators, where valve-point effects are considered. The input data for the three-generator system are given in Table 15.6. Note that the values of parameters are set as $\omega_{max} = 0.5$, $\omega_{min} = 0.1$, $c_1 = c_2 = 2$, and $\Delta = 0.31$ as applied to the test system with quadratic cost functions.

Here, the total demand for the three-generator is set as 850 MW. It was reported in Ref. 16 that the global optimum solution found for the three-generator system is

TABLE 15.4 Cost Coefficient of Test System with Quadratic Cost Function

Unit	a_i	b_i	γ_i	$P_{i,min}$	$P_{i,max}$
1	561.0	7.92	0.001562	150.0	600.0
2	310.0	7.85	0.00194	100.0	400.0
3	78.0	7.97	0.00482	50.0	200.0

TABLE 15.5 Comparison of Simulation Results of Each Method

Unit	NM	MHNN	IEP (pop = 10)	MPSO (par = 10)
1	393.170	393.800	393.170	393.170
2	334.604	333.100	334.603	334.604
3	122.226	122.300	122.227	122.226
TP	850.000	849.200	850.00000	850.00000
TC	8194.35612	8187.00000	8194.35614	8194.35612

pop, population size; par, number of particles; TP, total power (MW); TC, total generation cost ($).

TABLE 15.6 Cost Coefficient of Test System with Valve-Point Effects

Unit	a_i	b_i	γ_i	e_i	f_i	$P_{i,min}$	$P_{i,max}$
1	561.0	7.92	0.001562	300.0	0.0315	100.0	600.0
2	310.0	7.85	0.00194	200.0	0.042	100.0	400.0
3	78.0	7.97	0.00482	150.0	0.063	50.0	200.0

TABLE 15.7 Comparison of Simulation Results of Each Method

Unit	GA	IEP (pop = 20)	EP	MPSO (par = 20)
1	300.00	300.23	300.26	300.27
2	400.00	400.00	400.00	400.00
3	150.00	149.77	149.74	149.73
TP	850.00	850.00	850.00	850.00
TC	8237.60	8234.09	8234.07	8234.07

pop, population size; par, number of particles; TP, total power (MW); TC, total generation cost ($).

$8234.07. The obtained results for the three-generator system are compared with those from GA [18], IEP [3], and EP [14] in Table 15.7. It shows that the MPSO has succeeded in finding the global solution, as presented in Ref. 16, by always satisfying the equality and inequality constraints.

Finally, the MPSO has also been applied to the ED problem with 10 generators where the multiple-fuel effects are considered. In this case, the objective function is represented as the piecewise quadratic cost function. The input data and related constraints of the test system are given in Refs. 3, 4, and 17. In this case, the total system demand is varied from 2400 MW to 2700 MW with 100-MW increments.

For these problems, parameters are $\omega_{max} = 0.5$, $\omega_{min} = 0.1$, $c_1 = c_2 = 2$, and $\Delta = 0.05$. The best results from the MPSO are compared with those of HM [4],

TABLE 15.8 **Comparison of Simulation Results of Each Method (Demand = 2400 MW)**

S	U	F	HM GEN	F	MHNN GEN	F	AHNN GEN	F	IEP (pop = 30) GEN	F	MPSO (par = 30) GEN
1	1	1	193.2	1	192.7	1	189.1	1	190.9	1	189.7
	2	1	204.1	1	203.8	1	202.0	1	202.3	1	202.3
	3	1	259.1	1	259.1	1	254.0	1	253.9	1	253.9
	4	3	234.3	2	195.1	3	233.0	3	233.9	3	233.0
2	5	1	249.0	1	248.7	1	241.7	1	243.8	1	241.8
	6	1	195.5	3	234.2	1	233.0	3	235.0	3	233.0
	7	1	260.1	1	260.3	1	254.1	1	253.2	1	253.3
3	8	3	234.3	3	234.2	3	232.9	3	232.8	3	233.0
	9	1	325.3	1	324.7	1	320.0	1	317.2	1	320.4
	10	1	246.3	1	246.8	1	240.3	1	237.0	1	239.4
TP			2401.2		2399.8		2400.0		2400.0		2400.0
TC			488.500		487.87		481.700		481.779		481.723

S, system; U, unit; F, fuel type; GEN, generation; TP, total power (MW); TC, total generation cost ($).

IEP [3], MHNN [5], and AHNN [17] and given in Tables 15.8 through 15.11. In this case, the global solution is not known, or it may be impossible to find the global solution with the numerical approach for piecewise quadratic cost functions.

As seen in Tables 15.8 to 15.11, the MPSO [19] has always provided better solutions than HM [4] (except for the 2600-MW case), IEP [3], and MHNN [5].

TABLE 15.9 **Comparison of Simulation Results of Each Method (Demand = 2500 MW)**

S	U	F	HM GEN	F	MHNN GEN	F	AHNN GEN	F	IEP (pop = 30) GEN	F	MPSO (par = 30) GEN
1	1	2	206.6	2	206.1	2	206.0	2	203.1	2	206.5
	2	1	206.5	1	206.3	1	206.3	1	207.2	1	206.5
	3	1	265.9	1	265.7	1	265.7	1	266.9	1	265.7
	4	3	236.0	3	235.7	3	235.9	3	234.6	3	236.0
2	5	1	258.2	1	258.2	1	257.9	1	259.9	1	258.0
	6	3	236.0	3	235.9	3	235.9	1	236.8	3	236.0
	7	1	269.0	1	269.1	1	269.6	1	270.8	1	268.9
3	8	3	236.0	3	235.9	3	235.9	3	234.4	3	235.9
	9	1	331.6	1	331.2	1	331.4	1	331.4	1	331.5
	10	1	255.2	1	255.7	1	255.4	1	254.9	1	255.1
TP			2501.1		2499.8		2500.0		2500.0		2500.0
TC			526.700		526.13		526.230		526.304		526.239

S, system; U, unit; F, fuel type; GEN, generation; TP, total power (MW); TC, total generation cost ($).

TABLE 15.10 Comparison of Simulation Results of Each Method (Demand = 2600 MW)

		HM		MHNN		AHNN		IEP (pop = 30)		MPSO (par = 30)	
S	U	F	GEN	F	GEN	F	GEN	F	GEN	F	GEN
1	1	2	216.4	2	215.3	2	215.8	2	213.0	2	216.5
	2	1	210.9	1	210.6	1	210.7	1	211.3	1	210.9
	3	1	278.5	1	278.9	1	279.1	1	283.1	1	278.5
	4	3	239.1	3	238.9	3	239.1	3	239.2	3	239.1
2	5	1	275.4	1	275.7	1	276.3	1	279.3	1	275.5
	6	3	239.1	3	239.1	3	239.1	1	239.5	3	239.1
	7	1	285.6	1	286.2	1	286.0	1	283.1	1	285.7
3	8	3	239.1	3	239.1	3	239.1	3	239.2	3	239.1
	9	1	343.3	1	343.5	1	342.8	1	340.5	1	343.5
	10	1	271.9	1	272.6	1	271.9	1	271.9	1	272.0
TP			2600.0		2599.8		2600.0		2600.0		2600.0
TC			574.030		574.26		574.370		574.473		574.381

S, system; U, unit; F, fuel type; GEN, generation; TP, total power (MW); TC, total generation cost ($).

TABLE 15.11 Comparison of Simulation Results of Each Method (Demand = 2700 MW)

		HM		MHNN		AHNN		IEP (pop = 30)		MPSO (par = 30)	
S	U	F	GEN	F	GEN	F	GEN	F	GEN	F	GEN
1	1	2	218.4	2	224.5	2	225.7	2	219.5	2	218.3
	2	1	211.8	1	215.0	1	215.2	1	211.4	1	211.7
	3	1	281.0	3	291.8	1	291.8	1	279.7	1	280.7
	4	3	239.7	3	242.2	3	242.3	3	240.3	3	239.6
2	5	1	279.0	1	293.3	1	293.7	1	276.5	1	278.5
	6	3	239.7	3	242.2	3	242.3	1	239.9	3	239.6
	7	1	289.0	1	303.1	1	302.8	1	289.0	1	288.6
3	8	3	239.7	3	242.2	3	242.3	3	241.3	3	239.6
	9	3	429.2	1	355.7	1	355.1	3	425.1	3	428.5
	10	1	275.2	1	289.5	1	288.8	1	277.2	1	274.9
TP			2702.2		2699.7		2700.0		2700.0		2700.0
TC			625.180		626.12		626.240		623.851		623.809

S, system; U, unit; F, fuel type; GEN, generation; TP, total power (MW); TC, total generation cost ($).

Furthermore, it has provided solutions satisfying the equality and inequality constraints, whereas HM [4] and MHNN [5] do not satisfy the equality constraint.

When compared with AHNN [17], the MPSO has provided a better solution for the demand of 2700 MW.

15.2.5 Summary

This section presents the fundamentals of GAs and PSO and their applications to ED problems in power systems. During the past decade, GAs have become one of the important tools for practical and robust optimization methods in power system problems including ED problems. However, to deal with the limitations of the algorithm, various improvements have been explored in terms of GAs implementation to ED such as encoding scheme, genetic operation variation, fitness function evaluation, and constraints handling.

Moreover, this study presents a new approach to nonsmooth ED problems based on the PSO algorithm. The strategies for handling constraints are incorporated in the PSO framework in order to provide the solutions satisfying the inequality constraints. The equality constraint in the ED problem is resolved by reducing the degree of freedom by one at random. The strategies for handling constraints are devised while preserving the dynamic process of the PSO algorithm. Additionally, the dynamic search-space reduction strategy is applied to accelerate the convergence speed.

Numerical examples are tested on the sample ED problem by using GAs, and the results demonstrate that GAs have shown considerable success in providing good solutions in the ED problem. In addition, we can obtain that the MPSO has provided the global solution satisfying the constraints with a very high probability for the ED problems with smooth cost functions. For the ED problems with nonsmooth cost functions due to the valve-point effects, the MPSO has also provided the global solution with a high probability for the three-generator system. In the case of the nonsmooth function problem due to multifuel effects, the MPSO has shown superiority to the conventional numerical method, the conventional Hopfield neural network, and the EP approach while providing very similar results with the MHNN.

15.3 MAINTENANCE SCHEDULING

15.3.1 Maintenance Scheduling Problem

Maintenance schedule is a preventive outage schedule in a power system that determines the timing and duration for scheduled planned maintenance overhauls for generating units in order to minimize annual fuel costs and maximize the reliability of the power system while satisfying all of the complex logical constraints. Maintenance scheduling becomes a complicated problem when the power system contains a large number of generating units with different specifications and when numerous constraints have to be taken into consideration to obtain a practical and feasible solution.

Maintenance scheduling determines the optimal starting time for each preventive maintenance outage in a weekly period for six months to one year in advance while satisfying various system constraints and ensuring system reliability.

The reliability of the power system ensures that the demand is met even if an outage occurs in the system. Generally, the utility provides a spinning reserve by allowing for some back-up energy production capacity that can be made available at a few minutes' notice. This improves the system's reliability at the expense of increased operating costs. To minimize the cost of operation as well as maintenance, a good maintenance schedule that considers the number of operating hours, the downtime of the units, the health of the engine by inspecting the specific fuel consumption and the crew as well as resource availability constraints is desirable. A well-planned schedule for maintenance and repair work should be prepared for each unit in the system, according to the period of maintenance activities, which are based on the operating hours and condition of the engines.

An efficient and practical maintenance schedule thus increases system reliability, reduces operation and maintenance costs, and extends the lifetime of the generators. Moreover, an easily revisable maintenance schedule is required to adapt to increasing load demand and number of generating units in a larger-scale power system.

Maintenance scheduling is essentially an optimization problem where the primary objective is to obtain the best schedule while meeting various hard and soft constraints. The objective functions in maintenance scheduling are reliability and cost. The economic objective function includes the maintenance cost and operation cost [1]. The reliability objective function can be either deterministic, where the aim is to maximize the system's net reserve (the system installed capacity minus the maximum load and the maintenance capacity during the period under examination), or random, where the goal is to minimize the risk level.

Several hard and soft constraints should be taken into account to reflect the actual operating conditions of the power system in order to make the maintenance plan feasible. Listed below are some constraints for maintenance scheduling:

1. Crew constraint.
2. Availability of resources.
3. Maintenance period during which each unit should be maintained.
4. Load demand.
5. Rated capacity.
6. Condition of the engine.
7. Spinning reserve requirement.

15.3.2 GA, PSO, and ES Implementation

The test data for this problem is taken from a real power system consisting of two industrial parks located in Bintan and Batam in Indonesia and comprising a total of 19 generating units. A planning horizon of 25 weeks is considered in the generator scheduling problem. In each week, a maximum of three generators can be taken off for maintenance due to crew and resource constraints. The maintenance of a generator must be done in consecutive weeks, and care must be taken to ensure that the solution provides a feasible schedule.

15.3.2.1 Optimization Function There are three main costs that need to be minimized in the optimal maintenance scheduling of generators: total operating cost, maintenance cost, and penalty cost. The following sections will look into the details of each type of cost as well as the constraints associated with each cost.

15.3.2.2 Total Operating Cost

The operating cost (*toc*) of a generator per hour can be approximated by the following equation:

$$toc = aP^2 + bP + c, \qquad (15.18)$$

where P is the MW output power of the generator with a, b, and c as constant coefficients.

Based on Table 15.12, the total operating cost based on the schedule obtained by the search process can be evaluated for those generators in operation during the 25-week schedule.

15.3.2.3 Maintenance Cost

In the schedule, the maintenance cost of each generator is determined by two main factors. The first factor is the downtime (D) of the generator or the maintenance duration in weeks. Another factor is the maintenance cost per week (V), which may vary for each generator. Table 15.13 shows the data used in the evaluation of maintenance cost.

15.3.2.4 Penalty Cost

There are two parts in the evaluation of the penalty cost:

1. *Inability to meet the power demand including the spinning reserve constraint.* The schedule should ensure that the total available power is greater

TABLE 15.12   Generator Ratings

Unit	Rating (MW)	a	b	c
1	6.1	0.0038	5.44	52.6
2	6.1	0.0038	5.44	52.6
3	6.1	0.0038	5.44	52.6
4	6.1	0.0038	5.44	52.6
5	6.1	0.0038	5.44	52.6
6	6.1	0.0038	5.44	52.6
7	6.1	0.0038	5.44	52.6
8	6.1	0.0046	6.34	53.7
9	6.1	0.0050	5.34	51.5
10	6.4	0.0050	5.34	51.5
11	6.4	0.0057	5.34	52.5
12	6.4	0.0057	5.34	52.5
13	8.0	0.0346	8.06	76.5
14	8.0	0.0346	8.06	76.5
15	2.1	0.0076	7.10	55.4
17	2.1	0.0076	6.95	55.4
18	2.1	0.0076	7.30	55.4
19	6.1	0.0079	7.10	59.3

TABLE 15.13 Downtime and Maintenance Cost per Week of Each Generator

Unit	Downtime (D)	V ($/week)
1	4	750
2	1	750
3	1	750
4	1	750
5	1	750
6	3	750
7	4	750
8	3	750
9	3	750
10	3	850
11	2	850
12	1	820
13	1	1500
14	2	1500
15	3	600
16	2	600
17	1	600
18	3	600
19	1	900

than the power demand for every week. In case a power outage occurs, a spinning reserve (S) of 15 MW is included as a precaution.

Hence, *PenaltyCost1* can be evaluated as follows:

$$P_a = P_c - S - P_y. \tag{15.19}$$

If $P_d > P_a$, then *PenaltyCost1* = $1,000,000; else *PenaltyCost1* = $0. P_a is the available power, P_c is the total installed capacity, P_y is the power capacity of the generator in maintenance based on the schedule generated, and P_d is the power demand.

Table 15.14 illustrates the forecast of load demand for a scheduling horizon of 25 weeks.

2. *Inability to meet crew and resources constraints.* Two important constraints in the maintenance scheduling problem are the number of maintenance crew and the availability of resources. The maximum number of generators that can be taken off for maintenance per week is limited to three in this study system. This is due to the fact that for one week only, a maximum crew of 12 personnel is provided in the power plant. The availability of resources specifies the equipment needed in the maintenance, which includes oil, filters, fuel pumps, and other engine components. From the experience of the power plant personnel, the resource constraint is represented with an integer number from 0 to 9.

TABLE 15.14   Load Demand Forecasting for 25 Weeks

T (weeks)	Demand (P_d)	Installed Capacity (P_c)	Spinning Reserve (S)	Gross Reserve
1	70.2	104.6	15	19.4
2	69.1	104.6	15	20.5
3	69.3	104.6	15	20.3
4	66.4	104.6	15	23.2
5	70.5	104.6	15	19.1
6	73.4	104.6	15	16.2
7	67.5	104.6	15	22.1
8	68.9	104.6	15	20.7
9	70.1	104.6	15	19.5
10	70.4	104.6	15	19.2
11	68.7	104.6	15	20.9
12	67.6	104.6	15	22.0
13	70.3	104.6	15	19.3
14	71.4	104.6	15	18.2
15	67.9	104.6	15	21.7
16	70.8	104.6	15	18.8
17	72.1	104.6	15	17.5
18	72.3	104.6	15	17.3
19	70.7	104.6	15	18.9
20	66.9	104.6	15	22.7
21	71.1	104.6	15	17.9
22	76.0	104.6	15	13.6
23	71.7	104.6	15	17.9
24	70.6	104.6	15	19.0
25	71.6	104.6	15	18.0

Resources are considered to be enough if assigned 9, whereas a shortage of resources is indicated by a number approaching 0. Table 15.15 illustrates the concept:

The crew and resource constraints are inputs to a fuzzy knowledge-based system. The output is a cost multiplier per week associated with these constraints. For example, the week with full availability of crew and resources will have a cost multiplier of zero. For a week with more constraints, the cost multiplier will increase correspondingly. The cost multiplier (m) is a number in the range of 0 to 3 depending on the availability of crew and resources.

If n represents the number of generators to be maintained in a respective week, *PenaltyCost2* can be approximated as follows:

If $n = 1$, *PenaltyCost2* $= m \times \$1000$

If $n = 2$, *PenaltyCost2* $= m \times \$1500$

If $n = 3$, *PenaltyCost2* $= m \times \$2000$

TABLE 15.15 Cost Multiplier for Crew and Resource Constraints

Week	Crews	Resources	Cost Multiplier
1	12	8	0.2445
2	12	8	0.2445
3	12	8	0.2445
4	12	9	0.0000
5	12	9	0.0000
6	10	9	0.3407
7	12	9	0.0000
8	8	7	1.0684
9	8	7	1.0684
10	12	7	0.4253
11	12	9	0.0000
12	12	9	0.0000
13	12	9	0.0000
14	10	9	0.3407
15	12	8	0.2445
16	8	8	0.8515
17	10	8	0.6831
18	12	6	0.5689
19	12	6	0.5689
20	12	9	0.0000
21	12	9	0.0000
22	12	9	0.0000
23	10	6	0.9689
24	8	6	1.2331
25	8	7	1.0684

15.3.2.5 *Overall Objective Function* The overall function to be minimized can be represented in a compact form as follows:

$$F = \sum_{t=1}^{T} \sum_{x=1}^{X} 168(a_x P_x^2 + b_x P_x + c_x) + \sum_{y=1}^{Y} V_y D + \sum_{t=1}^{T} PenaltyCost, \qquad (15.20)$$

where X is the unit in operation for that week, Y is the unit in maintenance, T is the length of the maintenance planning schedule (weeks), P_x is the generator output (MW) of operation unit, a_i, b_i, c_i are the fuel cost coefficients, V_y is the maintenance cost per week ($/week), and D is the downtime (weeks).

In a week, there are a total of 168 hours (24 hours $\times$ 7 days). Because the equation for the operating cost of the generator (15.18) is given on a per hour basis, a multiplier of 168 is required to get the total cost for operation in one week.

15.3.2.6 Problem Representation Proper care has to be taken in the initial random generation of the candidate solutions because of the following constraints:

1. Each generator should be taken off for maintenance in consecutive weeks according to its downtime.
2. In each week, the number of generators that can be maintained is limited to three due to resources and crew constraints.
3. A generator can only be taken off once within a week. Hence, there should be no repeated values within a respective week of the schedule, except for the number 0, where it represents no generator to be maintained.

After much consideration, a useful representation of the candidate solution is in the form of a two-dimensional matrix. The rows of the matrix represent the number of weeks in the schedule and the columns represent the index of the generators to be taken off for maintenance.

As an illustration, consider one candidate solution as follows:

12 18 0   In week 1, generator units 12 and 18 to be maintained
7 18 0   In week 2, generator units 7 and 18 to be maintained
18 7 15   And so on....

Take note that unit 18 has to appear in three consecutive weeks because its downtime is 3, whereas a "0" represents no generator to be taken off.

Pseudocode to generate a feasible candidate solution that satisfies the constraints is as follows:

1. Extract the number of generators (*num_gen*) to be maintained.
2. Extract the downtimes (*D*) associated with each generator.
3. Initially create a two-dimensional solution matrix of zeros.

   Row = num_weeks in schedule (25 weeks)
   Col = max generators that can be taken off (3)

4. Insert the generator units into the matrix according to the downtimes.

   counter = 1;
   while (counter < =num_gen)

   a. Look for zero entries in the solution matrix.
   b. Randomly select a row entry to start inserting generator index.
   c. Check if the index can be inserted in consecutive order (*D*).
   d. If yes, insert and increment counter; else repeat b and c end.

GA, PSO, and ES have been used as optimization techniques for solving the maintenance scheduling problem. The flowchart for the GA for maintenance scheduling is given in Fig. 15.9.

Figure 15.10 shows the flowchart for an ES-based approach for this generator maintenance scheduling problem, which uses mutation as the main search operator.

Further improvements to the algorithm were made by employing two heuristics to the standard ES algorithm for this problem, which have resulted in better schedules with lower cost.

1. A new variable called *multiple_mutations* is introduced in the generation of the offspring solution matrix to model the global search ability in the beginning and refined search toward the end.

2. Because selection is based on the best individuals, it is highly possible that there may be repeated solutions with the same fitness. In order to increase diversity and at the same time not to discard other potential solutions, these repeated solutions are discarded so that all the best individuals are distinct from each other to ensure diversity in the next mutation process.

The third algorithm uses PSO for this problem. In recent years, PSO has proved to be a viable method in function optimization [20, 21]. Although simple and elegant in nature, it is often difficult to apply PSO to a real problem with many constraints. The two main equations in PSO are the velocity vector and position vector associated with each particle, which although useful in function optimization, cannot be directly applied. Several modifications, as described below, have therefore been made to adapt the algorithm to effectively solve the maintenance scheduling problem. The pseudocode for this algorithm is as follows:

1. Initialize population with particles.
2. Calculate fitness for each particle.
3. For each particle, update its own *Pbest* value if there is improvement in its fitness.
4. For each particle, update its position by moving toward the *Gbest* particle or its own *Pbest* position.

From Eq. (4.1), one can infer that the particle will update its velocity of flying by either moving toward the *Gbest* particle or moving toward its own *Pbest* position that it stored in memory. Although PSO has no explicit crossover and mutation search operation, one can infer from (4.1) that PSO has combined these two concepts in one single operation.

Crossover operation is implicitly implemented with the *Gbest* particle sharing global information with the rest of the particles. Also, each particle can move toward its *Pbest* position in memory; hence, crossover processes within particle itself can take place, which may lead to a faster convergence.

The implicit mutation process is represented by the inertia weight w, which randomly updates the status of the particle.

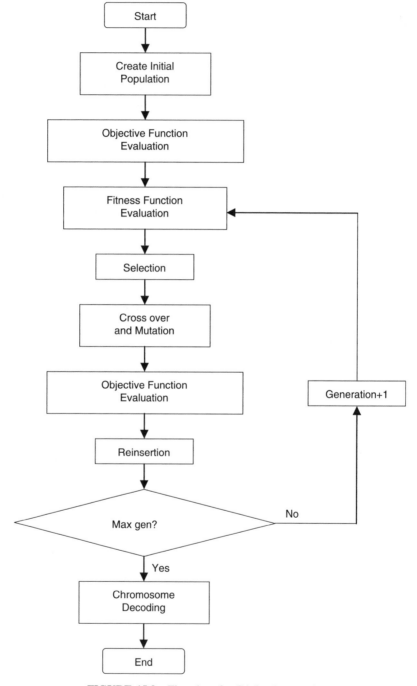

FIGURE 15.9 Flowchart for GA implementation.

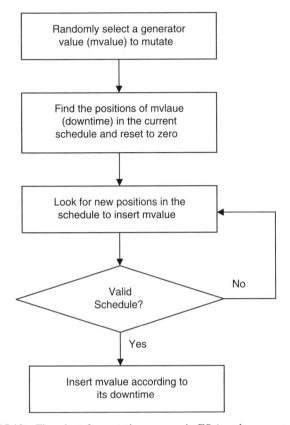

FIGURE 15.10 Flowchart for mutation process in ES (mvalue = mutation value).

A program to stimulate the crossover process in PSO has to be carefully written to obtain a feasible schedule that meets the various constraints described earlier. In addition, it should ensure that the following are satisfied:

1. Each generator should be taken off for maintenance in consecutive weeks according to its downtime.
2. In each week, the number of generators that can be maintained is limited to three.
3. A generator can only be taken off once within a week. Hence, there should be no repeated values within a respective week of the schedule, except for the number 0, where it represents no generator to be maintained.

The algorithm starts with two input schedules: *Gbest/Pbest* and *current* schedules, and for every generator in the schedule, it randomly extracts the maintenance weeks of that generator from either the *Gbest/Pbest* or *current* schedule. A new schedule is then created with the extracted information while all constraints are satisfied.

A variable *pmutate* (with probability set to 0.1) has been introduced to model the random mutation created by the inertia weight *w* in the PSO equation. The value has been set low so that most of the time, the schedule will be updated with crossover information from either *Gbest/Pbest* or *current* with low mutation rate. This means that in the event when no random mutation takes place, if *Gbest/Pbest* and *current* are exactly the same, the schedule will remain the same even after crossover (Fig. 15.11).

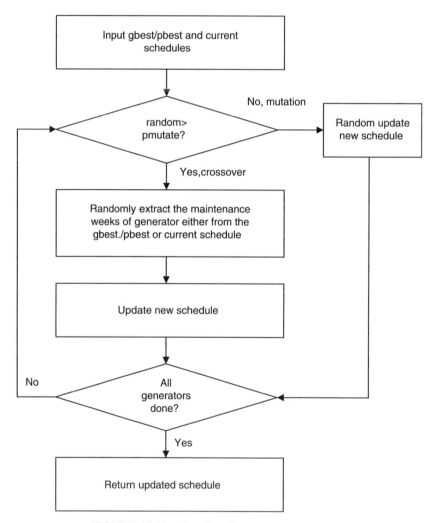

FIGURE 15.11 Flowchart for PSO implementation.

The following heuristics can be applied to increase the chance of finding a better schedule:

1. If the current schedule is the *Gbest* schedule itself, then randomly change the position of a generator value in the *Gbest* schedule. This heuristic refines the search within the current best solution.
2. Because of the fast convergence rate in PSO, a multiple change variable is introduced. This variable ensures thorough mixing of the candidate solutions by performing crossover and sometimes mutation over multiple times.

15.3.3 Simulation Results

The maintenance schedule is currently obtained by power plant engineers through a heuristic method based on trial and error. However, the schedule does not utilize the resources optimally. Furthermore, the engineers currently take approximately 1 week to prepare the maintenance schedule for 19 generating units for a 6-month planning horizon. In this section, the results from the three evolutionary approaches presented earlier are compared with the results obtained from the currently used heuristic approach for this problem. The fuzzy knowledge-based system is used as a common framework to represent all modes of imprecision in the generator scheduling problem. This fuzzy knowledge-based system evaluates the downtimes and cost multiplier values by considering the operating hours and crew and resource constraints, respectively. The results from the three optimization approaches are compared in this section. The results indicate that the schedules obtained by the PSO technique and ES result are superior to those by GA and heuristic approach.

The GA schedule was obtained with a maximum iteration of 400 with a population size of 100. For a fair comparison, the parameters for ES are set to be the same as GA with a mutation rate of 1. For PSO, the population size has no significant influence on the result and is usually small. A population size of 20 was chosen and the maximum iteration was set to 1000.

All the variants of PSO and ES obtained better results than GA. It was observed that although a pure PSO was time efficient, the average cost and schedule obtained are inferior when compared with PSO with multiple changes introduced. Similarly, ES with multiple mutations in the candidate solutions generally performed better than pure ES, which employs a single mutation in the candidate solutions. Modified ES with multiple mutations also resulted in greater diversity and minimum solution cost. Finally, another variant of ES that introduced crossover operation was tested with probability of crossover set to 0.1. The average cost of ES with crossover is generally lower as crossover increases the gene pool where offspring may benefit from the parents and in theory allows a more thorough search of the solution space. The results are given in Table 15.16.

TABLE 15.16 Simulation Results

Techniques	Average Cost Over 10 Runs (US$)	Optimized Cost (US$)	Approximate Time per Run
Heuristic	—	4,533,609.8487	1 week
GA	—	4,533,435.2287	40 min
PSO	4,496,457.2152	4,495,655.5439	23 min
PSO with multiple changes	4,496,217.4586	4,495,528.8839	52 min
ES	4,496,024.3566	4,495,625.0839	45 min
ES with multiple mutations	4,495,773.7109	4,495,511.2439	51 min
ES with multiple mutations and crossover	4,495,705.4489	4,495,625.0839	58 min

15.3.4 Summary

Maintenance schedule, which is the schedule to determine which generating units should be switched off, becomes a complicated problem due to the fact that there are many factors to be considered. This work considers various evolutionary techniques for solving this problem.

In this work, a fuzzy knowledge-based system evaluates the downtimes and cost multiplier values by considering the operating hours and crew and resource constraints, respectively. Three evolutionary techniques (GA, PSO, and ES) are then used to obtain the least cost of maintenance and operation while satisfying all constraints and at the same time improving system reliability by providing the spinning reserve and meeting the forecast load demands. It was observed that the evolutionary methods generated flexible maintenance schedules in a very short time. A comparison of results illustrates the potential of these approaches for obtaining optimum solutions to this problem for practical systems.

Because PSO and ES have been shown to perform better than GA, a hybrid approach to combine concepts from both PSO and ES is worth investigating. Different mutation rates for ES should also be investigated and compared with other evolutionary algorithms that use subpopulations.

15.4 COGENERATION SCHEDULING

Cogeneration systems [22] have been constructed and operated in many countries since the world oil crisis in the early 1970s. As a cogeneration system can supply both thermal and electric energies simultaneously, it is known as the total energy system. A cogeneration plant is usually built in an urban area in order to supply heat economically, so it has been used continuously as the distributed electric energy source in electric utilities. The main advantages of cogeneration are higher efficiency, quicker startup, lower environmental pollution, shorter construction time, and smaller site space. However, the facilities in the cogeneration system have to be well coordinated with the electric utilities to get more energy savings.

Many papers about cogeneration systems have been published. The modeling of cogeneration systems has been reported in Refs. 23 and 24. Cogeneration ED and planning have been presented in Refs. 25–28. The scheduling of a cogeneration system with auxiliary devices has been reported in Refs. [28–31].

Thus far, little work has been done on the optimal operation scheduling of cogeneration systems with heat storage tanks, auxiliary boilers, or batteries. This is an area of growing importance. However, the optimization process of scheduling such cogeneration systems using conventional linear or nonlinear programming techniques usually results in one of the many local minimum points; which are induced from operating so many different facilities in multiple time intervals, especially with heat storage, tanks and battery storage, which are time-lag facilities.

This section presents an application of a GA to solve such a problem. GA is an artificial intelligence method that is an optimization algorithm based on the mechanics of natural selections. The GA searches from a population of points and uses payoff (fitness or objective functions) information directly for the search direction, not derivatives or other auxiliary knowledge. GAs use probabilistic transition rules to select generations, not deterministic rules [32]. For the first time, the authors have used a simple GA for scheduling in a system with one load level only [33]. However, the simple GA was not good enough to give consistent results when we tried to solve the problem with any load levels. This section uses an improved GA (IGA) to solve such problems and gives consistent results for any reasonable load levels that are much better than the results from the simple GA. IGA uses only a fraction of the time as compared with GA to solve the same problem. The IGA given in this section is more suitable for practical cogeneration systems.

The objective of scheduling is to minimize the total operation cost while satisfying system constraints. This section deals with the operation scheduling problem on an industrial cogeneration system currently operated in Korea that consists of cogeneration units, auxiliary boilers, heat storage tanks, independent generators, and electricity chargers and batteries. It is a bottoming cycle cogeneration system, which mainly supplies thermal power but uses the remaining heat to produce electric power. The operational performance of the auxiliary devices is carefully studied. Different prices for buying and selling the electricity between the cogeneration system and electric utility have been considered. The efficiency equation for this cogeneration system is obtained by the least squares method from data taken over a period of 6 months.

A list of symbols used in this section is provided in Section 15.4.5.

15.4.1 Cogeneration Scheduling Problem

The cogeneration plant has two different units and four kinds of auxiliary devices. The energy flow of the proposed multicogeneration system is illustrated in Fig. 15.12. Electricity and heat flow through the electricity bus and the heat bus are simplified as a single bus.

In this section, the thermal and electrical loads during the operation period are given in the information from the industrial plants. The efficiencies of independent generators and auxiliary boilers are assumed to be constant. The electrical transmission loss generated between the electric utility and the industrial plant is charged to the owner of the cogeneration plants. Based on the above assumptions,

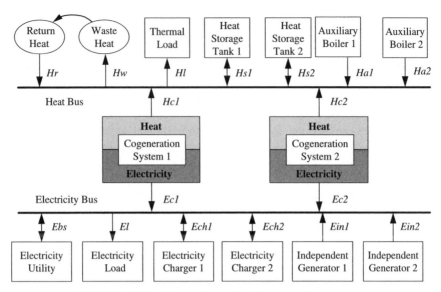

FIGURE 15.12 Energy flow of multicogeneration systems.

the objective function used to minimize the total operation is formulated. The total operation cost is the sum of fuel cost of the cogeneration system, auxiliary boilers, and independent generators, and the buying and selling cost of electricity between the cogeneration system and electric utility.

The objective function is represented by Eq. (15.21). The first term represents the buying and selling cost of electricity, and the second term is the total fuel cost of the cogeneration system, auxiliary boilers, and independent generator.

$$
\min \sum_{k=1}^{N} t \left(C_e(k) \begin{cases} \dfrac{1}{\gamma} E_{bs}(k) & \text{if } E_{bs}(k) > 0 \\ \gamma E_{bs}(k) & \text{otherwise} \end{cases} + \sum_{i=1}^{2} \dfrac{F_{ci}}{\eta_{ci}} H_{ci}(k) \right.
$$

$$
\left. + \sum_{i=1}^{2} \dfrac{F_{ai}}{\eta_{ai}} H_{ai}(k) + \sum_{i=1}^{2} \dfrac{F_{gi}}{\eta_{gi}} E_{ini}(k) \right)
$$

(15.21)

The system is subject to the following constraints:

$$
H_l(k) = \left(\sum_{i=1}^{2} (H_{ci}(k) + H_{ai}(k) - H_{hi}(k)) \right) - H_w(k) + H_r(k)
$$

$$
E_l(k) = \left(\sum_{i=1}^{2} (E_{ci}(k) + E_{ini}(k) - E_{chi}(k)) \right) + E_{bs}(k)
$$

(15.22)

$$
\underline{H}_{ci} \le H_{ci}(k) \le \overline{H}_{ci}
$$

$$
\underline{R}_{sei} H_{ci}(k) \le E_{ci}(k) \le \overline{R}_{sei} H_{ci}(k)
$$

$$\delta_{ai}\underline{H}_{ai} \leq H_{ai}(k) \leq \delta_{ai}\overline{H}_{ai}$$

$$\underline{E}_{ini} \leq E_{ini}(k) \leq \overline{E}_{ini},$$

where the over- and under-bars of the variables represent their maximum and minimum values; the efficiencies of cogeneration systems 1 and 2 vary according to the production of thermal and electric energy. Each efficiency was obtained as a quadratic equation by the least squares method as below:

$$\eta_{ci}(k) = \frac{1}{100}(a_i + b_i E_{ci} + c_i E_{ci}^2 + d_i H_{ci} + e_i H_{ci}^2). \tag{15.23}$$

The rules of electricity charging, heat transferred to the heat storage tank, and heat wasted and returned used in (15.22) are given in (15.24–15.28).

The electricity is charged according to the following equation:

$$E_{si}(k+1) = t\left\{\begin{matrix} \eta_{chi}E_{chi}(k) & E_{chi} > 0 \\ \dfrac{1}{\eta_{di}}E_{chi}(k) & \text{otherwise} \end{matrix}\right\} + E_{si}(k)(1 - \mu_{ei})^t \tag{15.24}$$

$$E_{si}(0) = E_{si0}.$$

The electric energy stored in the battery is restricted according to the following equation:

$$\underline{E}_{si} \leq E_{si}(k) \leq \overline{E}_{si}. \tag{15.25}$$

The thermal energy transferred to the heat storage tank is represented by the following equation:

$$H_{si}(k+1) = t\left\{\begin{matrix} \delta_{hi}H_{hi}(k) & H_{hi}(k) > 0 \\ \dfrac{1}{\delta_{hi}}H_{hi}(k) & \text{otherwise} \end{matrix}\right\} + H_{si}(k)(1 - \mu_{ti})^t \tag{15.26}$$

$$H_{si}(0) = H_{si0}.$$

The thermal energy stored in the heat storage tank is limited between the upper and lower boundaries and is shown in (15.27):

$$\underline{H}_{si} \leq H_{si}(k) \leq \overline{H}_{si}. \tag{15.27}$$

A part of the thermal energy produced in the cogeneration system dissipates as the thermal loss while the other part of the thermal energy, used as process heat, returns to the cogeneration system. Therefore, the waste heat and return heat can be represented by the following equation:

$$H_w(k) = \eta_w \left[H_l(k) + \sum_{i=1}^{2} (H_{hi}(k) - H_{ai}(k)) \right]$$

$$H_r(k) = \eta_r \left[H_l(k) + \sum_{i=1}^{2} (H_{hi}(k) - H_{ai}(k)) \right]. \tag{15.28}$$

15.4.2 IGA Implementation

GAs are search algorithms based on the mechanics of natural genetics and natural selection. The GA used here is the binary-coded one. Each control variable is encoded into a series of binary bits. Each bit simulates a gene. The series of binary bits of all control variables compose a string, which simulates a chromosome. Each chromosome represents a candidate solution consisting of electric power and thermal power in all intervals. GA is a population search method. A population of strings is kept in each generation. The next generation is produced by the simulation of the natural process of reproduction, gene crossover, and mutation. GA belongs to random search algorithms, but it is not a simple random walk. It effectively uses the information of the current population to direct the next search.

To make GA practical in the real-life large-scale systems, IGA is given in this section. The procedure of IGA is briefly as follows:

Initialization: The initial population of strings, s_i, $i = 1, 2, \ldots, m$, where m is the population size, is randomly selected in the binary-coded domain of control variables. Each s_i will be decoded back to the control variables to compute its fitness score f_i.

Statistics: The maximum fitness, minimum fitness, sum of fitnesses, and average fitness of this generation are calculated as follows:

$$\bar{f} = \left\{ f_i \middle| f_i \geq f_j \quad \forall f_j, j = 1, \ldots, m \right\}$$

$$\underline{f} = \left\{ f_i \middle| f_i \leq f_j \quad \forall f_j, j = 1, \ldots, m \right\}$$

$$f_\Sigma = \sum_{i=1}^{m} f_i \tag{15.29}$$

$$f_{\text{avg}} = \frac{f_\Sigma}{m}.$$

Reproduction: The strings of the same number as the population size will be copied into a "mating pool" according to their fitness values. The higher the fitness value, the greater the number of copies the string would probably have in the "mating pool." A simulated weighted roulette wheel is used to select mates. Each string in the current population has its sector slot in the wheel. The ratio of slot area of s_i to the whole wheel area is the ratio of f_i / f_Σ. Simulated m coin tosses on the roulette wheel select m mates.

Mutation: Every string in the "mating pool" may be mutated with the given mutation probability. For the string that is undergoing the mutation, a number will be randomly selected from a uniformly distributed [0, 1] domain. If this number is less than the mutation probability, the bits in a string will be changed from 1 to 0, or vice versa. The mutation operator produces a new string.

In general GAs, mutation probability is fixed throughout the whole search processing. However, in practical applications, a small fixed mutation probability could result in premature convergence, whereas the search with a large fixed mutation probability will not converge. An adaptive mutation is given in IGA to solve the problem as follows:

$$
p_m(k+1) = \begin{cases} p_m(k) - p_{m\text{Step}} & \text{if } \bar{f}(k) \text{ unchanged} \\ p_m(k) & \text{if } \bar{f}(k) \text{ increased} \\ p_{m\text{Final}} & \text{if } p_m(k) - p_{m\text{Step}} < p_{m\text{Final}} \end{cases} \tag{15.30}
$$

$$
p_m(0) = p_{m\text{Init}},
$$

where k is the generation number, and $p_{m\text{Init}}$, $p_{m\text{Final}}$, and $p_{m\text{Step}}$ are fixed numbers. $p_{m\text{Init}}$ would be around 1 and $p_{m\text{Final}}$ would be 0.005. $p_{m\text{Step}}$ depends on the maximum generation number.

Crossover: The strings in the "mating pool" are grouped in couples. Each couple of strings swap their bits according to the crossover probability. The uniform crossover is used here, which is better than one-point crossover and two-point crossover. The crossover operation would happen if a number for crossover, also randomly selected from a uniformly distributed [0, 1] domain, is less than the given crossover probability. A mask string is set up with randomly selected bits. The bits of the couple strings that correspond to 1's bits of the mask string will swap, whereas others will stay.

After all strings finish mutation and crossover, a new generation is reproduced. Each string is decoded back to the control variables to compute its fitness score. The statistics will compute the new maximum fitness, minimum fitness, sum of fitnesses, and average fitness. The convergence condition will be checked. The program will stop if the condition is met; otherwise, a new cycle of reproduction, mutation, and crossover will start.

For the sake of both speed and memory, IGA codes each variable in a fixed bit-length substring. For the cogeneration study, a 4-bit substring is used for any control variables. There are 15 steps of change for each variable ($2^4 - 1 = 15$). It would be long enough for some discrete variables. However, the length would not be enough to code some continuous variable. A dynamic hierarchy of coding system is developed in IGA to code the large number of control variables in a real-life system with a reasonable-length string without losing the resolution of the result. The search process will be divided into several stages with different change-steps for control variable vector V as follows:

Search stage 1

$$R_1 = [\underline{V}, \overline{V}]$$

$$V_{s1} = \frac{\overline{V} - \underline{V}}{15}. \tag{15.31}$$

Search stage 2

$$R_2 = \begin{cases} [V_{op1} - V_{s1}, \overline{V}] & \text{if } V_{op1} + V_{s1} > \overline{V} \\ [\underline{V}, V_{op1} + V_{s1}] & \text{if } V_{op1} - V_{s1} < \underline{V} \\ [V_{op1} - V_{s1}, V_{op1} + V_{s1}] & \text{otherwise} \end{cases} \tag{15.32}$$

$$V_{s2} = \frac{2V_{s1}}{15},$$

where V_{s1} and V_{s2} are change steps for the variable vector V in stage 1 and stage 2, respectively; V_{op1} is the control variable vector after the first stage search; and R_1 and R_2 are the search range for stages 1 and 2, respectively. The process would have several stages. However, two stages are enough in the cogeneration problem. In the cogeneration problem presented in this section, there are nine independent variables, that is, control variables, $\{H_{c1}, H_{c2}, H_{a1}, H_{a2}, E_{c1}, E_{c2}, E_{in1}, E_{in2}, E_{bs}\}$, and eight time intervals, so there are a total of 72 variables that will be accommodated in a 288-bit string. The population size is 30, so there will be 30 such strings in the process.

The objective of the cogeneration problem is to minimize the objective function. The minimization of the objective of cogeneration has to be changed to the maximization of fitness to be used in the simulated roulette wheel as follows:

$$fitness_i = \overline{f} - f_i, \tag{15.33}$$

where $\overline{f}$ is calculated in statistics.

After all strings finish mutation and crossover, a new generation is reproduced. Each string is decoded back to the control variables to compute its fitness score. The statistics will compute the new maximum fitness, minimum fitness, sum of fitness, and average fitness. The convergence condition will be checked. The program will stop if the condition is met; otherwise, a new cycle of reproduction, mutation, and crossover will start.

15.4.3 Case Study

The simulated cogeneration plant is shown schematically in Fig. 15.12. There are two cogeneration systems, two auxiliary boilers, two heat storage tanks, two independent generators, and two electricity chargers in the plant. The plant supplies both thermal and electric loads and is connected to the electricity utility. An eight-time-interval scheduling is used for the daily operation of the system, with a 3-hour period in each interval. The cogeneration system parameters are listed in Table 15.17. Three cases with different load levels are studied in the section. The light, base, and heavy loads and electricity buying-selling prices are given in Table 15.18. The coefficients of the efficiency equation for the cogeneration systems are given in Table 15.19, which are obtained by the least squares method based on the measurement data from the cogeneration plant of Buksan Energy Company, Korea. The electricity buying-selling prices are also provided by the same company.

TABLE 15.17 Cogeneration System Parameters

$\overline{E}_{s1}$	5.0	H_{s20}	1.0	$\overline{H}_{c1}$	25.0	μ_{e2}	0.05	δ_{a1}	0.98
$\underline{E}_{s1}$	1.0	$\overline{H}_{a1}$	5.0	$\overline{H}_{c2}$	30.0	η_{a1}	0.7	δ_{a2}	0.98
$\overline{E}_{s2}$	6.0	$\overline{H}_{a2}$	7.0	F_{a1}	19.55	η_{a2}	0.7	δ_{h1}	0.98
$\underline{E}_{s2}$	1.0	$\underline{H}_{a1}$	1.0	F_{a2}	19.55	η_{g1}	0.5	δ_{h2}	0.98
E_{s10}	1.0	$\underline{H}_{a2}$	1.0	F_{c1}	19.55	η_{g2}	0.5	$\overline{R}_{se1}$	0.85
E_{s20}	1.0	$\overline{E}_{in1}$	1.0	F_{c2}	19.55	η_w	0.3	$\underline{R}_{se1}$	0.65
$\overline{H}_{s1}$	6.0	$\overline{E}_{in2}$	1.0	F_{g1}	25.75	η_r	0.15	$\overline{R}_{se2}$	0.9
$\underline{H}_{s1}$	1.0	$\underline{E}_{in1}$	5.0	F_{g2}	25.75	η_{ch1}	0.9	$\underline{R}_{se2}$	0.7
$\overline{H}_{s2}$	7.0	$\underline{E}_{in2}$	7.0	μ_{t1}	0.05	η_{ch2}	0.9	t	6.0
$\underline{H}_{s2}$	1.0	$\overline{H}_{c1}$	5.0	μ_{t2}	0.05	η_{d1}	0.9	γ	0.98
H_{s10}	1.0	$\underline{H}_{c2}$	7.0	μ_{e1}	0.05	η_{d2}	0.0		

TABLE 15.18 Loads and Electricity Buying-Selling Prices

	k	0	1	2	3	4	5	6	7
Light	$H_l(k)$	43.06	45.06	50.22	50.22	47.33	47.33	45.06	43.06
Load	$E_l(k)$	30	42	52	55	52	50	42	33
Base	$H_l(k)$	48.06	50.06	55.22	55.22	54.33	52.33	50.06	48.06
Load	$E_l(k)$	40	50	60	60	62	55	50	45
Heavy	$H_l(k)$	53.06	55.10	57.18	58.22	57.22	55.20	55.10	53.06
Load	$E_l(k)$	45	55	63	67	67	58	53	48
$C_e(k)$	$C_e(k)$	31.54	31.54	82.44	82.44	82.44	82.44	58.21	58.21

TABLE 15.19 Coefficients for Cogeneration

	System Efficiency Equation				
i	a_i	b_i	c_i	d_i	e_i
1	50.0	0.0083	−0.00000029	0.90	−0.0032
2	49.0	0.0085	−0.00000031	0.85	−0.0038

The initial feasible variable settings (provided by the company) of the light load level case are listed in Tables 15.20 and 15.21. The units for all variables are MW, except for $H_{s1}(k)$, $H_{s2}(k)$, $E_{s1}(k)$, and $E_{s2}(k)$, whose units are MWh. The optimal variable-setting results for the same load level after the IGA search are given in Tables 15.22 and 15.23. The variable settings of the other two cases are omitted due to limited space. The production costs for all three cases, plus the costs for buying electricity from the electric utility and minus the profits for selling electricity to it, are given in Table 15.24. The savings are also given in Table 15.24.

The population size is 30, and each string is 288-bits long. The maximum generation is given as 200. The convergence tolerance is given as the difference between the maximum fitness and minimum fitness being less than 0.001 of the average fitness. IGA usually converges between 100 and 150 generations. The constraint functions are dealt with penalty functions. The penalty loops for the violations of the constraints are 5, 7, and 10 for the light, base, and heavy loads, respectively, with the penalty factors increased after each loop. The CPU time is 3 minutes 12 seconds, 4 minutes 17 seconds, and 5 minutes 28 seconds for three load levels, respectively, when a 166-MHz Pentium processor is used.

The simple GA used in Ref. 33, without the improvement mentioned in this section, has been used to compute these three loads, too. The population size is the same, but the string is 576-bits long. The maximum generation is given as 400, and the convergence is the same as above. However, in most loops, the procedures have to reach the maximum generation to stop. The penalty loops for the violation of the constraints are 9, 12, and 17 for these three load levels, respectively. The CPU times are 10 minutes 37 seconds, 14 minutes 52 seconds, and 20 minutes 48 seconds each in the same computer. The cost savings are also given in Table 15.24. Compared with the results the IGA has achieved, it can be seen that IGA has greatly enhanced the performance of GA.

With the IGA approach, it can be seen that very attractive savings for the multiple time interval scheduling of the daily operation of the two cogeneration systems and multiauxiliary devices have been achieved for different loads. However, it should be noted that the optimal allocation of thermal and electrical power depends on the load. It can also be seen that during the light load period, with a higher ability to regulate, a higher saving (34.38%) has been achieved.

15.4.4 Summary

This section has proposed the application of an IGA to the multiple time interval scheduling for cogeneration system operation for different loading conditions. The IGA could easily solve such complex scheduling problems with multi-variables in the multi-time intervals. The simulation results show a high potential for the proposed cogeneration model and the operation scheduling method to be applied to the industrial multi-cogeneration system, which usually has several cogeneration units and multi-auxiliary devices. It is believed that the same approach could be used for the trigeneration case [34].

TABLE 15.20 Initial Thermal Power Results for Light Load Level Case

k	$H_{c1}(k)$	$H_{c2}(k)$	$H_{a1}(k)$	$H_{a2}(k)$	$H_{h1}(k)$	$H_{s1}(k+1)$	$H_{h2}(k)$	$H_{s2}(k+1)$	$H_w(k)$	$H_r(k)$
0	25	14	4.9	6.86	1.7492	6.0	0.8639	3.3971	10.1739	5.0870
1	25	15	4.9	6.86	0.2911	6.0	1.1915	6.4157	10.4348	5.2174
2	25	19	4.9	6.86	−0.1991	4.5347	0.0	5.5007	11.4783	5.7391
3	25	20	4.9	6.86	0.6704	5.8590	0.0	4.7161	11.7491	5.8696
4	25	17	4.9	6.86	0.3322	6.0	0.6195	5.8649	10.9565	5.4783
5	25	16	4.9	6.86	0.0822	5.3858	0.0	5.0285	10.6957	5.3478
6	25	14	4.9	6.86	0.4702	6.0	0.1429	4.7313	10.1739	5.0870
7	25	11	4.9	6.86	0.0043	5.1570	0.0	4.0565	9.3913	4.6957

TABLE 15.21 Initial Electrical Power Results for Light Load Level Case

k	$E_{c1}(k)$	$E_{c2}(k)$	$E_{in1}(k)$	$E_{in2}(k)$	$E_{bs}(k)$	$E_{ch1}(k)$	$E_{s1}(k+1)$	$E_{ch2}(k)$	$E_{s2}(k+1)$
0	21.25	12.6	5	7	−13	1.5343	5.0	1.3157	4.4098
1	21.25	13.5	5	7	−4	0.2641	5.0	0.4859	5.0927
2	21.25	17.1	5	7	2	0.2641	5.0	0.0859	4.5982
3	21.25	18	5	7	4	0.25	4.9619	0.0	3.9424
4	21.25	15.3	5	7	4	0.2762	5.0	0.2738	4.1193
5	21.25	14.4	5	7	3	0.2641	5.0	0.3859	4.5737
6	21.25	12.6	5	7	−3	0.2641	5.0	0.5859	5.5032
7	21.25	9.9	5	7	−10	0.15	4.6919	0.0	4.7183

TABLE 15.22 Optimal Thermal Power Results for Light Load Level Case

k	$H_{c1}(k)$	$H_{c2}(k)$	$H_{a1}(k)$	$H_{a2}(k)$	$H_{h1}(k)$	$H_{s1}(k+1)$	$H_{h2}(k)$	$H_{s2}(k+1)$	$H_w(k)$	$H_r(k)$
0	23.218	30	0	0	1.7492	6.0	1.4674	5.1714	13.8830	6.9415
1	23.2773	29.3274	0	0	0.2911	6.0	0.3922	5.5870	13.7230	6.8615
2	25	29.8911	0	0	-1.3538	1.0	-1.1348	1.3163	14.3194	7.1597
3	24.7256	29.8713	4.79862	0	1.7492	6.0	0.3050	2.0253	14.2427	7.1213
4	24.6719	28.0253	0	0	-1.3538	1.0	-0.1525	1.2694	13.7471	6.8736
5	24.7808	29.9333	0	0	0.2474	1.5848	0.0	1.0884	14.2732	7.1366
6	23.1219	29.4372	0	1.3334	1.5787	6.0	0.3983	2.1043	13.7111	6.8555
7	23.0276	28.8382	0	0	0.2911	6.0	1.7496	6.9480	13.5302	6.7651

TABLE 15.23   Optimal Electrical Power Results for Light Load Level Case

k	$E_{c1}(k)$	$E_{c2}(k)$	$E_{in1}(k)$	$E_{in2}(k)$	$E_{bs}(k)$	$E_{ch1}(k)$	$E_{s1}(k+1)$	$E_{ch2}(k)$	$E_{s2}(k+1)$
0	19.7353	27	0	0	−13.403	1.5343	5.0	1.7981	5.7122
1	19.7857	26.3947	0	0	−5.86732	−0.9861	1.0	−0.7007	2.5617
2	21.25	26.902	3.93556	0	−0.28487	0.0428	1.0	−0.2401	1.3961
3	21.0168	26.8842	4.51259	5.1144	−0.81059	1.5343	5.0	0.1830	1.6912
4	20.9711	25.2228	4.99983	7	−5.90296	0.2641	5.0	0.0266	1.5219
5	21.0636	26.94	4.6869	4.5255	−6.84391	0.2641	5.0	0.108	1.5964
6	19.6536	26.4935	0	6.6762	−10.0149	0.2641	5.0	0.5443	2.8384
7	19.5734	25.9544	0	0	−13.4763	−0.9485	1.1253	0.0	2.4335

TABLE 15.24   Production Costs for All Three Load Level Cases

		Light Load	Base Load	Heavy Load
Production cost ($)	Initial	50744.4	59120.4	61225.9
	Optimal	33299.7	51003.8	57598.7
Savings (%)		34.38	13.73	5.92
Savings (%) with GA in Ref. 12 for comparison		15.56	8.17	3.35

15.4.5 Nomenclature

15.4.5.1 Thermal and Electric Variables

H_{ci} heat output from cogeneration system (MW)

H_l heat sending to thermal load (MW)

H_{hi} heat transferring to heat storage tank (MW)

H_{ai} heat output from auxiliary boiler (MW)

H_w waste heat (MW)

H_r return heat (MW)

E_{ci} electricity output from cogeneration system (MW)

E_{bs} electric power buying from or selling to electric utility (MW) ($+$: buying from utility; $-$: selling to utility)

E_l electric power sending to load (MW)

E_{chi} electricity charging to battery (MW) ($+$: charging; $-$: discharging)

E_{ini} electricity outputs from independent generator (MW)

E_{si} electric energy stored in battery (MWh)

H_{si} heat energy stored in heat storage tank (MWh)

15.4.5.2 Data

N total number of time intervals

F_{ci} fuel cost of cogeneration system ($/MWh)

F_{ai} fuel cost of auxiliary boiler ($/MWh)

F_{gi} fuel cost of independent generator ($/MWh)

C_e price of electricity buying from or selling to the electricity utility ($/MWh)

t length of each time interval (h)

15.4.5.3 Parameters

$\overline{R}_{sei}$ maximum ratio of steam to electricity of cogeneration system

$\underline{R}_{sei}$ minimum ratios of steam to electricity of cogeneration system

γ transmission efficiency between the electric utility and the cogeneration plant

η_{ci} efficiency of cogeneration system

η_{ai} efficiency of auxiliary boiler

η_{gi} efficiency of independent generator

η_w waste heat constant

η_r return heat constant

η_{chi} charge efficiency of battery

η_{di} discharge efficiency of battery

μ_{ei} natural discharge rate

μ_{ti} heat leakage rate

δ_{ai} heat transfer efficiency of auxiliary boiler

δ_{hi} heat transfer efficiency of heat storage tank

15.5 SHORT-TERM GENERATION SCHEDULING OF THERMAL UNITS

In order to supply high-quality electric energy to the consumer in a secure and economic manner, electric utilities face many economic and technical problems in operation, planning, and control of electric energy systems. One of the major problems is to determine the most economic and secure way of short-term generation scheduling and dispatch under given constraints. Various approaches were proposed for solving the short-term generation scheduling problems [35].

15.5.1 Short-Term Generation Scheduling Problem

The main objective of the short-term generation scheduling problem is to determine the output of thermal units so as to obtain a minimum total cost over a period of 24 hours subject to a set of constraints that arise from the system security requirements and restrictions on the operation of the units. The objective function to be minimized can be written as

$$\text{Min } F = \sum_{j=1}^{T} \left[\sum_{i=1}^{G} f_i(P_{ij}) + C(j,k) \right], \tag{15.34}$$

where G is the number of generating units, T is the time horizon of interest (24 hours), which is divided into 24 stages with each stage being 1 hour, f_i is the operation cost function for the ith unit, which can be defined as

$$f_i = a_i + b_i P_{ij} + c_i P_{ij}^2 \tag{15.35}$$

where a_i, b_i, and c_i are the coefficients that are determined by the characteristics of the ith unit, P_{ij} is the real power output of the ith unit in the jth stage, $C(i,k)$ is the additional cost relating to transition and penalty,

$$C(j, k) \in \{C_T(j, k), C_P(j, k)\},$$

where $C_T(j, k)$, $C_P(j, k)$ are transition and penalty costs from stage j to stage k. The constraints that the problem (15.34) are subject to are

(i) *Real power balance constraint:*

$$\sum_{i=1}^{G} P_{ij} = P_{Dj} + P_{Lj} \quad j = 1, 2, \ldots, T, \tag{15.36}$$

where P_{Dj} and P_{Lj} are the total demand and transmission loss in the area at the jth stage.

(ii) *Real power operating limits of generating units:*

$$P_i^{\min} \le P_{ij} \le P_i^{\max} \quad i = 1, 2, \ldots, G, \quad j = 1, 2, \ldots, T \tag{15.37}$$

where $P_i^{\min}$, $P_i^{\max}$ are the minimum and maximum real power outputs of the ith unit.

(iii) *Spinning reserve constraint:*

$$\left(\sum_{i=1}^{G} u_{ij} P_i^{\max} - P_{Dj} \right) \Big/ P_{Dj} \ge 0.1 \quad j = 1, 2, \ldots, T, \tag{15.38}$$

where $u_{i,j}$ is the status index of the ith unit at the jth stage (1 for up and 0 for down).

(iv) *Minimum uptime of units:*

$$(u_{ij} - u_{i,j-1})(w_{ij} - \tau h_i) \le 0 \quad i = 1, 2, \ldots, G, \quad j = 1, 2, \ldots, T, \tag{15.39}$$

where τh_i is the minimum uptime of the ith unit and

$$w_{ij} = u_{ij}(w_{i,j-1} + 1). \tag{15.40}$$

(v) *Minimum downtime of units:*

$$(u_{ij} - u_{i,j-1})(q_{ij} - \tau l_i) \le 0 \quad i = 1, 2, \ldots, G, \quad j = 1, 2, \ldots, T, \tag{15.41}$$

where τl_i is the minimum downtime of the ith unit and

$$q_{ij} = (1 - u_{ij})(q_{i,j-1} + 1). \tag{15.42}$$

(vi) *Maximum operating time of units:*

$$u_{ij}(v_{i,j-1} - \tau u_i) \le 0 \quad i = 1, 2, \ldots, G, \quad j = 1, 2, \ldots, T, \tag{15.43}$$

where τu_i is the maximum operating time of the ith unit and

$$v_{ij} = u_{ij}(v_{i,j-1} + 1). \tag{15.44}$$

To consider all the constraints mentioned above, the generation scheduling problem could be expressed in the form of a dynamic process as

$$F_j({}^jU^l) = \text{Min}\left\{\Phi_j^l({}^{j-1}U^k, {}^jU^l)\right\} \quad j = 1, 2, \ldots, T \tag{15.45}$$

subject to constraints (i)–(vi), where

$$\Phi_j^l({}^{j-1}U^k, {}^jU^l) = \sum_{i=1}^{G} f_i(P_{ij}) + \sum_{i=1}^{G} SC_i(q_{i,j-1}) + \sum_{i=1}^{G} DC_i$$
$$+ C_p(k,l) + F_{j-1}({}^{j-1}U^k). \tag{15.46}$$

Equation (15.45) is the minimal total operational cost to arrive at the state $({}^jU^l)$ from $({}^{j-1}U^k)$. In (15.46), the first, second, and third terms represent the total production fuel cost of a state and start-up and shut-down costs of units, respectively. The fourth term represents the penalty cost imposed when any of the transition constraints are violated, and the last term is the minimum total accumulated cost to reach to the state $({}^jU^l)$ from the initial stage. The constraints represented by (i)–(vi) will be treated in different ways. The operational constraints (i), (ii), and (iii) are handled using a conventional economic load dispatch module for each state while the search space is formed, and the transition constraints (iv), (v), and (vi) will be considered during the process of state transition by the dynamic programming (DP)-based conventional method to get reference results and the ant colony search algorithm (ACSA)-based technique to obtain final solution results. The penalty cost will also be applied for the violated transition constraints in the same process. Here, the solution procedure should be slightly modified so that the ACSA can be adopted easily.

15.5.2 ACSA Implementation

The ACSA for solving the general scheduling problem (GSP) is one of ACSAs described in Chapter 5. It can be described as follows:

- Build a construction graph that is similar to the one for the traveling salesman problem [36] for the GSP.
- m ants are initially positioned on n states chosen according to some initialization rules.
- Each ant builds a solution incrementally by repeatedly applying the state transition rule.

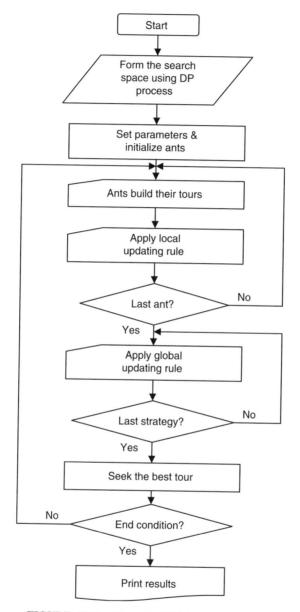

FIGURE 15.13 Flow of ACSA-based technique.

- While constructing its solution, an ant changes the amount of pheromone on the visited edges by applying the local updating rule.
- Once all ants have completed their solutions, the amount of pheromone is modified again by applying the global updating rule.

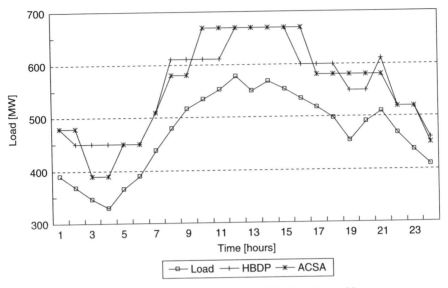

FIGURE 15.14 Comparison of scheduled total capacities.

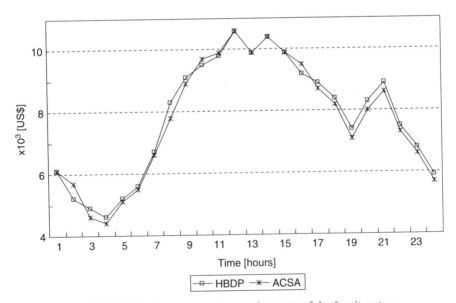

FIGURE 15.15 Comparison of generation costs of the 6-unit system.

- Seek the best solution using the solution process, in which ants are guided in building their solutions by both heuristic information and by pheromone information. An edge with a high amount of pheromone is a very desirable choice.
- The pheromone updating rules are designed so that they tend to give more pheromone to edges that should be visited by ants.

The overall flow of the proposed ACSA-based technique for GSP is briefly given in Fig. 15.13.

15.5.3 Experimental results

This proposed algorithm deals with a 24-hour generation scheduling or allocation problem. The test system is a 6-unit model power system. The system data are given in Ref. 37.

Figure 15.14 shows the scheduled total capacity of the 6-unit system obtained by the two algorithms, hybrid dynamic programming (HBDP) and ACSA. The difference in the generation cost is illustrated in Fig. 15.15, and the total generation costs of the two algorithms are $187,116.70 for HBDP and $184,841.50 for ACSA, respectively. All the results show that ACSA can achieve almost the same results as the ones obtained by the HBDP.

15.6 CONSTRAINED LOAD FLOW PROBLEM

15.6.1 Constrained Load Flow Problem

The constrained load flow (CLF) problem deals with the off-line adjustment of the power system control variables, such as the transformer tap-settings, and VAr (Volt-Ampere-Reactive) compensation blocks, in order to achieve optimum voltage values at the nodes of a power system subject to the physical and operating constraints.

The load flow problem can be expressed by the following two nonlinear equations:

$$Y = g(X, U) \tag{15.47}$$

$$Z = h(X, U) \tag{15.48}$$

where Y is a vector of nodal power injections, Z is a vector of constrained variables (power flows, reactive powers of PV (Power-Voltage) buses, etc.), X is a state vector (voltage angles and magnitudes), and U is a control vector (transformer taping settings, shunt compensation, voltage and power at PV buses, etc.).

The objective of the constrained load flow is to maintain some or all elements of X and Z vectors within given operating limits under the uncertainty of generating units' availability and load uncertainty.

15.6.2 Heuristic Ant Colony Search Algorithm Implementation

The challenge for the methods to solve this problem is that the methods should be able to provide optimal off-line settings of control variables for the entire planning period. A number of traditional algorithms, such as the Jacobian matrix-based algorithms [38, 39], the evolutionary computation techniques [40, 41] and optimal power flow method [42–45], are inefficient in accomplishing these requirements. More recently, the probabilistic CLF formulation [46] and heuristic reinforcement learning (RL) method [47–50] have been used for tackling the CLF problem with some success. In this case study, we focus on the application of a heuristic ACSA to solve the CLF problem [55]. The results obtained by using the ACSA will be compared with the ones obtained by using the Q-learning algorithm.

15.6.2.1 *Problem Formulation*   The ACSA formulates the CLF problem as a combinatorial optimization problem, $\zeta = (S, f, \Omega)$, which can be described as follows:

- A set of control variables, such as the transformer tap-settings and the VAr compensation blocks, are represented by U_i with discrete values $u_i \in D_i = \{d_1^i, d_2^i, \ldots, d_{|D_i|}^i\}$, $i = 1, 2, \ldots, n$, where d_j^i, $1 \leq j \leq |D_i|$, is a candidate setting for variable U_i, the subscript $|D_i|$ indicates the number of such candidate settings for the variable U_i, and n is the number of control variables in the CLF problem;
- A set, Ω, of constraints represented by the upper and lower bounds of control variables.
- An objective function

$$f = \frac{1}{p} \sum_{j=1}^{p} \left| \frac{2z_j - z_{j,\max} - z_{j,\min}}{z_{j,\max} - z_{j,\min}} \right| \tag{15.49}$$

where p expresses the number of constrained variables, such as the power flows, reactive power of PV buses; z_j is the value of jth constrained variable; and $(z_{j,\min}, z_{j,\max})$ are its lower and upper limits, respectively. All the constrained variables are combined as a vector—the vector of constrained variables, Z, which can be related to the state vector X and control vector U by (15.48). Especially, the objective function takes a similar formulation to the Q-learning reward function in the RL method for the ease of comparison of results.

The set of all feasible settings of control variables, as shown below, forms the search space.

$$S = \{s = \{(U_1, u_1), (U_2, u_2), \ldots, (U_n, u_n)\} | u_i \in D_i, i \in [1, n], \text{s.t. } \Omega\} \tag{15.50}$$

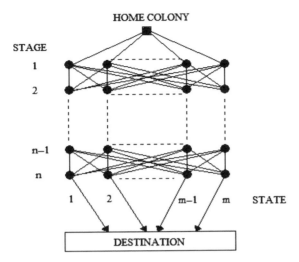

FIGURE 15.16   Construction graph (AS-graph) for the CLF problem.

15.6.2.2 *Construction Graph (AS-graph)*

The above combinatorial optimization problem is mapped on the construction graph for the CLF problem, which we refer to as the AS-graph in this work. The AS-graph is shown in Fig. 15.16, in which each control variable U_i, is represented by one stage; in other words, the ith stage corresponds with the ith control variable where $i = 1, 2, \ldots, n$, and n is the number of control variables. All possible candidate discrete settings for the ith control variable form the set D_i, each element in this set represents a state at the ith stage in the AS-graph. Here m is set to be the maximum number of elements of $D_j, j = 1, 2, \ldots, n$, for a better performance. Each ant will start its tour from HOME COLONY and stop at DESTINATION.

15.6.2.3 *ACSA for the CLF Problem*

An operating point comprising a load and generation pattern (operating point of the whole planning period of the system) is randomly created. For this operating point, first of all the AS-graph is created, and all paths receive an amount of pheromone that corresponds with an estimation of the best solution so that the ants test all paths in the initial iterations.

Then, ant k chooses the next states to go to in accordance with the transition probability calculated by

$$p(r, s) = \frac{\gamma(r, s)}{\sum_l \gamma(r, \ell)} \quad s, \ell \in N_r^k, \tag{15.51}$$

where matrix $\gamma(r, s)$ represents the amount of the pheromone trail, pheromone intensity, between points r and s.

When ant k moves from one stage to the next, the state of each stage will be recorded in a location list J^k. After the tour of ant k is completed, its location list is used to compute its current solution. Then the pheromone trails composed by nodes of location list J^k are updated by using

$$\gamma(r, s) = \alpha \cdot \gamma(r, s) + \Delta\gamma^k(r, s), \tag{15.52}$$

where α with $0 < \alpha < 1$ is the persistence of the pheromone trail, $(1 - \alpha)$ with representing the evaporation and $\Delta\gamma^k(r, s)$ the amount of pheromone that ant k puts on the trail (r, s). For the purpose of this research, the pheromone update $\Delta\gamma^k(r, s)$ is chosen as

$$\Delta\gamma^k(r, s) = \frac{1}{Q \cdot f}, \tag{15.53}$$

where f is the objective function and Q is a large positive constant.

In order to exploit the iteration in finding the best solution, the next two steps are considered:

(a) When all ants complete their tours, load flow is run and the objective function (15.49) is calculated for each run. Then, the pheromone trails (r, s) of the best ant tour [ant with minimum objective function (15.49)] is updated (global update) as:

$$\gamma(r, s) = \alpha \cdot \gamma(r, s) + \frac{R}{f_{best}} \quad r, s \in J^k_{best}, \tag{15.54}$$

where R is a large positive constant. Both Q in (15.53) and R are arbitrarily large numbers. Empirical tests have shown that the ACSA converges faster when Q is almost equal to R.

(b) To avoid search stagnation (the situation where all the ants follow the same path, that is, they construct the same solution [51]), the allowed range of the pheromone trail strengths is limited to:

$$\gamma(r, s) = \begin{cases} \tau_{min} & \text{if } \gamma(r, s) \le \tau_{min} \\ \tau_{max} & \text{if } \gamma(r, s) \ge \tau_{max} \end{cases}. \tag{15.55}$$

In this case study, the limits are chosen as

$$\tau_{max} = \frac{1}{\alpha \cdot f_{Gbest}}, \tag{15.56}$$

where f_{Gbest} is the global best solution (best over the whole past iterations), and

$$\tau_{min} = \frac{\tau_{max}}{M^2}, \tag{15.57}$$

where M is the number of ants.

The procedure is repeated for a large number of operating points covering the whole planning period. Once we have the set of optimal control settings for a large number of operating points, the one that minimizes the sum of multiobjective function (*mtf*) over the whole planning period is defined as a greedy-optimum control setting:

$$
mtf = \min\left\{\sum_{i=1}^{T} f_i\right\},
\tag{15.58}
$$

where T is the number of operating points sampled in the entire planning period.

The execution steps of the ACS-based algorithm applied to the CLF problem can be described as follows:

1. Create the AS-graph (search space) that represents the discrete settings (states) of the control variables (stages);

2. Insert the pheromone matrix $\gamma(m, n)$ according to nodes of AS-graph, where n is the number of stages and m the number of states.

3. Initialize the pheromone matrix $\gamma(m, n) = \gamma_0(m, n) = \tau_{max}$ (in (15.56), in this case f_{gBest} is an initial estimation of the best solution).

4. Repeat for a given number of operating points over the whole planning period.

   ▷ Repeat until the system convergence or iteration is less than a given maximum number.

   • Place randomly M ants on the states of the 1st stage ($i = 1$).

   • For $k=1$ to M

     • For $i = 2$ to n

       • When the ant-k has selected the r-state of the (i-1)-stage, it currently chooses the s-state of the (i)-stage in which will move according to transition rule (15.51).

       • Move the ant-k to s-state of i-stage.

       • Record s to J^k, and set $r = s$.

   • Run power flow for each ant

   • Calculate the objective function (15.49) for each ant.

   • Update the pheromone of (r, s)-trails for each

ant, using the local pheromone update formulae (15.50), (15.51).

- Update the pheromone of (r, s)-trails belonging to best ant tour (f_{best}), using the pheromone update formula (15.54).
  - Enforce the limits (15.55)-(15.57) in order to avoid the ants stagnations.
- Enforce each of the best control settings over the whole planning period and calculate (15.58).

5. Choose as a greedy-optimum control setting the one that minimizes (15.58).

15.6.3 Test Examples

Example 15.1 Reactive Power Control Problem for the IEEE 14-Bus System

A. Test System

The ACS algorithm is applied to adjust reactive control variables in the IEEE 14-bus test system as shown in Fig. 15.17.

The test system consists of the slack bus (node 1), 3 PV (nodes 2, 3, and 6), 10 PQ (Active-Reactive Power) buses, and 20 branches. The network data and load probabilistic data are the same as used in Ref. 46. They comprise six discrete distributions for the active load (at nodes 3, 6, 9, 10, 11, and 14), four discrete distributions for the reactive load (at nodes 9, 10, 11, and 14), with three to five impulses each and eight normal distributions for active and reactive loads at the remaining buses. The total

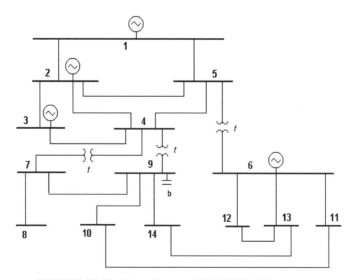

FIGURE 15.17 Line diagram of the IEEE 14-bus system.

**TABLE 15.25   Limits and Discretization of Actions and
Limits of Constrained Variables**

Control Actions	Umin	Umax	Step
t56	0.90	1.05	0.01
t49	0.90	1.05	0.01
t47	0.90	1.05	0.01
b9	0.00	0.24	0.03

installed capacity is equal to 4.9 p.u. (per-unit) and comprises 14 capacitor banks at node 1, 4 banks at node 2, 2 banks at node 3, and 2 banks at node 6. The voltage at all PV buses is taken equal to 1.0 p.u. and the slack bus voltage equal to 1.02 p.u. A fixed network topology is assumed. The control variables comprise all transformer taps (t) and reactive compensation (b) at bus 9 (Fig. 15.17).

Table 15.25 shows the limits of the control variables $Umin$ and $Umax$ and the discrete steps in variation. The transformer taps ($t56$, $t49$, and $t47$) are in 16 steps, and the reactive compensation ($b9$) is in nine steps. Therefore, the last step of $b9$ ($b9 = 0.24$) is repeated for the next seven steps, for steps 10 through 16. This makes the pheromone matrix $\gamma(m, n)$ of the AS-graph well-defined for all stages and states. Table 15.26 shows the upper and lower limits of all constrained variables $Wmax$ and $Wmin$.

B. Implementation of ACSA

For the IEEE 14-bus test system of this example, $T = 41$ operating points are selected from the entire planning period. These full correlated operating points are sampled

**TABLE 15.26   Limits and Discretization of Actions and
Limits of Constrained Variables**

Constrained Variables	Wmin	Wmax
Qg2	0.00	0.30
Qg3	0.00	0.70
Qg6	0.00	0.45
T23	0.00	0.75
T56	0.00	0.50
V4	0.96	1.05
V5	0.96	1.05
V7	0.96	1.05
V8	0.96	1.05
V9	0.96	1.05
V10	0.96	1.05
V11	0.96	1.05
V12	0.96	1.05
V13	0.96	1.05
V14	0.96	1.05

TABLE 15.27   ACS Parameters

Parameter	Value	Note
M	100	Number of ants in the colony.
N	4	Number of control variables, $t56$, $t49$, $t47$, and $b9$.
M	16	Number of states for all control variables. $b9$ has nine states, use 16 states for better performance.
Q	1,000,000	An arbitrary large number in (15.53).
R	1,000,000	An arbitrary large number in (15.54).
A	0.9865	Persistence of pheromone trails. Our experience shows that any value in the range [0.88–0.999] works well.

uniformly from the curves of normal and discrete distribution probabilities as follows:

$$\text{Load step} = \left(\mu \pm k \times \frac{3\sigma}{20} \right) \cdot 100\%, \tag{15.59}$$

where $k = 0, 1, 2, 3, \ldots, 20$. The μ and σ are the average values and the standard deviations for normal distributions of loads given in Ref. 52 and can be calculated using the formulae given by Ref. 53.

The ACS parameters used in the implementation are listed in Table 15.27.

The search will be terminated if one of the following criteria is satisfied: (a) the number of iterations since the last change of the best solution is greater than 1000 iterations, or (b) the number of iterations reaches 3000 iterations.

C. Performance

Performance of the ACSA is illustrated by the convergence curve (i.e., the best objective function values versus the iteration number during the ACS procedure). Figure 15.18 depicts the convergence curves of the ACS algorithm in a minimum value of (15.49), achieving the optimum control settings for each of the three operating points corresponding with the heavy, light, and nominal loads. The nominal load corresponds with the mean values of the load. The heavy and light loads correspond with the 1% confidence limit that all load values are lower and higher than these values, respectively.

Among the 41 optimal control settings, the *greedy-optimum* control settings are those that provide the minimum total function (15.58). Test results show that the greedy-optimum control settings that achieve the minimum total function (*mtf*) (15.58) over the whole planning period are the optimal control settings obtained for the nominal load (Fig. 15.18a). In this case, convergence of the ACSA took 1730 iterations.

Table 15.28 shows the results including the *mtf* value, the greedy-optimum control settings, and the operating space of constrained variables, obtained by the ACS-based algorithm with the results of the Q-learning (RL) [50] and the probabilistic CLF method [46], obtained for the same network.

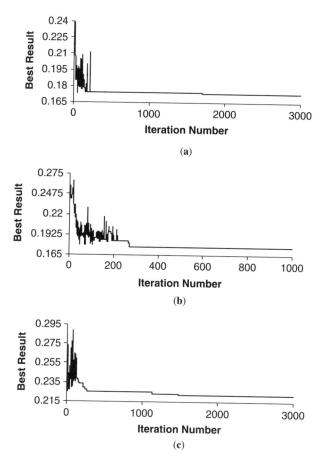

FIGURE 15.18 Performance of the ACSA in the (a) nominal, (b) heavy, and (c) light loads.

It can be seen that even with the greedy-optimum control settings in, reactive production at node 2 (Qg2) violates its limits in the ACS procedure. The Q-learning algorithm [50] provides slightly worse results, as Qg2 violates its limits and the voltage at node 14 violates its lower limit, too. The absolute value of maximum total reward (*mtr*) [50] in this case was calculated at 0.804, which is greater than the corresponding index of the ACSA (*mtf* = 0.732). The Q-learning algorithm [50] took about 38,800 iterations to find the greedy-optimum control settings. In the case of probabilistic CLF [46], the upper limit of reactive production at node 2 (Qg2) and the lower limit of voltage at node 14 as well as the upper limit of the apparent flow (T23) in lines 2–3 are violated.

In order to fully achieve the objective of the constrained load flow, which is to maintain some or all elements of *X* and *Z* vectors when given operating limits for the entire planning period, the violation of limits of Qg2 has to be fixed. One possible

TABLE 15.28 Comparison of Results Obtained by Using the ACS, RI, and Probabilistic Load Flow on the IEEE 14-Bus System

	ACSA		Q-Learning Algorithm		Probabilistic Load Flow	
	$mtf = 0.732$		$mtr = -0.804$			
Control Variables	Greedy-Optimal Settings		Greedy-Optimal Settings (a^*)		Optimal Settings	
t56	1.01		1.03		0.94	
t49	0.91		0.97		0.97	
t47	0.99		0.90		0.98	
b9	0.18		0.12		0.12	
Constrained Variables	Wmin	Wmax	Wmin	Wmax	Wmin	Wmax
Qg2	−0.5508*	−0.0795*	−0.5250*	−0.0527*	0.2069	0.3160*
Qg3	0.1554	0.5978	0.1751	0.6183	0.6420	0.6812
Qg6	0.0688	0.3606	0.0891	0.3787	0.3065	0.4161
T23	0.2273	0.6847	0.2284	0.6868	0.7223	0.7799*
T56	0.0731	0.2920	0.1675	0.4528	0.4144	0.4931
V4	0.9777	0.9984	0.9742	0.9951	0.9654	0.9731
V5	0.9876	1.0042	0.9864	1.0030	0.9682	0.9744
V7	0.9864	1.0235	0.9901	1.0274	0.9710	0.9857
V8	0.9864	1.0235	0.9901	1.0274	0.9710	0.9833
V9	0.9871	1.0316	0.9840	1.0284	0.9656	0.9833
V10	0.9819	1.0237	0.9775	1.0193	0.9644	0.9803
V11	0.9903	1.0139	0.9831	1.0067	0.9828	0.9912
V12	0.9899	1.0011	0.9802	0.9915	0.9907	0.9936
V13	0.9818	0.9994	0.9725	0.9903	0.9832	0.9878
V14	0.9600	1.0036	0.9541*	0.9977	0.9508*	0.9663

*Indicates the value exceeded a limit.

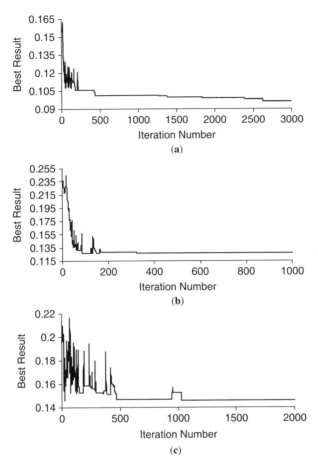

FIGURE 15.19 Performance of the ACSA when Qg2 is cut off in the (a) nominal, (b) heavy, and (c) light loads.

way is to relax the constant voltage limit at node 2 by considering it as a PQ bus and allowing the voltages at nodes 6 and 1 to be set at 1.021 p.u. and 1.03 p.u., respectively [46]. Rerunning the ACSA under these new considerations, the best solutions are obtained and shown in Fig. 15.19 for the three operating points corresponding with the nominal, heavy, and light loads, respectively. Among them, the greedy-optimum control settings that achieve the *mtf* (15.58) over the whole planning period are once more the optimal control settings obtained for the nominal load (Fig. 15.19a). In this case, convergence of the ACSA took 2645 iterations.

Table 15.29 shows results similar to the ones shown in Table 15.28 under the new consideration. In the case of the Q-learning algorithm [50], the corresponding absolute value (*mtr*) was calculated at 0.565, which is almost equal to *mtf* (*mtf* = 0.557). It must be underscored that the Q-learning algorithm took about 42,800 iterations to find the greedy-optimum control settings [50].

TABLE 15.29 Comparison of Results Between ACS, RI, and Probabilistic Load Flow on the IEEE 14-Bus System (Qg2 is Cutoff)

	ACSA		Q-Learning Algorithm		Probabilistic Load Flow	
	mtf = 0.557		mtr = −0.565		—	
Control Variables	Optimal Settings		Greedy-Optimal Settings (a*)		Optimal Settings	
t56	0.99		0.99		0.94	
t49	0.90		0.91		0.97	
t47	0.99		0.97		0.98	
b9	0.06		0.03		0.12	
V6	1.021		1.021		1.021	
V1	1.030		1.030		1.030	
Constrained Variables	Wmin	Wmax	Wmin	Wmax	Wmin	Wmax
Qg3	0.0132	0.5475	0.0100	0.5595	0.5120	0.5695
Qg6	0.0434	0.3770	0.0462	0.3794	0.3066	0.4214
T23	0.2182	0.6758	0.2180	0.6767	0.7063	0.7665*
T56	0.0332	0.2577	0.0330	0.2569	0.4126	0.4936
V4	0.9788	1.0060	0.9772	1.0044	0.9726	0.9822
V5	0.9898	1.0128	0.9888	1.0118	0.9763	0.9843
V7	0.9822	1.0241	0.9891	1.0311	0.9797	0.9954
V8	0.9822	1.0241	0.9891	1.0311	0.9797	0.9954
V9	0.9801	1.0284	0.9807	1.0289	0.9749	0.9933
V10	0.9780	1.0230	0.9786	1.0235	0.9737	0.9903
V11	0.9938	1.0191	0.9940	1.0192	0.9926	1.0013
V12	0.9997	1.0112	0.9998	1.0112	1.0008	1.0038
V13	0.9904	1.0086	0.9905	1.0086	0.9932	0.9979
V14	0.9600	1.0058	0.9603	1.0062	0.9605	0.9765

*Indicates the value exceeded a limit.

TABLE 15.30 Operating Space of Constrained Variables when the Criterion is Minimum VAr Compensation at Bus 9

Actions	Optimal Settings			
	$k = -2\ (\mu - 6\sigma/20)$		$k = -1\ (\mu - 3\sigma/20)$	
t56	1.01		0.99	
t49	0.94		0.92	
t47	0.93		0.93	
b9	0.00		0.00	
V6	1.021		1.021	
V1	1.03		1.03	
Constrained Variables	Wmin	Wmax	Wmin	Wmax
Qg3	0.0179	0.6201	0.0967	0.6377
Qg6	0.0628	0.4530	0.0871	0.3778
T23	0.2278	0.6858	0.2289	0.6878
T56	0.1113	0.2875	0.0556	0.2580
V4	0.9739	0.9944	0.9709	0.9914
V5	0.9876	1.0039	0.9837	1.0000
V7	1.0052	1.0421	1.0062	1.0431
V8	1.0052	1.0421	1.0062	1.0431
V9	0.9845	1.0280	0.9876	1.0312
V10	0.9817	1.0228	0.9843	1.0254
V11	0.9957	1.0189	0.9970	1.0202
V12	1.0000	1.0111	1.0003	1.014
V13	0.9911	1.0085	0.9915	1.0090
V14	0.9627	1.0056	0.9647	1.0076

The ACS and Q-learning algorithms provide the optimal results, rather than the near-optimal results given by the probabilistic CLF method [54], as all constraints including the upper limit of apparent flow (T23) on lines 2–3 are satisfied.

A key advantage of the proposed ACSA is its flexibility in providing control actions that can satisfy additional criteria and thus solve multicriteria optimization problems. For example, if the cost of VAr compensation should be taken into account, then as greedy-optimal action, the action that minimizes compensation at node 9 could be chosen.

Table 15.30 shows the operating space of constrained variables when the new optimal settings are enforced over the whole planning period. The convergence of ACSA to optimal actions in the case of minimum VAr compensation corresponding with operating points $k = -2$ and $k = -1$ takes 2563 and 2021 iterations, respectively. These results show that the ACSA provides control settings for the whole planning period and can be more effective than the probabilistic CLF method [46] as it satisfies constraints with minimum VAr compensation.

Example 15.2   Reactive Power Control Problem for the IEEE 136-Bus System

A. Test System

The ACSA is also applied to the reactive power control problem for the IEEE 136-bus system. This system consists of 136 buses (33 PV and 103 load buses), 199 lines, 24 transformers, and 17 reactive compensations.

The control variables selected comprise voltages at PV buses 4 and 21 (discrete variations 0.99 to 1.02, in step 0.01), taps at transformers 28, 41, and 176 (discrete variation of 0.92 to 1.00, in steps of 0.02), and reactive compensation (*b*) at buses 3 and 52 (discrete variation of six blocks). The total number of actions is $5^2 \times 5^3 \times 6^2 = 112,500$. The constrained variables include voltages at all PQ buses (from 0.96 to 1.05 p.u.) and three apparent power-flows at the most heavily loaded lines, 156 and 177 (upper limit 4.6 p.u.) and 179 (upper limit 3.4 p.u.).

B. Implementation of ACSA

For the IEEE 136-bus test system, $T = 41$ operating points are selected from the entire planning period similar to the IEEE 14-bus case. The ACS parameters used in the implementation and the termination criteria are the same as those in the IEEE 14-bus case.

C. Performance

The initial control settings violate the power flow limits of all the three most heavily loaded lines (156, 177, and 179), and upper limit of the voltages of buses 18, 19, and 23. The ACSA learns the greedy-optimal control action resulting in the satisfaction of the limits of constrained variables over the whole planning period as shown in Table 15.31.

TABLE 15.31   Results of ACS and Q-Learning Algorithms on the IEEE 136-Bus

	ACSA		Q-Learning Algorithm	
	$mtf = 0.628$		$mtr = -0.640$	
Control Actions	Optimal Settings		Greedy-Optimal Settings (a*) (p.u.)	
V4	1.01		1.00	
V21	0.99		0.99	
t28	0.94		0.92	
t41	0.92		0.92	
t176	0.92		0.92	
b3	0.17		0.15	
b52	0.17		0.17	
Constrained Variables	*Wmin*	*Wmax*	*Wmin*	*Wmax*
V18	0.990	1.027	0.987	1.021
V19	0.998	1.029	0.998	1.028
V23	1.012	1.050	1.010	1.049
T156	3.986	4.500	3.987	4.500
T177	3.948	4.501	3.948	4.501
T179	2.561	3.223	2.675	3.234

*Indicates the value exceeded a limit.

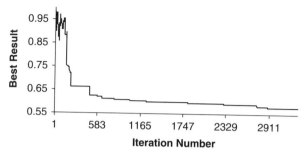

FIGURE 15.20 Performance of the ACSA in the average values of loads of IEEE-136 bus system.

The ants found the optimum control action at the average load values after about 2910 iterations as shown in Fig. 15.20 in contrast with 112,980 iterations of the Q-learning algorithm. The total computing time is about 8 seconds on a 1.4-GHz Pentium IV PC, compared with the 160 seconds achieved by Q-learning [50].

15.6.4 Summary

In this study case, the ACS approach is applied to the solution of the CLF problem. The CLF problem is modified as a combinatorial optimization problem. An iterative ACSA is implemented, providing the optimal off-line control settings over a planning period satisfying all operating limits of the constrained variables. Our algorithm consists of mapping the solution space, expressed by an objective function of the combinatorial optimization CLF problem on the space of control settings, where artificial ants walk. Test results show that the ACSA can be used to find optimum solutions within a reasonable time. The results of the proposed algorithm are also compared with those obtained by the Q-learning and probabilistic CLF methods. ACS and the Q-learning algorithms provide the optimal results, rather than the near-optimal results provided by the probabilistic CLF method [54]. A key advantage of the proposed ACSA is its flexibility in providing control actions that can accommodate additional criteria and thus solve multicriteria optimization problems (e.g., assigning priorities to control actions). The main advantage of the ACSA in comparison with the Q-learning algorithm is the better results in a far less number of iterations.

REFERENCES

1. Wood AJ, Wollenberg BF. Power generation, operation and control. New York: John Wiley & Sons; 1984.

2. Chowdhury BH, Rahman S. A review of recent advances in economic dispatch. IEEE Trans Power Systems 1990; 5(4):1248–1259.

3. Park YM, Won JR, Park JB. A new approach to economic load dispatch based on improved evolutionary programming. Eng Intelligent Systems Electrical Eng Commun 1998; 6(2):103–110.

4. Lin CE, Viviani GL. Hierarchical economic dispatch for piecewise quadratic cost functions. IEEE Trans Power Apparatus Systems 1984; 103(6):1170–1175.

5. Park JH, Kim YS, Eom IK, Lee KY. Economic load dispatch for piecewise quadratic cost function using Hopfield neural network. IEEE Trans Power Systems 1993; 8(3):1030–1038.

6. Walters DC, Sheble GB. Genetic algorithm solution of economic dispatch with valve point loading. IEEE Trans Power Systems 1993; 8(3): 325–1332.

7. Park JB, Won JR, Park YM, Lee KY. Economic dispatch solutions using an improved genetic algorithm based on the multi-stage method and directional crossover. Eng Intelligent Systems Electrical Eng Commun 1999; 7(4):219–225.

8. Shebel GB, Brittig K. Refined genetic algorithm—economic dispatch example. IEEE Trans Power Systems 1995; 10(1):117–124.

9. Bakirtzis A, Petridis V, Kazarlis S. Genetic algorithm solution to the economic dispatch problem. IEE Proc Gener Transm Distrib 1994; 141(4):377–382.

10. Song YH, Chou CSV. Advanced engineered-conditioning genetic approach to power economic dispatch. IEE Proc Gener Transm Distrib 1997; 144(3):285–292.

11. Orero SO, Irving MR. Economic dispatch of generators with prohibited operating zones: a genetic algorithm approach. IEE Proc Gener Transm Distrib 1996; 143(6):529–534.

12. Goldberg DE. Genetic algorithms in search, optimization, and machine learning. Reading, MA: Addison-Wesley; 1989.

13. Potts JC, Giddens TD, Yadav SB. The development and evaluation of an improved genetic algorithm based on migration and artificial selection. IEEE Trans Systems Man and Cybernetics 1994; 24(1):73–86.

14. Yang HT, Yang PC, Huang CL. Evolutionary programming based economic dispatch for units with non-smooth fuel cost functions. IEEE Trans Power Systems 1996; 11(1):112–118.

15. Sinha N, Chakrabarti R. Chattopadhyay PK. Evolutionary programming techniques for economic load dispatch. IEEE Trans Evol Computat 2003; 7(1):83–94.

16. Lin WM, Cheng FS, Tsay MT. An improved tabu search for economic dispatch with multiple minima. IEEE Trans Power Systems 2002; 17(1):108–112.

17. Lee KY, Sode-Yome A, Park JH. Adaptive Hopfield neural network for economic load dispatch. IEEE Trans Power Systems 1998; 13(2):519–526.

18. Walters DC, Sheble GB. Genetic algorithm solution of economic dispatch with the valve point loading. IEEE Trans Power Systems 1993; 8(3):1325–1332.

19. Park JB, Lee KS, Shin JR, Lee KY. A particle swarm optimization for economic dispatch with non-smooth cost functions. IEEE Trans Power Systems 2005; 20(1):34–42.

20. Fonseca N, Miranda V. EPSO—best-of-two-worlds meta-heuristic applied to power system problems. CEC '02. Proceedings of the 2002 Congress on Evolutionary Computation, Honolulu, Hawaii, 2002; 2:1080–1085.

21. Abrao PJ, Alves da Silva AP. Applications of evolutionary computation in electric power systems. CEC '02. Proceedings of the 2002 Congress on Evolutionary Computation, Honolulu, Hawaii, 2002; 2:1057–1062.

22. Verbruggen A. An introduction to CHP issues. Global Energy Issues 1996; 8(4):301–318.

23. Puttgen HB et al. Optimum scheduling procedure for cogenerating small power producing facilities. IEEE Trans Power Systems 1989; 4(3):957–964.

24. Ghoudjehbaklou H, et al. Optimisation topics related to small power producing facilities: operating under energy spot pricing polices. IEEE Trans Power Systems 1987; Vol. PWRS-2, No. 2.

25. Rooijers F J, van Amerongen RAM. Static economic dispatch for co-generation systems. IEEE Trans Power Systems 1994; 9(3):1392–1398.

26. Wang SM, Liu CC, Luu S. A negotiation methodology and its application to cogeneration planning. IEEE Trans Power Systems 1994; 9(2):869–875.

27. Horii S, et al. Optimal planning of gas turbine cogeneration plants based on mixed-integer linear programming. Int J Energy Res 1987; 11:507–518.

28. Moslehi K, et al. Optimization of multiplant cogeneration system operation including electric and steam networks. IEEE 1990 Summer Meeting, Paper No. 90, SM 459–8, PWRS.

29. Baughman ML, et al. Optimizing combined cogeneration and thermal storage systems: an engineering economics approach. IEEE Trans Power Systems 1989; 4(3):974–980.

30. Lee JB, Lyu SH, Kim JH. A daily operation scheduling on cogeneration system with thermal storage tank. Trans IEE Japan 1994; 114-B:1295–1302.

31. Chen B-W, Hong C-C. Optimum operation for a back-pressure cogeneration system under time-of-use rates. IEEE/PES Summer Meeting, Paper No. 95, SM 593-4 PWRS, July 1995.

32. Ma JT, Lai LL. Improved genetic algorithm for reactive power planning. Proceedings of the 12th Power Systems Computation Conference, PSCC, Dresden, Germany, Aug. 1996. 499–505.

33. Ma JT, Lai LL, Lee JB. Multi-time interval scheduling for daily operation of a two-cogeneration system with genetic algorithm. Proceedings of the IEEE/CSEE International Conference on Electrical Power Engineering, Beijing, China, Aug. 1996. 1348–1352.

34. Editorial. Recent novel developments in heat integration—total site, trigeneration, utility systems and cost-effective decarbonisation. Case studies waste thermal processing, pulp and paper and fuel cells. Appl Thermal Eng 2005; 25:953–960.

35. Sheble GB, Fahd GN. Unit commitment literature synopsis. IEEE Trans Power Systems 1994; Vol. PWRS-9, No. 1. 128–135.

36. Dorigo M, Gambardella LM. Ant colony system: a cooperative learning approach to the traveling salesman problem. IEEE Trans Evol Computat 1997; 1(1):53–66.

37. Yu IK, Song YH. A novel short-term generation scheduling technique of thermal units using ant colony search algorithms. Electric Power Energy Systems 2001; 23:471–479.

38. Kellermann W, El-Din HM, Graham, CE, Maria GA. Optimization of fixed tap transformer setting in bulk electric systems. IEEE Trans Power Systems 1991; 6(3):1126–1132.

39. Vlachogiannis JG. Control adjustments in fast decoupled load flow. Electric Power Systems Res 1994; 31(3):185–194.

40. Yoshida H, Kawata K, Fukuyama Y, Takayama S, Nakanishi Y. A particle swarm optimization for reactive power and voltage control considering voltage security assessment. IEEE Trans Power Systems 2000; 15(4):1232–1239.

41. Satpathy PK, Das D, Dutta Gupta PB. A novel fuzzy index for steady state voltage stability analysis and identification of critical busbars. Electric Power Systems Res 2002; 63(2):127–140.

42. Cova B, Losignore N, Marannino P, Montagna M. Contingency constrained optimal reactive power flow procedures for voltage control in planning and operation. IEEE Trans Power Systems 1995; 10(2):602–608.

43. Vaahedi E, Tamby J, Mansour Y, Li W, Sun D. Large scale voltage stability constrained optimal VAr planning and voltage stability applications using existing OPF/Optimal VAr planning tools. IEEE Trans Power Systems 1999; 14(1):65–74.

44. Gan D, Thomas RJ, Zimmerman RD. Stability constrained optimal power flow. IEEE Trans Power Systems 2000; 15(2):535–540.

45. Jabr RA, Coonick AH. Homogeneous interior point method for constrained power scheduling. IEE Proc Gener Transm Distrib 2000; 147(4):239–244.

46. Karakatsanis TS, Hatziargyriou ND. Probabilistic constrained load flow based on sensitivity analysis. IEEE Trans Power Systems 1994; 9(4):1853–1860.

47. Watkins CJCH, Dayan P. Q-learning. Machine Learning 1992; 8(3):279–292.

48. Kaelbling LP, Littman ML, Moore AW. Reinforcement learning: a survey. J Artif Intelligence Res 1996; 4:237–285.

49. Sutton RS, Barto AG. Reinforcement learning: An introduction, adaptive computations and machine learning. Cambridge, MA: MIT Press, 1998.

50. Vlachogiannis JG, Hatziargyriou ND. Reinforcement learning for reactive power control. IEEE Trans Power Systems 2004; 19(3):1317–1325.

51. Dorigo M, Maniezzo V, Colorni A. The ant system: optimisation by a colony of co-operating agents. IEEE Trans Systems, Man and Cybernetics Part B 1996; 26(1):29–41.

52. Allan RN, Al-Shakarchi MRG. Probabilistic techniques in a.c. load-flow analysis. IEE Proc Gener Transm Distrib 1977; 124(2):154–160.

53. Kenney JF, Keeping ES. Mathematics in statistics. Princeton, NJ: Nostrand, 1962.

54. Carpaneto E, Chicco G. Ant-colony search-based minimum losses reconfiguration of distribution systems. Proceedings of the 12th IEEE Mediterranean Electrotechnical Conference, Dubrovinik, Croatia, 12–15 May 2004 (MELECON 2004), Vol. 3. 971–974.

55. Vlachogiannis JG, Hatziargyriou ND, Lee KY. Ant colony system-based algorithm for constrained load flow problem. IEEE Trans Power Systems 2005; 20(3):1241–1249.

CHAPTER 16

Power System Controls

YOSHIKAZU FUKUYAMA, HAMID GHEZELAYAGH, KWANG Y. LEE,
CHEN-CHING LIU, YONG-HUA SONG, and YING XIAO

16.1 INTRODUCTION

In this chapter, the case studies of modern heuristic optimization techniques for control of power plants and power systems will be presented. In Section 16.2, the particle swarm optimization (PSO) technique is applied to power system control. One of the important operating tasks of power utilities is to keep voltage within an allowable range for high-quality customer services. Electric power loads vary from hour to hour, and voltage can be varied by a change of the power load. Power utility operators in control centers handle various equipment such as generators, transformers, static condensers (SCs), and shunt reactors (ShRs), so that they can inject reactive power and control voltage directly into target power systems in order to follow the load change. Voltage and reactive power control (Volt/Var Control; VVC) determines an on-line control strategy for keeping voltage of target power systems considering the load change and reactive power balance in target power systems.

In Section 16.3, genetic algorithm (GA) is applied to design controllers for a power plant. It is well-known that automatic control is deemed a necessary condition for safe operation, which minimizes material fatigue, the number of staff, and enables efficient plant management. However, the design of controllers for power plants is not a trivial task. The challenge in controller design for these plants exists because they are typically nonlinear and multivariable with multiple control objectives. Whereas conventional controls such as proportional-integral-derivative (PID) compensators yield an acceptable response, they do not have the flexibility necessary to provide a good performance over a wide range of operation. Application of modern optimal control techniques yields system performance that is optimal at only one operating point. Recent application of robust control to power plant models yielded quite favorable results. These robust controllers perform very well over a wide range of operation. However, these robust control design methodologies require frequency response information in

Modern Heuristic Optimization Techniques. Edited by K. Y. Lee and M. A. El-Sharkawi

addition to a linearized model of the plant. A goal of modern intelligent control design is to obtain a controller based on input/output information only, and the modern heuristic optimization techniques hold the promise for such a control system design.

In Section 16.4, a combination of different intelligent system techniques—neural networks, fuzzy systems, GA, and evolutionary programming (EP)—is used to design an intelligent predictive control system for power plants. Power plant control has been a subject of many classic and modern control studies. Dealing with the multiinput, multioutput (MIMO) characteristic of a power plant is a challenge in adaptive and optimal control approaches. In addition, a power plant has a wide range of nonlinear operation that makes the linear model-based controllers complicated. An intelligent control system is able to cover a wide range of operation and plant variation by using learning algorithms. The plant is modeled by a neuro-fuzzy system, and the connection weights are determined by GA. The predictive control has been optimized with evolutionary programming.

Finally, Section 16.5 focuses on the application of a multiobjective (MO) optimization approach to optimal operation of power systems incorporating flexible alternating current transmission systems (FACTS) devices. First, taking economic system operation and attaining specific local power flow control targets as objectives, a MO optimization model for FACTS control is formulated. For most prevalent MO optimization algorithms, it is very difficult to handle problems with these noncommensurable objectives; therefore, based on compromise programming (CP), a displaced worst compromise programming (DWCP) is introduced to solve the problem in an interactive procedure.

16.2 POWER SYSTEM CONTROLS: PARTICLE SWARM TECHNIQUE

One of the important operating tasks of power utilities is to keep voltage within an allowable range for high-quality customer services. Electric power loads vary from hour to hour, and voltage can be varied by a change of the power load. Power utility operators in control centers handle various equipment such as generators, transformers, SCs, and ShRs, so that they can inject reactive power and control voltage directly in target power systems into order to follow the load change. VVC determines an on-line control strategy for keeping voltage of target power systems considering the load change and reactive power balance in target power systems.

Current practical VVC in control centers is often realized based on power flow sensitivity analysis of the operation point using limited execution time and available data from the actual target power system. Reduction of power generation cost is one of the current issues of interest to power utilities. Therefore, an optimal control to minimize power transmission loss is required for VVC instead of simple power flow sensitivity analysis. Because many voltage collapse accidents have occurred over the past three decades [1], voltage security problems have dominated, and consideration of the problem has been required in VVC problems [2, 3]. Two evaluations should be performed to consider voltage security. The first one is to calculate the distance between the current operating point and the voltage collapse point. The calculation can be realized by drawing a P-V (Power-Voltage) curve using the continuation power flow (CPFLOW) technique [4]. The authors have developed a practical CPFLOW and verified it with practical power systems [5]. The second one is to suppose various faults for the current operating point in the target power system and calculate the distance between the post-fault operating points and voltage collapse points for each contingency. The calculation is called voltage contingency analysis [1]. If a sufficient distance can be kept for both calculations, the new operating condition calculated by VVC can be evaluated as a secure one. Thus, the advanced VVC requires optimal control strategy considering power loss minimization and voltage security.

VVC can be formulated as a mixed-integer nonlinear programming (MINLP) with continuous state variables such as automatic voltage regulator (AVR) operating values and discrete state variables such as on-load tap changer (OLTC) tap positions and the amount of reactive power compensation equipment such as SCs and ShRs. The objective function can be varied according to the power system condition. For example, the function can be the minimization of power transmission loss of the target power system for the normal operating condition as described above. Conventionally, the methods for the VVC problem have been developed using various methods such as fuzzy, expert system, mathematical programming, and sensitivity analysis [6–11]. However, a practical method for a VVC problem formulated as a MINLP with continuous and discrete state variables has been eagerly awaited.

PSO is one of the evolutionary computation (EC) techniques [12]. The method is improved and applied to various problems [13–16]. The original method is able to handle continuous state variables easily. Moreover, the method can be expanded to handle both continuous and discrete variables easily. Therefore, the method can be applicable to VVC formulated as a MINLP. Various methods have been developed for a MINLP such as generalized benders decomposition (GBD) [17] and outer approximation with equality relaxation (OA/ER) [18]. Using the conventional methods, the whole problem is usually divided into subproblems, and various methods are utilized for solving each subproblem. On the contrary, PSO can handle the whole MINLP easily and naturally, and it is easy to apply to various problems compared with the conventional methods. Moreover, VVC requires various constraints that prove difficult for mathematical methods to handle. PSO is expected to be suitable for VVC because it can handle such constraints easily.

This section presents a PSO for VVC formulated as a MINLP considering voltage security assessment (VSA). VSA is considered using a CPFLOW technique and a fast voltage contingency selection method. The feasibility of the proposed method for VVC is demonstrated and compared with the reactive tabu search (RTS) [19, 20] and the enumeration method on practical system models with promising results.

16.2.1 Problem Formulation of VVC

16.2.1.1 State Variables The following control equipment is considered in the VVC problem:

(a) AVR operating values (*continuous* variable).

(b) OLTC tap position (*discrete* variable).

(c) The amount of reactive power compensation equipment (*discrete* variable).

The above state variables are treated in load flow calculation as follows: AVR operating values are treated as voltage specification values. OLTC tap positions are treated as a tap ratio to each tap position. The amount of reactive power compensation equipment is treated as corresponding susceptance values.

16.2.1.2 *Problem Formulation* VVC for a normal power system condition can be formulated as follows:

$$\text{Minimize } f_c(x, y) = \sum_{i=1}^{n} Loss_i, \tag{16.1}$$

where n is the number of branches, x is *continuous* variables, y is *discrete* variables, and $Loss_i$ is the active power loss (ploss) at branch i, subject to

(a) *Voltage constraint*: Voltage magnitude at each node must lie within its permissible range to maintain power quality.

(b) *Power flow constraint*: Power flow of each branch must lie within its permissible range.

(c) *Voltage security*: The determined VVC strategy should keep voltage security of the target power system.

Ploss of the target power system is calculated for a certain VVC strategy using load flow calculation with both continuous variables (AVR operating values) and discrete variables (OLTC tap positions and the amount of reactive power compensation equipment). Voltage and power flow constraints can be checked at the load flow calculation, and penalty values are added if the constraints are violated.

P-V curves for the determined VVC strategy and various contingencies are generated and then checked to see if the VVC candidate is able to keep sufficient voltage security margins. The constant power load model is used because the load model is the severest to the voltage security problem. However, if a more complicated load model is required, the proposed method can be easily expanded using a ZIP (Impedance-Current-Power) load model [21].

16.2.2 Expansion of PSO for MINLP

PSO was originally developed for nonlinear optimization problems with continuous state variables. However, lots of engineering problems have to handle both discrete and continuous variables using nonlinear objective functions. PSO can be expanded for handling MINLP as follows. Discrete numbers instead of continuous numbers can be used to express the current position and velocity. Namely, discrete random number is used for *rand* in (4.1) and the whole calculation of the right-hand side of (4.1) is discretized to the existing discrete number. Using this modification for discrete numbers, both continuous and discrete numbers can be handled in the algorithm with no inconsistency.

16.2.3 Voltage Security Assessment

The static P-V curve represents the relation between load increase and voltage drop. Namely, the P-V curve can be calculated by increasing total loads in the target power system gradually and plotting the dropped voltage. CPFLOW utilizes power system loads as parameters and calculates the P-V curve by modification of the parameters using a continuation method. The continuation method is one of the methods in applied mathematics, and it calculates transition of equilibrium points (e.g., P-V curve) by modification of parameters. In order to avoid the ill condition around the saddle node bifurcation point (nose point), an arclength along the P-V curve is introduced as an additional state variable, and the power flow equation is expanded. The continuation method is applied to the expanded power flow equation and the P-V curve can be generated rapidly without ill condition around the nose point. CPFLOW can generate a P-V curve automatically and can be easily applied to large-scale power systems [4, 5].

The proposed method generates a P-V curve using the CPFLOW technique and calculates a megawatt (MW) margin, distance between the current operating point and the nose point, for the determined control strategy. The proposed method also utilizes the fast voltage contingency analysis method using CPFLOW [22]. Then, the method checks whether the MW margin is enough or not compared with the predetermined value. The procedure for voltage security assessment can be expressed as follows:

Step 1. Evaluation of the control strategy. The new power system condition is checked after applying the current control strategy to see if there is enough MW margin or not.

Step 2. Evaluation of various contingencies. Several severe contingencies for the new power system condition after applying the current control strategy are selected by the fast voltage contingency analysis method. The MW margin for only the severe contingencies are calculated using CPFLOW.

If the MW margins for the current control strategy and the severe contingencies are large enough, the current control strategy is selected. Otherwise, it is not selected. Using the procedure, the method checks whether the target power system can keep voltage security by the control or not.

VSA can be composed of static and dynamic VSA. The proposed method only considers the static VSA because of the limited calculation time for on-line VVC. If the dynamic VSA is still required, the VSA used in the proposed method can be replaced with a dynamic VSA tool such as Quasi Steady-State (QSS) described in Ref. 1. However, in such a case, we have to face the problem of execution time and we may have to develop a parallel computation method for the VSA based on distributed memory tools such as the parallel virtual machine (PVM) [23] and the message-passing interface (MPI) [24] or shared memory tools such as OpenMP [25].

16.2.4 VVC Using PSO

16.2.4.1 Treatment of State Variables Each variable is treated in PSO as follows: Initial AVR operating values are generated randomly between upper and lower bounds of the voltage specification values. The value is also modified in the search procedure between the bounds. OLTC tap position is initially generated randomly between the minimum and maximum tap positions. The value is modified in the search procedure among existing tap positions. Then, the corresponding impedance of the transformer is calculated for the load flow calculation. The amount of reactive power compensation equipment is also generated from 0 to the amount of existing equipment at the substation initially. The value is also modified in the search procedure between 0 and the amount of existing equipment.

16.2.4.2 VVC Algorithm Using PSO The proposed VVC algorithm using PSO can be expressed as follows:

Step 1. Initial searching points and velocities of agents are generated using the above-mentioned state variables randomly.

Step 2. Ploss to the searching points for each agent is calculated using the load flow calculation. If the constraints are violated, the penalty is added to the loss (evaluation value of agent).

Step 3. Pbest is set to each initial searching point. The initial best evaluated value (loss with penalty) among pbests is set to gbest.

Step 4. New velocities are calculated using velocity equations for continuous and discrete state variables.

Step 5. New searching points are calculated using state modification equations for continuous and discrete state variables.

Step 6. Ploss to the new searching points and the evaluation values are calculated.

Step 7. If the evaluation value of each agent is better than the previous pbest, the value is set to pbest. If the best pbest is better than gbest, the value is set to gbest. All gbests are stored as candidates for the final control strategy.

Step 8. If the iteration number reaches the maximum iteration number, then go to step 9. Otherwise, go to step 4.

Step 9. P-V curves for the control candidates and various contingencies are generated using the best gbest among the stored gbests (candidates). If the MW margin is larger than the predetermined value, the control is determined as the final solution. Otherwise, select the next gbest and repeat the VSA procedure mentioned above.

If the voltage and power flow constraints are violated, the absolute violated value from the maximum and minimum boundaries is largely weighted and added to the objective function (16.1). The maximum iteration number should be determined by presimulation. As mentioned below, PSO requires less than 100 iterations to obtain a good solution even for large-scale problems. There are several ways to formulate VVC considering VSA. Maximization of MW margin instead of loss minimization is one option. However, the purpose of the paper is to develop a VVC algorithm for steady-state operation. In such a case, enough voltage stability margins can usually be kept. Therefore, the authors decided to utilize only loss minimization as the objective function and to check whether the control strategy has enough voltage stability margins or not after loss minimization. Moreover, evaluation for each state is extremely time-consuming considering VSA during optimization procedure, and it is difficult to realize on-line VVC. Considering the trade-off between the optimal control and the execution time, the proposed method selected the way to handle the contingencies after generation of the optimal control candidates. If maximization of MW margin is required as the objective function, an approximation method such as the look-ahead method with parallel computation should be used during the search procedure for on-line VVC.

16.2.5 Numerical Examples

The proposed method has been applied to several power system models compared with RTS and the enumeration method. Our target VVC problem is formulated as a MINLP with discrete and continuous variables. Optimal power flow (OPF) basically only handles continuous variables and in some papers, such as Ref. 26, tried to handle discrete variables in OPF formulation. Unfortunately, the authors do not have an OPF program with such treatment. Therefore, the proposed method is compared with available combinatorial optimization software in the simulation.

16.2.5.1 IEEE 14 Bus System

1. **Simulation conditions:** Figure 16.1 shows a modified IEEE 14 bus system. Table 16.1 shows the operating condition of the system.

    The followings are control variables.

    (a) *Continuous* AVR operating values of nodes 2, 3, 6, and 8: Upper and lower bounds are 0.9 and 1.1 (pu (per-unit)).

    (b) *Discrete* tap positions of transformers between nodes 4–7, 4–9, and 5–6: These transformers are assumed to have 20 tap positions.

    (c) *Discrete* number of installed SC in nodes 9 and 14: Each node is assumed to have three 0.06 (pu) SC.

    The proposed method tries to generate an optimal control for the operating condition. Ploss of the original system is 0.1349 (pu). Generation of the VVC candidates (Steps 1–7 in the proposed VVC algorithm) by the proposed PSO-based

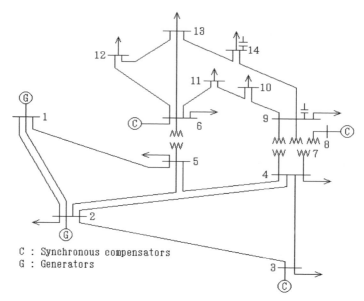

FIGURE 16.1   A modified IEEE 14 bus system.

TABLE 16.1   Operating Conditions of IEEE 14 Bus System

Bus No.	Vol. (pu)	Node Specification		SC (pu)
		P (pu)	Q (pu)	
1[a]	1.060	–	–	0.0
2[b]	1.045	−0.183	0.127	0.0
3[b]	1.010	0.942	0.190	0.0
4		0.478	−0.039	0.0
5		0.076	0.016	0.0
6[b]	1.070	0.112	0.075	0.0
7		0.000	0.000	0.0
8[b]	1.090	0.000	0.000	0.0
9		0.295	0.166	0.18[c]
10		0.090	0.058	0.0
11		0.035	0.018	0.0
12		0.061	0.016	0.0
13		0.135	0.058	0.0
14		0.149	0.050	0.18[c]

[a]Node 1 is slack.
[b]PV specification node.
[c]0.06 (pu) ∗ 3 SC.

method, RTS, and the enumeration method is compared in the simulation. The following parameters are utilized in the simulation according to the presimulation. The inertia weights approach (IWA) (4.2) is utilized for the coefficient function w of the velocity equation.

c_1 and c_2 of the velocity equation are set to 2.0. w_{max} and w_{min} of (4.2) are set to 0.9 and 0.4 according to the presimulation as shown below. The number of agents for PSO is 10. The parameters for RTS are also determined to appropriate values through presimulation. The initial tabu length is 10, and increase/decrease rate for tabu length is 0.2 for RTS in the simulation. The results are compared with 300 searching iterations. RTS and the enumeration method utilize discretized AVR operating values, and the interval is 0.01 (pu). The interval corresponds with 5 (kV) in a 500 (kV) system. The formulation as the combinatorial optimization problem (COP) has about 10^9 combinations in the problem. The system has been developed using C language (egsc ver. 1.1.1), and all simulation is performed using Engineering Work Station (SPECint95: 12.3).

2. *Simulation results:* Table 16.2 shows the best results by the proposed method, RTS, and the enumeration method. Table 16.3 shows the loss values and calculation time of the results. The best result by RTS is similar to that by the enumeration method (the optimal result formulated as a COP). However, the loss value calculated by PSO is smaller than the optimal value and a tap position is different between the results. When VVC is formulated as a COP, only solutions to discrete values are searched and the objective function shape between the discritized intervals is of no concern. Therefore, as is usually pointed out, the optimal solution formulated as a MINLP and a COP is different. The results indicate the necessity for the formulation of the VVC as a MINLP. PSO can generate smaller loss values than RTS with 15% possibility. The calculation time by PSO is about 15% faster than that of RTS. Table 16.4 shows the parameter sensitivity analysis of PSO. In the simulation, w_{max} and

TABLE 16.2 The Optimal Control for IEEE 14 Bus System

Control Variables	Methods		
	PSO	RTS	Enumeration Method
AVR 2	1.0463	1.05	1.05
AVR 3	1.0165	1.02	1.02
AVR 6	1.1000	1.10	1.10
AVR 8	1.1000	1.10	1.10
Tap 4–7	0.94	0.95	0.95
Tap 4–9	0.93	0.93	0.93
Tap 5–6	0.97	0.97	0.97
SC 9	0.18	0.18	0.18
SC 14	0.06	0.06	0.06

AVR 2: AVR operating values (pu) at node 2.
Tap 4–7: Tap ratio between node 4 and 7.
SC 9: Susceptance (pu) at node 9.

TABLE 16.3 Summary of Calculation Results by the Proposed Method and Reactive Tabu Search

Method	Compared Items	IEEE 14 Bus System	112 Bus System
PSO	Minimum loss value	0.1332276	0.1134947
	Average loss value	0.1335090	0.1175230
	Calculated time	16.5	54.2
RTS	Minimum loss value	0.1323657	0.1208179
	Calculated time	19.5	220.3

Loss value: active power loss (pu).
Calculated time: average calculation time (s).

w_{min} of Table 16.4 and c_i of the velocity equation are changed. The average and minimum Ploss with 100 searching iterations in 100 trials for each case are shown in the table. The results reveal that the appropriate values for w_{max} and w_{min} are 0.9 and 0.4. The appropriate value for c_i is 1.5. However, the minimum Ploss for 1.5, 2.0, and 2.5 are similar, and 2.0 is utilized in the simulation according to the suggested value in Ref. [13]. Consequently, the appropriate parameter values for the problem are the same as the ones suggested in Ref. 13.

The proposed method generates a P-V curve for the optimal control strategy using the CPFLOW technique and performs the voltage contingency analysis. It is verified that the strategy can maintain voltage security when the load margin to achieve 0.95 (pu) voltage is larger than 10% load increase in the simulation. The evaluation criteria depend on the target power system and should be determined for each system through presimulation. Figure 16.2 shows an example of a P-V curve for node 12 with the optimal control strategy.

16.2.5.2 Practical 112 Bus Model System

1. *Simulation conditions:* The proposed method is applied to a practical model system with 112 buses. The system models the extremly high voltage (EHV) system of Kansai Electric Practical system. The model system has 11 generators

TABLE 16.4 Parameter Sensitivity Analysis for IEEE 14 Bus System (100 trials)

w_{max} w_{min}		c_i						
		0.5	1.0	1.5	2.0	2.5	3.0	4.0
0.9	Ave.	0.133693	0.133573	0.133763	**0.133567**	0.133765	0.133986	0.134504
0.4	Min.	0.133012	0.133012	0.133012	**0.133012**	0.133012	0.133073	0.133076
2.0	Ave.	0.135519	0.135689	0.136362	0.136324	0.135763	0.136425	0.136245
0.9	Min.	0.133074	0.133073	0.133073	0.133121	0.133083	0.133125	0.133315
2.0	Ave.	0.134987	0.135226	0.13604	0.135661	0.135457	0.135795	0.136435
0.4	Min.	0.133015	0.133012	0.133012	0.133073	0.133014	0.133075	0.133115

Bold numbers indicate the minimum sensitivities.

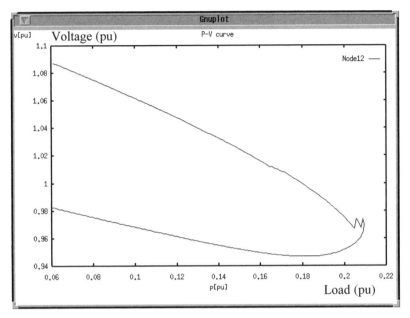

FIGURE 16.2 A P-V curve of the optimal control (node 12) for the IEEE 14 bus system.

for AVR control, 47 OLTCs with 9 to 27 tap positions, and 13 SC installed buses with 33 SCs for VVC. Therefore, the dimension of the problem is 91 (11 + 47 + 33) with continuous and discrete variables. The number of agents for PSO is set to 30 in order to get a high-quality solution within 1 (min). PSO and RTS are compared in 100 searching iterations. The same parameters for the IEEE 14 bus system except the above values are utilized in the simulation.

2. *Simulation results:* Figure 16.3 shows the statistical evaluation results by the proposed method in 100 trials. Table 16.3 shows the loss values and calculation time of the results. The average loss value by the proposed method is smaller than the best result by RTS. PSO generates a better solution than RTS with 96% possibility. Figure 16.4 shows typical convergence characteristics (Ploss transition of gbest by PSO and the best result by RTS). It is clear from the figure that the solution by PSO is converged to high-quality solutions at the early iterations (about 20 iterations). The average iteration number to the best result by the proposed method is 31.7. On the contrary, RTS reaches the best result gradually. The average calculation time by PSO is about 4 times faster than that by RTS. RTS generates neighboring solutions (candidates for the next searching point) in the solution space. It performs load flow calculation for each candidate and evaluates violation of operating constraints and tabu status for all candidates. Therefore, candidates that should be evaluated are increased exponentially as the dimension of the problem increases. On the contrary, PSO only calculates the velocity and state equations for each agent, and

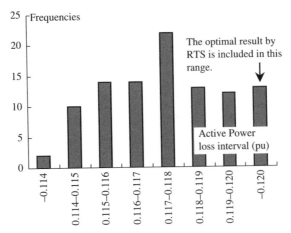

FIGURE 16.3 Statistical results by PSO (100 trials) for practical 112 bus system.

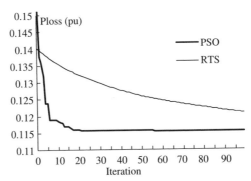

FIGURE 16.4 Convergence characteristics by PSO and RTS for practical 112 bus system.

the number of load flow calculation is the same for IEEE 14 and practical 112 bus system if the same number of agents are utilized for the simulation. The characteristic of PSO is suitable for the application to practical system. The determined VVC strategy candidate is evaluated as a secure one using a CPFLOW technique. Voltage contingency analysis is also performed for the candidate, and it is evaluated as secure. The calculation time for voltage contingency ranking is 11.0 (s) (112 contingencies) and the time for one CPFLOW calculation is 2.0 (s) for the 112 bus model system. Therefore, for example, the total calculation time for voltage security assessment is 19.0 (s) if CPFLOW is performed for the severest three contingencies (one CPFLOW calculation, contingency ranking, and three CPFLOW calculations). As described above, a large penalty is added at the evaluation of the objective function if the voltage and power flow constraints are violated. Therefore, all of the best solutions by both PSO and RTS within 100 searching iterations are feasible solutions without voltage and power flow constraints violation in the simulation. Although the best VVC strategy is evaluated as secure in the model system, voltage security assessment can become more important when the utilization rate of power equipment is increased and in the deregulation environment.

16.2.5.3 Large-Scale 1217 Bus Model System

1. *Simulation conditions:* The proposed method has been developed to apply the practical EHV system. Therefore, the applicability of the proposed method to the target system is already evaluated with the 112 bus system. However, in order to evaluate the applicability of the proposed method to large-scale systems, it has been applied to a 1217 bus system. The model system is composed by doubling the full-scale Kansai Electric power system. The system has 84 generators for AVR control, 388 OLTCs, and 82 SCs for VVC. Therefore, the dimension of the problem is 554(84 + 388 + 82) with continuous and discrete variables. The parameters for evaluated methods are the same as those utilized for the 112 bus model system.

2. *Simulation results:* Convergence characteristics for the 1217 bus system by RTS and PSO are the same as in Fig. 16.4. RTS requires about 7.6 (h) for 100 iterations. On the contrary, the average execution time for obtaining the optimal results (the average number of iterations for that is 27.5) by PSO is about 230 (s). Figure 16.5 shows the number of states to be evaluated at each iteration by RTS and PSO. The figure assumes that the number of agents is 30 in all cases. The number by RTS is the number of neighboring states of the current state at each iteration. Therefore, it increases drastically as the dimension of the problem increases. On the contrary, the number by PSO corresponds with the number of agents. Therefore, it is the same even for large-dimensional problems. Consequently, although PSO only evaluates the limited number of states using the velocity and state equations, the evaluation is efficient even for the large-scale problems and realizes a quick convergence characteristic to suboptimal solutions. The characteristic indicates the applicability of PSO to large-scale problems.

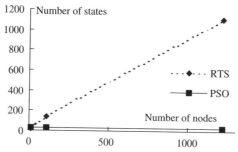

FIGURE 16.5   The number of states evaluated at each iteration by RTS and PSO.

The calculation time for evaluation of one state is increasing as the dimension of the problem increases. Therefore, if speed-up of the whole execution time has to be realized, parallel computation methods based on distributed memory tools such as PVM [23] and MPI [24] or shared memory tools such as OpenMP [25] can be utilized for the optimization part in a similar manner as for the VSA part.

16.2.6 Summary

This section presents a PSO for VVC considering VSA. The proposed method formulates VVC as a MINLP and determines a control strategy with continuous and discrete control variables such as AVR operating values, OLTC tap positions, and the amount of reactive power compensation equipment. The method also considers voltage security using CPFLOW and a voltage contingency analysis technique. The feasibility of the proposed method for VVC is demonstrated on practical power systems with promising results. The results can be summarized as follows:

(a) This section shows the practical applicability of PSO to a MINLP and suitability of PSO for application to large-scale VVC problems. PSO has several parameters. According to the simulation results, severe parameter tuning is not required and especially, PSO only requires <50 iterations for obtaining suboptimal solutions even for large-scale systems.

(b) Many power system problems can be formulated as a MINLP, and the results indicate the possibility of PSO as a practical tool for various MINLPs of power system operation and planning.

(c) VVC is sometimes formulated as a combinatorial optimization problem. However, discrete variables of the optimal result formulated as a MINLP and those formulated as a combinatorial optimization problem are different. Therefore, it indicates the efficiency of the formulation of VVC as a MINLP.

(d) Consideration of VSA is one of the important practical functions of VVC. The results reveal the possibility of treatment of the security by the proposed PSO-based method in VVC.

In addition to the proposed method, the following additional features make the proposed VVC more practical.

(e) Avoidance of control concentration to specific equipment.

(f) Tracking to load change.

(g) Look-ahead control using load forecast.

Especially, for handling (e) and (f), an optimal control strategy in several control intervals should be considered simultaneously. Improvement of the proposed method considering the above features and parallel computation is part of our future research.

16.3 POWER PLANT CONTROLLER DESIGN WITH GA

A practical control problem that has received a great deal of attention lately is the robust control of power plants. It is well-known that automatic control is deemed a necessary condition for safe operation, which minimizes material fatigue, the number of staff, and enables efficient plant management [27]. However, the design of controllers for power plants is not a trivial task. The challenge in controller design for these plants exists because they are typically nonlinear and multivariable with multiple control objectives. Whereas conventional controls such as PID compensators yield an acceptable response, they do not have the flexibility necessary to provide a good performance over a wide range of operation. Application of modern optimal control techniques yields system performances that are optimal at only one operating point.

Recent application of robust control to power plant models yielded quite favorable results [28]. These robust controllers perform very well over a wide range of operation. However, these robust control design methodologies require frequency response information in addition to a linearized model of the plant. A goal of modern intelligent control design is to obtain a controller based on input/output information only. The GA is a method that holds promise for such a control system design. The GA is a search technique based on the mechanics of natural genetics and survival of the fittest. Touted as an efficient and general method of searching a complex space [29], the GA has had success in many areas. The GA has proved successful in obtaining a solution to the traveling salesmen [30], optimal control of an aircraft autopilot system [30], multivariate curve fitting, and game-playing [31]. A number of these achievements suggest the potential utility of the GA as a method for control system design [32, 33]. In this section, we introduce an application of GA in power plant control: boiler-turbine control system design [32].

16.3.1 Overview of the GA

The GA is a parallel, global probabilistic search technique based on the principle of population genetics. Generally, the GA technique consists of three steps (illustrated in Fig. 16.6) and each step is concisely explained below [34, 35]:

1. *Encoding and initialization:* This first step needs to encode the parameters or solution space and initialize the population of the first generation.
   - *Encoding:* Before we use the GA for designing a control system, we have to encode the solution space using a suitable representation scheme. It is standard to translate the parameters into binary digit (bit) strings, that is, strings of 1's and 0's. Such strings can be lengthened to provide more resolution for the parameter representation. Using this scheme of representation, various components of a solution are represented by binary strings that are then

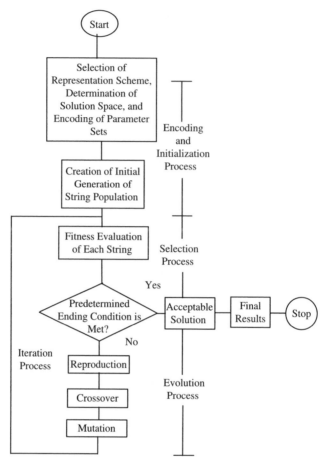

FIGURE 16.6 Flowchart of GA method.

concatenated to form a single binary string called a chromosome. Additionally, there are some forms of representations as well. For simplicity, we will use the binary representation in this paper.

- *Initialization of population:* Once a suitable representation has been selected, the next step is to initialize the population. This is usually done by a random generation of binary strings representing the chromosomes. A uniform representation of the solution space is ensured this way.

2. *Selection process:* This second step in the GA provides a measure of fitness for each candidate solution for the given problem.

  - *Fitness evaluation.* The strings in the current generation are decoded to be their decimal equivalents. Then, they are judged with some objective functions and assigned individually with fitness values.

3. *Genetic operations:* There are three fundamental operators in this process.
   - *Reproduction operation:* The strings with larger fitness values have a higher a probability of producing a larger number of their copies in the new generation by the reproduction operation. In elitist reproduction, the current best string is guaranteed a long life from generation to generation by making a copy of itself directly into the next generation. The reproduced strings are then placed in *a mating pool* for further use.
   - *Crossover operation:* The strings can exchange information in a probabilistic way by the crossover operation. Two strings are chosen from the mating pool and arranged to exchange their corresponding portions of the binary strings at a randomly selected position along them. This process can combine good qualities among the preferred strings.
   - *Mutation operation:* The strings can change their structure abruptly by the mutation operation through an occasional alternation of a value at a randomly selected bit position along them. The mutation process may quickly generate those strings that might not be conveniently produced by the reproduction and crossover operations. Because it may also spoil the opportunity of the appropriate generation, mutation usually occurs with a small probability.

Steps 2 and 3 usually need to be iterated before the final global optimal solution is found. The GA runs iteratively repeating the process until it arrives at a predetermined ending condition. Finally, the acceptable solution is obtained and decoded into its original pattern.

16.3.2 The Boiler-Turbine Model

The essential dynamics of a boiler-turbine have been remarkably captured for a 160-MW oil-fired drum-type boiler-turbine-generator unit in a third-order MIMO nonlinear model for overall wide-range simulations in Ref. 36. The inputs are the positions of valve actuators that control the mass flow rates of fuel (u_1 in pu), steam to the turbine (u_2 in pu), and feedwater to the drum (u_3 in pu). The three outputs are electric power (E in MW), drum steam pressure (P in kg/cm^2), and drum water level deviation (L in m). The three state variables are electric power, drum steam pressure, and fluid (steam-water) density (ρ_f). The state equations are

$$\frac{dP}{dt} = 0.9u_1 - 0.0018u_2P^{9/8} - 0.15u_3 \tag{16.2a}$$

$$\frac{dP}{dt} = 0.9u_1 - 0.0018u_2P^{9/8} - 0.15u_3 \tag{16.2b}$$

$$\frac{dE}{dt} = ((0.73u_2 - 0.16)P^{9/8} - E)/10. \tag{16.2c}$$

The drum water level output is calculated using the following algebraic equations:

$$q_e = (0.85u_2 - 0.14)P + 45.59u_1 - 2.51u_3 - 2.09 \tag{16.3a}$$

$$\alpha_s = (1/\rho_f - 0.0015)/(1/(0.8P - 25.6) - 0.0015) \tag{16.3b}$$

$$L = 50(0.13\rho_f + 60\alpha_s + 0.11q_e - 65.5), \tag{16.3c}$$

where α_s is the steam quality and q_e is the evaporation rate (kg/s). Positions of valve actuators are constrained to [0, 1], and their rates of change (pu/s) are limited to:

$$-0.007 \le du_1/dt \le 0.007 \tag{16.4a}$$

$$-2.0 \le du_2/dt \le 0.02 \tag{16.4b}$$

$$-2.0 \le du_2/dt \le 0.02. \tag{16.4c}$$

16.3.3 The GA Control System Design

The GA was used in two different controller designs: GA-based propotional-integral (PI) controller and GA-based linear quadratic regulator (LQR) controller designs [32]. In the design of the PI control system, the parameters are the 12 proportional, integral, and cross-coupling gains in the coupled controller illustrated in Fig. 16.7. In the LQR controller, the parameters are the nine state feedback gains in the standard LQR configuration. In each of the control designs, the quadratic performance index is selected:

$$J = \int_{t_0}^{t_f} \left((y - y^{ref})^T Q(y - y^{ref}) + u^T R u \right) dt, \tag{16.5a}$$

and a fitness measure is designated as

$$f = \frac{1}{1 + J}. \tag{16.5b}$$

16.3.3.1 PI Controller Design In the coupled PI controller design, the gains are trained in five stages. The first consists of training the proportional gains in the PI controller only and leaving the others fixed at zero. The initial gains are selected at random between some coarse upper and lower bounds and tuned through each genetic iteration. Once some prespecified convergence criterion has been achieved, the best-fit triplet of proportional gains is designated as the result for the stage. Stage 2 extends the strings in the population to include the next three integral

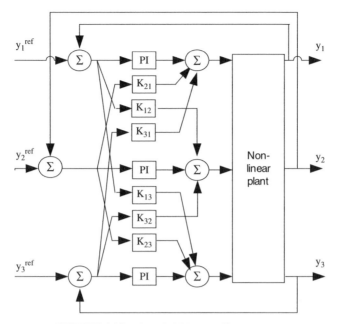

FIGURE 16.7 Coupled PI controller structure.

gains. The bounds on the first three values in the string (the proportional gains) are between $\pm 25\%$ of the result from the previous stage. This constrains the first three gains to be around their previous best value and yet allows fine-tuning with the introduction of the new integral gains. The integral gains are set initially to random values between some coarse upper and lower bounds and tuned through each genetic iteration. Again, this is executed until some convergence criterion has been attained. This procedure is repeated in the same manner with the string lengthened by two more bits to represent the coupling of the first input error to the other two. This is done two more times to train the final four coupling gains. The training cycle is tabulated in Table 16.5.

TABLE 16.5 Training Schedule for Coupled PI Gains

Stage	Gains to Train
I	K_{p1}, K_{p2}, K_{p3}
II	K_{i1}, K_{i2}, K_{i3}
III	K_{12}, K_{13}
IV	K_{21}, K_{23}
V	K_{31}, K_{32}

Each stage of training entails an alternating reference demand change in pressure and power. After the system has been trained to perform well with one reference change, it is then trained to perform well with a different reference demand change. This is done until some convergence criterion has been attained. By training the controller gains in this manner, the gains are tuned to some average performance between two high-performance results. The goal of this method is a controller that performs well at different operating points. A pleasure demand change between 100% and 120% of the nominal operating point is used in the first training cycle while the other inputs are held fixed at their nominal operating points.

The goal of the controller is to track step demands in power and pressure. To achieve this, the following performance index is selected to be minimized:

$$J = J_0 + J_{ss},$$ (16.6a)

where J_0 is defined by (16.5a). In addition, we penalize the steady-state error through computation of

$$J_{ss} = \sum_{i=1}^{3} e_i(t_f),$$ (16.6b)

where

$$e_i(t) = y_i(t) - y_i^{ref}(t).$$ (16.6c)

The weighting matrices are selected as

$$Q = I_{3\times3}, \qquad R = \begin{bmatrix} 724 & 0 & 0 \\ 0 & 141 & 0 \\ 0 & 0 & 5 \end{bmatrix}.$$ (16.6d)

The values of these matrices can be determined from the relative importance of the variables and from the relationships of the outputs to the inputs in the steady-state. Because the order of the system has been increased by including the integral action in the controller, we weigh the steady-state error heavily in the hope of achieving perfect steady-state tracking.

The following GA parameters are selected for the training cycle:

Population size	64
Crossover rate	0.9
Mutation rate	0.001
Parameter resolution	9 bits per substring

Each training cycle is executed until either 30 generations have elapsed or 95% convergence has been achieved. In Ref. 32, 95% convergence was the condition that was met first. Finally, the fitness function used in the training cycle is $10^7/J$.

16.3.3.2 LQR Controller Design

The goal of the GA is to determine the matrix gains in the feedback path to ensure tracking of the reference signal over a wide operating range. The controller structure is simply a state feedback matrix in the feedback path and a feedforward gain matrix to ensure tracking. This feed-forward matrix is found using the transfer function from the input to the output [37]. To obtain reasonable tracking over a wide operating range, the GA trains the gains of the level as well as a large load change. We select a demand change in power while attempting to maintain drum pressure and drum water level.

The GA is implemented with a number of different genetic parameters (i.e., cross-over rate, mutation rate, population number, and so on). In an effort to improve convergence, the individual matrix elements' resolutions are all increased at a specified generation of the run. At this specified time, all of the substrings are lengthened by two bits. The old matrix element values are retained but at a greater precision. We select this lengthening of the bits to occur one-third of the way through the optimiz-ation run. We choose the following parameters:

Population size	100
Crossover rate	0.8
Mutation rate	0.2
Terminating generation	30
Matrix element interval	$[-1, 1]$
Initial resolution	15 bits per substring
Resolution after 10 generations	17 bits per substring

In addition to these parameters, the initial population is seeded with three members from a previous run's "best fit" population with the remainder of the population con-structed at random. This is another attempt to obtain quicker convergence.

In the GA optimization runs, the fitness function is selected as $100/(1 + J)$, where J is the performance index (16.6a). In addition, the weighting matrices are identical to those in the PI controller design.

16.3.4 GA Design Results

The performance of each of the control systems is tested with step inputs in pressure and power while the drum water level is held constant.

16.3.4.1 GA/PI Controller Results

The PI gains at each training stage are summarized in Table 16.6. The performance of the system with the GA-designed coupled PI controller is shown in Figs. 16.8–16.10. The drum pressure is very oscil-latory with a large overshoot. This response settles down with a steady-state error of

0.175. The power response is good during the first change in pressure at 200 seconds. As can be seen, the change has little influence on power. When the demand in power steps up to 120 MW, the power output overshoots by about 5 MW for a duration of <200 seconds. This overshoot can be attributed to the saturation of the control valve for that period of time (not shown). Again, the change in power has little effect on pressure at 600 seconds. The steady-state error of the power is 0.0004 MW. The drum water level deviation is acceptable during both transitions in pressure and power. The largest deviation of the level occurs during the power demand change at 200 seconds when the level exceeds 0.15 m from the nominal level. The steady-state error for the level is 0.000243 m.

TABLE 16.6   Gains at Each Training Stage

	I	II	III	IV	V
K_{p1}	11.3883	8.9675	10.2678	11.7566	11.1185
K_{p2}	0.0793	0.0009	0.0007	0.0005	0.0004
K_{p3}	1.2051	0.9579	1.0423	1.1341	1.1631
K_{i1}	—	0.0041	0.0038	0.0036	0.0033
K_{i2}	—	0.0191	0.0149	0.0117	0.0093
K_{i3}	—	0.0269	0.0238	0.0211	0.0186
K_{12}	—	—	0.0405	0.0386	0.0292
K_{13}	—	—	0.0979	0.1214	0.1344
K_{21}	—	—	—	0.0381	0.0468
K_{23}	—	—	—	0.0950	0.0875
K_{31}	—	—	—	—	0.0842
K_{32}	—	—	—	—	0.0699

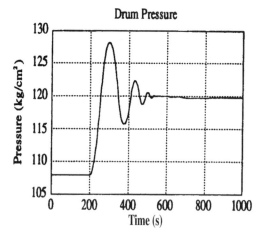

FIGURE 16.8   GA/PI pressure response.

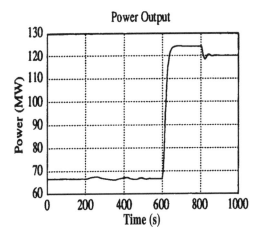

FIGURE 16.9   GA/PI power response.

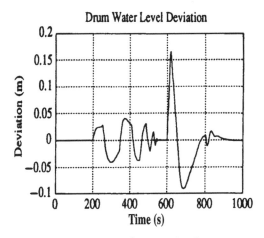

FIGURE 16.10   GA/PI drum level response.

16.3.4.2 GA/LQR Controller Results   The performance of the system with the GA-designed LQR controller is shown in Figs. 16.11–16.13. The result of the GA/LQR design is the state feedback matrix

$$K = \begin{bmatrix} 0.0354 & 0.2236 & -0.1451 \\ -0.0430 & 0.0588 & 0.0039 \\ 0.1609 & -0.6316 & 0.4275 \end{bmatrix}.$$

It should be noted that, to ensure steady-state tracking, a feedforward gain matrix is added to the system. This is obtained by calculating the system transfer function and

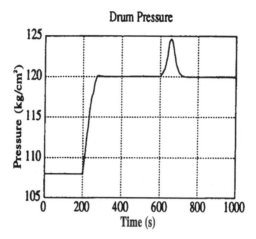

FIGURE 16.11 GA/LQR pressure response.

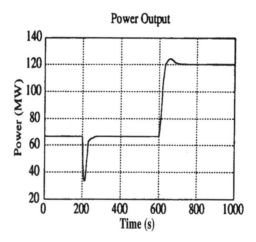

FIGURE 16.12 GA/LQR power response.

finding a suitable matrix, F, so that the system tracks step demands in steady state. The values of the matrix are found to be [37]

$$F = \begin{bmatrix} 0.0341 & 0.0018 & 0.6567 \\ -0.0051 & 0.0903 & -1.3537 \\ -0.4234 & -0.4088 & 94.3416 \end{bmatrix}.$$

The pressure tracks the initial demand change well, and there is a $5\,\text{kg/cm}^2$ pressure bump when the power demand is stepped up. The steady-state error in pressure is $0.252\,\text{kg/cm}^2$. The power pressure shows a large drop (25 MW) when the pressure demand is stepped up at 200 seconds. The power tracks the step demand at 600 seconds well, achieving a steady-state error of 0.68 MW. The drum

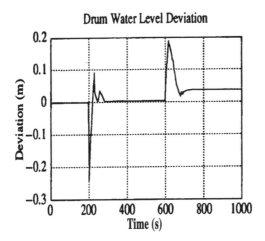

FIGURE 16.13   GA/LQR drum level response.

water level deviation undergoes relatively large deflections at each demand change in pressure and power. The largest of these is a level drop of 0.25 m. The steady-state error in level is 0.0425 m, due in large part to the deviation from the nominal operating point.

16.3.4.3 Summary A control system design methodology for a boiler-turbine plant is presented. The nonlinear MIMO plant was controlled by GA-designed PI and LQR controllers. The GA/PI control system achieved good steady-state tracking but oscillations due to the integral action were prevalent. The GA/LQR control system performed well but at the cost of small but finite steady-state error. In order to ensure the steady-state tracking, a feedforward gain matrix was added to the system. This matrix, however, was obtained by calculating the system transfer function of a linearized system. Often, the plant model is not available, and, moreover, the linearization is valid only at one operating point. This calls for another application of GA in determining the feedforward controller, which is illustrated in Ref. 33.

16.4 EVOLUTIONARY PROGRAMMING OPTIMIZER AND APPLICATION IN INTELLIGENT PREDICTIVE CONTROL

Power plant control has been a subject of many classic and modern control studies. Dealing with the MIMO characteristic of a power plant is a challenge in adaptive and optimal control approaches. In addition, a power plant has a wide range of nonlinear operation that makes the linear model-based controllers complicated. In recent years, intelligent control systems have been used extensively to solve the control problem of nonlinear systems. An intelligent control system is able to cover a wide range of operation and plant variation by using learning algorithms. The model reference controller is a widely used intelligent control structure [38] that has a complex implementation in the MIMO case. Fuzzy control approach is one solution to overcome the practical limitations of controller design [39]. Predictive control has been applied in power plant and process control extensively [40, 41]. Computational optimization methods, such as GA and evolutionary programming (EP), have begun a new era in training and adapting control systems to the plant variation.

In this section, development of an intelligent predictive controller system to control a boiler-turbine unit is introduced. A self-organized neuro-fuzzy identifier system performs as the plant model to anticipate the output response in a prediction time interval. The GA and error back-propagation training methods perform fuzzy rule extraction and membership function tuning, respectively. This predictive control scheme uses EP in the optimization of the control inputs to minimize errors of the predicted outputs and reference set-points. This section starts with a description of the structure of the intelligent predictive controller in a power plant application. The EP optimization formulation is expressed next. Neuro-fuzzy identifier and its training are explained briefly. Finally, the performance of the controller is reviewed.

16.4.1   Structure of the Intelligent Predictive Controller

A predictive controller anticipates the plant response for a sequence of control actions in a future time interval that is known as the *prediction horizon* [42]. The control action in this prediction horizon should be determined by an optimization method to minimize the difference between set-point and predicted response. Model-based predictive control (MPC) deploys this prediction of plant outputs and control inputs. The MPC uses a plant model to evaluate how control strategies will affect the future behavior of the plant. After finding a good strategy, MPC pursues that strategy for one control step and then reevaluates its strategy based on the plant's response. The MPC was originally developed for the linear model of a plant to provide the prediction formulation. Afterward, the predictive control method was extended limited classes of nonlinear systems [43]. Development of an adequate nonlinear empirical model for a complex system such as a power plant is a challenging problem. Intelligent systems such as neural network and fuzzy systems may replace the empirical model of the plant in predictive control methodology.

The structure of an intelligent predictive control system is shown in Fig. 16.14 for controlling a boiler-turbine power unit. A non-model-based identifier predicts the response of the actual plant in the prediction horizon. This identifier is a neuro-fuzzy system that is trained with appropriate training data to estimate a nonlinear system. The optimization block finds the sequence of inputs to minimize a cost function for future time, but only the first value of this sequence is applied to the plant. Figure 16.15 depicts the input/output prediction time horizon in the predictive control approach. The fuzzy rules and membership functions of this identifier are trained off-line by the actual measured data of the power unit. Then, the tuning process continues on-line to adapt the identifier with the plant's output while the control process performs. This training process improves the performance of the control system in the prediction interval.

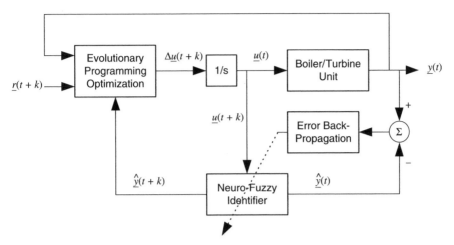

FIGURE 16.14   Intelligent predictive control of a power unit.

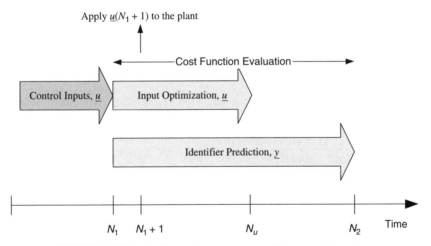

FIGURE 16.15   Prediction horizon in a predictive control system.

The prediction system is formed by an intelligent identifier to generate the anticipated plant output for a future time window of $N_1 \leq t \leq N_2$. The future control variable for this prediction stage is determined in an optimization algorithm for the time interval of $N_1 \leq t \leq N_u$, such that $N_u \leq N_2$. The optimal control inputs are determined by minimizing the following cost function:

$$J = \sum_{k=N_1}^{N_2} \left\| \underline{r}(t+k) - \hat{\underline{y}}(t+k) \right\|_R^2 + \sum_{k=N_1}^{N_u} \left\| \Delta \underline{u}(t+k) \right\|_Q^2 + \left\| \underline{r}(t) - \hat{\underline{y}}(t) \right\|_R^2, \quad (16.7)$$

where $\hat{y}(t+k)$ is the predicted plant output vector, which is determined by neuro-fuzzy identifier for time horizon of $N_1 \leq t \leq N_2$, $r(t+k)$ is the desired reference vector, $\Delta u(t+k)$ is the predicted input changes in $N_1 \leq t \leq N_u$ time range, and R and Q are weighting vectors.

16.4.2 Power Plant Model

In this study, the power plant simulation is performed by a model that was developed by Bell and Åström [44]. This is a third-order nonlinear model, derived by physical and empirical methods, as in the following:

$$\frac{dP}{dt} = -0.0018u_2 P^{9/8} + 0.9u_1 - 0.15u_3, \tag{16.8}$$

$$\frac{dE}{dt} = \frac{1}{10}((0.73u_2 - 0.16)P^{9/8} - E), \tag{16.9}$$

$$\frac{d\rho_f}{dt} = \frac{1}{85}(141u_3 - w_s), \tag{16.10}$$

$$w_s = (1.1u_2 - 0.19)P, \tag{16.11}$$

$$L = 0.05(0.13073\rho_f + 60\alpha_{cs} + 0.11q_e - 67.975), \tag{16.12}$$

$$\alpha_{cs} = \frac{(1 - 0.001538\rho_f)(0.8p - 25.6)}{\rho_f(1.0394 - 0.0012304p)}, \tag{16.13}$$

$$q_e = (0.85u_2 - 0.147)p + 45.6u_1 - 2.5u_3 - 2.1, \tag{16.14}$$

where p is drum steam pressure (kg/cm^2), E is electrical power (MW), w_s is steam mass flow rate (kg/s), L is water level deviation about mean (m), ρ_f is fluid density (kg/m^3), u_1, u_2, and u_3 are normalized fuel, steam, and feedwater valve positions, α_s is steam quality (mass ratio), and q_e is evaporation rate (kg/s).

In addition, the actuator dynamics of control valves are also modeled by

$$|du_1/dt| \leq 0.007 \quad (\text{s}^{-1}) \quad 0.0 \leq u_1 \leq 1.0, \tag{16.15}$$

$$-2.0 \leq du_2/dt \leq 0.02 \quad (\text{s}^{-1}) \quad 0.0 \leq u_2 \leq 1.0, \tag{16.16}$$

$$|du_3/dt| \leq 0.05 \quad (\text{s}^{-1}) \quad 0.0 \leq u_3 \leq 1.0, \tag{16.17}$$

to limit the rate of change in valve positions.

16.4.3 Control Input Optimization

The intelligent predictive control system does not depend on the mathematical model of the plant. Therefore, the optimization cannot be implemented by conventional methods in MPC. The search engine based on EP [45] is used to determine the optimized control variables for a finite future time interval. The EP performs a competition search in a population and its mutated offspring. The members of each population are the input vector deviations that are initialized randomly. The mutation and competition continue making new generations to minimize value of a cost function. The output of the optimizer block is the control valve deviations that are integrated and applied to the identifier and power unit.

The EP population consists of the individuals to present the deviation of control inputs. This population is represented by the following set:

$$U_n = \left\{ \underline{\Delta U}_{1,n}, \underline{\Delta U}_{2,n}, \ldots, \underline{\Delta U}_{n_p,n} \right\}, \tag{16.18}$$

such that U_n is the nth generation of population, and n_p is the population size. The ith individual in the nth generation is a matrix that is formed by input deviation vectors, expressed by

$$\underline{\Delta U}_{i,n} = \begin{bmatrix} \underline{\Delta u}_1^{i,n} & \underline{\Delta u}_2^{i,n} & \cdots & \underline{\Delta u}_m^{i,n} \end{bmatrix}, \quad \text{for } i = 1, 2, \ldots, n_p, \tag{16.19}$$

where m is the number of control inputs (or control valves in the power plant application). The $\underline{\Delta u}_j^{i,n}$ is the jth vector of the ith individual in the nth generation as in the following:

$$\underline{\Delta u}_j^{i,n} = \begin{bmatrix} \Delta u_j^{i,n}(1) & \Delta u_j^{i,n}(2) & \cdots & \Delta u_j^{i,n}(n_u) \end{bmatrix}^T, \quad \text{for } i = 1, 2, \ldots, m, \tag{16.20}$$

such that n_u is the number of steps in the discrete-time horizon for the power unit input estimation that is defined by

$$n_u = N_u - N_1. \tag{16.21}$$

where N_1 is the start time of prediction horizon, and N_u is the end time of the input prediction. The individuals of input deviation vector belong to a limited range of real numbers $\Re$:

$$\Delta u_j^{i,n}(\cdot) \in \left[\Delta u_{j,\min}, \Delta u_{j,\max} \right]. \tag{16.22}$$

In the beginning of the EP algorithm, population is initialized with randomly chosen individuals. Each initial individual is selected with uniform distribution from the above range of corresponding input.

The EP with adaptive mutation scale has shown a good performance in locating the global minima. Therefore, this method is used as it is formulated in Ref. 46. The flowchart of the EP process is shown in Fig. 16.16. The fitness value of each population is determined with a cost function to consider the error of predicted input and output in the prediction time window. The cost function of the ith individual in a population is defined by

$$f_{i,n} = \sum_{k=1}^{n_y} \left\| \underline{r}(t+k) - \hat{\underline{y}}_{i,n}(t+k) \right\|_R^2 + \sum_{k=1}^{n_u} \left\| \Delta \underline{U}_{i,n}(k) \right\|_Q^2 + \left\| \underline{r}(t) - \underline{y}(t) \right\|_R^2, \quad (16.23)$$

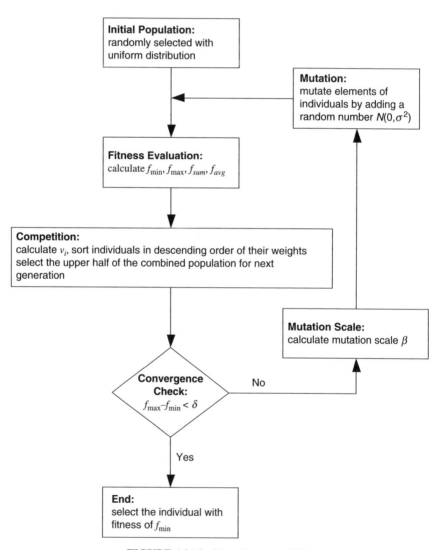

FIGURE 16.16 Flow diagram of EP.

where $r(t + k)$ is the desired reference set-points at a sampled time of $t + k$, and $\hat{y}_{i,n}(t + k)$ is the discrete predicted plant output vector, which is determined by applying $\Delta U_{i,n}(k)$ into the neuro-fuzzy identifier for time horizon of $n_y = N_2 - N_1$. The input deviation vector is determined in a smaller time window of n_u as in (16.21) such that $n_u \leq n_y$. The inputs of the identifier stay constant after $t + n_u$. Therefore, $\Delta U_{i,n}(k)$ is the kth row of the matrix in (16.19) for the ith individual of the nth generation and is expressed by

$$\Delta U_{i,n}(k) = \left[\Delta u_1^{i,n}(k) \quad \Delta u_2^{i,n}(k) \quad \cdots \quad \Delta u_m^{i,n}(k) \right], \qquad \text{for } i = 1, 2, \ldots, n_p. \quad (16.24)$$

The maximum, minimum, sum, and average of the individual fitness in the nth generation should be calculated for further statistical process by

$$f_{max}|_n = \{ f_{i,n} | f_{i,n} \geq f_{j,n} \quad \forall f_{j,n}, j = 1, 2, \ldots, n_p \}, \quad (16.25)$$

$$f_{min}|_n = \{ f_{i,n} | f_{i,n} \leq f_{j,n} \quad \forall f_{j,n}, j = 1, 2, \ldots, n_p \}, \quad (16.26)$$

$$f_{sum}|_n = \sum_{i=1}^{n_p} f_{i,n}, \quad (16.27)$$

$$f_{avg}|_n = \frac{f_{sum}|_n}{n_p}. \quad (16.28)$$

After determining the fitness values of a population, the mutation operator performs on the individuals to make a new offspring population. In mutation, each element of the parent individual as in (16.20) provides a new element by adding a random number, such as

$$\Delta u_j^{i+n_p,\, n}(k) = \Delta u_j^{i,n}(k) + N\left(\mu, \sigma_{i,j}^2(n) \right),$$

$$\text{for} \quad i = 1, 2, \ldots, n_p \quad j = 1, 2, \ldots, m \quad k = 1, 2, \ldots, n_u, \quad (16.29)$$

such that $N(\mu, \sigma_{i,j}^2(n))$ is a Gaussian random variable with mean μ and a variance of $\sigma_{i,j}^2(n)$. The variance of the random variable in (16.29) is chosen to be

$$\mu = 0,$$

$$\sigma_{i,j}^2(n) = \beta(n)(\Delta u_{j,\,max} - \Delta u_{j,\,min}) \frac{f_{i,n}}{f_{max}|_n}, \quad (16.30)$$

where $\beta(n)$ is the mutation scale of the nth population such that $0 < \beta(n) \leq 1$. After mutation, the fitness of offspring individuals is evaluated and assigned to them.

The generated new individuals and old individuals produce a new combined population with a size of $2n_p$. Each member of the combined population competes with some other members to determine which one is valued to survive to the next generation. For this purpose, the ith individual of the nth generation $\underline{\Delta U}_{i,n}$ competes with the jth individual $\underline{\Delta U}_{j,n}$, such that $j = 1, 2, \ldots, p$. The number of individuals to compete with is a fixed number p. Each of the p individuals is selected randomly with uniform distribution. The result of this competition is a binary number $v_{ij,n} \in \{0, 1\}$ to represent *lose* or *win* and is determined by

$$v_{ij,n} = \begin{cases} 1 & \text{if } \lambda_{j,n} < \dfrac{f_{j,n}}{f_{j,n} + f_{i,n}}, \\ 0 & \text{otherwise} \end{cases} \tag{16.31}$$

such that $\lambda_{j,n} \in [0, 1]$ is a randomly selected number with uniform distribution, and $f_{j,n}$ is the fitness of the jth selected individual in the nth generation. The value of $v_{ij,n}$ will be set to 1 if according to (16.31) the fitness of the ith individual is relatively smaller than the fitness of the jth individual, and that means the ith individual wins the competition. To select the survived individual, a weight value is assigned to each individual by

$$w_{i,n} = \sum_{k=1}^{p} v_{ij,n}, \qquad \text{for } i = 1, 2, \ldots, n_p, \tag{16.32}$$

where $w_{i,n}$ is the weight value of the ith individual in the nth generation. The first n_p individuals with higher weight values are the winners in the competition step and can survive to the next generation. The rest of the individuals with lower weight values are discarded.

The n_p individuals with the highest competition weight $w_{i,n}$ are selected to form the $(n + 1)$th generation. This newly formed generation participates in the next iteration. To determine convergence of the process, the difference of maximum and minimum fitness of the population is checked against a desired small number $\varepsilon > 0$ as in

$$f_{\max}|_n - f_{\min}|_n \leq \varepsilon. \tag{16.33}$$

If this convergence condition is met, then the individual with the lowest fitness is selected as a sequence of n_u input vectors for the future time horizon. The first vector is applied to the plant, and the time window shifts to the next prediction step.

Before starting the new iteration, the mutation scale changes according to the newly formed population. If the mutation scale is kept as a small fixed number, EP may have a premature result. In addition, a large fixed mutation scale will raise the possibility of having a nonconvergence process. An adaptive mutation scale provides a change of mutation probability according to the minimum fitness value of n_p individuals in the $(n + 1)$th generation. The mutation scale for the next generation is determined by

$$\beta(n+1) = \begin{cases} \beta(n) - \beta_{step} & \text{if } f_{min}|_n = f_{min}|_{n+1} \\ \beta(n) & \text{if } f_{min}|_n < f_{min}|_{n+1} \\ \beta_{final} & \text{if } \beta(n) - \beta_{step} < \beta_{final} \end{cases}, \qquad (16.34)$$

where n is generation number, and β_{step} is the predefined possible step change of the mutation scale in each iteration. A decreasing of the $\beta(n)$ should be limited such that $\beta(\cdot) \geq \beta_{final}$, always where β_{final} is a fixed positive number. The mutation scale does not change if the iterations of the EP keep converging to the final result. If the minimum fitness of the new generation does not change, then the mutation scale decreases with one step value. The initial value of $\beta(1)$ may be selected as large as 1 and then decreased by $0 < \beta_{step} < 1$. The practical values of initial mutation scale, β_{step} and β_{final}, are problem dependent, related to the number of generations and the complexity of the system.

16.4.4 Self-Organized Neuro-Fuzzy Identifier

The identification process may not perform desirably if it does not include the input/output interaction. For this purpose, series-parallel configuration [47] is chosen as it is drawn in Fig. 16.17. This identification structure considers the past output states in conjunction with the present inputs to determine the present output. The identifier with augmented inputs is represented by

$$\hat{y}(k+1) = \hat{f}\Big(\underline{y}(k), \ldots, \underline{y}(k-i); \underline{u}(k), \ldots, \underline{u}(k-j)\Big), \qquad (16.35)$$

such that $\hat{y}(k)$ is the estimated output at time step k, $\hat{f}(\cdot)$ is the identifier function, and $\underline{u}(k)$ and $\underline{y}(k)$ are plant input and output, respectively, at time step k. Neuro-fuzzy identifier technique is chosen for identification in predictive control loop [48].

The identifier is a neuro-fuzzy system that can be trained with appropriate algorithms. The structure of a MIMO neuro-fuzzy identifier, with m inputs and n outputs, is shown in Fig. 16.18. We assume that the ith input and the jth output retain p_i and q_j

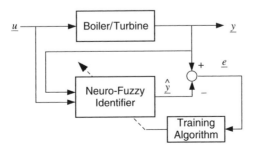

FIGURE 16.17   Series-parallel identification.

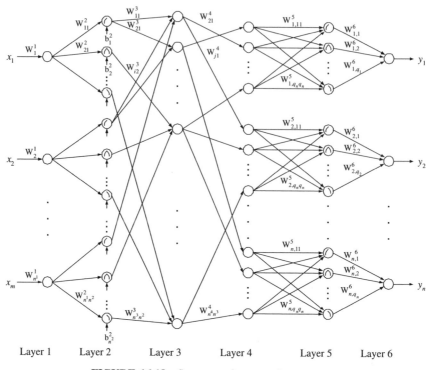

FIGURE 16.18 Structure of a neuro-fuzzy system.

bell-shaped membership functions, respectively. The number of possible rules, with *IF-THEN* format, is derived by

$$\eta = \prod_{i=1}^{m} p_i. \tag{16.36}$$

The first layer represents the input stage that provides the crisp inputs for the fuzzification step. The second layer performs fuzzification. The weights and biases respectively represent the widths and means of input membership functions. Number of neurons is $n^2 = \sum_{i=1}^{m} p_i$. The weighting matrix $\underline{W}^2$ is a block diagonal matrix containing the width of the input bell-shaped membership functions. The second layer has a bias vector $\underline{B}^2$ that contains the mean of the input membership functions. Using the exponential activation function, the outputs of the second layer neurons are the fuzzified system inputs. The number of the third layer neurons n^3 is equal to η. The weighting matrix of the third layer input represents antecedent parts of rules and is called the *premise matrix*. Each row of the premise matrix presents a fuzzy rule. The number of 1's in each row is equal to m. Column positions of 1's determine the assigned input membership function of a rule. The following $n^3 \times n^2$ matrix shows the possible combinations in the rule premises:

$$
\underline{W}^3 = \begin{bmatrix} 1 & 0 & \cdots & 0 & \cdots & 1 & 0 & \cdots & 0 \\ 1 & 0 & \cdots & 0 & \cdots & 0 & 1 & \cdots & 0 \\ & \vdots & & & \ddots & & \vdots & \\ 0 & 0 & \cdots & 1 & \cdots & 0 & 0 & \cdots & 1 \end{bmatrix}. \tag{16.37}
$$

The position of the 1's in the first row shows the existence of the first rule with the following premise:

$$
R^1\text{: IF } x_1 \text{ is } T^1_{x_1} \text{ AND} \ldots \text{AND } x_m \text{ is } T^1_{x_m} \text{ THEN} \ldots ,
$$

where $T^j_{x_i}$ is the jth linguistic term of the ith input. Other rows present the remaining rules in a similar manner. The neuron output is determined by the *min* composition that provides the firing strength of the rules. The fourth layer consists of separate sections for every system output. Each section represents consequent parts of rules for an output. The total number of neurons in this layer will be

$$
n^4 = \sum_{i=1}^{n} n_i^4 = \sum_{i=1}^{n} q_i, \tag{16.38}
$$

where n_i^4 or q_i is the number of membership functions for the ith output. The weighting matrix of the fourth layer defines consequences of rules and is called the *consequence matrix*. Because of multiple system outputs, it consists of n blocks:

$$
\underline{W}^4 = \begin{bmatrix} \underline{W}_1^4 & | & \underline{W}_2^4 & | & \cdots & | & \underline{W}_n^4 \end{bmatrix}^T. \tag{16.39}
$$

Each consequence matrix block is a $n_i^4 \times n^3$ matrix that consists of 1 and 0, e.g., the ith block can be

$$
\underline{W}_i^4 = \begin{bmatrix} 0 & 1 & \cdots & 0 \\ 1 & 0 & & 0 \\ \vdots & & \ddots & \vdots \\ 0 & 0 & \cdots & 1 \end{bmatrix}. \tag{16.40}
$$

Each row presents the output membership function term in the consequence part of a rule that is defined by a column number. For example, in (16.40), the position of 1 in the first row shows that the second rule (column 2) has the first membership function of the ith output term (row 1) in the consequence part, as in the following:

$$R^2: IF \dots THEN \dots y_i \text{ is } T^1_{y_i} \dots,$$

where $T^j_{y_i}$ is the jth linguistic term of the ith output. In each row, several 1's may exist, because consequences of several rules may have similar output semantic terms. Because a rule can have only one output term, the number of 1's in a column is not more than one. The fourth layer neuron output is the summation of firing strength of rules with similar consequences. Layer 5 makes the output membership functions. Combination of the fifth and sixth layers provides the defuzzification method, that is, approximation of the center of gravity, with the following equation for the ith output:

$$y_i = \frac{\sum\limits_{j=1}^{n_i^4} m^j_{y_i} \sigma^j_{y_i} \alpha^j}{\sum\limits_{j=1}^{n_i^4} \sigma^j_{y_i} \alpha^j}, \tag{16.41}$$

where $m^j_{y_i}$ and $\sigma^j_{y_i}$ are the mean and width of the jth output membership function for the ith system output. The weighting matrix of the fifth layer for the ith output section $\underline{W}^5_i$ is a diagonal $n_i^5 \times n_i^4$ matrix that contains the width of the output membership functions. The activation value of each neuron provides one term of summation in the defuzzified output. The linear activation function determines the output of the ith section. The sixth layer completes the defuzzification and provides a crisp output. The weighting vector of the ith neuron $\underline{W}^6_i$ contains the means of the ith output membership functions. The ith crisp output is derived by using an activation function similar to (16.41).

16.4.5 Rule Generation and Tuning

The neuro-fuzzy identifier can be configured as an intelligent system to be adapted to the boiler-turbine system. The training procedure can dedicate this to the identifier. Two steps of training are considered to configure the identifier. The first step is automatic rule generation. For this purpose, GA training is chosen [49], because of the specific structure of the neuro-fuzzy identifier. The GA finds the best fit for the fuzzy rules that perform the best error of estimated outputs. The second step of training takes advantage of the error back-propagation to tune the fuzzy membership function to match the boiler data. Figure 16.19 shows the implementation of this training. In the beginning, the identifier is loaded with the input/output membership functions and initial fuzzy rules. This data can be either chosen randomly or selected by a plant expert's decisions for the best result. The fuzzy rules are defined in the weighting matrices of the third and fourth layers. Positions of 1's in these premise and consequence weighting matrices define fuzzy rules. The position-related structure is suitable for using a GA in the rule generation. The encoded weighting matrices of the third and fourth layers define the GA chromosome.

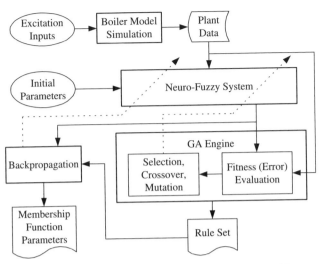

FIGURE 16.19 Training process of power unit identifier.

Because the neuro-fuzzy system is considered in a MIMO case, we can recall from (16.39) that the fourth layer weighting matrix consists of several blocks for every output. Therefore, the result of encoding is a compound chromosome that consists of n sections, corresponding with the number of system outputs. The number of genes in each section is equal to the rule number η. Each section encodes the premise and consequence matrices with nonbinary elements. Therefore, it is suitable to use GA with the nonbinary alphabet [50, 51]. The alphabet size of each section is equal to the membership function number of the corresponding output. A chromosome is expressed by

$$\underline{c} = \left[\theta_1^1 \theta_2^1 \cdots \theta_\eta^1 \theta_1^2 \theta_2^2 \cdots \theta_\eta^2 \cdots \theta_1^n \theta_2^n \cdots \theta_\eta^n \right], \tag{16.42}$$

such that n is the number of system outputs, and θ_j^i is a gene with *allele* values that belong to the alphabet set of

$$\theta_j^i \in A = \{0, 1, \ldots, q_i\}, \qquad \text{for } j = 1, \ldots, \eta. \tag{16.43}$$

The value of θ_j^i represents the row position of 1 in each column of the consequence matrix. This compound chromosome is considered as a set of subchromosomes. The GA will act separately on these subchromosomes to find the best fit. Having a set of experimental input/output plant data points, GA can be applied to find the optimal set of fuzzy rules. To declare the GA training problem, assume a MIMO neuro-fuzzy system $F: \mathfrak{R}^m \to \mathfrak{R}^n$, with n unknown vectors, representing encoded premise and consequence matrices:

$$F(\underline{\theta}_1, \underline{\theta}_2, \ldots, \underline{\theta}_n, x_1, x_2, \ldots, x_m) \in \mathfrak{R}^n, \tag{16.44}$$

where x_j is a system input, and $\underline{\theta}_i$ is the unknown parameter vector for the ith output. Input and parameter vectors are

$$\underline{x} = [x_1 \quad x_2 \quad \cdots \quad x_m]^T \in \mathfrak{R}^m, \tag{16.45}$$

$$\underline{\theta}_i = \left[\theta_1^i, \theta_2^i, \ldots, \theta_\eta^i\right]^T \in N_i^\eta, \tag{16.46}$$

for $N_i = \{1, 2, \ldots, q_i\}$. D is a set of training data as a collection of l pairs of input/output vector points

$$D = \{(\beta_{1i}, \ldots, \beta_{mi}, \gamma_{1i}, \ldots, \gamma_{ni})| \quad \forall i = 1, \ldots, l\}. \tag{16.47}$$

GA needs to find integer values r_j^i for $j = 1, \ldots, \eta$ and $i = 1, \ldots, n$ as unknown parameters, such that the evaluation of F at the set of D, expressed by $F(\underline{r}_1, \ldots, \underline{r}_n, \beta_{1i}, \ldots, \beta_{mi})$, will be the nearest to $\underline{\gamma}_i = [\gamma_{1i} \quad \gamma_{2i} \quad \cdots \quad \gamma_{ni}]^T$, where $\underline{r}_i = [r_1^i \quad r_2^i \quad \cdots \quad r_\eta^i]^T$. The fitness function of GA to solve this problem provides evaluation of population individuals and is based on the least squares principle. The fuzzy function F loaded with parameters of an individual $\underline{c}$ is called $F_{\underline{c}}$ and is evaluated on every data point of D. The sum of squares of differences between evaluated values and γ_i in (16.47) provides the error value of that individual as in

$$error(\underline{c}) = \sum_{i=1}^{L} \left(F_{\underline{c}}(\underline{r}, \underline{\beta}_i) - \underline{\gamma}_i\right)^2. \tag{16.48}$$

Subtracting this error function from the largest error value of the generation makes the fitness function:

$$f(\underline{c}) = error_{max} - error(\underline{c}). \tag{16.49}$$

The weighted roulette wheel assigns a weighted slot to each individual [52]. The proportion of selecting the ith individual of λ strings in the kth generation is

$$P_i(k) = \frac{f_i(k)}{\sum\limits_{j=1}^{\lambda} f_j(k)}. \tag{16.50}$$

The *crossover operator* generates two offspring strings from each pair of parent strings, chosen with probability of p_c. Crossover takes place in every subchromosome of parents. Crossover point is determined randomly with uniform distribution in the ith position:

$$\underline{a}_{\text{parent}} = [a_1^1 a_2^1 \cdots a_i^1 a_{i+1}^1 \cdots a_\eta^1 \cdots a_1^n a_2^n \cdots a_i^n a_{i+1}^n \cdots a_\eta^n],$$
$$\underline{b}_{\text{parent}} = [b_1^1 b_2^1 \cdots b_i^1 b_{i+1}^1 \cdots b_\eta^1 \cdots b_1^n b_2^n \cdots b_i^n b_{i+1}^n \cdots b_\eta^n].$$

(16.51)

Therefore, the offspring strings will be

$$\underline{a}_{\text{parent}} = [a_1^1 a_2^1 \cdots a_i^1 b_{i+1}^1 \cdots b_\eta^1 \cdots a_1^n a_2^n \cdots a_i^n b_{i+1}^n \cdots b_\eta^n],$$
$$\underline{b}_{\text{parent}} = [b_1^1 b_2^1 \cdots b_i^1 a_{i+1}^1 \cdots a_\eta^1 \cdots b_1^n b_2^n \cdots b_i^n a_{i+1}^n \cdots a_\eta^n].$$

(16.52)

Mutation method changes the value of a gene position with a frequency equal to the mutation rate, defined by p_m.

Tuning the parameters of fuzzy membership functions completes the training of the neuro-fuzzy system. Adjusting the membership function increases the accuracy of the identifier, as the initial membership functions have been chosen in the beginning. Error back-propagation is used for training neural networks and self-organized fuzzy logic systems. Let $D_k = \{\beta(k), \gamma(k)\}$ be a set of a given pair of desired system input/output, such that $\underline{\beta}(k) \subset \mathbb{R}^n$, $\underline{\gamma}(k) \subset \mathbb{R}^m$. If $\underline{\hat{y}}(k)$ is the output of the neuro-fuzzy system in response to the input $\underline{\beta}(k)$, the squared error is defined as the following

$$E = \frac{1}{2} [\underline{\gamma}(k) - \underline{\hat{y}}(k)]^T [\underline{\gamma}(k) - \underline{\hat{y}}(k)].$$

(16.53)

The training set contains the mean and width of the input/output membership functions. The sixth layer weighting element $w_{i,j}^6$ is updated by

$$w_{i,j}^6(k+1) = w_{i,j}^6(k) - v(\partial E / \partial w_{i,j}^6)\big|_k,$$

(16.54)

where v is the learning rate. The error rate is derived from (16.53) as in the following:

$$\frac{\partial E}{\partial w_{i,j}^6} = -[y_i(k) - \hat{y}_i(k)](\partial \hat{y}_i / \partial w_{i,j}^6)\big|_k,$$

(16.55)

such that rate of the neuro-fuzzy output is derived by

$$\frac{\partial \hat{y}_i}{\partial w_{i,j}^6} = \frac{I_{i,j}^6}{\sum_{l=1}^{n_i^5} I_{i,l}^6}.$$

(16.56)

Similar to derivation of (16.53)–(16.56), the width and mean of the membership functions in other layers can be derived. The learning rates of error back-propagation should be chosen appropriately. Training ends after achieving specified error or reaching the maximum iteration number.

16.4.6 Controller Implementation

An object oriented programming (OOP) environment is used in the implementation of the controller. The structural class definition of this programming technique is appropriate for the network architect of the neuro-fuzzy identifier. The identifier is trained and initialized before the control action starts. The training data is obtained from the mathematical model of the power unit. A set of valve position inputs is chosen to excite the power unit model in a staircase of electrical power operating points. The input vector of the identifier includes control valve positions, delayed drum pressure, and level deviation. Every input and output of the neuro-fuzzy identifier has seven membership functions. The GA training engine evaluates rule set candidates and converges to the final set. The crossover and mutation rates of GA are chosen to be $p_c = 0.7$, $p_m = 0.005$ while the population size is 10. In the second phase of training, error back-propagation performs membership function tuning.

After the initial training, the identifier is engaged in the closed loop of the predictive control as in Fig. 16.14. The parameters of prediction horizon are selected to be $n_y = 100$ and $n_u = 70$, with time step of $\Delta t = 10$ s. Population size is chosen to be $n_p = 10$. In the first simulation test, the electrical power is increased from low to medium power (15.26–66.65 MW). Both steam pressure and electrical power set-points are ramped up in this test. The transient response of the drum pressure P and electrical power E are shown in Figs. 16.20 and 16.21. In addition, the identifier is replaced by the mathematical model of the plant, and the response of this set up is also illustrated in these figures for comparison. As can be seen in these figures, steam pressure starts after the ramp variation with a fast slope, and that is due to the

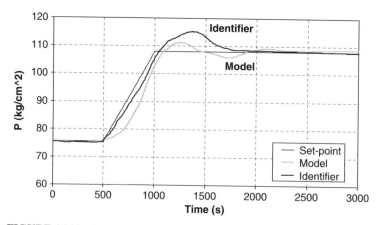

FIGURE 16.20   Response of drum pressure in power and pressure ramp test.

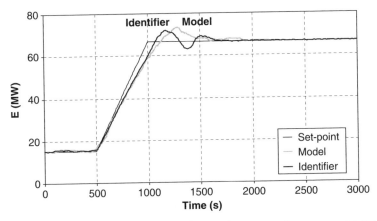

FIGURE 16.21   Response of electrical power in power and pressure ramp test.

prediction horizon. The pressure response takes an overshoot and settles to constant reference input after 75 prediction steps. The electrical power performs faster and makes a ringing before settling down. The on-line training of the identifier explains the ripples in steady state. To improve the transient response, one may consider a larger prediction time or a larger population size in EP.

Figures 16.22 and 16.23 show the transient of the drum pressure and electrical power in response to step change set-points. The electrical power and drum pressure are stepped up from medium power to the high power operating points (66.65 MW to 128.9 MW). The pressure response is slower than the electrical power transient and settles after a ringing in 150 prediction steps. However, the electrical power shows a transition that reflects the discrete nature of the control actions. The envelope of this power transient is similar to an overdamped transient response, settling in 100 prediction steps.

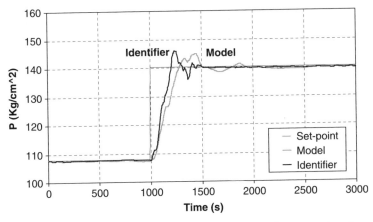

FIGURE 16.22   Response of drum pressure in power and pressure step change.

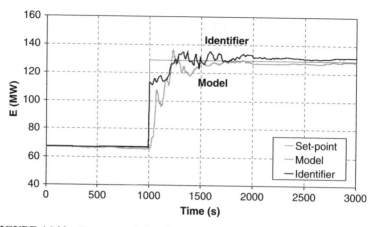

FIGURE 16.23 Response of electrical power in power and pressure step change.

16.4.7 Summary

Implementation of an intelligent predictive control is introduced for a boiler-turbine power unit. This controller uses a neuro-fuzzy identifier to predict the plant outputs in response to a sequence of valve positions in a future time window. This sequence of plant inputs is optimized with EP. Simulation of this predictive control in power/pressure step and ramp tests shows a fast transient response to the input changes. The GA and error back-propagation training of the neuro-fuzzy identifier is also formulated for automatic rule generation and membership function tuning. The obtained identifier is trained for local and wide-range training data. Overall, the intelligent predictive control is suitable for nonlinear MIMO systems with slow varying dynamics such as complex as power plants, where the plant's mathematical model is difficult to obtain.

16.5 AN INTERACTIVE COMPROMISE PROGRAMMING-BASED MO APPROACH TO FACTS CONTROL

MO optimization methodologies for decision aids have received considerable development over the past 30 years. In most power system applications, minimizing both operation costs and transmission losses or minimizing both operation and emission costs are taken as objectives. With the introduction of free competition to the market, besides economic operation, reliability and quality of electricity supply has received more attention than ever before. These are the major reasons why MO optimization techniques for decision support have assumed a new strategic importance in the current restructured electric power industry.

As already recognized, increased loading of power systems, environmental restrictions, combined with worldwide power-industry restructuring, particularly the unbundling of power generation from power transmission and a large quantity of bilateral contracts and wheeling services in the electricity market, requires a more flexible and effective means for power flow control (PFC) and for taking full advantage of existing power resources. On the other hand, by the use of line impedance, magnitude and angle of nodal voltage to redistribute power flow, the FACTS concept has opened up brand new diagrams for power system control with particular respect to thermal, voltage, and dynamics constraints. In the scope of steady-state power system analysis, a host of technical performances of FACTS applications in power networks have been identified and evaluated in a voluminous literature, including boosting available transfer capability [53], increasing the profit of auction of financial transmission rights (FTR) [54], improving network loadability [55], loop flow prevention and parallel flow elimination [56], providing secure tie line connections to neighboring regions for spinning reserve sharing [57], and so on. Practically, FACTS devices are to be applied to facilitate economic system operation and power transactions, and to improve system security and reliability. Meanwhile, they are also in demand to fulfill various tasks by local PFC to improve transmission services. Therefore, in order to take full advantage of FACTS devices, it will be of great benefit to design a concerted and deliberate control strategy with consideration of the major applications.

However, so far there are very few works published to consider optimal power system operation incorporating FACTS devices with multiple goals. In the OPF model of Ref. 58, with minimization of system operation costs as an objective, PFC targets of FACTS devices are formulated as a set of constraints. However, inappropriate setting of the constraints may cause infeasibility of the problem. Because more than one conflicting goal is involved in the problem, a single solution does not exist that can optimize all the objectives. For system operators, the most important task is to determine a satisfactory compromise solution by trading off all the objectives. Apparently, it is necessary to use MO optimization methods to deal with such a problem.

This section focuses on the application of a MO optimization approach to optimal operation of power systems incorporating FACTS devices. First, taking economic system operation and attainment of specific local PFC targets as objectives, a MO optimization model for FACTS control is formulated. For most prevalent MO optimization algorithms, it is very difficult to handle problems with these noncommensurable objectives. Therefore, based on CP, a DWCP is introduced to solve the problem in an interactive procedure. Different from normal interactive approaches (usually beset with the difficulties of determining the step and direction of weight or aspiration level correction), by using this method, it is very efficient to obtain the reference point (to reflect decision makers' (DMs') preference changes during the interactive process), without any further analysis or extra programming. Considering imprecise and vague objectives and preferences in the decision environment, fuzzy set theory is used to simulate them with the appropriate membership functions. Moreover, these membership functions can also conveniently represent relative deviations in the CP. A reduced practical power system is employed to demonstrate the performance of the proposed approach.

16.5.1   Review of MO Optimization Techniques

As the name implies, the MO optimization technique is one that can be used to achieve an optimal solution with multiple objectives simultaneously. Shown in (16.57), the general model of MO optimization problems has the same expression as that of conventional problems, except that the objective vector has a rank larger than 1.

$$\text{Minimize: } F(C, X)$$

subject to:

$$G(C, X) = 0$$

$$H(C, X) \leq 0, \tag{16.57}$$

where X is a state vector that consists of dependent variables and fixed system parameters; C is a control vector that consists of control variables; $F(C, X)$ is the objective(s) to be optimized; $G(C, X)$ are nonlinear equality constraints; and $H(C, X)$ are nonlinear inequality constraints.

In practice, for optimization models with multiple objectives, which sometimes are conflicting and noncommensurable, it usually becomes impossible to find an appropriate solution C^* to obtain the optimum values of all the objectives simultaneously. Based on this reality, a concept of optimality (called Pareto optimality), is used to define a domain of nondominated solutions on the boundary of the achieved domain [59]. The definition states that C^* is Pareto optimal if there exists no other feasible vector C that would improve some objective values without worsening at least one criterion simultaneously. In this sense, optimization actually means the generation of a range of noninferior solutions that show trade-offs among objectives.

The prevalent MO optimization algorithms are introduced briefly in the following subsections.

16.5.1.1   *Weighting Method*   Proposed by Zadeh in 1963, the basic idea of the method is to linearly combine the various objectives into a single-objective (SO) function by simply assigning relative weight to each objective. Thus, the transformed SO optimization problem can be easily solved by any conventional methodology. If the weights are interpreted as representing the relative preference of some DMs, then the solution to the SO problem is equivalent to the best-compromise solution of the corresponding MO problem.

Because of its simple implementation, the weighting method has been identified as one of the most widely used MO optimization approaches in power system applications [60–62]. However, it is not free from criticism:

- Reasonable weight assigning is a thorny point for any practical application, especially for the problem with noncommensurable objectives.

- Because of the incorporation of multiple noncommensurable factors into a single function, the objective may lose significance.
- It is a frequent observation that an evenly distributed set of weights fails to produce an even distribution of points for a good approximation of the Pareto surface.

16.5.1.2 *Goal Programming*

First described in 1961 by Charnes and Cooper [63], goal programming (GP) represents one of the most frequently used approaches in the literature of power systems. In a practical MO problem, the DM's aims are often to achieve a satisfactory level for all objectives (termed aspiration levels or goals). Based on this fact, the purpose of GP is to minimize the derivations between the desired goals and the actual results.

The GP has been used to calculate environmental marginal cost for the problem of environmentally constrained optimal generation dispatch [64, 65]. Fan et al. [66] applied GP to solve the real-time economic dispatch problem. The approach is also employed to solve weekly maintenance scheduling of generators under economic and reliability criteria in Ref. 67. However, generally the priority order of objectives is hard to presume. It may cause a relatively long execution time. Above all, the major shortcoming of GP is that inappropriate construction of goals can result in inferior solution [59].

16.5.1.3 *ε-Constraint Method*

Proposed in 1967 by Marglin, the ε-constraint method is regarded as the most intuitively appealing MO programming [68]. In the approach, one of the criteria is chosen as the objective function while all of the others are treated as inequality constraints by introducing the limit level ε. Thereby, the MO problem is transformed into a SO problem. Calculating the problem repeatedly for various values of ε, this method allows the DM to get all Pareto solutions of the problem. However, this approach can result in a considerable computation burden, due to the large quantity of ε combinations.

In Ref. 69, a two-stage solution methodology based on the ε-constraint method and simulated annealing (SA) technique is developed to solve the distribution network reconfiguration problem together with consideration of power loss minimization and relieving branch overload.

16.5.1.4 *Compromise Programming*

Technically, the ideal solution is defined as a collection of all objective values, which is preferred to all others among all achievable scores for any objective. With this concept, a compromise programming (CP), proposed by Zeleny in 1973, is implemented through replacing *a priori* defined goals of GP by the actually achievable ideal solution [59]. By the use of the CP, the best compromise solution among the objectives that is the closest one to the ideal point can be achieved. Most importantly, the CP provides an excellent base for an interactive procedure.

In 1995, Chattopadhyay, Banerjee and Parikh [70] applied the CP to evaluate demand-side management (DSM) options, which simultaneously take into account capital and operating cost savings, reduced emissions, and improved reliability of electricity supply. In 1999, Chattopadhyay and Momoh [71] used the CP to solve a generation-transmission operation planning problem.

16.5.1.5 Fuzzy Set Theory Applications In addition to the conventional MO optimization algorithms, applications of fuzzy set theory in MO problem solving have been drawing the attention of many researchers. In an attempt to bridge the gap between the conventional mathematics and natural discourse, in 1965 Zadeh introduced the concept of fuzzy quantifiers [72]. Given a collection of objects Y, a fuzzy set A is defined as

$$A = \{(y, \mu_A(y))|y \in Y\}, \tag{16.58}$$

where $\mu_A(y)$ is the membership function of y, representing the degree that y belongs to A. In contrast with a normal crisp set, where the membership value could only be either 0 or 1, the value in fuzzy sets may range from 0 to 1 continuously.

For problems with multiple conflicting objectives, because it is impossible to attain all the ideal values, the best compromise solution that can satisfy the DM is sought instead. Consequently, the motivations of the DM are always described using "as much as possible, at least larger than, about," and so forth, linguistically, which is difficult to in represent any conventional mathematics form. Fuzzy set theory can be used to translate these imprecise and vague preferences into a clearcut model. In addition, in typical fuzzy programming, assuming the full symmetry relationship between objective functions and constraints, the optimal solution can be achieved by seeking the maximum degree of "overall satisfaction" [73], that is, maximizing the minimum value of all the membership values of objectives and constraints included. Therefore, fuzzy set theory leads to computationally efficient MO algorithms; it has been playing a unique role in the MO optimization field for a long time now.

Thus far, in the scope of power system operation and planning, the fuzzy method has been proposed to deal with contingency constrained OPF [74], hydrothermal coordination [75], and transmission planning [76], and so on.

16.5.1.6 Genetic Algorithm Genetic algorithms were developed by John Holland and his colleagues in 1975. Based on the mechanics of natural selection and natural genetics, the GA starts with a population of strings that represent the possible solutions and generates successive populations of strings by combining survival of the fittest among string structures.

Proposed originally by Schaffer in 1984, and motivated by a suggestion of a nondominated GA outlined by Goldberg in 1989 [77], a number of studies on MO GAs have been undertaken and the implementations have been successfully applied in the various MO fields [78]. Different from normal optimization and search procedures, a GA searches from a population of solution points. Therefore, multiple noninferior solutions can be found in a single run of GA. In 1998, Das and Patvardhan [79] presented a GA-based approach to deal with the emission-constrained economic load dispatch problem.

16.5.1.7 Interactive Procedure   Actually, there are two participants involved in MO decision-making: the system analyst and the DM. With respect to the preference preassigned or the aspiration levels or objectives of the DM, the aim of the analyst is to provide efficient information and appropriate alternatives for supporting the DM. However, in practice, due to the characteristics of noncommensurability and conflict of objectives, vagueness and inconsistency of human desires, especially the DM's possible unfamiliarity with system optimization potential, the complete a *priori* knowledge of preference or aspiration levels of objective performance with precision is not always available. Consequently, both sides of the decision process (i.e., analysis of what is available and of what is desirable) are to be carried out jointly and dependent on each other. Therefore, it is necessary to introduce an interactive process between analysts and DMs, where objectives evolve by DMs according to available alternatives proposed by analysts, which are, in turn, adjusted and generated in accord with existing objectives.

After the reasonable aspiration changes as a result of learning and experience, this method has the advantage of allowing DMs to gain a greater understanding of the problem. More importantly, without forcing them into an exhaustive examination of all Pareto solutions, the procedure can guide DMs to obtain what they consider the best compromise.

In Ref. 80, an interactive method is applied to solve a reactive power planning problem, taking both economy and security of the system operation into account. Generally speaking, most of the current interactive methods focus on determining the step and direction of weight or aspiration level change, which are usually undertaken by sensitivity analysis [81, 82], asking DMs to obtain a series of paired comparisons between objectives [60], or even trial-and-error [83, 84]. The whole process could be quite difficult and unreliable. Besides that, it has to be pointed out that sometimes the DM's behavior is not rational or the reference points designated by the DM keep on changing frequently. This will result in a time-consuming interactive process.

16.5.2 Formulated MO Optimization Model

Generally speaking, among a variety of common objectives of the power system optimization, the major one is economy. For power systems, particularly heavily loaded systems and systems with loop flow or parallel flow, optimal generation dispatch is often restricted by stressed transmission circuits or (and) relatively low voltage nodes. Physically, this situation also limits bulk power transactions between long-distance electricity market participants. Therefore, by relieving restriction imposed by transmission network resources, essentially, FACTS can provide considerable economic benefits by allowing for the use of more cost-efficient generators.

Meanwhile, it is widely recognized that by controlling active, reactive power and voltage individually or simultaneously, FACTS devices are able to accomplish power flow control flexibly. From the steady-state power system operation viewpoint, by local power flow control to eliminate violation of voltage and thermal limits, so as to maintain a certain quantity of network transfer capacity demanded for security and further utilization, use of FACTS devices could also have a potentially beneficial effect in this regard.

16.5.2.1 Formulated MO Optimization Model for FACTS Control Based on the analysis above, a MO optimization problem is formulated as (16.59):

$$\underset{C}{\text{Min}}\, F(X, C)$$

$$\text{s.t.} \quad G(X,\, C) = 0$$

$$H(X, C) \le 0 \tag{16.59}$$

Minimization of system operation cost and local power flow control target fulfilling are taken as the objectives, which are denoted by the vector F. Operation cost is represented approximately by $\alpha + \beta P_G + \gamma P_G^2$. Therefore, for a power system with N_G generators, the operating cost can be further expressed as:

$$F_G = \sum_{i=1}^{N_G} \alpha_i + \beta_i P_G + \gamma_i P_G^2 \tag{16.60}$$

X represents a vector of state variables including magnitude and angle of nodal voltages. In addition to the control variables of FACTS devices, the control vector C includes generator output and voltage of PV node. Power flow equations are included as equality constraints and are denoted by G. The inequality constraints H are imposed on limitations of nodal voltage magnitude, line transfer limit, generator output, and FACTS device capacity.

Here, a power injection model (PIM)-based FACTS model and a power flow control approach is used to represent FACTS devices and to deal with the local PFC problem, which has been presented in detail in Ref. 85. For clarity and integrity of the chapter, the related model will be introduced briefly.

16.5.3 Power Flow Control Model of FACTS Devices

16.5.3.1 Control Variables In PIM, because active and (or) reactive power injections represent all the features of the steady-state FACTS device, it is rational to take the power injections as independent control variables for power flow control. According to the derivation of the PIM in Ref. 86, line flow control of FACTS devices can be considered as an additional controllable power flow through the line, which is superimposed on the "natural" power flow. Taking the unified power flow controller (UPFC) as an example, it is implemented by two back-to-back converter-based synchronous voltage sources (SVSs), which can generate or absorb reactive power independently. Therefore, for the UPFC installed on line

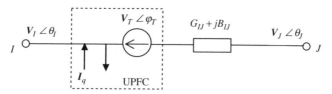

FIGURE 16.24 Voltage source model of UPFC.

L, $I - J$, near bus I, as shown in Figs. 16.24 and 16.25, besides the active power injection, $P_{I(\text{inj})}$, by the series SVS, there are two independent reactive power injections involved. One is $Q_{II(\text{inj})}$ for regulating voltage, which is injected by the shunt SVS directly to bus I as I_q. Generated by the series SVS, the other flows via line L for reactive power flow control, which is represented as $Q_{IL(\text{inj})}$. That is, in the model, the total reactive power injection to bus I, $Q_{I(\text{inj})} = Q_{II(\text{inj})} + Q_{IL(\text{inj})}$. The control vector is written as $C = [P_{I(\text{inj})} \ Q_{IL(\text{inj})} \ Q_{II(\text{inj})}]$. For active and reactive power flow control of the UPFC, with $P_{I(\text{inj})}$ and $Q_{IL(\text{inj})}$ as control variable, the controlled power flow of line L, P_L and Q_L, can be expressed as power flow P_{L0} and Q_{L0} with the effect of the controller, plus $P_{I(\text{inj})}$ and $Q_{IL(\text{inj})}$, which flow along line L, as shown in Fig. 16.25 clearly. As a result, for UPFC control, the relationship between the power injections and the controlled power flow can be expressed as:

$$P_L(X_L, P_{I(\text{inj})}) = P_{L0}(X_L) - P_{I(\text{inj})} \tag{16.61}$$

$$Q_L(X_L, Q_{IL(\text{inj})}) = Q_{L0}(X_L) - Q_{IL(\text{inj})} \tag{16.62}$$

$$V_I = V_I(Q_{II(\text{inj})}), \tag{16.63}$$

where X_L respesents voltage and phase angle of buses I and J, V_I, V_J, θ_I, θ_J.

As the power injections are only an interim result, once the required power injections are obtained, the original control parameters, including magnitude V_T, phase angle φ_T of the series inserted voltage, and magnitude of the current I_q, can be derived easily according to the voltage source model of the UPFC.

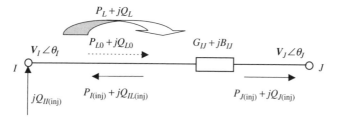

FIGURE 16.25 Power injection model of FACTS devices for power flow control.

TABLE 16.7 Control Variables for Available Control Targets and Control Parameters of FACTS Devices

Classification	Steady-State Functions	Control Targets	Power Injections		FACTS Devices	Control Parameters
			Control Variable(s)	Noncontrol Power Injection(s)		
Shunt controller	Voltage control	V_I	$Q_{II(inj)}$	—	SVC STATCOM	I_S
Series controller	Active power flow control	P_L	$P_{I(inj)}$	$Q_{IL(inj)}$, $Q_{J(inj)}$	TCPS TCSC SSSC	σ x_c V_C
Unified controller	Active, reactive power flow control, voltage control	P_L, Q_L, V_I	$P_{I(inj)}$, $Q_{IL(inj)}$, $Q_{II(inj)}$	$Q_{J(inj)}$	UPFC	V_T, φ_T, I_q

As a unified controller, the UPFC possesses the functions of both series controller and shunt controller. Series controllers, such as the thyristor-controlled series capacitor (TCSC), thyristor-controlled phase-shifter (TCPS), and static synchronous series compensator (SSSC), are characterized as active line flow control. Therefore, there is only one control variable, $P_{I(inj)}$, involved. The relationship between the power injections and the controlled power flow can be represented by (16.61). As shunt controllers, including static var compensator (SVC) and STATic synchronous COMpensator (STATCOM), are only used for voltage regulation, the relationship between the control variable, $Q_{II(inj)}$, and the control target, V_I, can be expressed as (16.63). Control models for all families of FACTS controllers are summarized in Table 16.7, where for TCPS, σ is the phase shifting angle, x_c represents the controllable reactance of TCSC, V_C denotes the series inserted voltage of SSSC, and I_S is the reactive current injected by SVC and STATCOM.

16.5.3.2 Applied Power Flow Control Model

In the power flow control model, minimization of the total mismatch of the preassigned control targets is taken as the objective. That is, for any FACTS device h that is installed on line $L, I - J$, as shown in Fig. 16.25, the objective function is represented by:

$$F_h = \mu_1 |P_L - P_{LT}| + \mu_2 |Q_L - Q_{LT}| + \mu_3 |V_I - V_{IT}|, \qquad (16.64)$$

where $P_{LT}, Q_{LT}, V_{IT}, P_L, Q_L$, and V_I are the control targets and the actual value of active, reactive power flow along line L and voltage of node I, respectively. According to the steady-state control functions and the corresponding classifications of FACTS devices, the model can be employed to accomplish the control of all prevalent FACTS devices flexibly, by suitably setting the coefficients μ_1, μ_2, and μ_3 as follows:

For the shunt controller: $\mu_1 = \mu_2 = 0$, $\mu_3 = 1$
For the series controller: $\mu_1 = 1$, $\mu_2 = \mu_3 = 0$
For the unified controller: $\mu_1 = \mu_2 = \mu_3 = 1$

For a power system with N_F FACTS devices installed to implement power flow control, the optimization objective is

$$\text{Min } F = \sum_{h=1}^{N_F} F_h.$$

16.5.4 Proposed Interactive DWCP Method

Inspired by the real-life interactive decision process, a DWCP is introduced to solve the MO problem by making the achieved goals as far as possible from the anti-ideal point, which is termed the worst compromise (WC) point [87]. Because the method is based on CP, for a better understanding, the applied CP model will be introduced briefly.

16.5.4.1 Applied Fuzzy Compromise Programming As introduced earlier, based on the concept of ideal point, the purpose of CP is to achieve the best compromise solution among objectives, which is the closest to the ideal point. Based on the fact that sometimes DMs strive to be as far as possible from the worst achievable objective values, which is referred to as an anti-ideal solution, CP can also be formulated to maximize the distance between the achieved objectives and the anti-ideal points [59]. This formulation of CP is applied here, as shown in (16.65):

$$\underset{C}{\text{Max}} \left(\sum_{N_{obj}} (F(X,C) - F^{AI})^2 \right)^{1/2}$$

$$\text{s.t.} \quad G(X,C) = 0$$

$$H(X,C) \leq 0, \tag{16.65}$$

where N_{obj} is the total number of objectives considered in the model, and F^{AI} is the corresponding anti-ideal value of each objective.

Practically, the operation and decision environment is usually vague. An important advantage of fuzzy formulation is its ability to handle elastic or ambiguous optimization objectives by introducing fuzzy membership functions. Particularly, local power flow control seems to be a promising field for the application of the concept, where not all control objectives can be stated and are therefore necessary to be reach precisely. Furthermore, in most applications of CP, in order to avoid any undesirable effects of the units of measurements on the objectives, relative deviations are usually adopted. They can be represented by fuzzy membership functions intuitively and conveniently. With these consequent benefits, application of fuzzy set theory together with CP will be more progressive and practical than conventional algorithms.

To reflect the satisfaction degree of a given solution, the objectives are reformulated by using fuzzy membership functions as follows:

16.5.4.2 Operation Cost Minimization

$$\underset{C}{\text{Max}}\,(\mu_{F_{EO}}(X, C)),$$

where the membership function of the objective of economic operation (EO) is

$$\mu_{F_{EO}}(X, C) = \begin{cases} 0 & F_{EO}(X, C) \geq F_{EO}^{WC} \\ \dfrac{F_{EO}^{WC} - F_{EO}(X, C)}{F_{EO}^{WC} - F_{EO}^{Ideal}} & F_{EO}^{Ideal} < F_{EO}(X, C) < F_{EO}^{WC} \\ 1 & F_{EO} \leq F_{EO}^{Ideal} \end{cases} \qquad (16.66)$$

F_{EO}^{Ideal} is an ideal value of the objective with membership value of 1. F_{EO}^{WC} represents the WC value of the objective with membership value of 0. According to the formulated linear membership function shown in Fig. 16.26, membership values of any solution between F_{EO}^{Ideal} and F_{EO}^{WC} can be determined easily.

16.5.4.3 Local Power Flow Control
Practically, nodal voltages are supposed to be within a specified range; any value within the range obtained is acceptable, which is not necessary to be fixed. Voltage control is included in the limitations of nodal voltages in H. Therefore, only active and reactive line flow controls are

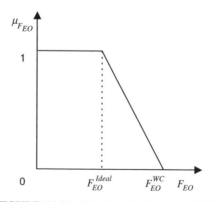

FIGURE 16.26 Membership function of EO.

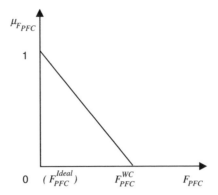

FIGURE 16.27 Membership function of PFC.

considered here. Because the membership functions of active and reactive power flow control have a similar feature, a general representation, including the fuzzy membership functions and the corresponding illustration curves, is shown in (16.67) and Fig. 16.27,

$$\max_{C} \; \min_{P,Q} (\mu_{F_{PFC}^{P}}(\boldsymbol{X},\boldsymbol{C}), \qquad \mu_{F_{PFC}^{Q}}(\boldsymbol{X},\boldsymbol{C})),$$

where

$$\mu_{F_{PFC}^{\Gamma}}(\boldsymbol{X},\boldsymbol{C}) = \begin{cases} 0 & F_{PFC}^{\Gamma}(\boldsymbol{X},\boldsymbol{C}) > F_{PFC}^{\Gamma WC} \\ 1 - \dfrac{F_{PFC}^{\Gamma}(\boldsymbol{X},\boldsymbol{C})}{F_{PFC}^{\Gamma WC}} & F_{PFC}^{\Gamma}(\boldsymbol{X},\boldsymbol{C}) \le F_{PFC}^{\Gamma WC}, \quad \Gamma = P, Q. \end{cases} \quad (16.67)$$

The interpretation of the membership function is the absolute value of deviation between the real value and the control target should be "as close to 0 as possible," at least not higher than F_{PFC}^{WC}, which is an assumed acceptable tolerance. The corresponding satisfaction degree will be decreased linearly in that tolerance range, as shown in Fig. 16.26.

It needs to be mentioned that, in the interactive process, the WC values of the objectives keep on changing to reflect the dynamic preferences of the DM, which will be demonstrated in Section 16.5.7.

16.5.5 Proposed Interactive Procedure with Worst Compromise Displacement

During the interactive process, in each interaction, a DM gives comments to the current solution of all objectives. As reference point to reflect the DM's preference changes, the WC points are to be adjusted according to these comments. They will be taken as the corresponding anti-ideal points of the model for the next iteration. Then, with the modified MO optimization model, another optimal solution can be calculated and presented to the DM. Thereby, the WC points keep on being displaced to capture the DM's intentions interactively, until he or she is satisfied with the solution provided. Generally, the procedure consists of three phases:

16.5.5.1 Phase 1: Model Formulation This phase is to formulate a model according to the DM's preference. If this is not the initial iteration, to reflect the DM's aspiration level changes, a new WC parameter for any objective can be determined following the rules in Table 16.8. Then, with the WC points displacing, a new optimization model is obtained with modified membership functions of the objectives.

16.5.5.2 Phase 2: Noninferior Solution Calculation With fuzzy membership functions formulated, the solution $S = [C, X]$ can be calculated as a result of the following problem using the CP:

$$\underset{C}{\text{Max}} \sum_{N_{obj}} (\mu_F(C, X))$$

$$\text{s.t.} \quad G(X, C) = 0$$

$$H(X, C) \le 0 \tag{16.68}$$

This conversion enables us to view the original MO optimization problem from a more practical angle; that is, rescheduling the control variables to achieve a scenario with the sum of satisfaction degrees of all the objectives as high as possible, without violating any rigid constraints.

16.5.5.3 Phase 3: Scenario Evaluation The resultant scenario is presented to the DM. If the DM is satisfied with the current solution, it means that the best compromise solution has been achieved. If not, the solution is evaluated to elicit his or her preferences. To all the objective values, the DM's opinions can be classified into three categories as listed in the left column of Table 16.8.

TABLE 16.8 Rules for New Worst Compromise Value Assignation

DM's comments on the objective values of iteration t	New worst compromise value assigned for iteration $t + 1$
Needs to be improved	Set at the current objective value attained
Retain at the current level	Keep at the current WC point
Allowed to be worsened	Is lowered to the nearest experienced point

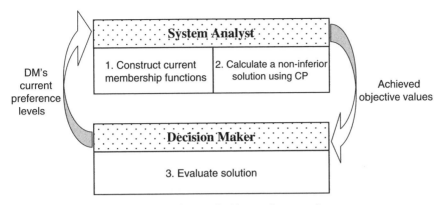

FIGURE 16.28 Applied interactive procedure.

With the information of the DM's preference obtained in phase 3, a new iteration can be started with phase 1. The whole procedure will not be terminated until the DM is satisfied with a decision alternative completely or the same decision alternative is computed in two consecutive sessions.

The structure of the applied interactive procedure is sketched in Fig. 16.28.

16.5.6 Implementation

For clarity, the whole interactive procedure of the proposed displaced worst compromise method for the problem is outlined in Fig. 16.29, where t_{max} is the pre-assigned maximum interaction number.

From the implementation perspective, there are two major modules involved in the proposed FACTS control approach: an optimization model and an AC power flow calculation. A predictor-corrector primal-dual interior point linear programming (PCPDIPLP), as proposed by Mehrotra in 1992 [88], is employed to solve the formulated OPF problem. Compared with other optimization algorithms, the method has advantages such as reliability of the optimization and high speed of the calculation, which are both very important for the interactive approach. Noting that conventional power flow algorithms are convergence failure-prone when applied to power systems with embedded FACTS devices [89, 90], an improved optimal multiplier Newton-Raphson (OMNR) power flow algorithm is adopted here, which has been presented in Ref. 91.

16.5.7 Numerical Results

In order to illustrate feasibility of the interactive MO approach proposed, a reduced practical system with 29 nodes and 69 branches is used. A single line diagram of the HV circuits of the test system is shown in Fig. 16.30. According to the data in Ref. [92], the cost coefficients of each generator are assumed. To demonstrate the impact of FACTS devices control more clearly, the system is slightly modified by

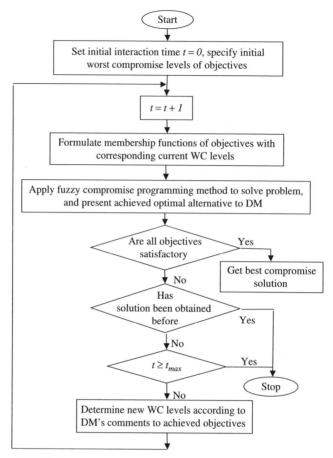

FIGURE 16.29   Flowchart of proposed interactive procedure.

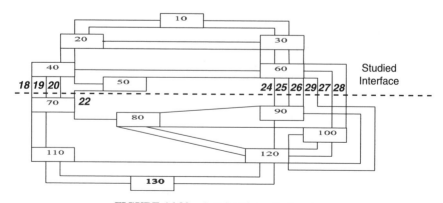

FIGURE 16.30   A reduced practical system.

increasing the original load and generation level by 10%. In this study, the per unit base is 100 MVA.

First, considering the objective of economy, a conventional optimal power flow is calculated only considering generation rescheduling and PV node voltage adjustment. The result is listed in Table 16.9 as case 1. The fact that there are much more economic generators in the North than in the south, together with the situation that generation center and load center are located in the north and in the south, respectively, results in an extremely stressed transfer interface. From the thermal burden of the interface branches of case 1, as illustrated in Fig. 16.31, it is evident that the serious situation of unbalanced line flow through the interface, particularly on lines 24, 25, 26, and 29, prevents system operation cost from further decreasing.

Aiming at gaining a better economy by adjusting power flow of the interface, four FACTS devices are applied. Besides those series controllers for line flow redistribution, due to the relatively low voltage of node 90 in case 1, a UPFC is installed on line 24 to support the node voltage. The installed FACTS devices, locations, and the derived control parameters are given in Table 16.10. As case 2, the resultant optimal solution is also shown in Table 16.9. From the comparisons of the operation costs between case 1 and case 2, which are summarized in Table 16.9, the potential benefit of FACTS devices on economic system operation support is clearly seen.

For the thermal burden of the interface branches in case 2, it is observed from Fig. 16.32 that, by control of FACTS devices, those controllable lines are all

TABLE 16.9 Scenarios With and Without FACTS Devices Under Cost Minimization Criterion

Case 1 (No FACTS) (pu)				Case 2 (with FACTS) (pu)			
Generator Output		PV Node Voltage		Generator Output		PV Node Voltage	
1	44.0193	1	1.0228	1	45.9099	1	1.0603
2	23.3952	2	1.0085	2	23.6438	2	1.0546
3	64.6059	3	1.0223	3	64.8031	3	1.0252
4	71.6372	4	1.0096	4	73.4717	4	1.0093
5	49.7671	5	1.0173	5	57.4298	5	1.0000
6	86.1291	6	1.0111	6	91.4889	6	1.0077
7	10.0000	7	1.0269	7	10.0000	7	1.0167
8	7.1523	8	1.0019	8	6.3785	8	0.9984
9	12.1736	9	0.9950	9	10.0000	9	0.9852
10	22.6943	10	1.0164	10	16.0150	10	1.0064
11	38.5423	11	1.0589	11	36.5183	11	1.0655
12	29.8411	12	1.0222	12	27.1509	12	1.0409
13	30.3545	13	1.0552	13	30.5214	13	1.0604
14	11.4169	14	1.0620	14	10.0000	14	1.0572
15	22.9201	15	1.0887	15	20.7512	15	1.0992
16	18.2779	—	—	16	16.5065	—	—
Operation cost = 840.24 ($\times 10^2$ £/hour)				Operation cost = 818.16 ($\times 10^2$ £/hour)			

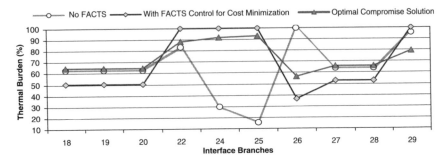

FIGURE 16.31 Thermal burden of interface branches in cases 1, 2, and 3.

TABLE 16.10 Derived Control Parameters of the FACTS Devices in Case 2

No.	Device	Line	First Node	Second Node	Control Parameters Derived for Cost Minimization in Case 2 (pu)
1	UPFC	24	90	60	$V_T = 0.8433$, $\varphi_T = 1.4305$, $I_q = 16.4293$
2	TCSC	25	60	90	$x_c = -0.0836$
3	SSSC	29	60	120	$V_C = -0.0036$
4	TCPS	22	50	70	$\sigma = 0.3695$

almost fully-loaded to allow better access to cheaper generations in the north. But, as a matter of fact, the increase in the power flow above a certain operating level will decrease the overall security of the system. Therefore, to operate the system securely, the level of power transfer between specific areas usually has to be limited. In view of

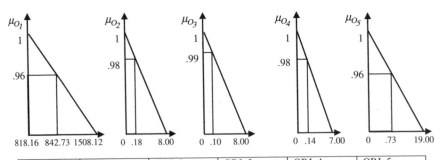

Objectives	OBJ. 1	OBJ. 2	OBJ. 3	OBJ. 4	OBJ. 5
Initial WC value	1508.12	8.00	8.00	7.00	19.00
DM's comments	Need to be improved	Allowed to be worsened	Allowed to be worsened	Allowed to be worsened	Allowed to be worsened
New WC value	842.73	8.00	8.00	7.00	19.00

FIGURE 16.32 First optimal alternative and DM's comments to achieved objectives.

this situation, besides improving system economy, these FACTS devices should also be controlled with the intention of alleviating heavy-load lines via local power flow control. However, owing to the current optimal generation pattern being interfered with, it will cause a relatively poor economy of system operation. The method to achieve a satisfactory compromise solution for this problem lies in trading off all the objectives considered; that is, to bring considerable savings in operating costs without jeopardizing the necessary level of system security. As case 3, the related study is carried out and the details are described as follows.

For the ideal value of the operation cost F_{EO}^{Ideal}, it is assumed to be equal to the cost of the system incorporating all the controllable measures under the cost minimization criterion. The operation cost of the conventional power flow without any control is taken as the initial value of F_{EO}^{WC}, which is 1508.12×10^2 £/hour. Under reasonable assumption, 50% of the power flow target value is taken as the relevant worst compromise value F_{PFC}^{WC}. In order to relieve the relevant lines of the interface for security and further utilization of the network, power flow control of lines 22, 24, 25, and 29 are considered. Based on thermal limits of the lines and thermal burden percentage of other interface branches, the corresponding targets are assigned with the aim of flattening the thermal burden curve. Therefore, including objective of economy, there are five objectives involved in the model.

The process of reaching a satisfactory solution is demonstrated in Figs. 16.32–16.34. The responses of the DM to the optimal alternatives and the corresponding new WC value assigned for the next iteration are also summarized in the figures. Because of the conflicting feature of the objectives, it is predictable that in any best compromise alternative, the objective value of economy should be higher than the ideal value of the objective. Therefore, for the economy objective, only the related part of the membership function curve is presented here.

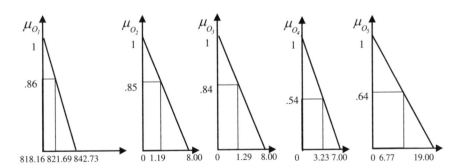

Objectives	OBJ. 1	OBJ. 2	OBJ. 3	OBJ. 4	OBJ. 5
DM's comments	Retain at the current level	Allowed to be worsened	Allowed to be worsened	Need to be improved	Need to be improved
New WC value	842.73	8.00	8.00	3.23	6.77

FIGURE 16.33 Second optimal alternative and DM's comments to achieved objectives.

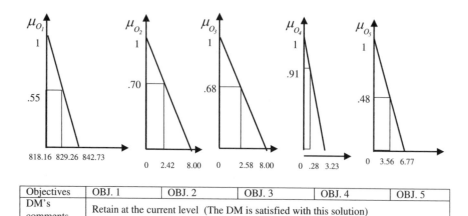

Objectives	OBJ. 1	OBJ. 2	OBJ. 3	OBJ. 4	OBJ. 5
DM's comments	Retain at the current level (The DM is satisfied with this solution)				

FIGURE 16.34 Third optimal alternative and DM's comments to achieved objectives.

The final satisfactory scenario is obtained in the third interaction. The settings of generators' output and PV nodes' voltage, as well as the control parameters of the FACTS devices, are given in Table 16.11. The thermal burden curve of the interface branches is also depicted in Fig. 16.31, from which it is clearly seen in the best compromise solution achieved that interface branches possess a relatively balanced thermal burden. This fact of the feasibility of the resultant scenario of case 3 highlights the importance of considering multiple objectives in the optimal operation of power systems incorporating FACTS devices.

16.5.8 Summary

Undoubtedly, flexible control of FACTS devices enables power systems to operate in a more robust, secure, and economic way. In order to make full use of the installed FACTS devices, it is desirable to take their major functions into account. In this chapter, an interactive compromise programming-based MO optimization approach has been proposed to determine FACTS control parameters for optimal operation of power systems in a coordinated way, with consideration of both economy and local power flow control simultaneously. It can give new insights into FACTS device functions and the corresponding control strategies. Moreover, a clearer picture of the situation of power systems with flexible and effective control means can also be presented.

For such a problem with multiple conflicting objectives, it is practical to seek an optimal compromise solution by trading off all the objectives. Based on CP, a DWCP is introduced to solve the problem in an interactive procedure. During the interactive process, as a reference point, the worst compromise point is displaced according to rules reflecting the DM's aspiration level changes. One of the salient features of this approach is that it is very efficient to determine the

TABLE 16.11 Final Optimal Scenario

Generator Output (pu)		PV Node		FACTS Devices (pu)				
		No.	Voltage (pu)	No.	Device	Control Targets	Achieved Targets	Control Parameters Derived
1	44.6354	1	1.0008					
2	23.3418	2	1.0514					
3	67.7259	3	0.9917					
4	75.8747	4	1.0048	1	UPFC	$P_L = 16$,	$P_L = 18.43$,	$V_T = 0.7287$, $\varphi_T = 1.4281$,
5	52.2182	5	1.0000			$Q_L = 0$	$Q_L = 0$	$I_q = 14.0510$
6	96.2635	6	1.0056					
7	10.0000	7	1.0205					
8	7.2068	8	1.0117	2	TCSC	$P_L = 16$	$P_L = 18.59$	$x_c = -0.0802$
9	10.0000	9	0.9991					
10	16.6947	10	1.0252	3	SSSC	$P_L = 14$	$P_L = 14.28$	$V_C = -0.0087$
11	37.7192	11	1.0496					
12	28.0036	12	0.9407	4	TCPS	$P_L = 38$	$P_L = 41.56$	$\sigma = 0.0512$
13	27.4389	13	1.0805					
14	10.0000	14	1.0845					
15	21.6590	15	1.0936		Operation cost $= 829.26 \times 10^2$ £/hour			
16	16.9966	—	—					

reference point without any further analysis or extra programming. With the introduction of fuzzy set theory, motivations of the DM can be reflected in a more realistic and credible way. Moreover, it also provides a very intuitive and convenient way to measure and represent the distances involved in the CP.

From the process of seeking the satisfactory solution in the illustrative example, it is evident that the proposed approach can lead to an efficient solution by solving the MO problem in an interactive procedure, without forcing the DM into an exhaustive examination of all the Pareto solutions.

Finally, it should be noted that, to achieve an optimal control scheme of FACTS devices, the major applications considered should not only be confined in economy improvement and local power flow control. And in terms of local power flow control, the example for load flow rebalance taken in this chapter is only one of many important applications that have been discussed in the first section of this chapter. Nevertheless, it is to be emphasized that, with the MO model and the solution method proposed in this section, it can be extended easily to deal with MO optimal operation problems of systems considering any other applications of FACTS devices by including more objectives.

REFERENCES

1. Van Cutsem T, Vournas C. Voltage stability of electric power systems. Dordrecht: Kluwer Academic Publishers; 1998.

2. Van Cutsem T. An approach to corrective control of voltage instability using simulation and sensitivity. IEEE Trans Power Systems 1995; 10(2):616–622.

3. Thukaram BD, Parthasarathty K. Optimal reactive power dispatch algorithm for voltage stability improvement. Int J Electrical Power & Energy Systems 1996; 18(7):461–468.

4. Chiang HD, Flueck AJ, Shah KS, Balu N. CPFLOW: a practical tool for tracing power system steady-state stationary behavior due to load and generation variations. IEEE Trans Power Systems 1995; 10(2):623–634.

5. Fukuyama Y, et al. A practical continuation power flow for large-scale power system analysis. Proc IEE Japan Annual Convention Record 1998; No. 1313 (in Japanese).

6. Tomsovic K. A fuzzy linear programming approach to the reactive power/voltage control problem. IEEE Trans Power Systems 1992; 7(1):287–293.

7. Cova B, Losignore N, Marannino P, Montagna M. Contingency constrained optimal reactive power flow procedures for voltage control in planning and operation. IEEE Trans Power Systems 1995; 10(2):602–608.

8. Ramos JLM, et al. A hybrid tool to assist the operator in reactive power/voltage control and operation. IEEE Trans Power Systems 1995; 10(2):760–768.

9. Wu QH, Ma JT. Power system optimal reactive power dispatch using evolutionary programming. IEEE Trans Power Systems 1995; 10(3):1243–1249.

10. Vu H, Pruvot P, Launay C, Harmand Y. An improved voltage control on large-scale power system. IEEE Trans Power Systems 1996; 11(3):1295–1303.

11. Le TL, Negnevitsky M, Piekutoski M. Network equivalents and expert system application for voltage and VAR control in large-scale power systems. IEEE Trans Power Systems 1997; 12(4):1440–1445.

12. Kennedy J, Eberhart R. Particle swarm optimization. Proceedings of IEEE International Conference on Neural Networks, Perth, Australia; 1995. Vol. IV. p. 1942–1948.

13. Shi Y, Eberhart R. A modified particle swarm optimizer. Proceedings of IEEE International Conference on Evolutionary Computation, Anchorage, Alaska, 4–9 May 1998. p. 69–73.

14. Zhenya H, et al. Extracting rules from fuzzy neural network by particle swarm optimization. Proceedings of IEEE International Conference on Evolutionary Computation, Anchorage, Alaska, 4–9 May 1998. p. 74–77.

15. Kennedy J, Spears W. Matching algorithm to problems: an experimental test of the particle swarm optimization and some genetic algorithms on the multimodal problem generator. Proceedings of IEEE International Conference on Evolutionary Computation, Anchorage, Alaska, 4–9 May 1998. p. 78–83.

16. Angeline P. Using selection to improve particle swarm optimization. Proceedings of IEEE International Conference on Evolutionary Computation, Anchorage, Alaska, 4–9 May 1998. p. 84–89.

17. Geoffrion AM. Generalized Benders decomposition. J Operation Theory Appl 1972; 10(4):237–260.

18. Kocis GR, Grossmann IE. Computational experience with DICOP solving MINLP problems in process systems engineering. Computer Chem Eng 1989; 13(3):307–315.

19. Battiti R. The reactive tabu search. ORSA J Computing 1994; 6(2):126–140.

20. Toune S, Fudo H, Genji T, Fukuyama Y, Nakanishi Y. Comparative study of modern heuristic algorithms to service restoration in distribution systems. IEEE Trans. Power Delivery 2002; 17(1):173–181.

21. Chiang HD, et al. The generation of ZIP-V curves for tracing power system steady state stationary behavior due to load and generation variations. Proc. of IEEE Power Engineering Society Summer Meeting, Edmonton, Alberta, Canada July 1999, Vol. 2, 647–651.

22. Chiang HD, Wang CS, Flueck AJ. Look-ahead voltage and load margin contingency selection functions for large-scale power systems. IEEE Trans Power Systems 1997; 12(1):173–180.

23. Dongarra J, et al. PMV 3 user's guide and reference manual. Oak Ridge, TN: National Laboratory; 1993.

24. Gropp W, et al. Using MPI. Cambridge, MA: The MIT Press; 1994.

25. OpenMP C and C++ application program interface version 1.0. OpenMP Architecture Review Board; 1998.

26. Liu WHE, Papalexopoulos AD, Tinney WF. Discrete shunt controls in a Newton optimal power flow. IEEE Trans Power Systems 1992; 7(4): 1509–1518.

27. Ben-Abdennour A, Lee KY, Edwards RM. Multivariable robust control of a power plant deaerator. IEEE Trans Energy Conversion 1993; 8(1):123–129.

28. Klefenz IG. Automatic control of steam power plants, 3rd ed. Bibliographisches Institut; 1986.

29. Golberg DE. Genetic algorithms in search, optimization, & machine learning. Reading, MA: Addison-Wesley; 1989.

30. Krichnakumar K, Goldberg DE. Control system optimization using genetic algorithms. J Guidance Control Dynamics 1992; 15(3):735–740.

31. Brady RM. Optimization strategies gleamed from biological evolution. Nature 1985; 317:804–806.

32. Dimeo R, Lee KY. Boiler-turbine control system design using a genetic algorithm. IEEE Trans Energy Conversion 1995; 10(4):752–759.

33. Zhao Y, Edwards RM, Lee KY. Hybrid feedforward and feedback controller design for nuclear steam generators over wide range operation using genetic algorithm. IEEE Trans Energy Conversion 1997; 12(1):100–106.

34. Stender J, Hillebrand E, Kingdon J. Genetic algorithms in optimization, simulation and modeling. Washington, DC: IOS Press; 1994.

35. Davis L. Genetic algorithms and simulated annealing. Los Altos, CA: Morgan Kaufmann Publishers; 1987.

36. Bell RD, Astrom KJ. Dynamic models for boiler-turbine-alternator units: data logs and parameter estimation for a 160 MW unit. TFRT-3192. Sweden: Lund Institute of Technology; 1987.

37. Kwakernaak H, Sivan R. Linear optimal control systems. New York: John Wiley & Sons; 1972.

38. Åstrom KJ, Wittenmark B. Adaptive control. Reading, MA: Addison-Wesley; 1989.

39. Garduno-Ramirez R, Lee KY. Wide range operation of a power unit via feedforward fuzzy control. IEEE Trans Energy Conversion 2000; 15(4):421–426.

40. Tolle H, Ersu E. Neurocontrol; learning control systems inspired by neuronal architectures and human problem solving. Lecture notes in control and information sciences. Berlin: Springer-Verlag; 1992.

41. Saint-Donat J, Bhat N, McAvoy TJ. Neural-net based model predictive control. In: Harris CJ, ed. Advances in intelligent control. London: Taylor & Francis; 1994.

42. Clarke D. Advances in model-based prediction control. New York: Oxford University Press; 1994.

43. Rawlings JB, Meadows ES, Muske KR. Nonlinear model predictive control: a tutorial and survey. Proceedings on ADCHEM '94, Kyoto, Japan, 1994.

44. Åstrom KJ, Bell RD. Dynamic models for boiler-turbine-alternator units: data logs and parameter estimation for a 160 MW unit. In: Report LUTFD2/(TFRT-3192)/1–137/ (1987). Lund, Sweden: Department of Automatic Control, Lund Institute of Technology; 1987.

45. Fogel LJ. The future of evolutionary programming. Proceedings of the 24th Asilomar Conference on Signals, Systems, and Computers. Pacific Grove, CA, 1991.

46. Lai LL. Intelligent system application in power engineering; evolutionary programming and neural networks. New York: John Wiley & Sons; 1998.

47. Narandra KS, Parthasarathy K. Identification and control of dynamical systems using neural networks. IEEE Trans Neural Networks 1990; 1(1):4–27.

48. Ghezelayagh H, Lee KY. Neuro-fuzzy identifier of a boiler system. Eng Intelligent Systems 1999; 4:227–231.

49. Goldberg DE. Genetic algorithm in search, optimization, and machine learning. Reading, MA: Addison-Wesley; 1989.

50. Mason AJ. Partition coefficients, static deception and deceptive problems for non-binary alphabets. Proceedings of the 4th International Conference on Genetic Algorithms; 1991. p. 210–214.

51. Antonisse J. A new interpretation of schema notation that overturns the binary encoding constraint. Proceedings of the 3rd International Conference on Genetic Algorithms; 1989. p. 86–91.

52. Goldberg DE, Deb K. A comparative analysis of selection schemes used in genetic algorithms. The Clearinghouse for Genetic Algorithms Tech. Report, No. 90007. Tuscaloosa, AL: The University of Alabama, Department of Engineering Mechanics; 1990.

53. Xiao Y, Song YH, Lui CC, Sun YZ. Available transfer capability enhancement using FACTS devices. IEEE Trans Power Systems 2003; 18(1):305–312.

54. Wang X, Song YH, Qiang Lu, Sun YZ. Optimal allocation of transmission rights in systems with FACTS devices. IEE Proc Generation, Transmission Distribution 2002; 149(3):359–366.

55. Galiana FD, Almeida K, Toussaint M, Griffin J, Atanackovic D, Ooi BT, Mcgillis DT. Assessment and control of the impact of FACTS devices on power system performance. IEEE Trans. Power Systems 1996; 11(4):1931–1936.

56. Baldick R, Kahn E. Paths contract, phase-shifters, and efficient electricity trade. IEEE Trans Power Systems 1997; 12(2):749–755.

57. Mutale J, Strbac G. Transmission network reinforcement versus FACTS: an economic assessment. IEEE Trans Power Systems 2000; 15(3):961–967.

58. Ge SY, Chung TS. Optimal active power flow incorporating power flow control needs in flexible ac transmission systems. IEEE Trans Power Systems 1999; 14(2):738–744.

59. Zeleny M. Multiple criteria decision making. New York: McGraw-Hill; 1982.

60. Brodsky SFJ, Hahn RW. Assessing the influence of power pools on emission constrained economic dispatch. IEEE Trans Power Systems 1986; 1(1):57–62.

61. Heslin JS, Hobbs BF. A multiobjective production costing model for analyzing emissions dispatching and fuel switching. IEEE Trans Power Systems 1989; 4(3):836–842.

62. El-Hawary ME, Ravindranath KM. Combining loss and cost objectives in daily hydro-thermal economic scheduling. IEEE Trans Power Systems 1991; 6(3):1106–1112.

63. Charnes A, Cooper WW. Management models and industrial applications of linear programming. Vol. 1. New York: Wiley, 1961.

64. Kermanshahi BS, Wu Y, Yasuda K, Yokoyama R. Environmental marginal cost evaluation by non-inferiority surface. IEEE Trans Power Systems 1990; 5(4):1151–1159.

65. Nanda J, Kothari DP, Linggamurthy KS. Economic load dispatch through goal programming techniques. IEEE Trans Power Systems 1988; 3(1):26–32.

66. Fan J-Y, Zhang L. Real-time economic dispatch with line flow and emission constraints using quadratic programming. IEEE Trans Power Systems 1998; 13(2):320–325.

67. Moro LM, Ramos A. Goal programming approach to maintenance scheduling of generating units in large scale power systems. IEEE Trans Power Systems 1999; 14(3):1021–1028.

68. Cohon JL. Multiobjective programming and planning. Mathematics in science and engineering, Vol. 140. New York: Academic Press; 1978.

69. Chiang H-D, Jean-Jumeau R. Optimal network reconfigurations in distribution systems. Part 1: A new formulation and a solution methodology. IEEE Trans Power Delivery 1990; 5(4):1902–1909.

70. Chattopadhyay D, Banerjee R, Parikh J. Integrating demand side options in electric utility planning: a multiobjective approach. IEEE Trans Power Systems 1995; 10(2):657–663.

71. Chattopadhyay D, Momoh J. A multiobjective operations planning model with unit commitment and transmission constraints. IEEE Trans Power Systems 1999; 14(3):1078–1084.

72. Zadeh LA. Fuzzy sets. Information and Control 1965; 8:338–353.

73. Zimmermann HJ. Fuzzy programming and linear programming with several objective functions. TIMS/studies in the management science 20. Amsterdam: Elsevier Science Publisher; 1984. p. 109–121.

74. Ramesh VC, Li X. A fuzzy multiobjective approach to contingency constrained OPF. IEEE Trans Power Systems 1997; 12(3):1348–1354.

75. Xiao Y, Song YH. Applying multi-objective fuzzy dynamic programming to hydro-thermal coordination including pumped-storage plant. Sixth UK Workshop on Fuzzy Systems, 8–9 September 1999. p. 241–246.

76. Sun H, Yu DC. A multiple-objective optimization model of transmission enhancement planning for Independent Transmission Company (ITC). IEEE 2000 Summer Meeting, Seattle, July 2000.

77. Goldberg D. Genetic algorithms in search, optimization & machine learning. Reading, MA: Addison-Wesley; 1998.

78. Kalyanmoy D. Multi-objective genetic algorithms: problem difficulties and construction of test problems. Evol Computat 1999; 7(3):205–230.

79. Das DB, Patvardhan C. New multi-objective stochastic search technique for economic load dispatch. IEE Proc Generation, Transmission Distribution 1998; 145(6):747–752.

80. Chen Y-L, Liu C. Optimal multi-objective VAR planning using an interactive satisfying method. IEEE Trans Power Systems 1995; 10(2):664–670.

81. Sakawa M, Nishizaki I, Uemura Y. Interactive decision making for large-scale multiobjective linear programs with fuzzy numbers. Fuzzy Sets Systems 1997; 88:161–172.

82. Yang J-B, Chen C, Zhang Z-J. The interactive step trade-off method (ISTM) for multiobjective optimization. IEEE Trans Systems Man Cybernetics 1990; 20(3):688–695.

83. Sakawa M, Nishizaki I, Uemura Y. Interactive fuzzy programming for two-level linear fractional programming with fuzzy parameters. Fuzzy Sets Systems 2000; 115:93–103.

84. Chen Y-L. An interactive fuzzy-norm satisfying method for multi-objective reactive power sources planning. IEEE Trans Power Systems 2000; 15(3):1154–1160.

85. Xiao Y, Song YH, Sun YZ. Power flow control approach to power systems with embedded FACTS devices. IEEE Trans Power Systems 2002; 17(4):943–950.

86. Han ZX. Phase shifter and power flow control. IEEE Trans PAS 1982; 101(10):3790–3795.

87. Wojtek M. Use of the displaced worst compromise in multiobjective programming. IEEE Trans Systems Man Cybernetics 1988; 18(3):472–477.

88. Mehrotra S. On the implementation of a primal-dual interior point method, SIAM J Optim 1992; 2:575–601.

89. Fuerte-Esquivel CR, Acha E. Newton-Raphson algorithm for the reliable solution of large power networks with embedded FACTS devices. Proc IEE, Pt. C, 1996; 143(5):447–454.

90. Hiskens IA. Analysis tools for power systems—contending with nonlinearities. Proc IEEE, 1995; 83(11):1573–1587.

91. Xiao Y, Song YH. Power flow studies of a large practical power network with embedded FACTS devices using improved optimal multiplier Newton-Raphson method. Euro Trans Electrical Power (ETEP) 2001; 11(4).

92. Wood AJ, Wollenberg, BF. Power generation, operation and control. 2nd ed. New York: John Wiley & Sons; 1996.

CHAPTER 17

Genetic Algorithms for Solving Optimal Power Flow Problems

LOI LEI LAI and NIDUL SINHA

17.1 INTRODUCTION

Historically, the solution of the economic load dispatch (ELD) by the equal incremental cost method was a precursor of optimal power flow (OPF). The arrival of OPF marked the end of the "classical" period of economic dispatch, which had developed for more than four decades. Both are optimization problems with the same minimum cost objective. However, ELD only considers real power generations and represents the electrical network by single equality constraints, the power balance equation, whereas the OPF in the simplest form is a general problem of real and reactive power flow, so as to minimize the instantaneous operating costs. It is a static optimization problem with a scalar objective function (cost function). It is to be noted that the objective function in OPF can take different forms like minimization of generation cost (optimal active and reactive power flow), minimizing electrical losses in transmission systems (optimal reactive power flow), and minimum shift of generation and other controls from an optimum operating point. The optimal power flow problem was introduced in the early 1960s by Carpentier and has grown into a powerful tool for power system operation and planning [1–16]. It has been extensively used by utility engineers for a number of applications starting from development of base cases, study of voltage instability, thermal limits, maximum transfer capability study, and contingency constrained OPF in power system networks provided with FACTS (Flexible AC Transmission Systems) devices, to mention a few.

After deregulation of the electricity industry, solving the optimal power flow problem is fundamental to the unbundling of transmission costs associated with transmission open access and is of increasing importance in power system operation. As open-access market principles are applied to power systems, significant changes in their operation and control are occurring. Market competition demands greater

Modern Heuristic Optimization Techniques. Edited by K. Y. Lee and M. A. El-Sharkawi
Copyright © 2008 the Institute of Electrical and Electronics Engineers, Inc.

attention to operating cost versus voltage stability, which in turn forces the operation of power systems under higher loading conditions. As a result, systems are being operated with reduced stability margins with the risk of voltage collapse events throughout the world. Because stability continues to be a basic requirement in the operation of any power system, new tools are being considered to analyze the effect of stability on the operating cost of the system, so that system stability can be incorporated into the costs of operating the system. Thus the incorporation of voltage stability criteria in the OPF has become essential. On the other hand, higher loading conditions of the power system make the solution of load flow with classical methods very difficult. Today's complicated system as well as its extensive exploitation demands for accurate models of different devices for accurate representation of power systems.

Classical methods rely on convexity to obtain the global optimum solution and are highly sensitive to starting points. Therefore, these methods frequently get trapped in local optimum or diverge and do not offer much freedom in choosing the objective function or the type of constraints that may be used. The OPF is in general nonconvex and as a result many local minima may exist. Especially after incorporation of a number of features in the objective function and in the constraints, it becomes a highly constrained and large-dimensional optimization problem that is very difficult, if not impossible, to solve with the classical method. Meta-heuristic methods like genetic algorithms (GAs) and evolutionary programming are the choices of solutions in such cases.

FACTS with the new technology of power electronics has given new control facilities in power systems in both the steady-state power flow control and dynamic stability control. In the steady-state operation of power systems, unwanted loop power flow and parallel power flow between utilities are problems in heavily interconnected bulk power systems. These two power flows sometimes are beyond the control of generators or it may cost too much with the generator regulations. However, with the phase shifter and/or other control facilities based on fast-reacted power electronic components in networks, the unwanted power flow will be easily regulated. Several papers have been published for dealing with power flow controls [17–20]. A unified power flow controller (UPFC) with voltage source converters can operate as a shunt compensator, tap-changer, series compensator, and phase shifter. It is possible to make use of circuit reactance and voltage angles as control facilities for power flows in the network. However, extra control facilities complicate the system operation. As the control facilities influence each other, a good coordination is required in order to bring all devices to work together without interfering with each other. Therefore, it becomes necessary to extend available system analysis tools, such as OPF, to represent FACTS controls. It has also been noted that the OPF problem with series compensation may be a nonconvex problem [17], which will lead the conventional optimization methods stuck into local minima.

In this chapter, the OPF problem will be treated starting with simple cases with the introduction of complexities like valve point loadings, voltage limits, transfer limits, and so forth, in steps ending with more complicated power system with FACTS. All problems are solved with GA.

17.2 GENETIC ALGORITHMS

The GAs are part of the evolutionary algorithms family, which are meta-heuristic computational methods. GAs are powerful stochastic search algorithms based on the mechanism of natural selection and natural genetics. GAs work with a population of binary strings, searching many peaks (hill climbing) in parallel. By employing genetic operators, they exchange information between the peaks, hence reducing the possibility of ending at a local optimum. GAs are more flexible than most search methods because they require only information concerning the quality of the solution produced by each parameter set (objective function values) and not like many traditional optimization methods that require derivative information or, worse yet, complete knowledge of the problem structure and parameters. GAs were invented by Holland [21] in the early 1970s and used in practical problems from the late 1980s [22]. GAs are search algorithms based on the mechanics of natural genetics and natural selection. GAs differ from other optimization methods in the following ways:

1. GAs search from a population of points, not a single point that equips them with inherent parallel computation ability. This population can climb over hills and move across valleys. GAs can therefore discover a globally optimal point.

2. GAs use payoff (fitness or objective functions) information directly for the search direction, not derivatives or other auxiliary knowledge. GAs therefore can deal with the nonsmooth, noncontinuous, and nondifferentiable functions that are experienced in real-life optimization problems. OPF is one such problem. This property also equips GAs to find more accurate solutions for highly nonlinear real-life optimization problems in contrast with traditional optimization methods, which make lots of approximations (hence incorporate error) in the model representation.

3. GAs use probabilistic transition rules to select generations, not deterministic rules, so they can reach a complicated and uncertain area to find the global optimum. GAs are more flexible and robust than conventional methods.

These features mean that GAs are robust and parallel search algorithms that adaptively search for the global optimal point.

17.2.1 Terms Used in GA

GA in essence, consists of a population of bit strings transformed by three genetic operators: *selection, crossover,* and *mutation*.

A population consists of individuals or chromosomes. The number of individuals in a population is called its population size.

17.2.1.1 Search Space   The space of all feasible solutions (the set of solutions among which the desired solution resides) is called *search space* (also state space).

17.2.1.2 Chromosome Points in the domain space of the search, usually real numbers over some range, are encoded as bit strings, called chromosomes. Each string (called chromosome) represents a possible solution to the problem being optimized.

17.2.1.3 Gene Each bit position in the string is called a *gene* and it represents a value for some variable of the problem.

17.2.1.4 Population Size Population size is the number of individuals (solutions) in a population. It is the collection of individuals/solutions with their respective fitness.

An example is used to illustrate the use of the terms. Let us assume that the objective is to minimize the operating cost of three generating units with the following production cost functions:

$$F_1 = 50 + 150P_{g1} + 10P_{g1}^2 \quad \$/h$$
$$F_2 = 30 + 110P_{g2} + 20P_{g2}^2 \quad \$/h \quad (17.1)$$
$$F_3 = 40 + 200P_{g3} + 40P_{g3}^2 \quad \$/h$$

subject to

$$P_{g1} + P_{g2} + P_{g3} = 300 \text{ MW} \quad \text{(equality constraint)} \quad (17.2)$$

$$\left. \begin{array}{l} 40 \le P_{g1} \le 100 \text{ MW} \\ 30 \le P_{g2} \le 110 \text{ MW} \\ 50 \le P_{g1} \le 150 \text{ MW} \end{array} \right\} \quad \text{(inequality constraints).} \quad (17.3)$$

Here the unit power outputs are the independent (decision) variables. A feasible solution is one, which satisfies all the constraints, and it may not be an optimum solution (but feasible). So, the different values of P_{gi}s (i.e., P_{g1}, P_{g2}, and P_{g3}) are chosen from within their respective unit-limits (to satisfy the inequality constraints), and costs of all the units are calculated for the set of values of P_{gi}s. The random choice of values decision variables may lead to violation of equality constraint. This violation is penalized and added to the cost of the solution. Then the fitness for the solution is calculated.

So, a population with five chromosomes is formed as a matrix as given below:

$$\text{Population} = \begin{bmatrix} P_{g1}^1 & P_{g2}^1 & P_3^1 & Fitness^1 \\ P_{g1}^2 & P_{g2}^2 & P_3^2 & Fitness^2 \\ P_{g1}^3 & P_{g2}^3 & P_3^3 & Fitness^3 \\ P_{g1}^4 & P_{g2}^4 & P_3^4 & Fitness^4 \\ P_{g1}^5 & P_{g2}^5 & P_3^5 & Fitness^5 \end{bmatrix}. \quad (17.4)$$

In this population matrix, each row (except the fitness column) is called a *chromo-some* (individual solution) and each decision variable (P_{g1}, P_{g2}, P_{g3}) is called a *gene*. Population size in this case is *five*.

17.2.1.5 Fitness The *fitness function* is a measure of how well a candidate solves the problem.

The fitness function, which may be the sum of objective function value and penalty for constraint violation, can be calculated for each individual of the parent population as

$$FIT_i = 1 \bigg/ \left(\sum_j f_i(x_{ij}) + \sum_{z=1}^{N_c} PF_z \right) \qquad (17.5)$$

and

$$PF_z = \lambda_z \times [VIOL_z]^2$$

where $VIOL_z$ is the violation of constraint z and λ_z is the penalty multiplier. N_c is the number of constraints.

A simple genetic algorithm follows the following steps:

1. Generate randomly a population of initial population within the feasible ranges of the decision variables.
2. Calculate the fitness for each string in the population.
3. Create offspring strings through reproduction, crossover, and mutation operation.
4. Evaluate the new strings and calculate the fitness for each string (chromosome).
5. If the search goal is achieved, or an allowable generation is attained, return the best chromosome as the solution; otherwise go to step 3.

In this chapter, only floating-point GA (FGA) is used in solving the problems, apart from the study with FACTS devices, as the use of FGA has a number of advantages over binary GA. The efficiency of the GA is increased as there is no need to convert the solution variables to the binary type. Moreover, less memory is required, and there is no loss in precision by discretization to binary or other values. Also, there is greater freedom to use different genetic operators.

17.2.1.6 Initialization An initial population of N individuals is generated. Each individual is taken as a pair of real-valued vectors x_i, $\forall_i \in \{1, 2, \ldots, N\}$, where x_is are the objective variables and determined by setting the jth component $X_j \sim U(X_{jmin}, X_{jmax})$, for $j = 1, 2, \ldots, n$. $X(X_{jmin}, X_{jmax})$ denotes a uniform random variable ranging over $[X_{jmin}, X_{jmax}]$.

Figure 17.1 shows the flowchart of a simple GA. In case of load flow problems in polar coordinates, the objective variables are the magnitudes of bus voltages and magnitudes of their phase angles.

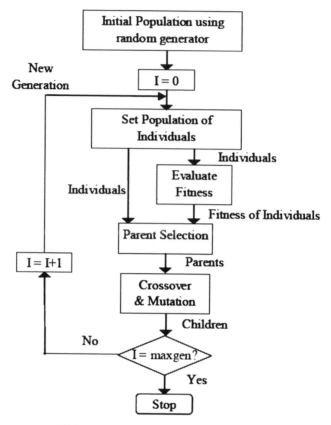

FIGURE 17.1 Flowchart of a simple GA.

The GA optimization toolbox (GAOT) in MATLAB proposed in Ref. 23 is used after minor modification for minimization for solving the load flow problem. The binary GA and FGA with different mutation and crossover functions available in the toolbox were tried on this problem. The FGA with heuristic crossover and nonuniform mutation appeared to perform better than FGAs with other types of crossover and mutation functions and hence considered for.

17.2.1.7 *Creation of Offspring* New solutions (i.e., offspring) are created by using two operators: (i) crossover and (ii) mutation. Only the heuristic crossover and nonuniform mutation will be described here. However, readers may refer to Ref. 23 for other types of crossover and mutation functions.

17.2.1.8 *Heuristic Crossover* This is a crossover operator that uses the fitness values of the two parent chromosomes to determine the direction of the search. The best and worst parents are determined by comparing the fitnesses. The offspring are created according to the following equations:

$$\text{Offspring1} = \text{BestParent} + r * (\text{BestParent} - \text{WorstParent}) \qquad (17.6)$$

$$\text{Offspring2} = \text{BestParent} \qquad (17.7)$$

where r is a uniform random number between 0 and 1.

It is possible that Offspring1 will not be feasible. This can happen if r is chosen such that one or more of its genes fall outside of the allowable upper or lower bounds. For this reason, heuristic crossover has a user settable parameter (n) for the number of times to try and find an r that results in a feasible chromosome. If a feasible chromosome is not produced after n tries, the WorstParent is returned as Offspring1.

17.2.1.9 Nonuniform Mutation Nonuniform mutation based on a nonuniform probability distribution, shrinks a single element of the selected vector chosen at random toward one of the bounds. This Gaussian distribution starts wide and narrows to a point distribution as the current generation approaches the maximum generation.

This mutation operator increases the probability that the amount of the mutation will be close to 0 as the generation number increases. Nonuniform mutation operates uniformly at an early stage and more locally as the current generation approaches the maximum generation. Thus, this mutation operator keeps the population from stagnating in the early stages of the evolution then allows the genetic algorithm to fine-tune the solution in the later stages of evolution.

The degree of nonuniformity is controlled with the shape parameter b. Nonuniformity is important for fine-tuning.

Nonuniform mutation randomly selects one variable, j, and sets it equal to a nonuniform random number:

$$x_i' = \begin{cases} x_i + (b_i - x_i)\, f(G) & \text{if } r_1 < 0.5 \\ x_i - (x_i + a_i)\, f(G) & \text{if } r_1 \geq 0.5, \\ x_i & \text{otherwise} \end{cases} \qquad (17.8)$$

where $f(G) = (r_2(1 - G/G_{\max}))^b$, r_1, r_2 are uniform random numbers between $(0, 1)$, i.e., $U(0, 1)$, G is current generation, $G_{\max}$ is maximum number of generations, and b is shape parameter.

17.2.1.10 Normalized Geometric Selection NormGeomSelect is a ranking selection function based on the normalized geometric distribution. Rank selection ranks the population first and then every chromosome receives a fitness value determined by this ranking. The worst will have the fitness 1, the second worst 2, and so forth, and the best will have fitness N (number of chromosomes in population).

Steps followed:

1. Rank individuals according to fitness.
2. Negative fitness is allowed.
3. P_i (selecting the ith individual) $= q(1 - q)^{(r-1)}/(1 - (1 - q))^N$, where q is the probability of selecting the best individual, r is the rank of the ith individual, and N is the population size.

17.2.1.11 *Crossover Probability* Crossover probability represents how often crossover will be performed. For better results, the crossover rate is taken to be much larger than the mutation rate (generally 20 times greater). The crossover rate generally ranges from 0.25 to 0.95.

17.2.1.12 *Mutation Probability* Mutation probability represents how often parts of a chromosome will be mutated. If there is no mutation, offspring are generated immediately after crossover (or directly copied) without any change. If mutation is performed, one or more parts of a chromosome are changed.

17.2.1.13 *Stopping Rule* The iterative procedure of generating new trials by selecting those with minimum function values from the competing pool is terminated when there is no significant improvement in the solution. It can also be terminated when a given maximum number of generations (iterations) are reached. In the current work, the latter method (i.e., a given number of iterations) is employed.

17.3 LOAD FLOW PROBLEM

The load flow equations can be developed either in polar coordinates or in rectangular coordinates. However, we have considered polar coordinates to enable us to make wiser initial assumptions because of the fact that the voltage magnitudes do not deviate much from unity.

The objective function results from the summation of squares of the power (both real and reactive power) mismatch and the violation of reactive power generation limits in generator buses whose minimum (ideally zero) coincides with the load flow solution

$$\min g(V, \theta) = \sum_{i \in N_g} \Delta P_i^2 + \sum_{i \in N_{PQ}} \Delta Q_i^2 + \sum_{i \in N_{Q_g^{\lim}}} \lambda_{Q_{gi}} (Q_{gi} - Q_{gi}^{\lim})^2, \qquad (17.9)$$

TABLE 17.1 Specific Generation, Loads, and Nodal Voltages for IEEE-14 Bus System (Light Load)

Bus No.	$|V_{sp}|$ (p.u.)	θ (deg.)	P_G (p.u.)	P_D (p.u.)	Q_G (p.u.)	Q_D (p.u.)	Bus Type	Q_{Gmin} (p.u.)	Q_{Gmax} (p.u.)
1	1.060	0.0	–	–	–	–	0	−0.4	0.5
2	1.045	–	0.40	0.217	–	0.127	1	0.0	0.4
3	1.010	–	0.0	0.942	–	0.19	1	0.0	0.0
4	–	–	0.0	0.4780	0.0	0.039	2	0.0	0.0
5	–	–	0.0	0.0760	0.0	0.016	2	0.0	0.0
6	1.070	–	0.0	0.112	–	0.075	1	−0.06	0.24
7	–	–	0.0	0.0	0.0	0.0	2	0.0	0.0
8	1.090	–	0.0	0.0	–	0.0	1	−0.06	0.24
9	–	–	0.0	0.295	0.0	0.166	2	0.0	0.0
10	–	–	0.0	0.090	0.0	0.058	2	0.0	0.0
11	–	–	0.0	0.035	0.0	0.018	2	0.0	0.0
12	–	–	0.0	0.061	0.0	0.016	2	0.0	0.0
13	–	–	0.0	0.135	0.0	0.058	2	0.0	0.0
14	–	–	0.0	0.149	0.0	0.050	2	0.0	0.0

Bus type: 0, slack bus; 1, voltage controlled bus; 2, load bus; base MVA, 100.
The tuned parameters of the FGA used are population size, 200; number of generations (iterations), 2000 (for light load), 10,000 (for heavy load).

where

$$\Delta P_i = |P_i^{sp} - P_i| \qquad i \in N_{PQ} + N_V$$
$$\Delta Q_i = |Q_i^{sp} - Q_i| \qquad i \in N_{PQ}$$
$$P_i^{sp} = P_{gi} - P_{di}$$
$$Q_i^{sp} = Q_{gi} - Q_{di}.$$

The decision variables for the load flow problem (IEEE-14 bus system) are

$$[V_1, V_2, \ldots V_{14}, \theta_1, \theta_2, \ldots \theta_{14}]$$

Assumptions are that transformer tap settings and reactive powers (compensations) are fixed. Table 17.1 shows the generation and loads for the IEEE-14 bus system under light load. Figure 17.2 shows the convergence of results. Table 17.2 shows the load flow result obtained by the FGA.

Table 17.3 shows generation and loads for the system under heavy load, and Fig. 17.3 shows convergence result derived with a GA. Load flow results are shown in Table 17.4.

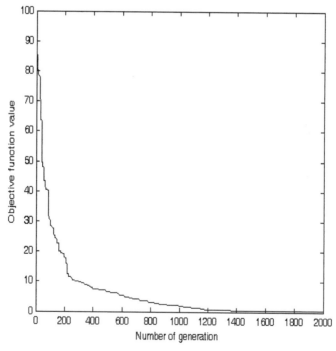

FIGURE 17.2   Result of GA-based load flow study, IEEE-14 bus system (light load).

TABLE 17.2 Results of Load Flow Study (Light Load) for IEEE-14 Bus System Using FGA

| Node No. | V_i (p.u.) | θ (deg.) | Mismatches (p.u.) | | Total Real Power Loss (p.u.) |
			$P_i^{sp} - P_i$ (Total)	$Q_i^{sp} - Q_i$ (Total)	
1	1.0600	0			
2	1.0450	−4.4439	0.0	−0.4481	
3	1.0100	−11.1475	−0.1567	−0.2106	
4	1.0120	−9.3650	−0.0005	0.0463	
5	1.0159	−7.8961	−0.0003	0.0504	
6	1.0700	−12.4499	0.0	−0.1562	0.0216
7	1.0560	−12.7262	0.0	0.0108	
8	1.0900	−13.5481	0.0937	−0.2111	
9	1.0483	−13.9370	0.0	0.0110	
10	1.0413	−13.8925	−0.0001	0.0360	
11	1.0498	−13.2510	−0.0003	0.0210	
12	1.0598	−12.8236	−0.0392	−0.0002	
13	1.0548	−12.9330	−0.1047	−0.0027	
14	1.0333	−14.5028	−0.0006	−0.0018	

TABLE 17.3 Specific Generations, Loads, and Nodal Voltages for IEEE-14 Bus System (Heavy Load)

Bus No.	$|V_{sp}|$ (p.u.)	θ (deg.)	P_G (p.u.)	P_D (p.u.)	Q_G (p.u.)	Q_D (p.u.)	Bus Type
1	1.060	0.0	–	–	–	–	0
2	1.045	–	4.50	1.217	–	1.50	1
3	1.010	–	1.00	3.500	–	2.00	1
4	–	–	0.0	4.500	0.0	1.00	2
5	–	–	0.0	2.500	0.0	1.00	2
6	1.070	–	2.00	3.50	–	2.00	1
7	–	–	0.0	0.50	0.0	0.50	2
8	1.090	–	2.50	2.0	–	0.00	1
9	–	–	0.0	2.50	0.0	2.00	2
10	–	–	0.0	2.50	0.0	1.00	2
11	–	–	0.0	1.50	0.0	1.00	2
12	–	–	0.0	2.50	0.0	0.10	2
13	–	–	0.0	1.50	0.0	0.58	2
14	–	–	0.0	2.40	0.0	0.05	2

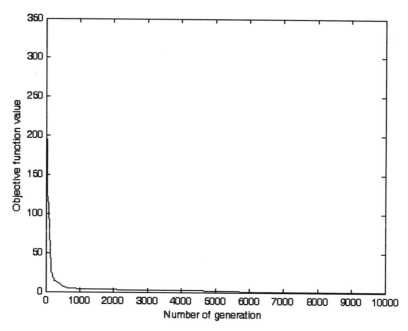

FIGURE 17.3   Result of GA-based load flow study, IEEE-14 bus system (heavy load).

TABLE 17.4   Results of Load Flow Study (Heavy Load) for IEEE-14 Bus System Using FGA

| Node No. | V_i (p.u.) | θ (deg.) | Mismatches (p.u.) | | Total Real Power Loss (p.u.) |
			$P_i^{\text{sp}} - P_i$ (Total)	$Q_i^{\text{sp}} - Q_i$ (Total)	
1	1.060	0			
2	1.045	−11.616			
3	1.010	−34.368			
4	0.4899	−40.662			
5	0.4504	−36.537			
6	1.070	−126.954	0.4044	0.1886	5.1610
7	0.5519	−97.51			
8	1.0900	−89.438			
9	0.4923	−124.003			
10	0.5226	−127.323			
11	0.7417	−128.785			
12	0.9586	−129.582			
13	0.9018	−128.35			
14	0.6433	−136.621			

17.4 OPTIMAL POWER FLOW PROBLEM

In general, the optimal power flow problem is a nonlinear programming (NLP) problem that is formulated as follows:

$$\text{Minimize} \sum_{i=1}^{N_g} F_i = \sum_{i=1}^{N_g} (a_i P_{gi}^2 + b_i P_{gi} + c_i) \quad \$/\text{hr} \tag{17.10}$$

subject to:

(a) Active power balance in the network

$$P_i(V, \theta) - P_{gi} + P_{di} = 0 \qquad (i = 1, 2, 3, \ldots N_b) \tag{17.11}$$

(b) Reactive power balance in the network

$$Q_i(V, \theta) - Q_{gi} + Q_{di} = 0 \qquad (i = N_v + 1, N_v + 2, \ldots N_b) \tag{17.12}$$

(c) Security related constraints (soft constraints)

(i) Limits on real power generations

$$P_{gi}^{\min} \le P_{gi} \le P_{gi}^{\max} \qquad (i = 1, 2, 3, \ldots N_g) \tag{17.13}$$

(ii) Limits on voltage magnitudes

$$V_i^{\min} \le V_i \le V_i^{\max} \qquad (i = N_v + 1, N_v + 2, \ldots N_b) \tag{17.14}$$

(iii) Limits on voltage angles

$$\theta_i^{\min} \le \theta_i \le \theta_i^{\max} \qquad (i = 1, 2, 3, \ldots N_b) \tag{17.15}$$

(d) Functional constraints, which is a function of control variables

(i) Limits on reactive power

$$Q_{gi}^{\min} \le Q_{gi} \le Q_{gi}^{\max} \qquad (i = 1, 2, 3, \ldots N_g) \tag{17.16}$$

(ii) Limits on active and reactive power flow on lines

$$|S_k| \le S_k^{\max} \qquad (k \in N_e). \tag{17.17}$$

Real power flow equations are

$$P_i(V, \theta) = V_i \sum_{i=1}^{N_b} V_j(G_{ij} \cos \theta_{ij} + B_{ij} \sin \theta_{ij}). \tag{17.18}$$

Reactive power flow equations are

$$Q_i(V, \theta) = V_i \sum_{i=1}^{N_b} V_j(G_{ij} \sin \theta_{ij} - B_{ij} \cos \theta_{ij}), \tag{17.19}$$

where N_g = number of generator buses, N_b = number of buses, N_v = number of voltage controlled buses, N_{PQ} = set of PQ bus numbers, N_e = set of numbers of total branches having limit on power flow, $N_{V_{PQ}^{\lim}}$ = set of numbers of PQ buses at which voltages violate the limits, $N_{Q_g^{\lim}}$ = set of numbers of buses at which reactive power generations violate the limits, P_i = active power injection into the buses, Q_i = reactive power injection into the buses, P_{di} = active load on bus i, Q_{di} = reactive load on bus i, P_{gi} = active power generation on bus i, Q_{gi} = reactive power generation on bus i, V_i = magnitude of voltage at bus i, θ_i = magnitude of phase angle at bus i, $\theta_{ij} = \theta_i - \theta_j$, $Y_{ij} = G_{ij} + B_{ij}$, and $\overline{V}_i = V_i e^{j\theta_i}$. The power flow equations are used as equality constraints; the active and reactive power generation restrictions, shunt capacitor/reactor reactive power restrictions, apparent power flow restrictions in branches, transformer tap-setting restrictions, and bus voltage restrictions are used as inequality constraints. The active power generation at the slack bus P_{gs}, load bus voltages V_{load}, reactive power generations Q_g, and branch apparent power flows are state variables, which are restricted by adding them as the quadratic penalty terms to the objective function to form a penalty function. Equation (17.10) is therefore changed to the following generalized objective function:

$$\min f = \sum_{i \in N_g} (a_i + b_i P_{gi} + c_i P_{gi}^2) + \lambda_{P_{gs}}(P_{gs} - P_{gs}^{\lim})^2 + \sum_{i \in N_{V_{PQ}^{\lim}}} \lambda_{V_i}(V_i - V_i^{\lim})^2$$

$$+ \sum_{i \in N_{Q_g^{\lim}}} \lambda_{Q_{gi}}(Q_{gi} - Q_{gi}^{\lim})^2 + \sum_{i \in N_{S_k^{\lim}}} \lambda_{S_k}(|S_k| - S_k^{\max})^2 \tag{17.20}$$

$$0 = P_i(V, \theta) - V_i \sum_{i=1}^{N_b} V_j(G_{ij} \cos \theta_{ij} + B_{ij} \sin \theta_{i \cdot})$$

$$\tag{17.21}$$

$$0 = Q_i(V, \theta) - V_i \sum_{i=1}^{N_b} V_j(G_{ij} \sin \theta_{ij} - B_{ij} \cos \theta_{ij})$$

$$\begin{aligned}
P_{gi}^{\min} \leq P_{gi} \leq P_{gi}^{\max} && i \in N_g, i \neq s \\
Q_{ci}^{\min} \leq Q_{ci} \leq Q_{ci}^{\max} && i \in N_c, \\
T_k^{\min} \leq T_k \leq T_k^{\max} && i \in N_t, \\
V_i^{\min} \leq V_i \leq V_i^{\max} && i \in N_g,
\end{aligned} \tag{17.22}$$

where s is the slack bus. $\lambda_{P_{gs}}$, λ_{V_i}, $\lambda_{Q_{gi}}$, and λ_{S_k} are the penalty factors. If all quantities are per unit values, the initial penalty factors will be given as the average value of c_i for the penalty terms to be comparable with the objective function. The inequalities are the only control variable constraints so they are self-restricted

$$P_{gs}^{\lim} = \begin{cases} P_{gs}^{\min} & \text{if } P_{gs} < P_{gs}^{\min} \\ P_{gs}^{\max} & \text{if } P_{gs} > P_{gs}^{\max} \end{cases}$$

$$V_i^{\lim} = \begin{cases} V_i^{\min} & \text{if } V_i < V_i^{\min} \\ V_i^{\max} & \text{if } V_i > V_i^{\max} \end{cases}$$

$$Q_{gi}^{\lim} = \begin{cases} Q_{gi}^{\min} & \text{if } Q_{gi} < Q_{gi}^{\min} \\ Q_{gi}^{\max} & \text{if } Q_{gi} > Q_{gi}^{\max} \end{cases}.$$

The objective of the problem is to minimize the fuel cost and to ensure security of the system under normal and contingent conditions.

17.4.1 Application Examples

The developed FGA has been implemented in the MATLAB command line and used to solve an OPF problem of IEEE-30 bus system under varying operating conditions as outlined in Table 17.5. The cost function is considered to be quadratic with cost parameters as given in Table 17.5. Figure 17.4 shows the variation of objective function values with respect to number of generation under cost minimization situation.

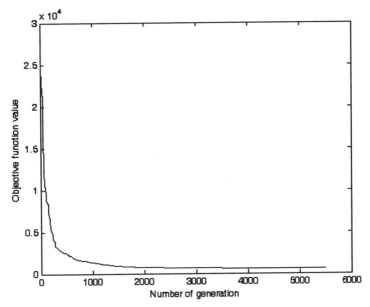

FIGURE 17.4 Result of GA-based OPF for IEEE-30 bus system cost minimization.

TABLE 17.5 Variable Limits and Generation Cost Parameters

	Power Generation Limits and Fuel Cost Parameters $(S_B = 100 \text{ MVA})$					
Bus	1	2	5	8	11	13
P_g^{max}	2	0.8	0.5	0.35	0.3	0.4
P_g^{min}	0.5	0.2	0.15	0.1	0.1	0.12
Q_g^{max}	2	1	0.8	0.6	0.5	0.6
Q_g^{min}	−0.2	−0.2	−0.15	−0.15	−0.1	−0.15
a	0	0	0	0	0	0
b	200	175	100	325	300	300
c	37.5	175	625	83.4	250	250

Bus Voltage Limits				Branch Apparent Power Limit S_k^{max}
V_g^{max}	V_g^{min}	V_{load}^{max}	V_{load}^{min}	Branch (8,28)
1.1	0.95	1.05	0.95	0.12

	Transformer Tap Setting Limits			
Branch	(6,9)	(6,10)	(4,12)	(28,27)
T_k^{max}	1.1	1.1	1.1	1.1
T_k^{min}	0.9	0.9	0.9	0.9

	Capacitor/Reactor Installation Limits								
Bus	10	12	15	17	20	21	23	24	29
Q_c^{max}	0.05	0.05	0.05	0.05	0.05	0.05	0.05	0.05	0.05
Q_c^{min}	0	0	0	0	0	0	0	0	0

The decision variables for the OPF problem are

$$[V_1, V_2, \ldots V_{30}, \theta_1, \theta_2, \ldots \theta_{30}, P_{g1}, P_{g2}, P_{g5}, P_{g8}, P_{g11}, P_{g13}, Q_{g1}, \ldots Q_{g13},$$
$$tr_1, \ldots tr_4, Q_{c1}, \ldots Q_{c9}].$$

The tuned parameters of the FGA used are

Population size: 200
Number of generations (iterations): 5500
Crossover: Heuristic crossover
Mutation: Nonuniform mutation

Variation of values of objective function with respect to an increase of generation is shown in Fig. 17.5. Variable limits and generation cost parameters are shown in Table 17.5. Different results are shown in Tables 17.6, 17.7 and 17.8 below for various constraints and limits.

TABLE 17.6   Optimal Active Power Generations, Fuel Costs (Quadratic), Generator Voltages, Transformer Taps and Shunt Capacitor/Reactor Compensations, Reactive Power Generations

Bus	Active Power Generations	Bus Voltage Magnitudes	Bus Voltage Phase Angles (deg.)	Shunt Capacitor Compensations	Reactive Power Generations
1	2.2581	0.952	0		2.35
2	0.2034	0.992	−3.5891		0.134
3		0.988	−5.166		
4		0.994	−6.5095		
5	0.151	1.072	−10.157		0.043
6		1.0153	−7.325		
7		1.035	−8.615		
8	0.1173	1.0186	−7.571		0.182
9		0.9969	−9.168		
10		1.0054	−11.646	0.0287	
11	0.1034	1.048	−7.092		0.309
12		1.020	−14.695	0.0436	
13	0.12083	1.047	−15.029		0.0867
14		0.992	−16.574		
15		1.002	−15.681	0.044	
16		1.005	−14.135		
17		1.0013	−12.608	0.0023	
18		0.998	−13.077		
19		0.999	−11.728		
20		1.003	−11.353		0.0029
21		1.0016	−12.660	0.045	
22		1.0022	−12.753		
23		0.993	−16.508	0.049	
24		0.984	−14.577	0.045	
25		0.967	−12.557		
26		0.951	−14.589		
27		0.966	−9.893		
28		1.0172	−7.451		
29		0.9552	−8.792	0.0009	
30		0.957	−9.707		

Branch	Transformer Tap Settings	Fuel Cost ($/hr)
(6,9)	0.958	827.88
(6,10)	1.0925	
(4,12)	1.031	
(28,27)	0.952	

Loss = 0.12944 p.u.

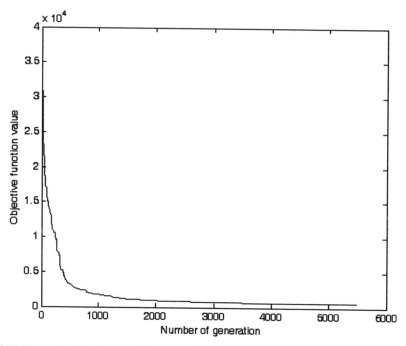

FIGURE 17.5   Result of GA-based OPF for IEEE-30 bus system under outage of line (6,28) and line flow limit in line (8,28).

17.4.1.1   Optimal Power Flow for Loss Minimization

To consider the accurate cost curve of each generating unit, the valve-point effects must be included in the cost model. Therefore, the sinusoidal function is incorporated into the quadratic function. The cost function addressing valve-point loadings of generating unit i is accurately represented as:

$$F_i(p_i) = a_i + b_i p_i + c_i p_i^2 + |e_i \sin(f_i(p_{i\,\min} - p_i))|. \qquad (17.23)$$

The parameters of the cost functions are as given in Table 17.9. A typical valve point effect of a generator and its impact on the convergence of the cost function are shown in Fig. 17.6 and Fig. 17.7 respectively. A set of result is shown in Table 17.10.

17.5   OPF WITH FACTS DEVICES

In this section, the UPFC is used as a power flow controller. The phase shifters and/ or series compensators based on UPFC are adopted as the power flow controlling facilities. Simulation studies have been carried out in a modified IEEE 30-bus system. The phase shifters and/or series compensators based on UPFC are adopted as the power flow control facilities. The active power loss is used as the objective

TABLE 17.7 Optimal Active Power Generations, Fuel Costs (Quadratic), Generator Voltages, Transformer Taps and Shunt Capacitor/Reactor Compensations, Reactive Power Generations, Loss, Line Flow Through Line (8,28)

Bus	Active Power Generations	Bus Voltage Magnitudes	Bus Voltage Phase Angles (deg.)	Shunt Capacitor Compensations	Reactive Power Generations
1	2.9180	0.951	0		4.987
2	0.2083	1.060	−5.510		0.5910
3		1.010	−6.165		
4		1.02	−7.618		
5	0.1522	1.086	−11.773		0.7889
6		1.028	−8.753		
7		1.046	−10.375		
8	0.1104	1.026	−8.654		0.2752
9		1.001	−13.452		
10		0.961	−16.181	0.0095	
11	0.1017	1.068	−12.881		0.1128
12		0.921	−14.675	0.0372	
13	0.1269	0.878	−14.128		0.0111
14		0.921	−16.729		
15		0.933	−16.260	0.0492	
16		0.922	−16.365		
17		0.940	−16.77	0.0039	
18		0.944	−15.521		
19		0.947	−15.246		
20		0.952	−15.215	0.0025	
21		0.950	−17.258	0.0378	
22		0.950	−17.143		
23		0.944	−16.812	0.0432	
24		0.941	−15.38	0.0342	
25		0.954	−7.208		
26		0.947	−4.421		
27		0.969	−4.884		
28		1.038	−6.816		
29		0.9258	−1.751	0.0119	
30		0.946	−2.593		

Branch	Transformer Tap Settings	Fuel Cost ($/hr)
(6,9)	1.0744	1088.71
(6,10)	1.0127	
(4,12)	1.0929	
(28,27)	1.0721	

Loss (real) = 0.3526 p.u. Line flow (8,28) = 0.1167 p.u.

TABLE 17.8 Optimal Active Power Generations, Fuel Costs (Quadratic), Generator Voltages, Transformer Taps and Shunt Capacitor/Reactor Compensations, Reactive Power Generations

Bus	Active Power Generations	Bus Voltage Magnitudes	Bus Voltage Phase Angles (deg.)	Shunt Capacitor Compensations	Reactive Power Generations
1	2.653	0.990	0		−1.1736
2	0.325	0.9987	−3.408		0.205
3		0.998	−5.082		
4		0.999	−5.727		
5	0.264	0.992	−9.459		0.543
6		0.993	−6.082		
7		0.985	−7.950		
8	0.3391	0.982	−5.877		0.352
9		0.990	−4.085		
10		1.01	−5.999	0.04968	
11	0.2695	1.0199	−0.234		0.2507
12		0.991	−8.074	0.0283	
13	0.3741	1.024	−5.090		−0.1018
14		0.982	−10.54		
15		0.9851	−11.588	0.0131	
16		1.0014	−6.32		
17		1.0088	−5.679	0.027	
18		0.990	−17.313		
19		0.996	−17.619		
20		1.001	−15.858	0.0495	
21		1.013	−5.302	0.02132	
22		1.0144	−5.3547		
23		0.9933	−10.458	0.0415	
24		0.9982	−8.388	0.0374	
25		0.993	−12.915		
26		0.9788	−14.853		
27		1.002	−14.481		
28		0.9842	−6.780		
29		0.9858	−19.296	0.0496	
30		0.9813	−20.941		

Branch	Transformer Tap Settings	Fuel Cost ($/hr)
(6,9)	1.0989	1306.00
(6,10)	1.0737	
(4,12)	0.9603	
(28,27)	1.04323	

Power loss (real) = 0.021131 p.u.

TABLE 17.9 Cost Parameters of the Generating Units with Valve Point Loading

Power Generation Bus	Cost Parameters				
	a	b	c	e	f
1	0	200	37.5	200	0.042
2	0	175	175	150	0.063
5	0	100	625	100	0.084
8	0	325	83.4	100	0.084
11	0	300	250	100	0.084
13	0	300	250	150	0.063

a, b, c, e and f are the coefficients.

function, and the active power flow limits in transmission lines during the contingency state are used to justify the UPFC control.

The objectives for optimal power flows are to minimize production cost, MW losses, or mega VAr (MVAr) losses and control variable shift. The control variables could be phase-shift transformer taps, switched capacitors, static VAr compensator (SVC) control setting, generator MW, voltage and MVAr control, load shedding and line switching. Simulation studies have been carried out in a modified IEEE 30-bus system to show the effectiveness of the power flow control in FACTS with the GA.

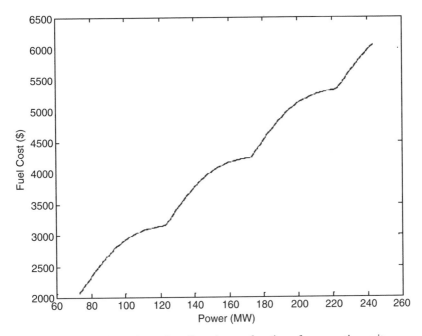

FIGURE 17.6 Valve point effects in cost function of a generating unit.

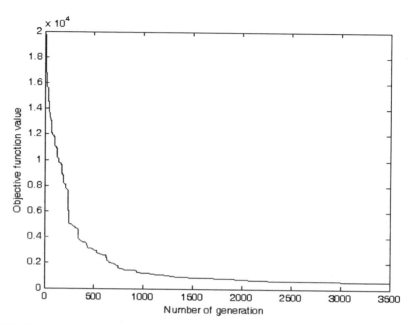

FIGURE 17.7   Convergence nature of FGA for IEEE-30 bus system with cost functions including valve point loadings (limits on slack bus power relaxed).

17.5.1   FACTS Model

17.5.1.1   Phase Shifter   The power injection model is used to represent the phase shifter, which is derived in Refs. 18–20 but is based on the phase-shifter transformer. A general model based on UPFC is given in this section. The tap change is independent of the phase shift so the angle and the magnitude of the transmission line voltage can be separately regulated. The equivalent circuit of the model is given in Fig. 17.8. The variable with a " $-$ " above represents a complex number.

The power flow equations of the branch can be derived as follows:

$$
\begin{aligned}
P_{ij} &= V_i^2 T_{ij}^2 g_{ij} - V_i V_j T_{ij}(g_{ij}\cos(\theta_{ij} + \Phi) + b_{ij}\sin(\theta_{ij} + \Phi)) \\
Q_{ij} &= -V_i^2 T_{ij}^2 b_{ij} - V_i V_j T_{ij}(g_{ij}\sin(\theta_{ij} + \Phi) - b_{ij}\cos(\theta_{ij} + \Phi)) \\
P_{ji} &= V_j^2 g_{ij} - V_i V_j T_{ij}(g_{ij}\cos(\theta_{ij} + \Phi) - b_{ij}\sin(\theta_{ij} + \Phi)) \\
Q_{ji} &= -V_j^2 b_{ij} + V_i V_j T_{ij}(g_{ij}\sin(\theta_{ij} + \Phi) + b_{ij}\cos(\theta_{ij} + \Phi)),
\end{aligned}
\tag{17.24}
$$

where

$$
\begin{aligned}
y_{ij} &= g_{ij} + jb_{ij} \\
\bar{V}_i &= V_i e^{j\theta_i} \\
\theta_{ij} &= \theta_i - \theta_j
\end{aligned}
\tag{17.25}
$$

TABLE 17.10 Optimal Active Power Generations, Fuel Costs (with Valve Point Loading), Generator Voltages, Transformer Taps and Shunt Capacitor/Reactor Compensations, Reactive Power Generations

Bus	Active Power Generations	Bus Voltage Magnitudes	Bus Voltage Phase Angles (deg.)	Shunt Capacitor Compensations	Reactive Power Generations
1	0.8348	0.9506	0		0
2	0.28191	0.9629	−1.256		0.13916
3		0.981	−2.393		
4		0.985	−3.329		
5	0.16198	1.002	−6.326		0.65050
6		1.010	−3.501		
7		1.002	−4.666		
8	0.18126	1.054	−3.844		0.55527
9		1.012	−4.176		
10		1.002	−7.527	0.00947	
11	0.11056	0.961	−0.609		0.35411
12		0.975	−13.734	0.00721	
13	0.12598	1.000	−14.78		0.41701
14		0.961	−17.255		
15		0.968	−16.146		
16		0.958	−10.113	0.0473	
17		0.981	−7.448		
18		0.975	−13.473	0.0054	
19		0.985	−11.051		
20		0.993	−9.827		
21		1.002	−9.111	0.0343	
22		1.002	−9.409	0.0456	
23		0.966	−17.350		
24		0.9754	−13.874	0.0387	
25		0.984	−13.396	0.0323	
26		0.9781	−15.723		
27		0.993	−10.781		
28		1.022	−3.638		
29		0.983	−13.182		
30		0.979	−14.373	0.0351	

Branch	Transformer Tap Settings	Fuel Cost ($/hr)	Power loss (real)
(6,9)	0.9943	433.068	0.1373365
(6,10)	0.9297		
(4,12)	1.0283		
(28,27)	0.9603		

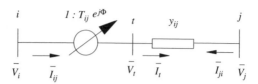

FIGURE 17.8   Equivalent circuit of power flow controller.

The effect of phase shifter can be represented by equivalent injected powers at both buses of the branch as in Fig. 17.9, and a general power injection model based on UPFC is given in (17.26). Considering g_{ij} is much less than b_{ij}, the following equation could be derived:

$$P_{is} = b_{ij}V_iV_j(T_{ij}\sin(\theta_{ij} + \Phi) - \sin\theta_{ij})$$
$$Q_{is} = b_{ij}(V_i^2(T_{ij}^2 - 1) - V_iV_j(T_{ij}(\cos(\theta_{ij} + \Phi) - \cos\theta_{ij})))$$
$$P_{js} = -b_{ij}V_iV_j(T_{ij}\sin(\theta_{ij} + \Phi) - \sin\theta_{ij})$$
$$Q_{js} = -b_{ij}V_iV_j(T_{ij}(\cos(\theta_{ij} + \Phi) - \cos\theta_{ij}).$$

$$(17.26)$$

These equivalent injected powers, together with a regular transmission branch as in Fig. 17.9, will be used in the power flow equations. The derivation in Ref. 19 shows that the sensitivities of the equivalent injected powers to the magnitudes and angles of voltages at the buses i and j are much smaller than the corresponding elements in the ordinary Jacobian matrix of power flow equations. Therefore, these injected powers can be treated as loads or generations at buses i and j. Within each iteration, the following equations hold:

$$0 = P_{gi} + P_{is} - P_{di} - V_i \sum_{j\in N_i} V_j(G_{ij}\cos\theta_{ij} + B_{ij}\sin\theta_{ij}) \quad i \in N_{B-1}$$
$$0 = Q_{gi} + Q_{is} - Q_{di} - V_i \sum_{j\in N_i} V_j(G_{ij}\sin\theta_{ij} - B_{ij}\cos\theta_{ij}) \quad i \in N_{PQ},$$

$$(17.27)$$

where N_i is the set of numbers of buses adjacent to bus i (including bus i), N_{B-1} is the set of numbers of total buses (excluding slack bus), and N_{PQ} is the set of PQ-bus numbers. P_{is} and Q_{is} will be computed after each iteration but will not be used in

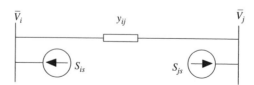

FIGURE 17.9   Power injection model.

deriving the Jacobian matrix, so the symmetry property of the bus admittance matrix is maintained and the P-Q decoupled power flow can be used without any modification.

17.5.1.2 Series Compensator The series compensator is shown in Fig. 17.10 as a controllable capacitor. The capacitor is just simply used as a changeable reactance to be implemented in the bus susceptance matrices.

17.5.2 Problem Formulation

The power flow controllers are used as phase shifters and series compensators to regulate the power flow. Because the active power loss is used as the objective function and the active power flow limits of transmission lines during the contingency states are used as the security constraints, the objective function of OPF in FACTS is therefore expressed as follows:

$$\min P_{\text{loss}} = \sum_{\substack{k \in N_E \\ k=(i,j)}} g_k \left(V_i^2 + V_j^2 - 2V_i V_j \begin{cases} \cos \theta_{ij} & \text{if } k \notin N_{\text{pfc}} \\ \cos(\theta_{ij} + \Phi_k) & \text{if } k \in N_{\text{pfc}} \end{cases} \right)$$

$$\text{s.t. } 0 = P_i - V_i \sum_{j \in N_i} V_j (G_{ij} \cos \theta_{ij} + B_{ij} \sin \theta_{ij}) \qquad i \in N_{B-1}$$

$$0 = Q_i - V_i \sum_{j \in N_i} V_j (G_{ij} \sin \theta_{ij} - B_{ij} \cos \theta_{ij}) \qquad i \in N_{PQ} \tag{17.28}$$

$$T_k^{\min} \le T_k \le T_k^{\max} \qquad k \in N_{\text{phs}}$$

$$\Phi_k^{\min} \le \Phi_k \le \Phi_k^{\max} \qquad k \in N_{\text{phs}}$$

$$0 \le x_{ck} \le x_{ck}^{\max} \qquad k \in N_{\text{sc}}$$

$$|P_k| \le P_k^{\max} \qquad k \in N_E^{\lim}$$

where N_E is the set of branch numbers, N_{phs} is the set of numbers of phase-shifter branches, N_{sc} is the set of numbers of series compensator branches, $N_E^{\lim}$ is the set of numbers of limited power flow branches, P_{loss} is the network real power loss, and g_k is the conductance of branch k.

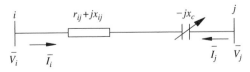

FIGURE 17.10 Series compensator line.

Power flow equations are used as equality constraints; phase-shifter angles, tap settings, series compensator settings, and active power flow of transmission lines are used as inequality constraints. The phase-shifter angles, tap settings, and series compensator settings are control variables so they are self-restricted. The active power flows of transmission lines are state variables, which are restricted by adding them as the quadratic penalty terms to the objective function to form a penalty function. Equation (17.28) is therefore changed to the following generalized objective function:

$$\min f = P_{\text{loss}} + \sum_{k \in N_E^{\text{lim}}} \lambda_k (|P_k| - P_k^{\text{max}})^2$$

$$\text{s.t.} \quad 0 = P_i - V_i \sum_{j \in N_i} V_j (G_{ij} \cos \theta_{ij} + B_{ij} \sin \theta_{ij}) \qquad i \in N_{B-1} \qquad (17.29)$$

$$0 = Q_i - V_i \sum_{j \in N_i} V_j (G_{ij} \sin \theta_{ij} - B_{ij} \cos \theta_{ij}) \qquad i \in N_{PQ},$$

where λ_k is the penalty factor, which can be increased in the optimization process.

17.5.3 Numerical Results

In this section, a modified IEEE 30-bus system has been used to show the effectiveness of the algorithm. The network is shown in Fig. 17.11 and the impedances, loads, and power generations, except the generation from the slack bus that is bus 1, are given in Ref. 24. All voltage and power quantities are per-unit values. The total loads are $P_{\text{load}} = 2.834$, $Q_{\text{load}} = 1.262$. The voltages at generator buses, P-V buses, and a slack bus are set to 1.0. Four branches, (6,10), (4,12), (10,22), and (28,27),

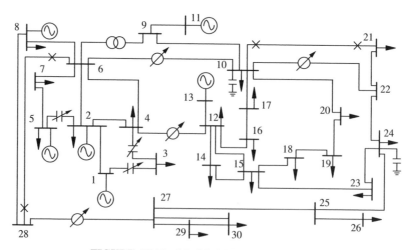

FIGURE 17.11 Modified IEEE 30-bus system.

TABLE 17.11   Series Compensator Limits

Branch	(1,3)	(3,4)	(2,5)
x_c^{max}	0.1	0.02	0.1
x_c^{max}/x_{ij} (%)	54	53	50

are installed with phase shifters. Three branches, (1,3), (3,4), and (2,5), are installed with series compensators. The phase-shifter limits are given as $-20° \le \Phi_k \le 20°$, and the series compensator limits are listed in Table 17.11.

Three cases have been studied. Case 1 is the normal operation state. Cases 2 and 3 are contingency states. Case 2 has one circuit outage of branch (6,28). Case 3 has two circuit outages of branches (6,28) and (10,21). Two branches, branches (10,22) and (8,28), are selected as limited power flow branches whose limits are set to 0.1. The initial phase shifter angles are $0°$ and the compensator reactance is 0. The optimal phase angles and series compensator reactance are listed in Table 17.12. Table 17.13 shows the optimization process. Generations and power losses are listed in Table 17.14. The power flows of branches (10,22) and (8,28) are regulated

TABLE 17.12   Optimization Process

				Overloads (%)	
	Outer Loop	Generation (Min = 40)	Time (min)	(10,22)	(8,28)
Case 1	Initial				
	1	107	2:42	0	0
	Total	107	2:42	0	0
Case 2	Initial				
	1	95	2:20	0	47
	2	84	2:01	0	24
	3	50	1:14	0	6.7
	4	40	0:59	0	1.5
	5	40	1:00	0	0.3
	Total	309	7:34	0	0
Case 3	Initial			131	58
	1	111	2:55	53	52
	2	126	3:36	19	20
	3	92	3:12	4.4	4.7
	4	73	2:09	1.0	1.2
	5	52	1:35	0.2	0.2
	6	40	1:14	0	0
	Total	494	14:41		

Note: 1. The population sizes are 30 for all cases.
2. Outer loop is used to increase the penalty factor when the overloads still exist after the optimization converges.

TABLE 17.13   Optimal Phase Angles and Compensators

	Branch										
	(6,10)		(4,12)		(10,22)		(28,27)		(1,3)	(3,4)	(2,5)
	$\Phi°$	T	$\Phi°$	T	$\Phi°$	T	$\Phi°$	T	x_c	x_c	x_c
Case 1	−7.87	1.00	−4.35	0.99	−0.04	1.00	−2.20	1.03	0.021	0.011	0.023
Case 2	0.23	1.04	−1.56	1.02	−0.10	1.00	−4.81	1.00	0.021	0.012	0.023
Case 3	−1.86	1.10	1.84	1.10	−10.76	0.94	−12.16	0.99	0.027	0.016	0.024

back into their limits in both cases 2 and 3 as shown in Table 17.15. After the optimization, there is a 0.00061 p.u. power saving in case 1. In the other two cases, there are even more active power losses. The regulations of phase angles are used to remove the overloads of branches after the circuit outages. Such regulations produce even more active power losses than the original power losses but effectively remove all line overloads. If there is no overload limits applied to the branches, the optimal power losses will be 0.06207 p.u. in case 2 and 0.06687 p.u. in case 3. Therefore, the power savings will be 0.0006 p.u. in case 2 and 0.00087 p.u. in case 3. It can also be seen that most iterations are used to eliminate the overloads after the circuit outages. Without the overload limits, the optimization would be completed within 3 minutes of CPU time (166 MHz PC). However, with the overload limits, it takes 7 minutes 34 seconds in case 2 and 14 minutes 41 seconds in case 3. The optimization converges in about 70 generations.

TABLE 17.14   Generations and Power Losses

	Initial Conditions				Optimal Results			
	P_g	Q_g	P_{loss}	Q_{loss}	P_g	Q_g	P_{loss}	Q_{loss}
Case 1	2.89388	0.98020	0.05988	−0.28180	2.89327	0.99244	0.05927	−0.26956
Case 2	2.89667	1.00151	0.06267	−0.26049	2.89849	0.99417	0.06450	−0.26783
Case 3	2.90174	1.01477	0.06774	−0.24722	2.92480	1.07876	0.09080	−0.18324

TABLE 17.15   Power Flows for Branches

	Branches			
	Initial Conditions		Optimal Results	
	(10,22)	(8,28)	(10,22)	(8,28)
Case 1	0.080	0.021	0.0766	0.0254
Case 2	0.086	0.147	0.100	0.100
Case 3	0.231	0.158	0.100	0.100

17.6 CONCLUSIONS

The potential for application of GA to OPF under various constraints and limits has been shown in this chapter. GA effectively finds out good results of the regulation with no need to differentiate the objective function and the constraint equations. Also, GA is the suitable optimization method to be used in power flow control of FACTS with different control facilities. It can also be seen that UPFC regulations would not give as much power savings as the regulations of generator bus voltages and/or VAr sources [25], but UPFC is a very effective control facility in neglecting the power flow to remove the overloads of transmission lines. Results on the application of hybrid GA to OPF can be found in Ref. 26.

REFERENCES

1. Carpentier J. Contibution to the economic dispatch problem. Bull Soc Francaise Elect 1962; 3(8):431–447 (in French).

2. Dommel HW, Tinney WF. Optimal power flow solutions. IEEE Trans Power Apparatus Systems 1968; PAS-87(10):1866–1876.

3. Lai LL, Ma JT. Improved genetic algorithms for optimal power flow under both normal and contingent operation states. Electrical Power & Energy Systems 1997; 19(5):287–292.

4. Sasson AM, Viloria F, Abotes F. Optimal load flow solution using the Hessian matrix. IEEE Trans Power Apparatus and Systems 1973; PAS-92:31–41.

5. Burchett RC, Happ HH, Wirgau KA. Large scale optimal power flow. IEEE Trans Power Apparatus and Systems 1982; PAS-101:3722–3732.

6. Huneault M, Galiana FD. A survey of the optimal power flow. IEEE Trans Power Systems 1991; 6(2):762–770.

7. Momoh JA et al. Challenges to optimal power flow. IEEE Trans Power Systems 1997; 12(1):444–455.

8. Bakirtzis AG, Biskas PN, Zoumas CE, Petridis V. Optimal power flow by enhanced genetic algorithm. IEEE Trans Power Systems 2002; 17(2):229–236.

9. Stott B, Alsac O. Fast decoupled load flow. IEEE Trans Power Apparatus Systems 1974; PAS-93:859–869.

10. Fuerte-Esquivel CR, Acha E. Unified power flow controller: a critical comparison of Newton-Raphson UPFC algorithms in power flow studies. IEE Proc Gener Trans Distrib 1997; 144(5):437–444.

11. Ge S, Chung TS. Optimal active power flow incorporating power flow control needs in flexible ac transmission systems. Electrical Power Energy System 1999; 14(2):738–744.

12. Walters DC, Sheble GB. Genetic algorithm solution of economic dispatch with valve point loading. IEEE Trans Power Systems 1993; 8(3):1325–1332.

13. Sinha Nidul, Chakrabarti R, Chattopadhyay PK. Evolutionary programming techniques for economic load dispatch. IEEE Trans Evol Computat 2003; 7(1):83–94.

14. Lin C-H, Lin S-Y, Lin S-S. Improvements on the duality based method used in solving optimal power flow problem. IEEE Trans Power Systems 2002; 17(2):315–323.

15. Yuryevich J, Wong KP. Evolutionary programming based optimal power flow algorithm. IEEE Trans Power Systems 1999; 14(4):1245–1250.

16. Kulworawanichpong T, Sujitjorn S. Optimal power flow using tabu search. IEEE Power Eng Rev 2002; 22(6):37–40.

17. Taranto GN, Pinto LMVG, Pereira MVF. Representation of FACTS devices in power system economic dispatch. IEEE Trans Power Systems 1992; PWRS-7(2):572–576.

18. Han ZX. Phase shifter and power flow control. IEEE Trans Power Apparatus Systems 1982; PAS-101(10):3790–3795.

19. Noroozian M, Andersson G. Power flow control by use of controllable series components. IEEE Trans Power Delivery 1993; PWRD-8(3):1420–1429.

20. Lai LL, Ma JT. Optimal power flow in facts using genetic algorithms. IEEE Stockholm Power Tech International Symposium on Electric Power Engineering, 1995, Stockholm, Sweden.

21. Holland JH. Adaptation in natural and artificial systems. Ann Arbor, MI: The University of Michigan Press; 1975.

22. Goldberg E. Genetic algorithms in search, optimization & machine learning. Reading, MA: Addison-Wesley; 1989.

23. Houck CR, Jones JA, Kay MG. A genetic algorithm for function optimization: A Matlab implementation. Technical Report NCSU-IE TR 95-09. North Carolina State University; 1995.

24. Lee KY, Park YM, Ortiz JL. A united approach to optimal real and reactive power dispatch. IEEE Trans Power Apparatus Systems 1985; PAS-104(5):1147–1153.

25. Ma JT, Lai LL. Application of genetic algorithm to optimal reactive power dispatch. 7th IFAC Symposium on Large Scale Systems: Theory and Applications (Postprint). Pergamon, UK, Elsevier Science; 1995, 383–388.

26. Chung TS, Li YZ. A hybrid GA approach for OPF with consideration of FACTS devices. IEEE Power Eng Rev 2001; February:47–57.

CHAPTER 18

An Interactive Compromise Programming-Based Multiobjective Approach to FACTS Control

YING XIAO, YONG-HUA SONG, and CHEN-CHING LIU

18.1 INTRODUCTION

Multiobjective (MO) optimization methodologies for decision aids have received a considerable amount of development during the past 30 years. In most power system applications, minimizing both operation costs and transmission losses or minimizing both operation and emission costs are taken as objectives. With the introduction of free competition to the market, besides economic operation, reliability and quality of supply of electricity has never been paid more attention. These are the main reasons why MO optimization techniques for decision support have assumed a new strategic importance in the current restructured electric power industry.

As already recognized, increased loading of power systems, environmental restrictions, combined with a worldwide power industry restructuring—particularly the unbundling of power generation from power transmission and a large quantity of bilateral contracts and wheeling services in the electricity market—require a more flexible and effective means for power flow control (PFC) and for taking full advantage of existing power resources. On the other hand, by the use of line impedance, magnitude and angle of nodal voltage to redistribute power flow, the FACTS (Flexible Alternating Current Transmission Systems) concept has opened up brand new avenues for power system control with particular respect to thermal, voltage, and dynamics constraints. In the scope of steady-state power system analysis, a host of technical uses for FACTS applications to the power network have been identified and evaluated in the now voluminous literature and includes boosting available transfer capability [1], increasing the profit of auction of financial transmission rights (FTR) [2], improving network loadability [3], loop flow prevention and parallel flow elimination [4], providing secure tie line connections to neighboring regions for

Modern Heuristic Optimization Techniques. Edited by K. Y. Lee and M. A. El-Sharkawi

spinning reserve sharing [5], and so on. Practically, installed FACTS devices are to be applied to facilitate economic system operation and power transactions and to improve system security and reliability. Meanwhile, they are also demanded to fulfill various tasks by local PFC to improve transmission services. Therefore, in order to take full advantage of FACTS devices, it will be of great benefit to design a concerted deliberate control strategy with consideration of the major applications.

However, so far there are very few works published to consider power system optimal operation incorporating FACTS devices with multiple goals. In the optimal power flow (OPF) model of Ref. 6, with minimization of system operation costs as an objective, PFC targets of FACTS devices are formulated as a set of constraints. However, inappropriate setting of the constraints may cause problem infeasibility. Because more than one conflicting goal is involved in the problem, there does not exist a single solution that can optimize all the objectives. For system operators, the most important task is to determine a satisfactory compromise solution by trading off all the objectives. Thus, it is necessary to use MO optimization methods to deal with such a problem.

This chapter focuses on the application of a MO optimization approach to optimal operation of power systems incorporating FACTS devices. First, taking economic system operation and attainment of specific local PFC targets as objectives, a MO optimization model for FACTS control is formulated. For most prevalent MO optimization algorithms, it is very difficult to handle problems with these noncommensurable objectives; therefore, based on compromise programming (CP), a displaced worst compromise programming (DWCP) is introduced to solve the problem in an interactive procedure. Different from normal interactive approaches, which are usually beset with the difficulties of determining the step and direction of weight or aspiration level correction, using this method, it is very efficient to get the reference point, which is to reflect decision-makers' (DMs') preference changes during the interactive process, without any further analysis or extra programming. Considering imprecise and vague objectives and preferences in the decision environment, fuzzy set theory is used to simulate them with appropriate membership functions. Moreover, these membership functions can also conveniently represent relative deviations in the CP. A reduced practical power system is employed to demonstrate the performance of the proposed approach.

The remainder of the chapter is organised as follows. First, the prevalent MO optimization methodologies and applications in power systems are reviewed briefly in Section 18.2. In Section 18.3, based on the analysis of major applications of FACTS technology to power network, a MO optimization model for FACTS control is presented in detail. An interactive DWCP is proposed in Section 18.4. In Section 18.5, aimed at dealing with the problem, the interactive procedure is introduced. In Section 18.6, the implementation of the interactive process is delineated using a flowchart. Then in Section 18.7, case studies with a reduced practical system are presented, with which the validity of the proposed approach is demonstrated. Finally, a brief summary is given in Section 18.8.

18.2 REVIEW OF MULTIOBJECTIVE OPTIMIZATION TECHNIQUES

As the name indicates, the MO optimization technique is one that can be used to achieve an optimal solution with multiple objectives simultaneously. Shown in (18.1), the general model of MO optimization problems has the same expression as that of conventional problems, except that the objective vector has a rank larger than 1.

Minimize: $\qquad\qquad\qquad F(C, X)$

subject to:

$$G(C, X) = 0$$

$$H(C, X) \leq 0 \qquad\qquad (18.1)$$

where X is the state vector (consisting of dependent variables and fixed system parameters), and C is the control vector consisting of the control variables:

$F(C, X)$: the objective(s) to be optimized;

$G(C, X)$: nonlinear equality constraints;

$H(C, X)$: nonlinear inequality constraints.

In practice, for optimization models with multiple objectives, which sometimes are conflicting and noncommensurable, it usually becomes impossible to find an appropriate solution C^* to obtain the optimum values of all the objectives simultaneously. Based on this reality, a concept of optimality, called Pareto optimality, is used to define a domain of nondominated solutions on the boundary of the achieved domain [7]. Verbally, the definition states that C^* is Pareto optimal if there exists no other feasible vector C that would improve some objective values without worsening at least one criterion simultaneously. In this sense, optimization actually means the generation of a range of noninferior solutions that show the trade-offs among objectives.

The prevalent MO optimization algorithms are introduced briefly in the following subsections.

18.2.1 Weighting Method

Proposed by Zadeh in 1963, the basic idea of the method is to linearly combine the various objectives into a single-objective (SO) function by simply assigning relative weight to each objective. Thus, the transformed SO optimization problem can be easily solved by any conventional methodologies. If the weights are interpreted as representing the relative preference of some DMs, then the solution to the SO problem is equivalent to the best-compromise solution of the corresponding MO problem.

Because of its simple implementation, the weighting method has been identified as one of the most widely used MO optimization approaches in power system applications [8–10]. However, it is not free from criticism:

- Reasonable weight assigning is a thorny point for any practical application, especially for the problem with noncommensurable objectives.

- Because of the incorporation of multiple noncommensurable factors into a single function, the objective may lose significance.
- It is a frequent observation that an evenly distributed set of weights fails to produce an even distribution of points for a good approximation of the Pareto surface.

18.2.2 Goal Programming

Goal programming (GP) represents one of the most frequently used approaches in the literature of power systems. In a practical MO problem, the DM's aims are often to achieve a satisfactory level for all objectives, which is termed aspiration levels or goals. Based on this fact, the purpose of GP is to minimize the derivations between the desired goals and the actual results; it was first described in 1961 by Charnes and Cooper [11].

The GP has been used to calculate environmental marginal cost for the problem of environment-constrained optimal generation dispatch [12, 13]. Fan et al. [14] applied GP to solve the real-time economic dispatch problem. The approach is also employed to solve weekly maintenance scheduling of generators under economic and reliability criteria in Ref. 15. However, generally the priority order of objectives is hard to presume. It may cause a relatively long execution time. Above all, the major short-coming of GP is that the inappropriate construction of goals can result in an inferior solution [7].

18.2.3 ε-Constraint Method

Proposed in 1967 by Marglin, the ε-constraint method is considered to be the most intuitively appealing MO programming approach [16]. In this approach, one of the criteria is chosen as the objective function while all of the others are treated as inequality constraints by introducing the limit level ε. Thereby, the MO problem is transformed into a SO problem. Calculating the problem repeatedly for various values of ε, this method allows the DM to get all Pareto solutions of the problem. However, this approach can result in a considerably heavy burden of computation, with a large quantity of combinations of ε.

In Ref. 17, a two-stage solution methodology based on the ε-constraint method and the simulated annealing (SA) technique is developed to solve the distribution network reconfiguration problem, taking into consideration the minimizing power losses and relieving the branch overload.

18.2.4 Compromise Programming

Technically, the ideal solution is defined as the collection of all objective values, which is preferred to all others among all achievable scores for any objective. With this concept, a CP, proposed by Zeleny in 1973, is implemented by replacing *a priori* defined goals of GP by the actual achievable ideal solution [7]. By the use of the CP, the best compromise solution among the objectives that is the closest one to the ideal point can be achieved. Most importantly, the CP provides an excellent base for the interactive procedure.

In 1995, Chattopadhyay [18] applied the CP to evaluate demand-side management (DSM) options, which take into account capital and operating cost savings, reduced emissions, and improved reliability of electricity supply simultaneously. In 1999, he used the CP to solve a generation-transmission operation planning problem [19].

18.2.5 Fuzzy Set Theory Applications

In addition to the conventional MO optimization algorithms, applications of fuzzy set theory in MO problem solving have been drawing the attention of many researchers. In an attempt to bridge the gap between the conventional mathematics system and nature discourse, in 1965 Zadeh introduced the concept of fuzzy quantifiers [20]. Given a collection of objects Y, a fuzzy set A is defined as:

$$A = \{(y, \ \mu_A(y)) \mid y \in Y\} \tag{18.2}$$

where $\mu_A(y)$ is the membership function of y, representing the degree that y belongs to A. In contrast with a normal crisp set, where the membership value could only be either 0 or 1, the value in fuzzy sets may range from 0 to 1 continuously.

For problems with multiple conflicting objectives, because it is impossible to attain all the ideal values, the best compromise solution that can satisfy the DM is sought instead. Consequently, the motivations of the DM are always described linguistically using "as much as possible, at least larger than, about," and so forth, which is difficult to represent in any conventional mathematics form. Fuzzy set theory can be used to translate these imprecise and vague preferences into a clearcut model. In addition, in typical fuzzy programming, assuming the full symmetry relationship between objective functions and constraints, the optimal solution can be achieved by seeking the maximum degree of "overall satisfaction" [21]; that is, maximizing the minimum value of all the membership values of objectives and constraints included. Therefore, fuzzy set theory leads to computationally efficient MO algorithms; it has been playing a unique role in the MO optimization field.

Thus far, in the scope of power system operation and planning, the fuzzy method has been proposed to deal with contingency constrained OPF [22], hydrothermal coordination [23], transmission planning [24], and so on.

18.2.6 Genetic Algorithm

Genetic algorithms (GAs) were developed by John Holland and his colleagues in 1975. Based on the mechanics of natural selection and natural genetics, the GA starts with a population of strings that represent the possible solutions and generates successive populations of strings by combining survival of the fittest among string structures.

Proposed originally by Schaffer in 1984, and motivated by a suggestion of a nondominated GA outlined in Goldberg in 1989 [25], a number of studies on MO GAs have been undertaken and the implementations have been successfully applied in various MO fields [26]. Different from normal optimization and search procedures, GA searches from a population of solution points. Therefore, multiple noninferior solutions can be found in a single run of GA. In 1998, Das and Patvardhan [27] presented a GA-based approach to deal with the emission-constrained economic load dispatch problem.

18.2.7 Interactive Procedure

Actually, there are two participants involved in MO decision-making: the system analyst and the DM. With respect to the preference preassigned or the aspiration levels of the objectives of the DM, the aim of the analyst is to provide efficient information and appropriate alternatives for supporting the DM. However, in practice, due to the characteristics of noncommensurability and conflict of objectives, the vagueness and inconsistency of human desires, especially the DM's possible unfamiliarity with system optimization potential, the complete *a priori* knowledge of preference or aspiration levels of objective performance with precision is not always available. Consequently, both sides of the decision process (i.e., analysis of what is available and of what is desirable) are to be carried out jointly and dependent on each other. Therefore, it is necessary to introduce an interactive process between analysts and DMs, where objectives evolve by DMs according to available alternatives proposed by analysts, which are, in turn, adjusted and generated in accord with existing objectives.

Following reasonable aspiration changes as a result of learning and experience, this method has the advantage of allowing DMs to gain a greater understanding of the problem. More importantly, without forcing them into an exhaustive examination of all Pareto solutions, the procedure can guide DMs to obtain what they consider the best compromise.

In Ref. 28, an interactive method is applied to solve a reactive power planning problem taking both economy and security of system operation into account. Generally speaking, most of the current interactive methods focus on determining the step and direction of weight or aspiration level change, which are usually undertaken by sensitivity analysis [29, 30], asking DMs to get a series paired comparisons between objectives [8], or even trial and error [31, 32]. The whole process could be quite difficult and unreliable. Besides that, it needs to be pointed out that sometimes the DM's behavior is not rational or the reference points designated by the DM keep on changing. This will result in a time-consuming interactive process.

18.3 FORMULATED MO OPTIMIZATION MODEL

Generally speaking, among a variety of common objectives of the power system optimization, the most underlying one is economy. For power systems, particularly heavily loaded systems and systems with loop flow or parallel flow, optimal generation dispatch is often restricted by stressed transmission circuits or (and) relatively low-voltage nodes. Physically, this situation also limits bulk power transactions between long-distance electricity market participants. Therefore, by relieving restrictions imposed by transmission network resources, essentially FACTS can provide considerable economic benefits by allowing for the use of more cost-efficient generators.

Meanwhile, it is well-known that by controlling active reactive power and voltage individually or simultaneously, FACTS devices are able to accomplish power flow control flexibly. From the viewpoint of steady-state power system operation by local power flow control to eliminate violation of voltage and thermal limits, so as to maintain a certain quantity of network transfer capacity demanded for security and further utilization, use of FACTS devices could also have a potentially beneficial effect in this regard.

18.3.1 Formulated MO Optimization Model for FACTS Control

Based on the analysis above, a MO optimization problem is formulated as:

$$\underset{C}{\text{Min}} \qquad F(X, C)$$

$$\text{s.t.} \quad G(X, C) = 0$$

$$H(X, C) \leq 0. \tag{18.3}$$

Minimization of system operation cost and fulfilling of local power flow control targets are taken as the objectives, which are denoted by the vector F. Operation cost is represented approximately by $\alpha + \beta P_G + \gamma P_G^2$. Therefore, for a power system with N_G generators, the operating cost can be further expressed as:

$$F_G = \sum_{i=1}^{N_G} \alpha_i + \beta_i P_G + \gamma_i P_G^2. \tag{18.4}$$

X represents the vector of state variables including magnitude and angle of nodal voltages. In addition to the control variables of FACTS devices, the control vector C includes generator output and voltage of the PV node. Power flow equations are included as equality constraints, which is denoted by G. The inequality constraints H are imposed on limitations of nodal voltage magnitude, line transfer limit, generator output, and FACTS device capacity.

Here, a power injection model (PIM)-based FACTS model and a power flow control approach is used to represent FACTS devices and to deal with the local PFC problem, which has been presented in detail in Ref. 33. For clarity and integrity of the chapter, the related model will be introduced briefly.

18.3.2 Power Flow Control Model of FACTS Devices

18.3.2.1 Control Variables In the PIM, because active and (or) reactive power injections represent all the features of the steady-state FACTS device, it is rational to take the power injections as independent control variables for power flow control. According to the derivation of the PIM in Ref. 34, line flow control of FACTS devices can be considered as an additional controllable power flow through the line, which is superimposed on the "natural" power flow. Taking the unified power flow controller (UPFC) as an example, it is implemented by two back-to-back converter-based synchronous voltage sources (SVSs), which can generate or absorb reactive power independently. Therefore, for the UPFC installed on line L, $I - J$, near bus I, as shown in Figs. 18.1 and 18.2, besides the active power injection, $P_{I(inj)}$, by the series SVS, there are two independent reactive power injections involved. One is $Q_{II(inj)}$ for regulating voltage, which is injected by the shunt SVS directly to bus I as I_q. Generated by the series SVS, the other flows via line L for reactive power flow control, which is represented as $Q_{IL(inj)}$. That is, in the model, the total reactive power injection to bus I, $Q_{I(inj)} = Q_{II(inj)} + Q_{IL(inj)}$. The control vector is written as $C = \begin{bmatrix} P_{I(inj)} & Q_{IL(inj)} & Q_{II(inj)} \end{bmatrix}$. For active and reactive power flow control of the UPFC, with $P_{I(inj)}$ and $Q_{IL(inj)}$ as control variable, the controlled power flow of line L, P_L and Q_L, can be expressed as power flow P_{L0} and Q_{L0} with the effect of the controller, plus $P_{I(inj)}$ and $Q_{IL(inj)}$ which flow along line L, as shown in Fig. 18.2 clearly. As a result, for UPFC control, the relationship between the power injections and the controlled power flow can be expressed as:

$$P_L(X_L, P_{I(inj)}) = P_{L0}(X_L) - P_{I(inj)} \tag{18.5}$$

$$Q_L(X_L, Q_{IL(inj)}) = Q_{L0}(X_L) - Q_{IL(inj)} \tag{18.6}$$

$$V_I = V_I(Q_{II(inj)}), \tag{18.7}$$

where X_L respesents voltage and phase angle of buses I and J, V_I, V_J, θ_I, θ_J.

As the power injections are only an interim result, once the required power injections are obtained, the original control parameters, including magnitude V_T, phase angle φ_T of the series inserted voltage, and magnitude of the current I_q, can be derived easily according to the voltage source model of the UPFC.

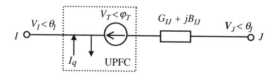

FIGURE 18.1   Voltage source model of UPFC.

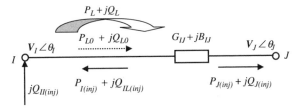

FIGURE 18.2 Power injection model of FACTS devices for power flow control.

As a unified controller, the UPFC possesses the functions of both series controller and shunt controller. Series controllers, such as thyristor-controlled series capacitor (TCSC), thyristor-controlled phase-shifter (TCPS), and static synchronous series compensator (SSSC), are characterized as active line flow control. Therefore, there is only one control variable, $P_{I(inj)}$, involved. The relationship between the power injections and the controlled power flow can be represented by (18.5). As shunt controllers, including the static var compensator (SVC) and STATic Synchronous COMpensator (STATCOM), are only used for voltage regulation, the relationship between the control variable, $Q_{II(inj)}$, and the control target, V_I, can be expressed as (18.7). Control models for all families of FACTS controllers are summarized in Table 18.1, where for TCPS, σ is the phase-shifting angle, x_c represents the controllable reactance of TCSC, V_C denotes the series inserted voltage of SSSC, and I_S is the reactive current injected by SVC and STATCOM.

18.3.2.2 Applied Power Flow Control Model

In the PFC model, minimization of the total mismatch of the preassigned control targets is taken as the objective. That is, for any FACTS device h that is installed on line $L, I-J$, as shown in Fig. 18.2, the objective function is represented by:

$$F_h = \mu_1|P_L - P_{LT}| + \mu_2|Q_L - Q_{LT}| + \mu_3|V_I - V_{IT}|, \tag{18.8}$$

where $P_{LT}, Q_{LT}, V_{IT}, P_L, Q_L,$ and V_I are the control targets and the actual value of active, reactive power flow along line L and voltage of node I, respectively. According to the steady-state control functions and the corresponding classifications of FACTS devices, the model can be employed to accomplish the control of all prevalent FACTS devices flexibly, by suitably setting the coefficients $\mu_1, \mu_2,$ and μ_3 as follows:

For the shunt controller: $\mu_1 = \mu_2 = 0, \mu_3 = 1$;
For the series controller: $\mu_1 = 1, \mu_2 = \mu_3 = 0$;
For the unified controller: $\mu_1 = \mu_2 = \mu_3 = 1$.

TABLE 18.1 Control Variables for Available Control Targets and Control Parameters of FACTS Devices

Classification	Steady-State Functions	Power Injections			FACTS Devices	Control Parameters
		Control Targets	Control Variable(s)	Noncontrol Power Injection(s)		
Shunt controller	Voltage control	V_I	$Q_{II(inj)}$	—	SVC STATCOM	I_S
Series controller	Active power flow control	P_L	$P_{I(inj)}$	$Q_{IL(inj)}, Q_{J(inj)}$	TCPS TCSC SSSC	σ x_c V_C
Unified controller	Active reactive power flow control, Voltage control	P_L, Q_L, V_I	$P_{I(inj)}, Q_{IL(inj)}, Q_{II(inj)}$	$Q_{J(inj)}$	UPFC	V_T, φ_T, I_q

For a power system with N_F FACTS devices installed to implement power flow control, the optimization objective is

$$\text{Min } F = \sum_{h=1}^{N_F} F_h.$$

18.4 PROPOSED INTERACTIVE DISPLACED WORST COMPROMISE PROGRAMMING METHOD

Inspired by the real-life interactive decision process, a DWCP is introduced to solve the MO problem by making the achieved goals as far as possible from the anti-ideal point, which is termed the worst compromise (WC) point [35]. Because the method is based on the CP, for a better understanding, the applied CP model will be introduced briefly.

18.4.1 Applied Fuzzy CP

As introduced earlier, based on the concept of ideal point, the purpose of CP is to achieve the best compromise solution among objectives, which is the closest to the ideal point. Based on the fact that sometimes DMs strive to be as far as possible from the worst achievable objective values, which is referred to as an anti-ideal solution, CP also can be formulated to maximize the distance between the achieved objectives and the anti-ideal points [7]. This formulation of CP is applied here, as shown in (18.9):

$$\underset{C}{\text{Max}} \quad \left(\sum_{N_{obj}} (F(X, C(- F^{AI})^2 \right)^{1/2}$$

$$\text{s.t.} \quad G(X, C) = 0$$

$$H(X, C) \leq 0, \tag{18.9}$$

where N_{obj} is the total number of objectives considered in the model, and F^{AI} is the corresponding anti-ideal value of each objective.

Practically, the operation and decision environment is usually vague. An important advantage of fuzzy formulation is its ability to handle elastic or ambiguous optimization objectives by introducing fuzzy membership functions. Particularly, local power flow control seems to be a promising field for the application of the concept, where not all control objectives can be stated, therefore are necessary to be reached precisely. Furthermore, in most applications of CP, in order to avoid any undesirable effects of the units of measurement on the objectives, relative deviations are usually adopted. They can be represented by fuzzy membership functions intuitively and conveniently. With these consequent benefits, application of fuzzy set theory together with CP will be more progressive and practical than conventional algorithms.

To reflect the satisfaction degree of a given solution, the objectives are reformulated by using fuzzy membership functions as follows.

18.4.2 Operation Cost Minimization

$$\underset{C}{\text{Max}} \ (\mu_{F_{EO}}(X, C)),$$

where the membership function of the objective of economic operation (EO) is

$$\mu_{F_{EO}}(X, C) = \begin{cases} 0 & F_{EO}(X, C) \geq F_{EO}^{WC} \\ \dfrac{F_{EO}^{WC} - F_{EO}(X, C)}{F_{EO}^{WC} - F_{EO}^{Ideal}} & F_{EO}^{Ideal} < F_{EO}(X, C) < F_{EO}^{WC}. \\ 1 & F_{EO} \leq F_{EO}^{Ideal} \end{cases} \qquad (18.10)$$

F_{EO}^{Ideal} is an ideal value of the objective with membership value of 1. F_{EO}^{WC} represents the WC value of the objective with membership value of 0. According to the formulated linear membership function shown in Fig. 18.3, membership values of any solution between F_{EO}^{Ideal} and F_{EO}^{WC} can be determined easily.

18.4.3 Local Power Flow Control

Practically, nodal voltages are supposed to be within a specified range; any value within the range obtained is acceptable, which is not necessary to be fixed. Voltage control is included in the limitations of nodal voltages in H. Therefore, only active and reactive line flow controls are considered here. Because the membership functions of active and reactive power flow control have a similar feature, a general representation, including the fuzzy membership functions and the corresponding illustration curves, is shown in (18.11) and Fig. 18.4.

$$\underset{C}{\text{max}} \ \underset{P,Q}{\text{min}} \ (\mu_{F_{PFC}^{P}}(X, C), \ \mu_{F_{PFC}^{Q}}(X, C)),$$

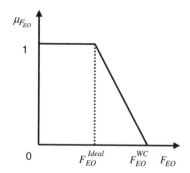

FIGURE 18.3 Membership function of EO.

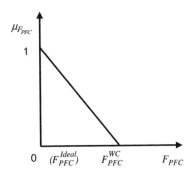

FIGURE 18.4   Membership function of PFC.

where

$$
\mu_{F_{PFC}^{\Gamma}}(X, C) = \begin{cases} 0 & F_{PFC}^{\Gamma}(X, C) > F_{PFC}^{\Gamma WC} \\ 1 - \dfrac{F_{PFC}^{\Gamma}(X, C)}{F_{PFC}^{\Gamma WC}} & F_{PFC}^{\Gamma}(X, C) \le F_{PFC}^{\Gamma WC}, \quad \Gamma = P, Q. \quad (18.11) \end{cases}
$$

The interpretation of the membership function is the absolute value of deviation between the real value and the control target should be "as close to 0 as possible," at least not higher than F_{PFC}^{WC}, which is an assumed acceptable tolerance. The corresponding satisfaction degree will be decreased linearly in that tolerance range, as shown in Fig. 18.3.

It needs to be mentioned that, in the interactive process, the WC values of the objectives keep on changing to reflect the dynamic preferences of the DM, which will be demonstrated in Section 18.5.

18.5   PROPOSED INTERACTIVE PROCEDURE WITH WC DISPLACEMENT

During the interactive process, in each interaction, a DM gives comments to the current solution of all objectives. As a reference point to reflect the DM's preference changes, the WC points are to be adjusted according to these comments. They will be taken as the corresponding anti-ideal points of the model for the next iteration. Then, with the modified MO optimization model, another optimal solution can be calculated and be presented to the DM. Thereby, the WC points keep on being displaced to capture the DM's intentions interactively, until he or she is satisfied with the solution provided. Generally, the procedure consists of three phases.

18.5.1   Phase 1: Model Formulation

This phase is to formulate a model according to the DM's preference. If this is not the initial iteration, to reflect the DM's aspiration level changes, a new WC parameter for

TABLE 18.2   Rules for New Worst Compromise Value Assignation

DM's comments on the objective values of iteration t	New worst compromise value assigned for iteration $t + 1$
Need to be improved	Set at the current objective value attained
Retain at the current level	Keep at the current WC point
Allowed to be worsened	Is lowered to the nearest experienced point

any objective can be determined following the rules in Table 18.2. Then, with the WC points displacing, a new optimization model is obtained with modified membership functions of the objectives.

18.5.2   Phase 2: Noninferior Solution Calculation

With fuzzy membership functions formulated, the solution $S = [C, X]$ can be calculated as a result of the following problem using the CP:

$$\underset{C}{\text{Max}} \sum_{N_{obj}} (\mu_F(C, X))$$

$$\text{s.t.} \quad G(X, C) = 0$$

$$H(X, C) \le 0. \tag{18.12}$$

This conversion enables us to view the original MO optimization problem from a more practical angle; that is, rescheduling the control variables to achieve a scenario with the sum of satisfaction degrees of all the objectives as high as possible, without violating any rigid constraints.

18.5.3   Phase 3: Scenario Evaluation

The resultant scenario is presented to the DM. If the DM is satisfied with the current solution, it means that the best compromise solution has been achieved. If not, the solution is evaluated to elicit his or her preferences. To all the objective values, the DM's opinions can be classified into three categories as listed in the left column of Table 18.2.

With the information of the DM's preference obtained in phase 3, a new iteration can be started with phase 1. The whole procedure will not be terminated until the DM is satisfied with a decision alternative completely or the same decision alternative is computed in two consecutive sessions.

The structure of the applied interactive procedure is sketched in Fig. 18.5.

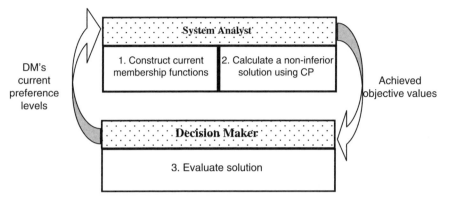

FIGURE 18.5 Applied interactive procedure.

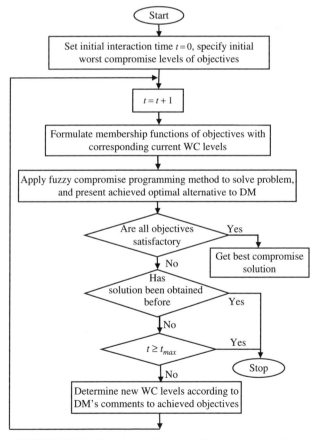

FIGURE 18.6 Flowchart of proposed interactive procedure.

18.6 IMPLEMENTATION

For clarity, the whole interactive procedure of the proposed displaced worst compromise method for the problem is outlined in Fig. 18.6, where t_{max} is the preassigned maximum interaction number.

From the implementation perspective, there are two major modules involved in the proposed FACTS control approach: an optimization model and an AC power flow calculation. A predictor-corrector primal-dual interior point linear programming (PCPDIPLP), as proposed by Mehrotra in 1992 [36], is employed to solve the formulated OPF problem. Compared with other optimization algorithms, the method has such advantages as reliability of the optimization and high speed of the calculation, which are both very important for the interactive approach. Noting that conventional power flow algorithms are convergence failure-prone when applied to power systems with embedded FACTS devices [37, 38], an improved optimal multiplier Newton-Raphson (OMNR) power flow algorithm is adopted here, which has been presented in Ref. 39.

18.7 NUMERICAL RESULTS

In order to illustrate the feasibility of the proposed interactive MO approach, a reduced practical system with 29 nodes and 69 branches is used. A single line diagram of the high voltage (HV) circuits of the test system is shown in Fig. 18.7. According to the data in Ref. 40, the cost coefficients of each generator are assumed. To demonstrate the impact of FACTS devices control more clearly, the system is slightly modified by increasing the original load and generation level by 10%. In this study, the per unit base is 100 MVA.

First, considering the objective of economy, a conventional optimal power flow is calculated only considering generation rescheduling and PV node voltage adjustment. The result is listed in Table 18.3 as case 1. The fact that there are much more

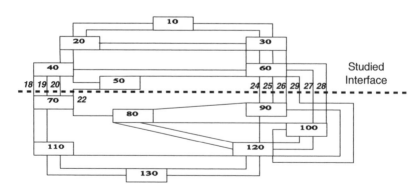

FIGURE 18.7 A reduced practical system.

**TABLE 18.3   Scenarios with and without FACTS Devices Under
Cost-Minimization Criterion**

Case 1 (No FACTS) (p.u.)				Case 2 (with FACTS) (p.u.)			
Generator Output		PV Node Voltage		Generator Output		PV Node Voltage	
1	44.0193	1	1.0228	1	45.9099	1	1.0603
2	23.3952	2	1.0085	2	23.6438	2	1.0546
3	64.6059	3	1.0223	3	64.8031	3	1.0252
4	71.6372	4	1.0096	4	73.4717	4	1.0093
5	49.7671	5	1.0173	5	57.4298	5	1.0000
6	86.1291	6	1.0111	6	91.4889	6	1.0077
7	10.0000	7	1.0269	7	10.0000	7	1.0167
8	7.1523	8	1.0019	8	6.3785	8	0.9984
9	12.1736	9	0.9950	9	10.0000	9	0.9852
10	22.6943	10	1.0164	10	16.0150	10	1.0064
11	38.5423	11	1.0589	11	36.5183	11	1.0655
12	29.8411	12	1.0222	12	27.1509	12	1.0409
13	30.3545	13	1.0552	13	30.5214	13	1.0604
14	11.4169	14	1.0620	14	10.0000	14	1.0572
15	22.9201	15	1.0887	15	20.7512	15	1.0992
16	18.2779	—	—	16	16.5065	—	—
Operation cost = 840.24 ($\times 10^2$ £/hour)				Operation cost = 818.16 ($\times 10^2$ £/hour)			

economic generators in the North than in the South, together with the situation that
generation center and load center are located in the North and in the South, respectively, results in an extremely stressed transfer interface. From the thermal burden of
the interface branches of case 1, as illustrated in Fig. 18.8, it is evident that the serious
situation of unbalanced line flow through the interface, particularly on lines 24, 25,
26, and 29, prevents system operation cost from further decreasing.

Aiming at gaining a better economy by adjusting power flow of the interface, four
FACTS devices are applied. Besides those series controllers for line flow redistribution, because of the relatively low voltage of node 90 in case 1, a UPFC is installed on

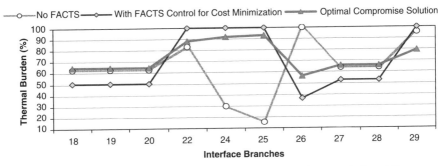

FIGURE 18.8   Thermal burdens of interface branches in cases 1, 2, and 3.

TABLE 18.4 Derived Control Parameters of the FACTS Devices in Case 2

No.	Device	Line	First Node	Second Node	Control Parameters Derived for Cost-Minimization in Case 2 (p.u.)
1	UPFC	24	90	60	$V_T = 0.8433$, $\varphi_T = 1.4305$, $I_q = 16.4293$
2	TCSC	25	60	90	$x_c = -0.0836$
3	SSSC	29	60	120	$V_C = -0.0036$
4	TCPS	22	50	70	$\sigma = 0.3695$

line 24 to support the node voltage. The installed FACTS devices, locations, and the derived control parameters are given in Table 18.4. As case 2, the resultant optimal solution is also shown in Table 18.3. From the comparisons of the operation costs between case 1 and case 2, which are summarized in Table 18.3, the potential benefit of FACTS devices on economic system operation supporting is clearly seen.

For thermal burden of the interface branches in case 2, it is observed from Fig. 18.9 that, by control of FACTS devices, those controllable lines are all almost full-loaded to allow better access to the cheaper generations of the North. But, as a matter of fact, increase in the power flow above a certain operating level will decrease the overall security of the system. Therefore, to operate the system securely, the level of power transfer between specific areas usually has to be limited. In view of this situation, besides improving system economy, these FACTS devices should also be controlled with an intention to alleviating heavy-load lines via local power flow control. However, owing to the current optimal generation pattern being interfered with, it will cause a relatively poor economy of system operation. The method to achieve a satisfactory compromise solution for this problem lies in trading off all the objectives considered, that is, to bring considerable savings in operating costs

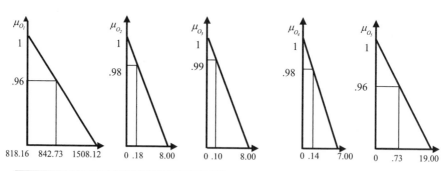

Objectives	OBJ. 1	OBJ. 2	OBJ. 3	OBJ. 4	OBJ. 5
Initial WC value	1508.12	8.00	8.00	7.00	19.00
DM's comments	Need to be improved	Allowed to be worsened	Allowed to be worsened	Allowed to be worsened	Allowed to be worsened
New WC value	842.73	8.00	8.00	7.00	19.00

FIGURE 18.9 First optimal alternative and DM's comments to achieved objectives.

without jeopardizing the necessary level of system security. As case 3, the related study is carried out, and the details are described as follows.

For the ideal value of the operation cost F_{EO}^{Ideal}, it is assumed to be equal to the cost of the system incorporating all the controllable measures under the cost-minimization criterion. The operation cost of the conventional power flow without any control is taken as the initial value of F_{EO}^{WC}, which is 1508.12×10^2 £/hour. Under reasonable assumption, 50% of the power flow target value is taken as the relevant worst compromise value F_{PFC}^{WC}. In order to relieve the relevant lines of the interface for security and further utilization of the network, power flow control of lines 22, 24, 25, and 29 is considered. Based on thermal limits of the lines and thermal burden percentage of other interface branches, the corresponding targets are assigned with an aim to flatten the thermal burden curve. Therefore, including objective of economy, there are five objectives involved in the model.

The process of reaching a satisfactory solution is demonstrated in Figs. 18.9 to 18.11. The responses of the DM to the optimal alternatives and the corresponding new WC value assigned for the next iteration are also summarized in the figures. Because of the conflicting feature of the objectives, it is predictable that in any best compromise alternative, the objective value of economy should be higher than the ideal value of the objective. Therefore for the economy objective, only the related part of the membership function curve is presented here.

The final satisfactory scenario is obtained in the third interaction. The settings of generators output and PV nodes voltage, as well as control parameters of the FACTS devices, are given in Table 18.5. Thermal burden curve of the interface branches is also depicted in Fig. 18.8, from which it is clearly seen in the best compromise solution achieved that interface branches possess a relatively balanced thermal burden. This fact that the feasibility of the resultant scenario of case 3 highlights the importance of considering multiple objectives in the optimal operation of power systems incorporating FACTS devices.

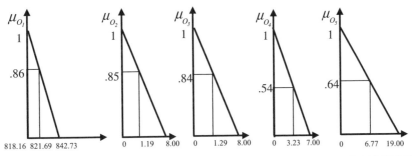

Objectives	OBJ. 1	OBJ. 2	OBJ. 3	OBJ. 4	OBJ. 5
DM's comments	Retain at the current level	Allowed to be worsened	Allowed to be worsened	Need to be improved	Need to be improved
New WC value	842.73	8.00	8.00	3.23	6.77

FIGURE 18.10 Second optimal alternative and DM's comments to achieved objectives.

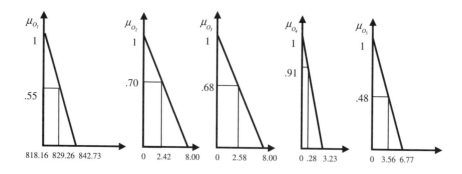

Objectives	OBJ. 1	OBJ. 2	OBJ. 3	OBJ. 4	OBJ. 5
DM's comments	Retain at the current level (The DM is satisfied with this solution)				

FIGURE 18.11 Third optimal alternative and DM's comments to achieved objectives.

TABLE 18.5   Final Optimal Scenario

Generator Output (p.u.)		PV Node Voltage (p.u.)		FACTS Devices (p.u.)				
				No.	Device	Control Targets	Achieved Targets	Control Parameters Derived
1	44.6354	1	1.0008					
2	23.3418	2	1.0514					
3	67.7259	3	0.9917					
4	75.8747	4	1.0048	1	UPFC	$P_L = 16$	$P_L = 18.43$	$V_T = 0.7287$
5	52.2182	5	1.0000			$Q_L = 0$	$Q_L = 0$	$\varphi_T = 1.4281$
6	96.2635	6	1.0056					$I_q = 14.0510$
7	10.0000	7	1.0205					
8	7.2068	8	1.0117	2	TCSC	$P_L = 16$	$P_L = 18.59$	$x_c = -0.0802$
9	10.0000	9	0.9991					
10	16.6947	10	1.0252	3	SSSC	$P_L = 14$	$P_L = 14.28$	$V_C = -0.0087$
11	37.7192	11	1.0496					
12	28.0036	12	0.9407	4	TCPS	$P_L = 38$	$P_L = 41.56$	$\sigma = 0.0512$
13	27.4389	13	1.0805					
14	10.0000	14	1.0845					
15	21.6590	15	1.0936		Operation cost = 829.26×10^2 £/hour			
16	16.9966	—	—					

18.8 CONCLUSIONS

Undoubtedly, flexible control of FACTS devices enables a power system to operate in a more robust, secure, and economic way. In order to take full advantage of FACTS devices installed, it would be desirable to take their major functions into account. In this chapter, an interactive compromise programming-based MO optimization approach has been proposed to determine FACTS control parameters for optimal operation of power systems in a coordinative way, with consideration of both economy and local power flow control simultaneously. It can give new insights into FACTS device functions and the corresponding control strategies. Moreover, a clearer picture of the situation of power systems with flexible and effective control means can also be presented.

For such a problem with multiple conflicting objectives, it is practical to seek an optimal compromise solution by trading off all the objectives. Based on compromise programming, a displaced worst compromise programming is introduced to solve the problem in an interactive procedure. During the interactive process, as a reference point, the worst compromise point is displaced according to rules reflecting the DM's aspiration level changes. One of the salient features of this approach is it is very efficient to determine the reference point without any further analysis or extra program. With the introduction of fuzzy set theory, motivations of the DM can be reflected in a more realistic and credible way. Moreover, it also provides a very intuitive and convenient way to measure and represent the distances involved in the CP.

From the seeking process of the satisfactory solution in the illustrative example, it is evident that the proposed approach can lead to an efficient solution by solving the MO problem in an interactive procedure without forcing the DM into an exhaustive examination of all the Pareto solutions.

Finally, it should be noted that to achieve an optimal control scheme of FACTS devices, the major applications considered should not only be confined to economy improvement and local power flow control. And in terms of local power flow control, the example for load flow rebalance taken in this chapter is only one of many important applications, which has been discussed in the first section of this chapter. Nevertheless, it is emphasized that, with the multiobjective model and the solution method proposed in this chapter, it can be extended easily to deal with MO optimal operation problems of systems considering any other applications of FACTS devices by including more objectives.

REFERENCES

1. Xiao Y, Song YH, Liu CC, Sun YZ. Available transfer capability enhancement using FACTS devices. IEEE Trans Power Systems 2003; 18(1):305–312.
2. Wang X, Song YH, Lu Q, Sun YZ. Optimal allocation of transmission rights in systems with FACTS devices. IEE Proc Generation Transmission Distribution 2002; 149(3):359–366.

3. Galiana FD, Almeida K, Toussaint M, Griffin J, Atanackovic D, Ooi BT, Mcgillis DT. Assessment and control of the impact of FACTS devices on power system performance. IEEE Trans Power Systems 1996; 11(4):1931–1936.

4. Baldick R, Kahn E. Paths contract, phase-shifters, and efficient electricity trade. IEEE Trans Power Systems 1997; 12(2):749–755.

5. Mutale J, Strbac G. Transmission network reinforcement versus FACTS: an economic assessment. IEEE Trans Power Systems 2000; 15(3):961–967.

6. Ge SY, Chung TS. Optimal active power flow incorporating power flow control needs in flexible ac transmission systems. IEEE Trans Power Systems 1999; 14(2):738–744.

7. Zeleny M. Multiple criteria decision making. New York: McGraw-Hill; 1982.

8. Brodsky SFJ, Hahn RW. Assessing the influence of power pools on emission constrained economic dispatch. IEEE Trans Power Systems 1986; 1(1):57–62.

9. Heslin JS, Hobbs BF. A multiobjective production costing model for analyzing emissions dispatching and fuel switching. IEEE Trans Power Systems 1989; 4(3):836–842.

10. El-Hawary ME, Ravindranath KM. Combining loss and cost objectives in daily hydro-thermal economic scheduling. IEEE Trans Power Systems 1991; 6(3):1106–1112.

11. Charnes A, Cooper WW. Management models and industrial applications of linear programming. Vol. 1. New York: John Wiley & Sons; 1961.

12. Kermanshahi BS, Wu Y, Yasuda K, Yokoyama R. Environmental marginal cost evaluation by non-inferiority surface. IEEE Trans Power Systems 1990; 5(4):1151–1159.

13. Nanda J, Kothari DP, Linggamurthy KS. Economic load dispatch through goal programming techniques. IEEE Trans Power Systems 1988; 3(1):26–32.

14. Fan J-Y, Zhang L. Real-time economic dispatch with line flow and emission constraints using quadratic programming. IEEE Trans Power Systems 1998; 13(2):320–325.

15. Moro LM, Ramos A. Goal programming approach to maintenance scheduling of generating units in large scale power systems. IEEE Trans Power Systems 1999; 14(3):1021–1028.

16. Cohon JL. Multiobjective programming and planning Mathematics in Science and Engineering Series. Vol. 140. New York: Academic Press; 1978.

17. Chiang H-D, Jean-Jumeau R. Optimal network reconfigurations in distribution systems: Part 1: A new formulation and a solution methodology. IEEE Trans Power Delivery 1990; 5(4):1902–1909.

18. Chattopadhyay D, Banerjee R, Parikh J. Integrating demand side options in electric utility planning: a multiobjective approach. IEEE Trans Power Systems 1995; 10(2):657–663.

19. Chattopadhyay D, Momoh J. A multiobjective operations planning model with unit commitment and transmission constraints. IEEE Trans Power Systems 1999; 14(3):1078–1084.

20. Zadeh LA. Fuzzy sets. Information and Control 1965; 8:338–353.

21. Zimmermann HJ. Fuzzy programming and linear programming with several objective functions. TIMS/Studies in the management science 20. Amsterdam: Elsevier Science Publisher; 1984. 109–121.

22. Ramesh VC, Xuan L. A fuzzy multiobjective approach to contingency constrained OPF. IEEE Trans Power Systems 1997; 12(3):1348–1354.

23. Ying X, Song YH. Applying multi-objective fuzzy dynamic programming to hydro-thermal coordination including pumped-storage plant. 6th UK Workshop on Fuzzy Systems, University of Sheffield, Sept. 8–9, 1999, 241–246.

24. Sun H, Yu DC. A multiple-objective optimization model of transmission enhancement planning for Independent Transmission Company (ITC). IEEE 2000 Summer Meeting, Seattle, July 2000.

25. Goldberg D. Genetic algorithms in search, optimization & machine learning. Reading, MA: Addison-Wesley; 1998.

26. Kalyanmoy D. Multi-objective genetic algorithms: problem difficulties and construction of test problems. Evol Computat 1999; 7(3):205–230.

27. Das DB, Patvardhan C. New multi-objective stochastic search technique for economic load dispatch. IEE Proc Generation Transmission Distribution 1998; 145(6):747–752.

28. Chen Y-L, Liu C. Optimal multi-objective VAR planning using an interactive satisfying method. IEEE Trans Power Systems 1995; 10(2):664–670.

29. Sakawa M, Nishizaki I, Uemura Y. Interactive decision making for large-scale multi-objective linear programs with fuzzy numbers. Fuzzy Sets Systems 1997; 88:161–172.

30. Yang J-B, Chen C, Zhang Z-J. The interactive step trade-off method (ISTM) for multi-objective optimization. IEEE Trans Systems Man Cybernetics 1990; 20(3):688–695.

31. Sakawa M, Nishizaki I, Uemura Y. Interactive fuzzy programming for two-level linear fractional programming with fuzzy parameters. Fuzzy Sets Systems 2000; 115:93–103.

32. Chen Y-L. An interactive fuzzy-norm satisfying method for multi-objective reactive power sources planning. IEEE Trans Power Systems 2000; 15(3):1154–1160.

33. Xiao Y, Song YH, Sun YZ. Power flow control approach to power systems with embedded FACTS devices. IEEE Trans Power Systems 2002; 17(4):943–950.

34. Han ZX. Phase shifter and power flow control. IEEE Trans PAS, 1982; 101(10):3790–3795.

35. Wojtek M. Use of the displaced worst compromise in multiobjective programming. IEEE Trans Systems Man Cybernetics 1988; 18(3):472–477.

36. Mehrotra S. On the implementation of a primal-dual interior point method. SIAM J Optim 1992; 2:575–601.

37. Fuerte-Esquivel CR, Acha E. Newton-Raphson algorithm for the reliable solution of large power networks with embedded FACTS devices. Proc IEE C 1996; 143(5):447–454.

38. Hiskens IA. Analysis tools for power systems—contending with nonlinearities. Proc IEEE 1995; 83(11):1573–1587.

39. Xiao Y, Song YH. Power flow studies of a large practical power network with embedded FACTS devices using improved optimal multiplier Newton-Raphson method. Eur Trans Electrical Power (ETEP) 2001; 11(4):247–256.

40. Wood AJ, Wollenberg BF. Power generation, operation and control, 2nd ed. New York: John Wiley & Sons; 1996.

CHAPTER 19

Hybrid Systems

VLADIMIRO MIRANDA

19.1 INTRODUCTION

This chapter serves several didactic purposes. The models that it addresses contain some hybrid solutions between evolutionary computation techniques and other modeling approaches, either classic and analytical or based on other branches of computational intelligence, such as the fuzzy set theory or particle swarm optimization (PSO). One wishes to underline the power of hybrid models and to discard from this book any accusation of fundamentalism or dogmatic belief in the superiority of meta-heuristics over any other technique. But, in places, there will be occasions for underlining particular aspects and suggesting new ways to be explored.

The reader will be confronted with three models. The first has a didactic flavor and is presented with as much simplicity as possible. The problem addressed is related to capacitor location and sizing, in networks, in order to reduce losses, control voltage, and at the same time optimize investment. The main topic underlined here is the successful combination of a genetic algorithm (GA) with information from a classic analytical power flow model. The lesson to be extracted is that one must not blindly trust in the blind force of an evolutionary process—especially when it is built in a naïve way by humans who expect it to find a particular optimum as quickly as possible. Therefore, one must "help," say, by introducing some Lamarckism in a fundamentally Darwinist process.

The second model is related to a real-world application. It is related to a unit commitment problem, especially designed for isolated systems with a high possible penetration of renewable energies, in particular of wind generation. An application written with such a model at its core is currently running in the Control Center of the island of Crete, Greece, operated by the Greek national utility PPC (Public Power Corporation). A first lesson to be extracted with this model is that there is no opposition between GA and fuzzy set theory, and they may be combined to get to desired solutions for some problems. A second lesson is that one must keep looking to biology and the most amazing findings of molecular genetics and

Modern Heuristic Optimization Techniques. Edited by K. Y. Lee and M. A. El-Sharkawi
Copyright © 2008 the Institute of Electrical and Electronics Engineers, Inc.

perhaps use some of the tricks that have been "selected" by evolution instead of only being confined to the development of naïve chromosome models.

In both these models, we can find a conceptually interesting perspective: in one way or the other, we are introducing (or making more evident) traits of Lamarckism in what is traditionally seen as a Darwinist process.

From a modern Darwinist point of view (and without entering the controversial ground that sometimes divides specialists over details), "evolution is the external and visible manifestation of the differential survival of alternative replicators" [1]. This survival is achieved through competition, namely at the phenotypic level, and the emergence of variation does not derive for any process of learning or instruction. Alternatively, Lamarckism is "a theory of directed variation" [2], that is, effects or procedures resulting from learning or sets of instructions assembled at the macroscopic level are translated back into genotypes, modifying them and thus modifying future phenotypes.

From what we currently know from molecular genetics, Lamarckism does not represent the biological mechanism of evolution. Lamarckism requires a one-to-one mapping between the phenotype and the genotype, so that a reverse translation "back into the genes" is possible. However, we have learned that this is not how biology works. The biological genotype contains a set of instructions on how to proceed to build a body, but it is not possible to find in the phenotype a code to build the genotype.

Under this light, we can now understand that most of the models proposed in GA are, in fact, full of Lamarckian potential. Although nature does not work like that, it seems that there is nothing wrong, in principle, in employing such principles in an artificial world of simulation. It would, anyway, be interesting if someone would demonstrate the evolutionary superiority of Darwinism over Lamarckism, which would be a surprising explanation of why we have in the world the former and not the latter . . . .

Discarding this tempting thought, we realize that most GA models exhibit precisely that property of a one-to-one mapping between genotype and phenotype. Therefore, one must seriously ask if learning cannot have a role in improving the genotypes of solutions, so that descendants may have more competitive traits and higher chances of survival. On the other hand, one might suspect that too much learning might trap the individuals in local optima, which is undesirable.

The third model we will discuss is also hybrid, but in a different sense—it shows how one can take advantage of ideas from PSO, to form a hybrid evolutionary algorithm where the operations of mutation/recombination are replaced by the rules of particle movement from this latter method. Experimental results show that in a number of problems, the combined push of the selection pressure and the particle movement rule add up to reinforce the algorithm and accelerate convergence. A real-world application will be mentioned related to voltage/VAR management and loss reduction in distribution networks.

In the following paragraphs, we will propose the examination of the three models under this light: although the actual problems addressed are in fact interesting examples, demonstrating the potential for practical application of the concepts, we wish to underline the importance of moving away from the naïve paradigm of the simple GA.

19.2 CAPACITOR SIZING AND LOCATION AND ANALYTICAL SENSITIVITIES

We will look first at a problem of locating capacitors in a system in order to reduce losses and control the voltage profile while at the same time minimizing the overall investment [3]. There have been interesting proposals of evolutionary models for this problem, but here we wish to discuss some didactic aspects of modeling. The general objective function will be built based on three different criteria—capacitor investment cost, energy cost of losses, voltage limit penalty—and subject to a number of constraints related to the problem.

We will consider two types of capacitors: fixed (with no possible switching but cheaper) and switched (allowing regulation of shunt capacity, but more expensive). A mix of these two types may be installed at any of a specified set of nodes in a network. In this case, we will assimilate fitness with objective, and therefore, it will be improved with the minimization of a numerical index. The formulation can be described in simple mathematical terms as follows:

$$\text{Minimize} \quad \text{OBJ} = [\text{CC} + \text{EC} + \text{VL}]$$

$$\text{Subject to:} \quad f(X, u) = 0$$

$$S_{min} \leq S_i \leq S_{max}$$

$$F_{min} \leq F_i \leq F_{max},$$

where CC is the capacitor investment function, EC is the energy cost of losses function, VL is the voltage limit penalty function, $f(X, u)$ is the equality constraints of the power flow problem, F_i is the quantity of fixed capacitors at b, and S_i is the quantity of switched capacitors at bus i.

Now admit that the investment function is calculated based on costs of capacitors and installed capacity. Different costs for fixed and switched capacitor types may be used, with higher value for switched capacitor banks, reflecting the incorporation of the costs of switching devices into the nominal cost of capacitor alone. In the following examples, linearized costs were used for installation cost of capacitors, assuming that the installation cost is directly proportional to installed capacity—although this does not imply any loss of generality of the model.

The capacitor investment function, given as the multiplication of total installation cost of capacitors by a scale factor (representing, among other things, the annualization of capital costs), is:

$$\text{CC} = \text{sf}_{\text{C}} \left[\left(\sum_{i=1}^{n} \text{NFC}_i \times \text{CF} \right) \times \left(\sum_{i=1}^{n} \text{NSC}_i \times \text{CS} \right) \right],$$

where sf_{C} is the scale factor of capacitor investment function, NFC_i is the quantity of fixed capacitor installed in bus i, NSC_i is the quantity of switched capacitor in bus i, CF is the per unit cost of fixed capacitors, and CS is the per unit cost of switched capacitors.

The total power losses of the system are related to the cost of losses function. The total energy loss is calculated from system power losses and time duration of the planning period. To have a more realistic model, the load curve is divided into a number m of load levels and approximated by a step model. The energy cost function formulation is in general as follows:

$$\text{EC} = \text{sf}_\text{E} \left[\sum_{i=1}^{m} \text{PL}_i \times T_i \times \text{MC} \right],$$

where sf_E is the scale factor of energy cost function, PL_i is the power loss of load level i, T_i is the planning time duration of load level i, and MC is the energy marginal cost of the system.

In the case of the voltage limit penalty function, a combination of approaches will be used to separate the priority of two different zones: a soft square penalty function when the voltage lies outside an internal preferred interval and a strongly sloped linear penalty function when the voltage is outside an externally nesting interval of still acceptable values,

$$\text{VL} = \text{sf}_\text{V} \left[\sum_{i=1}^{n} \left(\sum_{k=1}^{m} \text{SF}_k^i + \text{sf}_\text{M} \sum_{k=1}^{m} \text{LF}_k^i \right) \right],$$

where sf_V, sf_M are scale factors,

$$\text{SF}_k^i = \begin{bmatrix} \left(V_k^i - \text{VN}_k^i \right)^2 & \text{if } V_k^i \text{ off internal limit} \\ 0 & \text{if } V_k^i \text{ inside limit} \end{bmatrix}$$

$$\text{LF}_k^i = \begin{bmatrix} \left(V_k^i - \text{VN}_k^i \right) & \text{if } V_k^i \text{ off external limit} \\ 0 & \text{if } V_k^i \text{ inside limit} \end{bmatrix},$$

V_k^i is the voltage at bus k in load level i, and VN_k^i is the nominal voltage of bus k in load level i.

19.2.1   From Darwin to Lamarck: Three Models

We will adopt a genetic modeling approach to try to solve this problem. As a general rule, we can say that the number of genes in each chromosome is taken as a multiple of the number of PQ buses times the number of periods associated with the load levels.

In order to illustrate the importance of (a) a good chromosome scheme and (b) Lamarckism, we have compared the performance of three models. Using the same fitness (objective) function mentioned above, three different algorithms, *simple genetic algorithm* (SGA), *genetic-based algorithm* (GBA), and *genetic-based algorithm with gradient search* (GBAGS), were created based on different strategies and different genetic control operators.

The SGA is a typical genetic algorithm based on a Darwinist process, where the replicator units are the genes and where the selection acts on their phenotypic expressions. In the SGA, as close as possible to the type described by Goldberg [4], we defined a non-overlapping population, where all individuals of each generation are entirely replaced by newly generated individuals. We built each chromosome as a binary string and adopted a flip mutator and single-point crossover. The binary genes were read, for each location and load level, in such a way as to give the number of capacitors to be placed, both fixed and switched.

The GBA is a model that retains some characteristics of genetic coding of solutions but that also acquires properties from evolutionary programming. In this model, the one-to-one mapping between genotype and phenotype is quite straightforward, so that mutations may be thought of as acting almost directly on the phenotype. In the GBA, instead of having binary genes, we defined each gene as an integer, representing the number of capacitors of a given amount of reactive power to be used in series, at each location and time period. We also adopted an overlapping population, where only some individuals were replaced by new ones coming out from genetic operations. We used uniform crossover (see Chapter 3) and modified Gaussian mutation: randomly generated values according to a Gaussian distribution are rounded to the closest integer value so that they can be used in problems with discrete domain.

The third model, GBAGS, retains the evolutionary programming feature of GBA, but we've added the possibility of improving the phenotype of selected individuals and propagating such improvement to their descendancy. These improvements are acquired based on the best knowledge available on what seems good to the phenotype, and this is Lamarckism. However, it is still an evolutionary process, because it is submitted to selection and elimination of the less fit. The same strategy and genetic operators mentioned above for GBA were used in GBAGS, but selected individuals in the current generation are repaired by a gradient-based method described below. This gradient method uses local information based on sensitivity coefficients derived from the classic power flow equations; thus, adjustments in the number of capacitors at certain nodes can be made, improving the solution.

However, the algorithm has been carefully designed so that an "excess of Lamarckism" would not, in the end, trap the procedure in local optima. The phenotypic improvement is only applied in certain generations and to certain selected individuals, as illustrated in the example to be presented further on. This process of acquired characteristics becomes useful because it contributes to maintain diversity in the gene pool, and diversity has been recognized as playing an important role in exploring the search space [5].

19.2.2 Building a Lamarckian Acquisition of Improvements

Because the reactive power is the control variable for minimizing losses, it is necessary to identify a relationship between changing reactive power and system losses. Changing of system losses can also be seen as changing the active power at the reference bus, given a traditional AC power flow model; load is a constant value in each load level (or time step), and the incremental change in losses can be expressed as

$$\Delta P_{loss} = \Delta P_G,$$

where ΔP_{loss} is the incremental change in system power losses, and ΔP_G is the incremental change in active power at the swing bus.

The gradient vector of losses with respect to reactive power changes at bus i is given as:

$$\frac{\partial P_{loss}}{\partial Q_i} = \sum_{i=1}^{n} \frac{\partial P_G}{\partial \delta_i} \frac{\partial \delta_i}{\partial Q_i} + \sum_{i=1}^{n} \frac{\partial P_G}{\partial V_i} \frac{\partial V_i}{\partial Q_i},$$

where δ_i is the angle at node i, V_i is the voltage magnitude at node i, Q_i is the injected reactive power at node i, and P_G is the injected power at the swing bus.

The partial derivatives are sensitivity coefficients derived from the nodal power injection equations and the inverse of the Jacobian of the Newton-Raphson method. A possible better reactive power setting is suggested by the gradient vector of losses relative to reactive power changes:

$$Q_i^{new} = Q_i^{old} - \alpha \frac{\Delta P_{loss}}{\Delta Q_i},$$

where Q_i^{old} is the reactive power at bus i, Q_i^{new} is the new reactive power setting at bus i after a gradient vector move, and α is the positive iteration step.

During the procedure, this gradient search is included when necessary to increase the diversity, to overcome a possible local optimum, and to move solutions in the (possible) right direction toward the global optimum without disturbing the evolutionary nature of the algorithm. The gradient search is only activated when the fitness of the current individuals is worse than a specified value (Fig. 19.1).

The best individual's fitness value of a previous generation is selected as the judging value to decide which individuals of the current generation are to be improved or repaired. The selection of such a threshold value depends on the trade-off between how-much-time-we-can-spend and how-good-results-we-want-to-achieve. If the best fitness value of such a previous generation, close to the current generation, is used as the threshold value for repairing individuals, the number of individuals needing to be repaired is likely to be much higher than using the best fitness value of a previous generation, more distant from the current generation. The amount of individuals needing to be repaired in the current generation depends on the threshold value selection as the algorithm is evolving and the average fitness values are improving generation by generation.

Although the gradient search was included in the algorithm, a full utilization of gradient iterations to find the new reactive power setting is not advisable:

- Rounding off the gradient vector is necessary because of the discrete nature of the reactive power setting; and
- Such a procedure risks getting trapped in local optima.

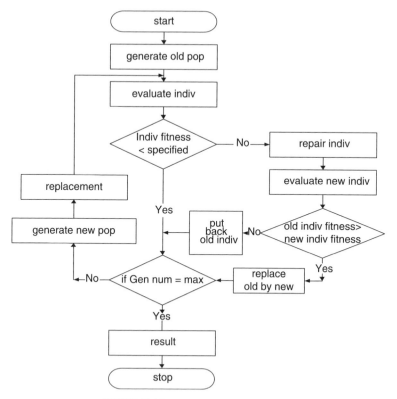

FIGURE 19.1 Flowchart of GBAGS.

19.2.3 Analysis of a Didactic Example

We will now illustrate the theoretical principles explained above with results from studies conducted on the system shown in Fig. 19.2. It is a simplified diagram from a real system in the Azores Islands.

As candidate buses to install capacitors, we selected only PQ buses, reflecting the idea that capacitor adding at the reference bus or at voltage-controlled buses would not affect power losses (one would then have to include in the study the influence of voltage control). A discrete step in capacitor size was defined, as a continuous formulation for capacitor sizes would not reflect the practical planning constraints.

Two different types of capacitor banks were adopted: fixed capacitor banks, which provide a base amount of reactive power injected in the system, and switched capacitor banks, which allow a policy of regulation, namely considering the changing in load curves.

Table 19.1. presents some data and control parameters used in the test studies.

Three different load levels (off-peak, peak, and normal) were used to reflect a realistic situation of a load curve. In the cases of peak and off-peak load levels, a shorter time duration was selected compared with time duration for the normal load level.

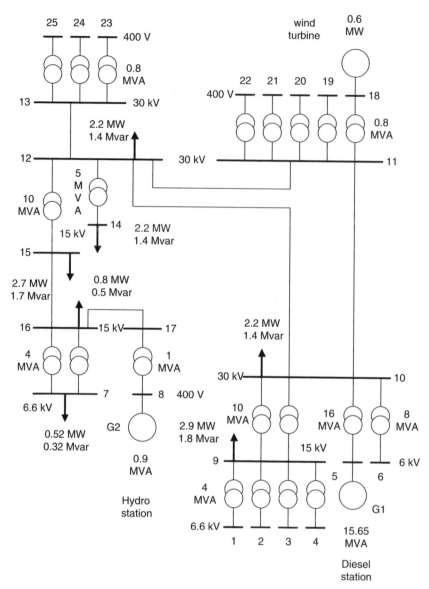

FIGURE 19.2 Single line diagram of the test system.

Both different and same load for three load levels were used in different exercises to test the ability of the algorithms. Satisfactory results were achieved from both types of tests. But the results from an example with the three load levels having the same load were selected for illustrative purposes, because they provide a quick check on the fact that fixed-type capacitor installation is preferable over switched type for the flat load situation. Furthermore, they also provide a quick

TABLE 19.1 Data and Control Parameters Used

Installation costs	
Fixed capacitors	US$90/100 kVAr
Switched capacitors	US$150/100 kVAr
Energy costs	
Avg. marginal value	US$0.06/kWh
Population	
30	
Crossover probability	
SGA, GBA, GBAGS	0.9
Mutation probability	
SGA	0.005
GBA, GBAGS	0.05
Max no. generations	
300	

check on the quality of the solutions, because the GA should provide the same solution for all periods if the load was the same for all of them. As we shall see, that is not always happening.

In this study and for GBAGS, the best individual of 10 generations before the current generation has been taken as a candidate to define the specified value for selecting individuals needing to be repaired. Each algorithm was run 10 times, and the fitness values were recorded for the 100th, 150th, 200th, 250th, and 300th generations. Other important results such as losses, installation costs, and voltage profile were also collected at the end of each run. Then, the average values of the results were calculated, providing the possible approximate results of the programs, and they were used to compare the performance of the algorithms.

The comparisons among the three algorithms were made based on fitness values, installation costs, costs of losses, and voltage profile at buses. Average values, calculated from the results, are mainly taken as the parameters to compare the performance of algorithms in this study except for the control setting of installed capacitors. The comparison of fitness values highlights the differences in overall performance of the algorithms. Figure 19.3 represents the average fitness value comparison of the three algorithms, and it clearly indicates gradual improvements from SGA to GBA to GBAGS. The comparison of the fitness value of the best run is shown in Fig. 19.4, displaying the same trend as in Fig. 19.5. This indicates that average results are enough to reflect the overall performance of the algorithms.

The performance of algorithms, decomposed in average installation costs and costs of losses can be compared in Fig. 19.5. Although the costs of losses have not significantly changed for the three algorithms, a great improvement is obtained by installing capacitors in the convenient nodes of the network, with a monthly reduction of 48%, from US$6283.3 to US$3254.8 between the cases of system without and with capacitors. Another significant improvement is seen on the installation cost of capacitors: it

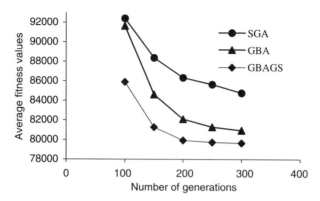

FIGURE 19.3 Average fitness value comparison.

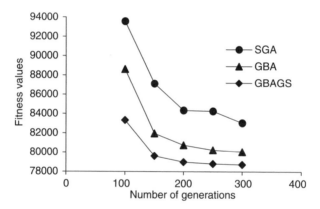

FIGURE 19.4 Fitness value comparison of best run.

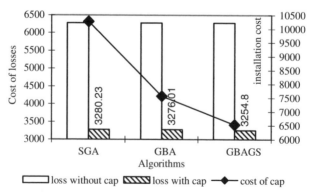

FIGURE 19.5 Economic performance of algorithms: monthly costs of losses and installation costs.

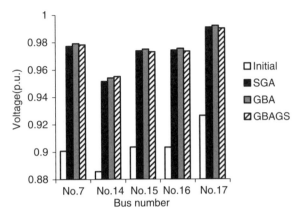

FIGURE 19.6 Voltage profile performance.

comes down to US$6531 from US$10,272, reducing the installation costs US$3741, a 37% reduction of the total installation cost between SGA and GBAGS.

The buses with voltage level outside limits in the initial situation (buses 7, 14, 15, 16, 17) were taken as the basis for the comparison of algorithms. The worst voltage profile, 0.8859 p.u., occurred at bus 14 in the initial system. In Fig. 19.6, we observe that even the worst voltage profile at bus 14 was improved up to within the specified limit of ± 5% of nominal value; besides, all three algorithms showed the ability to improve the voltage profiles into the acceptable limits defined, with good voltage profile results achieved.

In Table 19.2, we find the costs of losses and installation costs for the best run in each trial. Furthermore, we identify that switched-type capacitors were proposed by the SGA, although we had defined a flat load situation. This shows that the results we achieved with SGA were not good. However, both GBA and GBAGS suggested instead only fixed-type capacitors, indicating that the solutions they achieved were logically correct.

We can make a more detailed analysis of the quality of the solutions proposed by SGA, GBA, and GBAGS. With this purpose, we have organized Table 19.3, where we can see the number of 100 kVAr units proposed at each bus for each period of the load curve, by the best run of each of the three algorithms.

TABLE 19.2 Performance Based on the Best Run of the Three Algorithms

Algorithm	Cost of Losses (US$)	Cost of Capacitors (US$)	Capacitor (MVA) (Switched)	Capacitor (MVA) (Fixed)
SGA	3278.50	9690	2.5	6.6
GBA	3282.14	7380	0	8.2
GBAGS	3231.68	6570	0	7.3

TABLE 19.3 Best Run Obtained with Each of the Three Algorithms, Indicating the Number of 100 kVAr Units Proposed at Each Bus for Each Period of the Load Curve

Bus No.	SGA			GBA			GBAGS		
	Off-Peak	Normal Hours	Peak Hours	Off-Peak	Normal Hours	Peak Hours	Off-Peak	Normal Hours	Peak Hours
1	1	2	1	1	1	1	2	2	2
2			2				1	1	1
3		2	1				1	1	1
4	1	1	1	1	1	1			
6	3	3	3	3	3	3			
7	3	3	2	2	2	2			
9	8	7	8	5	5	5	4	4	4
10	2		4	6	6	6	6	6	6
11	3	4	3	8	8	8			
12	3	3	5	7	7	7	9	9	9
13	6	5	5	3	3	3	4	4	4
14	2	1					9	9	9
15	3	1	3	4	4	4			
16	15	14	14	9	9	9	13	13	13
17	15	15	14	15	15	15	15	15	15
18	11	13	12	15	15	15	9	9	9
19									
20	2	1	2	3	3	3			
21									
22									
23									
24									
25	1								

Remember that we have run all tests with a flat load profile (i.e., with equal value for the load at the three periods defined: off peak, normal and peak hours). Therefore, we do not expect the need for switching capacitors, and this will be represented, at any node, by a constant number of kVAr proposed for all time periods. In order to freely test the capacity of the algorithms to generate structurally correct solutions, we have included no penalties at all in the fitness function that would force the solution to have an equal number of capacitors at all time periods. However, we may notice that the solution proposed by the SGA presents an odd aspect, because we find in several buses distinct kVAr values at different hours. For instance, at bus 18, we must conclude that the SGA suggests 11×100 kVAr of fixed capacitors and 2×100 kVAr of switched capacitors.

This reflects the weakness of the solutions proposed by the SGA, which is much more important than an eventual higher cost. In fact, this demonstrates that the evolutionary process embedded in the SGA has difficulties in generating correct solutions. The chromosome scheme seems to be in trouble to either develop or maintain building blocks with the necessary structure.

The main cause of diversity on the SGA is crossover. Crossover clearly did not work too well in this case. It is as if we had joined together, in one run, three separate evolutionary subprocesses: one for off-peak hours, one for normal, and one for peak hours. The results seem to suggest that these subprocesses propose suboptimal solutions, which in fact are close but have difficulty in converging to each other. In order for this to happen, perhaps one would have to depend on very lucky mutations.

It is evident that the coding scheme of GBA behaved more robustly and discovered solutions with good internal structure. This may seem surprising, because the type of crossover adopted was uniform crossover, which has a much greater chance of disrupting any building blocks. However, we must remember that we adopted a distinct strategy for changing the generations of individuals, allowing selection to act upon the overlapping of parent and son generations. Therefore, we may conclude that the diversity originated but the uniform crossover process plus the higher mutation rate, acting upon the "phenotypic" chromosome scheme now adopted, played a crucial role in finding good structures for the solutions.

However, it is also obvious that the Lamarckism effort of improving solutions under the pressure of "necessity" and passing acquired characteristics to the descendants gave good results. It has been, however, a true evolutionary process, because selection was allowed to act upon the new individuals and eventually eliminate the "unfit."

This is conceptually very different from the interesting work published by Miu et al. [6], who also proposed the use of sensitivities in an evolutionary process to solve the same type of problem. However, these authors organized their approach in a two-phase model, with the GA acting in a first stage and a sensitivity-based heuristic performing a sort of postoptimization. As we define it, therefore, this is not a true hybrid algorithm, and it does not benefit from Lamarckism to accelerate evolution.

19.3 UNIT COMMITMENT, FUZZY SETS, AND CLEVERER CHROMOSOMES

19.3.1 The Deceptive Characteristics of Unit Commitment Problems

One of the difficulties that one has to deal with when adopting genetic algorithms arises with problems having a large amount of integer or discrete variables, creating a large number of local optima and giving the problem some deceptive flavor. Many researchers have found that, although GAs approach relatively easily a zone of near-optimal costs, they have some difficulty in finding the exact optimum, especially in problems of this type. Because of the discrete nature of the domain, landing on the optimum becomes very much dependent on chance—the chance of generating the exact chromosome composition needed when neighboring compositions do not suggest that this would improve the overall fitness.

What we have been saying applies to unit commitment (UC) problems in general. Some classic mathematical models make it very clear: they have the form of fixed cost-minimization models, which give place to a nonconvex objective function. One possible way of representing such models is given by

$$\min F = \sum_{i \in \text{Gen Set}} a_i \mu_i + b_i P_i$$

$$\text{subject to} \quad \begin{cases} \mu_i = 0 \text{ if } P_i = 0 \\ \mu_i = 1 \text{ if } P_i > 0 \end{cases},$$

$$\sum_{i \in \text{Gen Set}} P_i = L$$

where P_i is the power generated by generator i, Gen Set is the set of all generators, and L is the total load. The parameters a_i and b_i represent fixed costs and marginal generation costs, and μ_i are binary variables whose action is to originate the "jump" at the origin in each generator cost function.

We all know that nonconvex functions may have many local optima. This can be illustrated in a problem with only two generators that must satisfy a load of 6 MW. The cost functions of each generator in this example are

$$G_1 = 20\mu_1 + 2P_1$$
$$G_1 = 3\mu_2 + 4P_2$$

In Fig. 19.7, we can immediately see why this problem may be classified as deceptive for a GA approach. The global optimum is found with $P_1 = 0$ (and therefore with $P_2 = 6$, with an objective function value of 27). However, the tendency of a canonical GA will be to push the solutions toward the other (local) minimum, at $P_1 = 6$, with objective function value of 32. This characteristic may appear in complex problems, and one will need to activate ways of overcoming it.

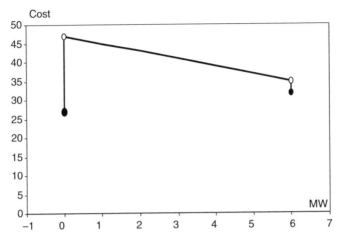

FIGURE 19.7 Variation of the objective function value with the variation of P_1 for the problem in the text. The deceptive characteristics of this function are evident here.

19.3.2 Similarity Between the Capacitor Placement and the Unit Commitment Problems

We will now discuss why one may find similarities in apparently unrelated problems, such as capacitor placement and unit commitment. In fact, imagine that one has a diversity of available generators of distinct technological principles and that one must define an hourly schedule for these generators over a 48-hour period.

One piece of data that one will need to consider is the load curve forecast. Except for exceptional days, the aggregated load curve of a power system evolves regularly with a periodic behavior in a 24-hour cycle. Therefore, unless extraordinary circumstances prevail, one will expect that the optimal unit commitment of generation will follow, in the second day, a pattern similar to the pattern of the first 24 hours.

The consideration of this cycling is unavoidable, even if one is seeking to obtain logical solutions for 24 hours only. Otherwise, one would have serious difficulties in taming the "tail effect" that would most probably occur—any algorithm ignoring such an effect would feel "free" to propose very asymmetrical solutions that would not connect well when considering the sequence of days. For instance, such an algorithm could be "tempted" to propose solutions where all available water in hydropower stations would be used, because it would not "see" the need for hydrogeneration on the following day. If one is going to model a 48-hour period, one must therefore consider that there are two 24-hour periods in succession that are not optimized independently.

Doesn't this sound familiar? Well, it certainly reminds us of the three-period division of the load curve in the capacitor placement problem in the flat load case—one of the characteristics of a good solution was the exact repetition of the structure in the location and size of capacitors for all three periods. Also in the unit commitment problem, one of the characteristics of a good solution is the (almost exact) repetition of the unit scheduling in successive days. And what did we learn with the capacitor placement problem? For one thing, that a simple GA would not perform well and would not lead to well-structured solutions, in the sense of having similar aspects in different time steps. And then, we learned that a touch of Lamarckism enhanced the efficiency of a GBA and allowed the discovery of good solutions.

It is therefore natural that one may believe that an evolutionary computing algorithm including procedures to ameliorate unit commitment solutions and translating such phenotypic changes back in the genotype of solutions, adding them to the gene pool, would be more efficient than a SGA. As we shall see, it is indeed so.

19.3.3 The Need for Cleverer Chromosomes

In order to develop a GA feasible procedure for the unit commitment problem, we will have to consider the problem of efficiently coding a solution in a chromosome.

The direct coding of an hourly schedule has obvious disadvantages of dimension. This coding strategy will represent, at each hour, a generator by a bit (1 if it is scheduled to operate, 0 if it is not). This means that, for example, for a problem of hourly unit commitment for a period of 48 hours, and considering a generation set of 25 units, one would require a chromosome with 1200 bits (48 × 25).

Long chromosomes have a negative effect: they define a much larger search space. This means that the whole process of convergence will require many more generations than for a limited search space. In order to cover such space regularly, one could randomly generate, at the beginning of the process, a larger number of chromosomes, but this will slow down the computation process, too, because of the need to perform the evaluation of the fitness function to all individuals. Also, such indiscriminate coverage of the search space is, in some sense, a waste, because it allows the representation of solutions that common sense, engineering judgment, and constraints indicate will be either unfeasible or unacceptable due to their characteristics or high cost. This would be the case, for instance, of a solution where generators would be connected and disconnected in succession every hour. Not only start-up and shutdown costs would render this scheme prohibitive, but also constraints such as minimum shutdown or start-up times would make them impossible to consider.

Therefore, one is led to imagine how to have a flexible representation of a solution, using a shorter chromosome but still covering all the important and interesting zones. The first step to be taken derives from the observation that generators scheduled to operate will remain so for a sequence of hours, usually. Furthermore, because the load curve usually presents flat regions and "plateaus," it is reasonable to assume that a scheme of generators some-on/some-off will last for a number of successive hourly periods before being disrupted and replaced by another scheme. If one could identify, by judgment or by analysis of past scheduling exercises, the location and duration of such periods, one would be able to code the generator schedule not for each hour but simply for each block of successive hours.

1 0 0 1 1	1 0 1 1 1	1 1 0 1 1	1 0 0 1 0
G1 G2 G3 G4 G5	G1 G2 G3 G4 G5	G1 G2 G3 G4 G5	G1 G2 G3 G4 G5
1st Block	2nd Block	3rd Block	4th Block
5 hours	8 hours	3 hours	8 hours

FIGURE 19.8 Example of coding with fixed length blocks for a problem with five generators and four blocks in 24 hours.

Thus, the first approach in reducing chromosome size is to divide the scheduling period into scheduling blocks of several hours each (the blocks do not need to be of equal length) and use 1 bit to code the status up-or-down of each generator. For instance, if the same problem of scheduling 25 generators for 48 hours could be divided in 10 blocks of homogeneous hours, the chromosome length required would be reduced to 250 bits (25 × 10). See another example in Fig. 19.8. This is a remarkable reduction in chromosome length and gives great hopes that the computing time may be considerably reduced as well. However, this is obviously a suboptimal procedure. The "*a priori*" definition of position and length of each block is inflexible and imposes a strict format to the problem that will be most likely conflict with the optimizing solution structure.

19.3.4 A Biological Touch: The Chromosome as a Program

The analysis of biological genomes has taught that genes do not code directly for phenotypic aspects of the individual. Instead, they code for proteins that act as enzymes. The genes must be read and translated, and their set can be interpreted as a program to build and operate an individual.

Some of the genes or some of the sequences code for the way in which the program must be read; there are stop signs, start-here signs, and even genes that disconnect other genes—this is how cell differentiation occurs. The activation or inactivation of specific genes changes the way in which the genome gets expressed. This has been the inspiration for the coding scheme devised:

- To add some bits that would introduce flexibility in the interpretation of the meaning of the chromosome.
- To add some bits that would switch on or off some parts of the chromosome, allowing the expression of new phenotypic structures.

The flexibility in chromosome interpretation has been achieved by adding a c number of bits to each block, denoting a possible anticipation or delay in the start of the block. The program for reading the chromosome was built by adding another bit to each block that would switch on or off its whole block, causing the extension of the coding of the previous block through the entire duration of the disconnected block. These two improvements receive a symbolic description in Fig. 19.9. From this figure, we can see that block 3 is disconnected and so the

1 0 0 1 0 0 1 1	1 0 0 1 0 1 1 1	0 1 1 1 1 0 1 1	1 1 1 1 0 0 1 0
S T1 T2 G1 G2 G3 G4 G5	S T1 T2 G1 G2 G3 G4 G5	S T1 T2 G1 G2 G3 G4 G5	S T1 T2 G1 G2 G3 G4 G5
1st Block	2nd Block	3rd Block	4th Block
5 hours	8 hours	3 hours	8 hours

FIGURE 19.9   Coding scheme where each block is controlled by a switching bit S and a time setting pair of bits T1 and T2.

schedule of block 2 extends throughout the period of block 3. Furthermore, bits T1 and T2 in the fourth block are on, with the possible meaning that the schedule of this block will start with an advancement of 3 hours.

It is a fact that the effect of disconnecting block 3 could be achieved in another way: by having the block switched on and having the schedule of the generators as a copy of the previous block. But this only means that there is not really a one-to-one mapping between genotype and phenotype, which makes the model closer to biological reality. It is also a fact that the existence of the switch bits S allows jumping into areas of the search space very quickly, instead of the painfully slow process of building the adequate or necessary sequence of bits.

Although this is still a simple or naïve program, it is certainly something more that the primitive coding process that would be represented by the simple GA with direct representation of the unit commitment schedule in the form of a bit per generator and per hour.

The experience with this model, applied to a real problem, demonstrated that it is indeed effective in finding good schedules, accepted by system operators and takes in account a number of constraints. Furthermore, all tests conducted also demonstrated that the SGA could not consistently generate such types of acceptable schedules. And the increase in chromosome length is quite acceptable. In fact, for the same problem of scheduling 25 generators for 48 hours divided in 10 blocks, the chromosome length required would be 325 bits ($25 \times (1 + 2 + 10)$). This is a very small price to pay for the additional flexibility obtained.

19.3.5   A Real-World Example: The CARE Model in Crete, Greece

The project CARE, supported by the European Union through the JOULE-THERMIE research program, had the following aim: to propose new concepts for control centers of weak or isolated networks that include wind generation. There are a number of these systems within the European Union (EU), mainly where there are numerous islands (e.g., in Greece or in the United Kingdom, Spain, Portugal, etc.) and in developing countries.

The EU projects such as CARE are typically delivered by a consortium of research institutions and companies from several countries; in this case, we have the participation of researchers from France (ARMINES), Greece (NTUA, Univ. Thessaloniki), Portugal (INESC Porto), the United Kingdom (UMIST), an industrial manufacturer (EFACEC, Portugal), and two utilities (PPC, the public national utility of Greece, and EDA, the public utility of the islands of the Azores, Portugal). The applicability of the tools developed within CARE were tested on the systems of Crete and of the Azores.

Among the modules to be developed in the CARE project, there was the need for a new unit commitment approach that would specifically take into account wind generation. The unit commitment module described in the following paragraphs has load and wind forecasts as inputs and provides a scheduling of generation for a certain period in the future. It admits any kind of generation, such as conventional thermal, diesel, hydro run-of-the-river, hydro from reservoirs, geothermal, and wind.

Wind cannot be considered a fully dispatchable generation source: the only possibility of control is the disconnection of generators. And especially in off-peak periods, this action may be necessary if the wind generation penetration is above some threshold that determines the definition of insecure operation states, namely for dynamic stability reasons. Both wind and hydro run-of-the-river generation depend on forecasts of the availability of their primary sources; but in general, the wind prediction is much more complicated and uncertain than the water inflow prediction.

The basic innovations brought up by this new model are the following:

1. The adoption of an evolutionary computing approach (a GA) with special features, such as coding compression, chromosome repair, and Lamarckism (improvement of solutions).
2. The ability to adopt load and wind forecasts in the form of intervals of uncertainty, instead of crisp numbers.
3. The adoption of fuzzy constraints to accommodate uncertainty and risk.
4. The ability to include a fuzzy constraint on wind penetration in order to reach a compromise between potential risk and the reduced costs of increasing wind generation.
5. The consideration of all kinds of costs, from running costs related to fuel consumption, ramping costs in thermal and diesel generators, and shutdown and start-up costs.
6. The consideration of all kinds of constraints: power balance, minimum shutdown and start-up times, minimum and maximum technical limits of generators, ramping limits of thermal units, start-up and warm-up programs, maximum wind penetration, spinning reserve requirements, and specific constraints expressed under the form of "rules" or "procedures" instead of a mathematical expression.

For the sake of clarity, in the following sections we will concentrate on the description of the model for a mixed thermal-wind generation system, its extension to other generation forms being quite simple. The concept developed in the CARE project is based on the following sequence of operations:

1. Long-term unit commitment for the next 48 hours, run on a hourly basis.
2. Ten-minute unit commitment for the next 8 hours, run on a 10-minute basis, having the results of the long-term problem as constraints for the last 10-minute periods.

Both algorithms are based on the same GA principles. However, they display distinct details of representation. For instance, in the long-term UC, generator ramping is represented by an approximate rate value, whereas in the 10-minute UC a more accurate start-up and warm-up schedule for thermal generators is followed, respecting the usual instructions given to operators. In the following paragraphs, we will not describe in full detail the complex model developed. The model has been converted as a module in an application, also called CARE, delivered to PPC and is currently being used in the Control Center of Crete, Greece.

19.3.5.1 General Comments
The unit commitment procedure developed within CARE searches for global optimal solutions considering the following costs and constraints:

- Unit shutdown and start-up costs
- Operation costs (usually, fuel costs)
- Ramping costs
- Load satisfaction
- Ramping constraints
- Spinning reserve criteria
- Maximum wind penetration
- Minimum unit downtimes
- Minimum start-up times (as function of downtime)

To achieve a practical procedure able to be used in real time, we have adopted a GA with some adaptations, where fitness is calculated as the sum of all costs and penalties.

19.3.5.2 Evolutionary Process Techniques: Chromosome Compression Technique
The coding technique adopted in the chromosomes is the one described earlier, with the adoption of 10 blocks, 2 bits for advancement or delay of each block, and 1 switch on/off bit for each block.

19.3.5.3 Evolutionary Process Techniques: Selection and Deterministic Crowding
Selection is directed under the deterministic crowding (DC) approach [7]. The DC will randomly pair all population elements in each generation. Each pair of parents will undergo mutation, crossover, and yield two children. Finally, children will compete with the parents, and the two fittest will enter the next generation. DC can successfully perform niching and preserve the diversity in the GA.

19.3.5.4 Evolutionary Process Techniques: Dynamic Mutation Rate
The mutation rate is increased whenever a certain number of generations pass without obtaining any improvement in the best solutions found; it is reset to

the original value once new improvements are found. These changes must be smooth to be beneficial; we have witnessed that a too large change of mutation rate does more harm than good to the convergence of the algorithm.

19.3.5.5 *Evolutionary Process Techniques: Crossover* The classic single-point random crossover has been adopted.

19.3.5.6 *Evolutionary Process Techniques: Chromosome Repair* Solutions generated during the evolutionary process are checked for feasibility, and, if verified unfeasible, an attempt is made to correct them and bring them into feasibility—their chromosomes are changed and added to the genetic pool.

19.3.5.7 *Evolutionary Process Techniques: A Lamarckist Adaptation* At certain generations, a process called "neighborhood digging" (ND) is launched. It tries to displace the start-up and shutdown times of generators in order to improve solutions (which are selected with a given probability). The general ND tests are

- Enabling/disabling some blocks.
- Shutting down or starting up generators within some blocks.
- Anticipating/postponing the start of a new block.
- Avoiding unnecessary shutdowns/start-ups of generators.

A set of rules has been derived to guide the ND process, for instance for postponement and anticipation of start-up and shutdown of generators, in order to improve the efficiency of this nevertheless time-consuming process. The decision to adopt changes in the solutions may come either from feasibility verification or from fitness calculation. If better solutions are found, their chromosome coding is added to the genetic pool. This process is again a form of Lamarckist adaptation. More interesting, because there is not a one-to-one mapping between phenotype and genotype, some decisions must be made in terms of which will be the preferred genetic representation of an improved individual.

19.3.5.8 *Fitness Evaluation: Dispatch with Fuzzy Wind Model* In each time step, an approximate dispatch cost is calculated with the fuzzy model described in Refs. 8 and 9, taking into account, if necessary, the constraint on wind penetration. The dispatch keeps the record of the discrete nature of the power dispatchable in wind parks (by disconnection of generators, if necessary). In each time step, we need to calculate the operation costs for a given scheduling alternative and, indirectly, to supply indications on the ramping limits for following time steps. This is achieved with a simple yet accurate enough dispatch model, based on a single bus model, and includes a distinct fuzzy constraint on how much wind generation is allowed into the system.

The basic mathematical model, for a time period or time step, is the following:

$$\min \Phi = \sum_i C_i(P_i) + C_w(W_{av} - W)$$

$$\sum_i P_i + W = L$$

subject to $W \tilde{\leq} \Pi$

$$P_i^{min} \leq P_i \leq P_i^{Max}$$
$$0 \leq W \leq W_{av}$$

where Φ is the operation cost for the time step, P_i is the constant power (MW) generated by generator i, $C_i(P_i)$ is the generation cost function of unit i ($), W is the wind power (MW), W_{av} is the maximum wind power available (upon prediction, MW), C_w is the penalty cost for not using all available wind, L is the load (upon prediction, MW), Π is the wind generation penetration (MW), and P_i^{min}, P_i^{Max} are the technical limits for each generator in the current time step (MW) and includes ramping constraints. The inequality sign with a tilde is used to represent a fuzzy constraint.

The penalty cost C_w is introduced because in many cases, by contract or by regulation, the dispatcher is forced to pay some compensation to private owners of wind parks if all available wind power is not utilized—meaning a loss for the private owners of such plants, based on unsold energy.

Whereas the first constraint (energy balance) is taken as crisp, the second one is defined as a fuzzy constraint: the wind generation penetration W defines a fuzzy border. Starting with two thresholds Π_{min} and Π_{Max}, one can calculate, in each time step, values in MW for Π_{min}, below which the operation is considered secure, relative to wind perturbations, and Π_{Max}, above which operation is not accepted due to dynamic stability problems in the case of wind perturbations.

The solution for this problem may be found according to the following algorithm, whose formal justification may be found in Ref. 8:

```
Solve two classical dispatch problems, replacing the
fuzzy constraint by W ≤ Π^min, to obtain Φ^Max, and by
W ≤ Π^Max to obtain Φ^min.
```

Start with a trial $\lambda = -\dfrac{\beta}{\alpha}$

```
REPEAT
```
$\quad$ Set all generators with $\dfrac{\partial C}{\partial P} > \lambda$ to P^{min}.

$\quad$ Set all generators with $\dfrac{\partial C}{\partial P} < \lambda$ to P^{Max}.

$\quad$ Calculate $\sum P_i$ and $\sum C_i(P_i)$ for the P_i values obtained with the current value of λ.

```
Calculate μ' and μ''
```

$$\mu' = \frac{\Phi^{Max} - \left[\sum_i C_i(P_i) + C_w\left(W_{av} - \left(L - \sum_i P_i\right)\right)\right]}{\Delta\Phi}$$

$$\mu'' = \frac{\Pi^{Max} - \left(L - \sum_i P_i\right)}{\Delta\Pi}$$

```
IF  μ' < μ''  THEN decrease λ.
IF  μ' > μ''  THEN increase λ.
UNTIL |μ'-μ''| < Tolerance specified.
```

19.3.5.9 *Fitness Evaluation: Ramping Rules* The limitations of thermal units in increasing generation beyond a limiting rate are represented according to adequate detail. For a long-term UC, a ramping rate in MW/h is adopted for each generator. In the 10-minute UC, the load pick-up schedule for each generator is taken in account, according to the instructions followed by operators, because the time needed to move from zero load to full load is usually much greater than 10 minutes. This means that a set of rules is specified for each type of generator, whose satisfaction of a greater or lesser degree will be translated in terms of a penalty to be included in the overall fitness evaluation.

19.3.6 Fitness Evaluation: Reliability (Spinning Reserve as a Fuzzy Constraint)

Two criteria for spinning reserve may be chosen: a percentage of the load in each time step or a probabilistic criterion, specifying a maximum probability of loss of load. The degree of satisfaction of these criteria is reflected as penalties in the fitness function of the GA. It must be said that the probabilistic criterion, based on the classic calculation of outage capacity probability tables, consumes much more computing time.

The calculation of the spinning reserve takes into account the ramping limitations of the generators. Also, that the wind generators do not contribute to the spinning reserve is considered. The constraints on spinning reserve are specified as flexible or fuzzy, so that a small degree of violation may still be accepted; however, beyond a certain threshold, a solution is definitely considered unfeasible.

19.3.7 Illustrative Results

We have extensively tested the model described and now illustrate the conclusions of such tests with some didactic results from an example based on real data from the island of Crete—a system composed of 17 different generators including six steam turbines, four diesel, seven gas turbines, plus three wind parks. The example will

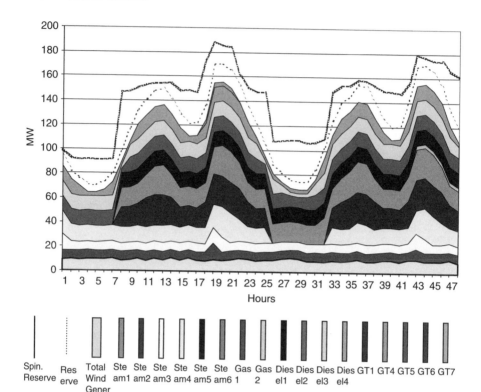

FIGURE 19.10 Solution for a long-term UC. The base strip corresponds with wind generation. The spinning reserve criterion is satisfied (as well as all other constraints).

be based on long-term UC runs, leading to the planning of a generation schedule over a period of 2 days. Figure 19.10 illustrates the result of a run, with an identification of the fractions of the load to be occupied by each scheduled generator in a 48-hour horizon. In this figure, one may also find the spinning reserve available at each hour (top dark line) and the requirement for spinning reserve specified (dashed line below).

The results in Fig. 19.10 are for a problem encoded in 16 blocks, with 2 bits for block flexibility. The population size was 200 and the maximum number of generations was set to 500.

We simulated load and wind prediction curves for 48 hours, according to the parameters used in CARE. The presented results verify a spinning reserve criterion of 10% of load and a maximum wind penetration of 20–25%. The penalties for violating constraints such as spinning reserve were decreased for the latest time steps of the scheduling period, given the increased uncertainty in load and wind predictions.

To assess the merits of the new model, we developed a series of tests to check on the effect and relevance of the ND procedure:

- StGA (straightforward genetic algorithm): running the algorithm without ND.
- NDlast: ND routine only in the last generation.

- ND150: ND routine every 40 generations, only until generation 160.
- ND350: ND routine only in the last 150 generations, every 40 generations.
- GAND: ND routine every 40 generations.

For each of these trials, we did 30 runs with random initialization of chromosomes. The best solution found was not the same in all runs, and the operation + scheduling cost varied to different degrees, depending on the trial set. Therefore, we have characterized the quality of the algorithms, not only as a function of the best run produced, but also as a function of its robustness (i.e., the variance or standard deviation of the results in successive trials). This is important in a real-world environment like a control center where there is no time to run the algorithm many times and thus the operators must have confidence that the procedure does not generate results deviating from a good (hopefully the best) solution. Table 19.4 illustrates the aggregate results obtained.

It is clear that the GAND configuration of the algorithm gives the best and more robust results: not only does it consistently detect better solutions than all other approaches, but also the variance of solutions offered in a number of runs is much narrower. The price to pay for this accuracy is an increase in computing time compared with the StGA alternative with no ND. However, the computing time (3 minutes in a 350 MHz Pentium) for GAND is still acceptable.

We can also extract some enlightening conclusions from Fig. 19.11, showing two samples for the evolution of the fitness function for a GAND and a StGA run. We can clearly see that it took GAND more or less half the number of generations (250) to reach solutions of the same quality of the ones obtained by SGA in 500 generations.

Thus, the difference in time becomes very small for the same level of solution costs—but then, the ND procedure allows the operator a bonus in quality of the proposed solution. It is not feasible to admit that one will do 30 runs in a real operation environment—but even so, the worst results of GAND will be expected to be better than the best of StGA.

One can see that the best run of GAND, in terms of cost, added with the standard deviation for GAND, still gives a lower cost than the best run of StGA. And besides, StGA has a much higher variance, so the risk of accepting a bad solution from StGA is much higher than from GAND. It can also be seen that the ND procedure had a positive influence on the algorithm outcome, even if in the first iterations, for some

TABLE 19.4 Comparison of Results Obtained with Different Coding Strategies

	Best Cost	Standard Deviation from the Mean	Standard Deviation (%)	Time (s)
StGA	59,320	1419	2.39	74.8
NDlast	58,864	1126	1.91	
ND150	59,139	1147	1.94	
ND350	58,757	1083	1.84	
GAND	58,255	795	1.36	181.8

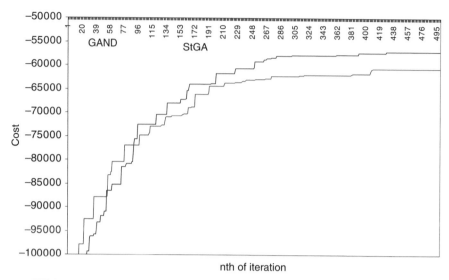

FIGURE 19.11 Evolution of fitness function: comparison of GAND with StGA.

reason possibly linked to random initialization of the chromosomes, the StGA showed to have some advance in fitness value.

We repeated the test for an 8-hour-ahead scheduling exercise, the 10-minute UC. We obtained a reduction in computing times, explained because the load profile is much smoother in this case—an average of 80 seconds for 30 runs with random initialization of chromosomes. The tests were conducted with a 350-MHz Pentium PC with Windows NT. The running times are absolutely compatible with the use of the technique in real time, in control centers of medium-sized systems or in isolated networks—one of the objectives of CARE.

19.4 VOLTAGE/VAR CONTROL AND LOSS REDUCTION IN DISTRIBUTION NETWORKS WITH AN EVOLUTIONARY SELF-ADAPTIVE PSO ALGORITHM: EPSO

19.4.1 Justifying a Hybrid Approach

Evolutionary algorithms (EAs) and PSO algorithms have been presented in many places as competing. It is not necessarily so. In this chapter, we will show how one may have a unified perspective over both methods and build a powerful algorithm that efficiently solves problems in many areas and, of course, in power systems. This algorithm has been named EPSO, for evolutionary self-adaptive particle swarm optimization [10]. Its discussion in the following sections assumes that the reader is familiar with the fundamentals of both families of methods: EA and PSO.

The classic PSO approach, with the concept of information sharing among particles, has proved to be efficient; however, it lacks the learning ability that EC methods display. The search in the decision space is always performed under the same rule and the same parameters, no matter what kind of problem or space topology is being attacked. On the contrary, self-adaptive evolution strategies (σSA-ES) [11] benefit from an evolutionary process to progressively select the most adequate parameters that condition the search.

A technique merging the two basic ideas, where the good traits of both techniques could add to each other instead of canceling, should lead to a more reliable and robust method. It is true that other attempts have been made to build a hybrid of swarm-type algorithms and evolutionary methods [12, 13], but EPSO may be fairly seen as the first true natural convergent evolution of both approaches.

19.4.2 The Principles of EPSO: Reproduction and Movement Rule

EPSO may be seen under two perspectives:

- As an EC method of the σSA-ES type, with a special rule for the replication of individuals instead of ordinary crossover and mutation of object parameters; or
- As a special PSO method where the weights that condition the movement in space undergo self-adaptive mutation.

Both perspectives are useful, but the EC paradigm may perhaps give insight on why the method works well and usually better than ES or PSO taken isolated. Seen as an evolution strategy, the EPSO algorithm may be described as follows. At a given iteration, consider a set of solutions or alternatives that we will keep calling particles (in the PSO tradition). The general scheme of EPSO is the following:

- *Replication*: each particle is replicated $r - 1$ times, creating a set of $r - 1$ clones.
- *Mutation*: each clone has its strategic parameters mutated.
- *Reproduction*: each particle (original plus clones) generates an offspring according to a rule similar to the particle movement rule of PSO.
- *Evaluation*: each offspring has its fitness evaluated.
- *Selection*: by stochastic tournament or other selection procedure, the best particles survive to form a new generation.

The reproduction (movement) rule for EPSO is the following: given a particle $\mathbf{X}_i$, a new particle $\mathbf{X}_i^{\text{new}}$ results from

$$\mathbf{X}_i^{(k+1)} = \mathbf{X}_i^{(k)} + \mathbf{V}_i^{(k+1)}$$

$$\text{©} = \mathbf{V}_i^{(k+1)} = w_{i0}^* \mathbf{V}_i^{(k)} + w_{i1}^* (\mathbf{b}_i - \mathbf{X}_i) + w_{i2}^* \mathbf{P}(\mathbf{b}_\mathbf{G}^* - \mathbf{X}_i),$$

where $\mathbf{b}_i$ is the best point found by particle i in its past life up to the current generation, $\mathbf{b}_g$ is the best overall point found by the swarm of particles in their past life up

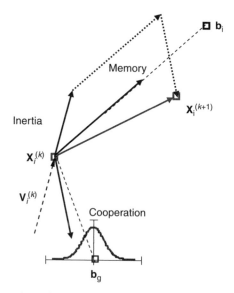

FIGURE 19.12   Illustration of EPSO particle reproduction: a particle $\mathbf{X}_i$ generates an offspring at a location commanded by the movement rule.

to the current generation, $\mathbf{X}_i^{(k)}$ is the location of particle i at generation k, $\mathbf{V}_i^{(k)} = \mathbf{X}_i^{(k)} - \mathbf{X}_i^{(k-1)}$ is the velocity of particle i at generation k, w_{i1} is the weight conditioning the *inertia* term (the particle tends to move in the same direction as the previous movement), w_{i2} is the weight conditioning the *memory* term (the particle is attracted to its previous best position), and w_{i3} is the weight conditioning the *cooperation* or *information exchange* term (the particle is attracted to the overall best-so-far found by the swarm). $\mathbf{P}$ is a diagonal matrix with elements of value 0 or 1 determined in each iteration by a communication probability p.

The p value externally fixed, controls the passage of information within the swarm and is 1 in classical formulations (communication topology called the *star*). This means that, for each dimension in the search space and in each iteration, there is a probability p that a particle will not receive (use) the information available on the best location found by the swarm—the particle evolves then only under the influence of inertia and memory components (a *stochastic star*).

The symbol * indicates that these parameters will undergo evolution under a mutation process to be explained. This rule is illustrated in Fig. 19.12.

19.4.3 Mutating Strategic Parameters

As in a σSA-evolution strategy, we distinguish, in each particle or solution representation, object parameters and strategic parameters. Object parameters are those giving the *phenotypic* description of a solution (its natural variables). Strategic parameters are those that condition the evolution of a given solution. The basic mutation rule for the first three strategic parameters is the following:

$$w_{ik}^* = w_{ik}[\log N(0, 1)]^\tau$$

where $k = 1$, 2, 3, $\log N(0, 1)$ is a random variable with lognormal distribution derived from the Gaussian distribution $N(0, 1)$ of 0 mean and variance 1, and τ is a learning parameter, fixed externally, controlling the amplitude of the mutations— smaller values of τ lead to a higher probability of having values close to 1.

The $\log N$ distribution is classically adopted for strategic parameters because, in this multiplicative form of mutation, the probability of having a new value multiplied by m is the same as having a value multiplied by $1/m$. Approximations to this scheme could be obtained by

$$w_{ik}^* = w_{ik}[1 + \tau N(0, 1)],$$

provided that τ is small and the outcome is controlled so that negative weights are ruled out. This scheme is preferable to additive mutations of the type

$$w_{ik}^* = w_{ik} + \tau N(0, 1),$$

because in this case, the absolute value of the mutation is insensitive to the value of w.

Thus, the mutations for weights w_{ik} for $k = 1, 2, 3$ are straightforward. As for the global best $\mathbf{b}_g$, it is randomly disturbed to give

$$\mathbf{b}_g^* = \mathbf{b}_g + w_{i4}^* N(0, 1),$$

where w_{i4} is the fourth strategic parameter associated with particle i. It controls the "size" of the neighborhood of $\mathbf{b}_g$ where it is more likely to find the real global best solution (assumed not found so far during the process) or, at least, a solution better that the current $\mathbf{b}_g$. This weight w_{i4} is mutated (signaled by $*$) according to the general mutation rule of strategic parameters, allowing the search to focus on a given point, if convenient.

19.4.4 The Merits of EPSO

Why does EPSO behave well and converge?

The basic answer is that EPSO is a method belonging to the self-adaptive ES family and, therefore, it must display such properties. In any ES, the generation of offspring is regulated by operations of mutation and crossover. There is a solid theoretical background giving insight on why ES achieves convergence and how a near-optimal progress rate is achieved [14]. The fact is that the operator selection, acting on random mutations (and transformations), provides a positive push toward the optimum.

The movement rule of PSO has proven to be, in itself, an average improver of solutions. In classic PSO we may say that we have a reproduction scheme but selection is trivial—each parent has one child and each child survives to its parent, in a sort of parallel $n \times (1, 1)$ES. However, the reproduction rule, by itself, ensures the progress to the optimum, meaning that, on average, each generation will be better than the preceding one.

EPSO adds this effect to the power of effective Darwinist selection. Therefore, we have two mechanisms acting in sequence, each one with its own probability of producing not just better individuals but an average better group. And it happens that selection acts on a generation that is already on average better than the preceding, so the effects are additive.

The fact that EPSO is self-adaptive adds another interest to the method: it avoids in a large scale the need for fine-tuning the parameters of the algorithm, because the procedure will hopefully learn (in the evolutionary sense) the characteristics of the search space and will self-tune the weights in order to produce an adequate rate of progress toward the optimum.

19.4.5 Experience with EPSO: Basic EPSO Model

Several EPSO models may be conceived. The basic model has the following characteristics:

- Each particle is only cloned once, resulting in a pair of particles at a location $(r = 2)$.
- The cloned particle sees its weights mutated.
- Both original and cloned particles are moved (reproduction).
- Selection acts on this pair of new particles to preserve the best and discard the worst.
- The process is repeated to all particles.

In the $(\mu, +\lambda)$ notation usual in ES, this forms, for each particle, a $(1, 2)\sigma$SA-evolution strategy. The global process may be seen as a multiple $p(1, 2)\sigma$SA-ES for all p particles, where the individual ES processes become coupled because they share information in the movement rule via the cooperation term.

19.4.6 EPSO in Test Functions

We have conducted a large number of experiments that have convinced us of the goodness of EPSO. Some of these experiments have been performed on classic test problems, such as Schaffer, Rosenbrock, Sphere, or Alpine functions, and have been reported in other publications [15, 16]. In general, we may summarize the results by saying that, in comparing EPSO with classic PSO (in 1000 runs with random initialization),

- EPSO displays a faster convergence.
- EPSO is more accurate (EPSO has an average best result superior to PSO).
- EPSO is more robust (the variance of the results is much smaller than for PSO).
- EPSO is more insensitive to weight initialization (the performance of PSO degrades very easily with the variation of parameters compared with EPSO).

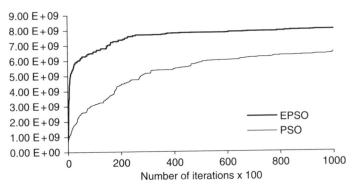

FIGURE 19.13 Comparison of the average fitness evolution in EPSO and classic PSO with increasing generations.

The following figures and results illustrate these points. They refer to the application of EPSO in a difficult test problem, usually known as the maximization of the Alpine function, which is given by

$$\max f_4(x) = \sin(x_1) \ldots \sin(x_n) \times \sqrt{x_1 \ldots x_n}.$$

It is a function with multiple local optima and the comparison is made with the performance of the classic PSO.

In one experiment, we used the algorithm to find the maximum of the Alpine 10-dimension function in a $[0, 100]^{10}$ search space. In Fig. 19.13, we can observe the average convergence in an experiment series of 20 simulations, with 40 particles. We verify a superior behavior of EPSO compared with the classic PSO. This has happened for all spaces, no matter how many dimensions.

We designed another experiment to verify the robustness and reliability of EPSO against classic PSO. In this experiment, we run 1000 trials with random initialization, each for 100 iterations. We employed EPSO with a stochastic tournament probability p_T such that $(1 - p_T) = 0.0375$. This experiment aimed at evaluating the dispersion of results generated by the two competing algorithms. For 1000 trials, the average result and the standard deviation of results given by EPSO and classical PSO are given in Table 19.5.

In order to make this clear, we have partitioned the fitness values into a set of intervals of fixed width and have represented, in Figs. 19.14 and 19.15, the histograms for the frequency of occurrence of the final solution within each interval.

TABLE 19.5 Results for the Alpine Function

Alpine 2-Dimension, 1000 Trials	Average Result	Standard Deviation
EPSO	98.69669	1.05910
Classic PSO	90.62233	6.81301

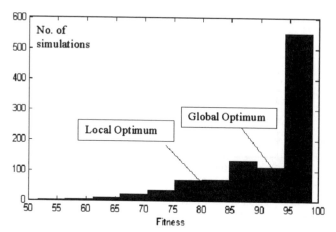

FIGURE 19.14 Histogram with the performance of classic PSO in 1000 simulations on the Alpine 2-dimension function: distribution in a partition with intervals of width 5.

It is obvious that EPSO reveals itself as a more robust and reliable algorithm. Independently of the random initialization of starting solutions (and weights for the self-adapting algorithm), EPSO almost always discovered the global optimum or stopped extremely close to it, whereas classic PSO was trapped in local optima on a significant number of occasions and had trouble in reaching close to the optimum. This performance of EPSO has been observed in other test functions, and the results of a diversity of experiments have been reported in a number of publications [15, 16].

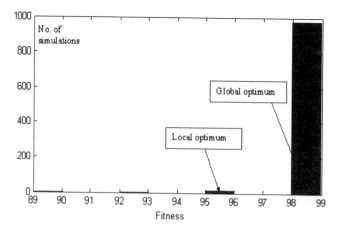

FIGURE 19.15 Histogram with the performance of EPSO in 1000 simulations on the Alpine 2-dimension function: distribution in a partition with intervals of width 1. Comparing with Fig. 19.4, notice that the x-axis scale is different.

19.4.7 EPSO in Loss Reduction and Voltage/VAR Control: Definition of the Problem

We have applied EPSO to problems of loss reduction and voltage/VAR control in power systems by acting in generator excitation, in transformer taps, and in capacitor banks [17].

The reactive power dispatch problem will be first defined as a multiple criteria problem:

> Minimize: Active losses: $\varphi_p(u, x)$
> Distance to an admissible voltage band: $\varphi_V(u, x)$
>
> Subject to: $H(u, x, p) \leq 0.$

Active losses $\varphi_p(u, x)$ are calculated from the AC power flow equations. The control variables u represent the following:

- Voltage values at PV buses where voltage control is possible.
- Tap position in tap-changing transformers.
- Amount of on-line capacity in switched capacitor banks.

The variables in the first set are continuous. However, the second and third sets of variables have a discrete nature: transformer taps have fixed discrete values and the amount of capacity switched on is defined in discrete steps. The admissible voltage band is defined as an interval around the nominal voltage $[V_{\min}, V_{\max}]$. If the point is inside the band, its distance $\varphi_V(u, x)$ is zero, otherwise this distance is measured according to a specified space metrics, not necessarily Euclidian. The constraints $H(u, x, p) \leq 0$ relate to the usual AC power flow model, plus branch flow limits and control variable limits.

In order to apply EPSO to this problem, a single objective function is built in the form

$$\text{Minimize} \quad K\varphi_p(u, x) + (1 - K)\varphi_V(u, x).$$

$K \in [0, 1]$ is a parameter whose control defines the type of problem to be solved. It may be defined:

- As a weight—and then compromise solutions between loss reduction and voltage violation are searched for; this allows for the calculation of trade-offs between the two objectives with an adequate variation of K.
- As a Boolean value that switches on or off the evaluation of losses whenever there is not or there is some voltage violation. This latter case means that the consideration of the multiple criteria is hierarchic: the first priority is to avoid voltage violations, and only if there is none, can one proceed to minimize losses.

19.4.8   Applying EPSO in the Management of Networks with Distributed Generation

Modern distribution systems display an increasing penetration of distributed gene-ration, of many sources: independent producers, industrial cogeneration, domestic cogeneration, mini-hydros, wind generation, and, in the foreseeable future, photovol-taic and fuel cells.

This phenomenon is highly visible in many countries in the European Union, where an active policy of incentives has been designed and applied. However, tech-nical regulations in many cases have not been adequate and have imposed practices that reduce the benefits from the existence of distributed generation for the system and its agents. A typical case happens with reactive power: in several countries, laws or regulations impose strict limits or obligations for reactive power generation from independent agents. For instance, in the case of wind generation with induction machines, some regulations impose the installation of heavy compensation in terms of capacitor banks, whose costs must be added to the global investment, reducing the window of feasibility for renewable energy sources.

Instead of imposing such regulations, a more convenient strategy could however be devised, if the operator of the distribution system could reach an agreement with the independent producers in order to use the reactive power resources to reduce losses in the system and obtain better voltage profiles. A study of this kind requires precisely a model such as the one described in the previous section.

EPSO has proved to be efficient in deriving strategies for voltage/VAR control in distribution networks. The practical importance of these results derives from the fact that, against common intuition, minimizing losses may lead to inadequate voltage values in some buses in a system, namely a distribution system. This is illustrated in Ref. 18, where EPSO was used in several types of studies, applied to the IEEE Reliability Test System with 24 branches and 36 lines, including 31 transmission lines, five transformers, 11 capacitor banks, and nine synchronous generators. The studies applied an EPSO algorithm to four cases:

- Minimizing the sum of deviations outside the admissible band (DV)
- Minimizing the sum of the square deviations (DV2)
- Minimizing the largest deviation (Min Max)
- Just minimizing power losses (Min loss)

The results for voltage levels in system buses are partially displayed in Fig. 19.16. They lead to a couple of interesting conclusions:

- Minimizing the sum of deviations or the sum of the square of deviations from an admissible voltage band does not lead to loss minimization and does not guarantee that, at certain buses, the voltage will not assume unacceptable values.
- Minimizing losses also may lead to unacceptable nodal voltages (see bus 5 in Fig. 19.16, where minimizing losses caused the worst voltage value of all four studies).

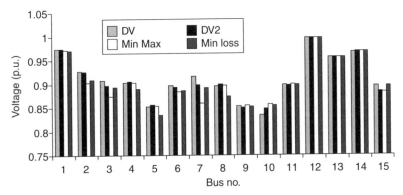

FIGURE 19.16 Bus voltages as a result of EPSO under four distinct criteria (from a study reported in Ref. 18).

- The best voltage profile, obtained by minimizing the largest deviation (a min-max objective), does not correspond with the solution with smaller loss levels.

The studies also showed that EPSO could work well under difficult objectives, such as the min-max type, making it a promising algorithm to deal with planning problems and risk analysis, for instance.

A practical application study applying EPSO has been successfully conducted by INESC Porto for EDP Distribuição, the distribution utility in Portugal. The studies on 60-kV and 30-kV networks were complex, taking in account a number of scenarios and possibilities of control by the utility of dispersed reactive power, but the final result was positive; that is, one could achieve simultaneously an improved voltage profile and a loss reduction if the imposing of strict regulation could be switched for a negotiated control. In order to achieve this result, it became apparent that reactive power generation capacity, imposed by law, should not be used at certain places, to the advantage of both the grid operator and the private investors. Details of this study may be found in Ref. 19. Of course, this example does not mean that the application of EPSO is restricted to distribution systems. More results of the application of EPSO to power systems may be found in Ref. 20.

19.5 CONCLUSIONS

This chapter was designed with the purpose of calling the attention of the reader to interesting aspects of the evolutionary principles that rule the algorithms discussed in this book. To illustrate the points raised, we've referred to three models, for optimal capacitor placement, for optimal unit commitment, and for voltage/VAR control and loss reduction the main purpose, however, has not been to discuss in detail the particular modeling of those real-world problems.

In fact, one of the main conclusions to be extracted is that our theoretical and engineering efforts are still in a relatively naïve state, and therefore much can be expected from further research in the topic. The chapter examples illustrate that Lamarckism may be a positive driving force toward evolution in a sense desired by model builders. However, they have also illustrated that there are advantages to be extracted from more complex coding procedures and from hybrid models.

There are many powerful mechanisms that act in nature to speed-up evolution that our evolutionary engineering models still do not consider. Geneticists see fitness, at the gene level, not as a value in itself but always as a comparative measure of performance of a gene against its alleles. But in our engineering language, we are still very much influenced by a form of language that measures fitness as an absolute value. In our models, we do not use the techniques that drive evolution in nature, such as "altruism," "parasitism," or "arms races."

Our individuals exist by themselves and do not depend on each other in any way, do not interact with each other, do not influence the evolution of each other. Also, our individuals receive a very naïve coding, more related to a Lamarckian perspective of evolution than to a Darwinian one. There is usually a one-to-one mapping between genotype and phenotype instead of having a chromosome as a program to build and operate individuals and genes competing against each other.

This may not be incorrect in itself, because in our engineering models we have been pretty much centered on the individual (the best fit individual), whereas in modern Darwinian currents of opinion, there is a recentering on the role of the genes as the real replicator units. Nature is all around us, showing its full potential to generate evolution. But nature's way of doing things, based on a chromosome with a program to be run, may be much more difficult to reproduce in engineering modeling. If this is so, then we should be able to make a better use of the inherent Lamarckian potential of our method of modeling systems.

The EPSO algorithm, on the other hand, shows that in engineering one does not need to be tied to the kind of solutions nature has developed with success (this applies also to the use of Lamarckism). The exploration of the space of solutions is achieved in nature as efficiently as possible with the limitations imposed by the biochemical mechanisms—but in engineering we may be as inventive as we like, and the generation of new solutions may be guided (by the particle movement rule) instead of random (with mutation or classic recombination schemes). The right amount of hybrid components is therefore the challenge.

REFERENCES

1. Dawkins R. Replicator selection and the extended phenotype. Zeitschrift für Tierpsychologie 1978; 47:61–76.

2. Gould SJ. Shades of Lamarck. Natural History 1979; 88(8):22–28.

3. Miranda V, Naing WO, Fidalgo JN. Experimenting in the optimal capacitor placement and control problem with hybrid mathematical-genetic algorithms. Proceedings of ISAP'01– Intelligent Systems Applications in Power Systems Conference, Budapest, Hungary, June 2001.

4. Goldberg DE. Genetic algorithms in search, optimization and machine learning. New York: Addison-Wesley; 1989.

5. Mahfoud SW. Niching methods for genetic algorithms. Ph.D. dissertation, Univ. of Illinois, Urbaba-Champaign 1995.

6. Miu KN, Chiang HD, Darling G. Capacitor placement, replacement and control in large-scale distribution systems by a GA-based two-stage algorithm. IEEE Trans Power Systems 1997; 12(3):1160–1166.

7. Mahfoud S. Niching methods for genetic algorithms. IlliGAL Report No. 95001. Illinois Genetic Algorithm Laboratory; 1995.

8. Miranda V, Pun HS. Economic dispatch model with fuzzy wind generation constraints and preference attitudes of the decision maker. Proceedings of the International Symposium on Electric Power Engineering at the Beginning of the Third Millenium, Naples-Capri, CD edition, University of Naples, Italy, May 12–18, 2000.

9. Miranda V, Pun HS. Unit commitment with a genetic algorithm, probabilistic spinning reserve and fuzzy wind dispatch–The CARE Project (EU Joule Programme). Proceedings of the XVI National Symposium on Security in the Operation of Power Systems, Bacau, Romania, September 22–23, 1999.

10. Miranda V, Fonseca N. EPSO–best-of-two-worlds meta-heuristic applied to power system problems. Proceedings of WCCI/CEC–World Conference on Computational Intelligence, Conference on Evolutionary Computation, Honolulu, Hawaii, June 2002.

11. Schwefel H-P. Evolution and optimum seeking. New York: Wiley; 1995.

12. Løvbjerg M, Rasmussen TK, Krink T.: Hybrid particle swarm optimiser with breeding and subpopulations. In: Spector L, Goodman ED, Wu A, Langdon W, Voigt H-M, Gen M, Sen S, Dorigo M, Pezeshk S, Garzon MH, Burke E. (eds): Proceedings of the Genetic and Evolutionary Computation Conference (GECCO-2001). San Francisco, Morgan Kaufmann 2001, 469–476.

13. Clerc M. The swarm and the queen: towards a deterministic and adaptive particle swarm optimization. In: Proceedings of the 1999 Congress of Evolutionary Computation, Vol. 3. Piscataway, NJ: IEEE Press; p. 1951–1957.

14. Beyer H-G. Toward a theory of evolution strategies: self-adaptation. Evolu Computat 1996; 3(3):311–347.

15. Miranda V, Fonseca N. EPSO–best-of-two-worlds meta-heuristic applied to power system problems. Proceedings of WCCI/CEC–World Conference on Computational Intelligence, Conference on Evolutionary Computation, Honolulu, Hawaii, June 2002.

16. Miranda V, Fonseca N. New evolutionary particle swarm algorithm (EPSO) applied to voltage/var control. Proceedings of PSCC'02–Power System Computation Conference, Seville, Spain, June 24–28, 2002.

17. Miranda V, Fonseca N. Reactive power dispatch with EPSO—evolutionary particle swarm optimization. Proceedings of PMAPS'2002–6th International Conference on Probabilistic Methods Applied to Power Systems, Naples, Italy, September 2002.

18. Miranda V, Fonseca N. EPSO - evolutionary particle swarm optimization, a new algorithm with applications in power systems. Proceedings of IEEE Transmission and Distribution Conference 2002–Asia-Pacific, Yokohama, Japan, October 2002.

19. Lopes P, Mendonça A, Pinto JS, Seca L, Fonseca N. Evaluation of the use of production in special regime (PRE) to voltage and VAR control—Assesment of the impact of PRE on HV and MV distribution networks. Report for EDP Distribuição, S.A., INESC Porto,

Power Systems Unit Porto, Portugal, March 2003 [in Portuguese, limited access by request].

20. Miranda V, Keko H and Jaramillo A. EPSO: Evolutionary particle swarms. In: Jain L, Palade V, Srinivasan D (eds.): Advances in evolutionary computing for system design, Volume 66. New York: Springer; 2007.

Modern Heuristic Optimization Techniques. Edited by K. Y. Lee and M. A. El-Sharkawi
Copyright © 2008 by the Institute of Electronics Engineers, Inc.